D0623317

Mechanical Engineering Series

Frederic F. Ling
Series Editor

Springer
New York
Berlin
Heidelberg
Hong Kong
London
Milan
Paris
Tokyo

Mechanical Engineering Series

J. Angeles, **Fundamentals of Robotic Mechanical Systems: Theory, Methods, and Algorithms, 2nd ed.**

P. Basu, C. Kefa, and L. Jestin, **Boilers and Burners: Design and Theory**

J.M. Berthelot, **Composite Materials: Mechanical Behavior and Structural Analysis**

I.J. Busch-Vishniac, **Electromechanical Sensors and Actuators**

J. Chakrabarty, **Applied Plasticity**

K.K. Choi and N.H. Kim, **Structural Sensitivity Analysis and Optimization 1: Linear Systems**

K.K. Choi and N.H. Kim, **Structural Sensitivity Analysis and Optimization 2: Nonlinear Systems and Applications**

G. Chryssolouris, **Laser Machining: Theory and Practice**

V.N. Constantinescu, **Laminar Viscous Flow**

G.A. Costello, **Theory of Wire Rope, 2nd ed.**

K. Czolczynski, **Rotordynamics of Gas-Lubricated Journal Bearing Systems**

M.S. Darlow, **Balancing of High-Speed Machinery**

J.F. Doyle, **Nonlinear Analysis of Thin-Walled Structures: Statics, Dynamics, and Stability**

J.F. Doyle, **Wave Propagation in Structures: Spectral Analysis Using Fast Discrete Fourier Transforms, 2nd ed.**

P.A. Engel, **Structural Analysis of Printed Circuit Board Systems**

A.C. Fischer-Cripps, **Introduction to Contact Mechanics**

A.C. Fischer-Cripps, **Nanoindentation, 2nd ed.**

J. García de Jalón and E. Bayo, **Kinematic and Dynamic Simulation of Multibody Systems: The Real-Time Challenge**

W.K. Gawronski, **Advanced Structural Dynamics and Active Control of Structures**

W.K. Gawronski, **Dynamics and Control of Structures: A Modal Approach**

G. Genta, **Dynamics of Rotating Systems**

K.C. Gupta, **Mechanics and Control of Robots**

J. Ida and J.P.A. Bastos, **Electromagnetics and Calculations of Fields**

(continued after index)

Kyung K. Choi
Nam H. Kim

Structural Sensitivity Analysis and Optimization 2

Nonlinear Systems and Applications

With 100 Figures

Kyung K. Choi
Department of Mechanical and
Industrial Engineering
The University of Iowa
Iowa City, IA 5224
USA
kkchoi@ccad.uiowa.edu

Nam Ho Kim
Department of Mechanical and
Aerospace Engineering
The University of Florida
POB 116250
Gainsville, FL 32611-6250
USA
nkim@ufl.edu

Series Editor
Frederick F. Ling
Ernest F. Gloyna Regents Chair in Engineering, Emeritus
Department of Mechanical Engineering
The University of Texas at Austin
Austin, TX 78712-1063, USA
and
William Howard Hart Professor Emeritus
Department of Mechanical Engineering,
Aeronautical Engineering and Mechanics
Renssalaer Polytechnic Institute
Troy, NY 12180-3590, USA

Library of Congress Cataloging-in-Publication Data
Choi, Kyung K.
Structural sensitivity analysis and optimization / Kyung K. Choi, Nam H. Kim.—1st ed.
p. cm. — (Mechanical engineering series)
Includes bibliographical references and index.
ISBN 0-387-23232-X (alk. paper) — ISBN 0-387-23336-9 (alk. paper)
1. Structural analysis (Engineering) I. Kim, Nam H. II. Title. III. Mechanical engineering series (Berlin, Germany)
TA645.C48 2005
624.1´71—dc22 2004062574

ISBN 0-387-23336-9 Printed on acid-free paper.

Printed in the United States of America. (EB)

9 8 7 6 5 4 3 2 1 SPIN 11012467

springeronline.com

◆

Dedicated to our wives

Ho-Youn

Jee-Hyun

◆

Preface

Structural design sensitivity analysis concerns the relationship between design variables available to the design engineer and structural responses determined by the laws of mechanics. The dependence of response measures such as displacement, stress, strain, natural frequency, buckling load, acoustic response, frequency response, noise-vibration-harshness (NVH), thermoelastic response, and fatigue life on the material property, sizing, component shape, and configuration design variables is implicitly defined through the governing equations of structural mechanics. In this text, first- and second-order design sensitivity analyses are presented for static and dynamics responses of both linear and nonlinear structural systems, including elastoplastic and frictional contact problems.

Prospective readers or users of the text are seniors and graduate students in mechanical, civil, biomedical, industrial, and engineering mechanics, aerospace, and mechatronics; graduate students in mathematics; researchers in these same fields; and structural design engineers in industry.

A substantial literature exists on the technical aspects of structural design sensitivity analysis. While some studies directly address the topic of design sensitivity, the vast majority of research is imbedded within texts and papers devoted to structural optimization. The premise of this text is that a comprehensive theory of structural design sensitivity analysis for linear and nonlinear structures can be treated in a unified way. The objective is therefore to provide a complete treatment of the theory and practical numerical methods of structural design sensitivity analysis. Design sensitivity supports optimality criteria methods of structural optimization and serves as the foundation for iterative methods of structural optimization. One of the most common structural design methods involves decisions made by the design engineer based on experience and intuition. This conventional mode of structural design can be substantially enhanced if the design engineer is provided with design sensitivity information that explains design change influences, without requiring a trial and error process.

Such advanced, state-of-the-art analysis methods as finite element analysis, boundary element analysis, and meshfree analysis provide reliable tools for the evaluation of the structural design. However, they give the design engineer little help in identifying ways to modify the design to either avoid problems or improve desired qualities. Using design sensitivity information generated by methods that exploit finite element, boundary element, or meshfree formulations, the design engineer can carry out systematic trade-off analysis and improve the design. This text presents design sensitivity analysis (DSA) theory and numerical implementation to create advanced design methodologies for mechanical systems and structural components, which will permit economical designs that are strong, stable, reliable, and have long service life. The design methodologies can be used by design engineers in the academia, industry, and government to obtain optimal structural designs for ground vehicles, aircraft, space systems, ships, heavy equipment, machinery, biomedical devices, etc. Extensive numerical methods for computing design sensitivity are included in this text for practical application and software development. More importantly, the numerical method allows seamless integration of CAD-FEA-DSA software tools, so that design optimization can be carried out using CAD geometric models instead of FEA models. This capability allows integration of CAD-CAE-CAM so that optimized designs can be manufactured effectively.

The book is organized into two volumes, four parts, and fourteen chapters. Parts I and II are in Volume 1: Linear Systems, and Parts III and IV are in Volume 2: Nonlinear Systems and Applications.

Part I introduces structural design concepts that include the CAD-based design model, design parameterization, performance measures, costs, and constraints. Based on the design model, an analysis model is introduced using finite element analysis. A broad overview of design sensitivity analysis methods is provided. By relying on energy principles to develop design sensitivity analysis theory, the design sensitivity method is developed without requiring highly sophisticated mathematics. The energy method is introduced in order to develop the variational equation and its relationship to the finite element method. Chapters 2 and 3 are essentially a review for students who have already learned energy methods in structural mechanics. The finite element method is explained as a technique based on a piecewise polynomial approximation of the displacement field and as an application of the variational method for approximating a solution to the governing boundary-value problem. In Part II, this relationship is successfully used in the development of discrete and continuum design sensitivity analysis methods and their relationships.

Part II treats design sensitivity analysis of linear structural systems. Both discrete and continuum design sensitivity analysis methods are explained. Chapter 4 describes finite-dimensional problems in which the structural response is a finite-dimensional vector of structural displacements, and the design variable is a finite-dimensional vector of design parameters. Governing structural equations are matrix equations. Direct design differentiation and adjoint variable methods of design sensitivity analysis are presented, along with the design derivatives of eigenvalues and eigenvectors. The computational aspects of implementing these methods are treated in some detail in conjunction with finite-element analysis codes. Chapters 5 through 7 treat continuum problems in which response and design variables are functions (displacement field and material distribution) and governing structural equations are the variational equations introduced in Chapters 2 and 3. Sizing, shape, and configuration design variables are treated separately in Chapters 5 through 7, respectively. Both the direct differentiation and adjoint variable method of design sensitivity analysis are developed, and design derivatives of eigenvalues are derived. Analytical solutions to simple examples and numerical solutions to more complex examples are presented. For both shape and configuration design variables, the material derivative concept is taken from continuum mechanics to predict the effect of design changes on the structural response. For a structural component with curvature, a more general configuration design theory is presented in Section 7.5 of Chapter 7. For shape design sensitivity, the adjoint variable method is used to derive expressions for differentials of the structural response, either as boundary integrals (the boundary method) or domain integrals (the domain method). A similar method is used for the shape design sensitivity of eigenvalues.

Part III treats design sensitivity analysis of nonlinear structural systems using continuum design sensitivity analysis methods. As with Chapters 2 and 3, the equilibrium equations for nonlinear structural systems are derived using the principles of virtual work from Chapter 8. Both geometric and material nonlinearities are treated. Nonlinear elasticity, buckling, hyperelasticity, elastoplasticity, nonlinear transient dynamics, and frictional contact problems are included. In nonlinear structural analysis, total and updated Lagrangian approaches have been introduced. The equilibrium equations are then linearized at the previously known configuration to yield incremental formulations for nonlinear analysis. The linearized equilibrium equation plays a key role in design sensitivity analysis in subsequent chapters, since the first-order variation with respect to the design parameter includes linearization of the energy form. The linearized form that appears during design sensitivity analysis is the same as the linearized form for nonlinear analysis. Sizing, shape, and configuration design variables are treated separately in Chapters 9 through 11, respectively. Both adjoint variable and direct differentiation methods are given for the nonlinear elastic problem. However, for nonlinear elastoplastic problems, only the direct differentiation method is used, since the design sensitivity is path-dependent. Analytical derivations of design sensitivity expressions for structural components are presented, along with numerical examples of sensitivity computations.

Part IV is devoted to practical design tools and applications: sizing and shape design parameterization, design velocity field computation, numerical implementation of the sensitivity for general-purpose code development, and various other practical design applications. In Chapter 12, sizing design parameterization for line and surface design components is introduced. For shape design parameterization, a three-step process is developed. One important aspect of shape design parameterization is the connection between the design parameterization and the computation of the design velocity field, as explained in Chapter 13. In Chapter 13, the computational aspects of design sensitivity analysis are considered, using the finite element method to solve the original governing and adjoint equations. The numerical method allows seamless integration of CAD-FEA-DSA, so that design optimization can be carried out using CAD models instead of FEA ones. Chapter 14 includes a number of practical design applications of linear and nonlinear structural systems with additional applications in thermoelastic analysis and fatigue design optimization to aid application-oriented readers.

A final comment on the notation used in this text. The structural design engineer may find that the notation conventionally used in structural mechanics has not always been adhered to. The field of design sensitivity analysis presents a dilemma regarding notation since it draws from fields as diverse as structural mechanics, differential calculus, calculus of variations, control theory, differential operator theory, and functional analysis. Unfortunately, the literature in each of these fields assigns a different meaning to the same symbol. Consequently, it is at times necessary to use symbols that look identical in an equation, but that come from very different notational systems. As a result, some notational compromise is required. The authors have adhered to standard notation except where ambiguity would arise, in which case the notation being used is indicated.

This book has been made possible due to contributions from the authors' former students, namely, Drs. R.-J. Yang, H.G. Lee, H.G. Seong, B. Dopker, T.-M. Yao, J.L.T. Santos, J.-S. Park, J. Lee, S.-L. Twu, K.-H. Chang, M. Godse, S.M. Wang, I. Shim, C.-J. Chen, Y.-H. Park, H.-Y. Hwang, X. Yu, W. Duan, S.-H. Cho, B.S. Choi, I. Grindeanu, J. Tu, B.-D. Youn, and Y. Yuan. Special appreciation is given to Professor K.-H. Chang at University of Oklahoma for his contributions to numerical methods and his examples in shape design sensitivity analysis and optimization. In addition, the authors value the contributions of colleagues Drs. J. Cea, B. Roussellet, J.P. Zolesio, R. Haftka, B.M. Kwak, G.W. Hou, H.L. Lam, and Y.M. Yoo. Finally, special thanks to Mr. R. Watkins for his outstanding work editing the manuscript.

Kyung K. Choi
Nam H. Kim
December 2004

Contents
1: Linear Systems

Contents
2: Nonlinear Systems and Applications

PART III
Design Sensitivity Analysis of Nonlinear Structural Systems

8
Nonlinear Structural Analysis

A nonlinear variational equation and an incremental solution procedure are introduced to be used for design sensitivity analysis purposes in subsequent chapters. Geometric nonlinearity is considered using the material and spatial descriptions of a deformation. Material nonlinearity includes hyperelasticity, rate-independent elastoplasticity, finite rotation using objective integration, and finite deformation elastoplasticity with the multiplicative decomposition of a deformation gradient. A contact/impact problem is sometimes called boundary nonlinearity, categorized by flexible-rigid and multibody contact/impact. In addition, frictional effect is important to certain applications. Transient dynamics significantly increase these three types of nonlinearities, and both implicit and explicit time integration methods are considered. Since the main purpose of this chapter is to introduce the basic procedure of nonlinear structural analysis, detailed discussions will refer to the literature for nonlinear structural analysis.

In nonlinear structural analysis, two approaches have been introduced: the total and the updated Lagrangian. The former refers to the initial configuration as a frame of reference, such that stress and strain are expressed in the initial domain of a structure, whereas the latter uses the current configuration as a frame of reference. In both formulations, equilibrium equations are obtained using the principle of virtual work. These equations are then linearized at the previously known configuration to yield incremental formulations for nonlinear analysis. As noted in [1] and [2], these two formulations are analytically equivalent.

To solve a nonlinear variational equation, the Newton-Raphson iterative method is frequently chosen. If a tangent operator of this nonlinear equation is exact, this method provides a quadratic convergence when the initial estimate is close enough to the solution. Even if the tangent operator is inexact, the response analysis may still converge after a greater number of iterations are performed. However, in sensitivity analysis, the inexact tangent operator produces an error in the sensitivity result.

8.1 Nonlinear Elastic Problems

Static and eigenvalue analyses of a nonlinear elastic problem are introduced in this section. The variational principle of a nonlinear elastic problem can be obtained from the minimum potential energy theory. The Newton-Raphson method is used to solve nonlinear variational problems. Through consistent linearization and transformation, it will be shown that the total and updated Lagrangian formulations are equivalent.

Based on the incremental equilibrium equation, the variational equation of a linear eigenvalue problem can be used for stability analysis in nonlinear structural systems. Various linear and nonlinear eigenvalue problems have been proposed to evaluate the stability status of nonlinear structural systems [3] through [8]. These problems differ in the assumption made between the critical load factor and the estimated critical load. The critical load factor of a nonlinear structural system can be evaluated by solving a linear eigenvalue problem at any prebuckling equilibrium configuration. The formulation of a stability analysis may include the effect of large displacements, large rotations, large strains, and material nonlinearities with appropriate kinematic and constitutive descriptions.

Hyperelasticity describes a rubberlike material with a strain energy density function, such that stress can be obtained by differentiating this strain energy function with respect to strain. In the case of an isotropic material, the strain energy function is expressed as a function of three invariants. The nearly incompressible properties of rubber can be considered by imposing a constraint condition between the displacement, calculated from the hydrostatic pressure, and the hydrostatic pressure obtained from pressure interpolation functions in the least-squares sense.

8.1.1 Static Problem

In the nonlinear problem, the assumption of infinitesimal deformation is not valid. Thus, the initial and deformed configurations have to be distinguished. Figure 8.1 shows the initial and deformed configurations of a structural domain, occupying ${}^{0}\Omega(\boldsymbol{X})$ initially and ${}^{n}\Omega(\boldsymbol{x})$ at the current configuration. The left superscript n is used to denote the configuration time t_n, and will be omitted unless needed for clarity. In static analysis, time is a convenient variable to denote different configurations during deformation, and to represent a different load. Let point $\boldsymbol{X}$ in the initial configuration move to point $\boldsymbol{x}$ in the deformed configuration. The total displacement of material point $\boldsymbol{X}$ is expressed as

$$\boldsymbol{z} = \boldsymbol{x} - \boldsymbol{X}. \tag{8.1}$$

Deformation gradient $\boldsymbol{F}$ is defined as a tensor, which associates infinitesimal vector $d\boldsymbol{X}$ at $\boldsymbol{X}$ with vector $d\boldsymbol{x}$ at $\boldsymbol{x}$ as

$$\boldsymbol{F} = \frac{\partial \boldsymbol{x}}{\partial \boldsymbol{X}} = \boldsymbol{1} + \frac{\partial \boldsymbol{z}}{\partial \boldsymbol{X}} = \boldsymbol{1} + \nabla_0 \boldsymbol{z}, \tag{8.2}$$

where $\boldsymbol{1}$ is the second-order identity tensor, and $\nabla_0 = \partial/\partial \boldsymbol{X}$ is the partial derivative in the initial configuration, such that $(\nabla_0 \boldsymbol{z})_{ij} = \partial z_i/\partial X_j$. The determinant of deformation gradient $\boldsymbol{F}$ is denoted by $J = det\ \boldsymbol{F}$.

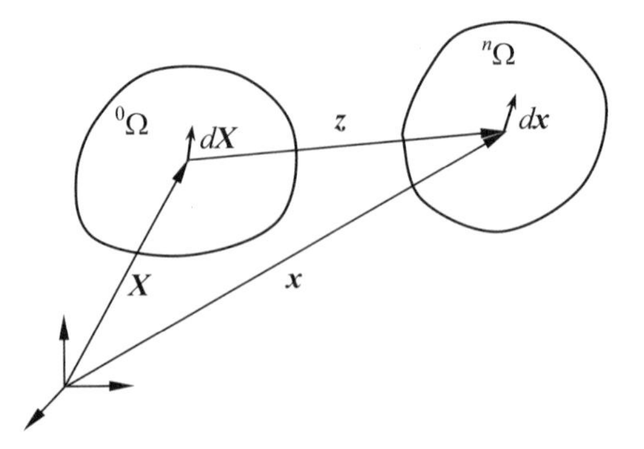

Figure 8.1. Configuration of a deformed body.

Total Lagrangian Formulation

The variational formulation of a nonlinear elastic problem can be obtained from the minimum total potential energy theory, assuming that the applied load is conservative. The total potential energy functional of an elastic problem is the difference between the stored strain energy Π^{int} and the work done by an external force Π^{ext}, and is given for a displacement formulation by

$$\begin{aligned}\Pi(\mathbf{z}) &= \Pi^{int}(\mathbf{z}) - \Pi^{ext}(\mathbf{z}) \\ &= \int_{{}^0\Omega} W(\boldsymbol{E})\, d\Omega - \int_{{}^0\Omega} \mathbf{z}^T \boldsymbol{f}^B\, d\Omega - \int_{{}^0\Gamma^S} \mathbf{z}^T \boldsymbol{f}^S\, d\Gamma,\end{aligned} \tag{8.3}$$

where $W(\boldsymbol{E})$ denotes the stored strain energy density function, $\boldsymbol{f}^B$ is the body force, $\boldsymbol{f}^S$ is the surface traction force on the boundary ${}^0\Gamma^S$, and $\boldsymbol{E}$ is the Green-Lagrange strain tensor, defined as

$$\boldsymbol{E} = \frac{1}{2}(\boldsymbol{F}^T\boldsymbol{F} - \boldsymbol{1}) = \frac{1}{2}(\nabla_0 \mathbf{z}^T + \nabla_0 \mathbf{z} + \nabla_0 \mathbf{z}^T \nabla_0 \mathbf{z}). \tag{8.4}$$

The minimum total potential energy theorem holds that displacement field $\mathbf{z} \in V$ minimizes (8.3). By taking the first-order variation of $\Pi(\mathbf{z})$ in the direction of $\bar{\mathbf{z}}$, and by equating $\delta\Pi(\mathbf{z})$ to zero, we obtain the variational equation

$$a(\mathbf{z}, \bar{\mathbf{z}}) = \ell(\bar{\mathbf{z}}), \quad \forall \bar{\mathbf{z}} \in Z, \tag{8.5}$$

where $a(\mathbf{z}, \bar{\mathbf{z}})$ is the structural energy form and $\ell(\bar{\mathbf{z}})$ is the load form, defined as

$$a(\mathbf{z}, \bar{\mathbf{z}}) = \int_{{}^0\Omega} \boldsymbol{S} : \bar{\boldsymbol{E}}\, d\Omega \tag{8.6}$$

$$\ell(\bar{\mathbf{z}}) = \int_{{}^0\Omega} \bar{\mathbf{z}}^T \boldsymbol{f}^B\, d\Omega + \int_{{}^0\Gamma^S} \bar{\mathbf{z}}^T \boldsymbol{f}^S\, d\Gamma. \tag{8.7}$$

$\boldsymbol{S} = \partial W / \partial \boldsymbol{E} = W_{,\boldsymbol{E}}$ is the second Piola-Kirchhoff stress tensor, while

$$\bar{\boldsymbol{E}} = sym(\nabla_0 \bar{\mathbf{z}}^T \boldsymbol{F}) = sym[\nabla_0 \bar{\mathbf{z}}^T (\nabla_0 \mathbf{z} + \boldsymbol{1})] \tag{8.8}$$

is the variation of the Green-Lagrange strain tensor presented in (8.4). The notation ":" is the contraction operator of tensors, such that $\boldsymbol{a} : \boldsymbol{b} = a_{ij}b_{ij}$, with summation in repeated indices, and $sym(\bullet)$ denotes the symmetric part of the tensor. In (8.5), solution space V and kinematically admissible displacement space Z are defined as

$$V = \left\{ \mathbf{z} \mid \mathbf{z} \in [H^1(\Omega)]^N, \mathbf{z}|_{\Gamma^g} = \boldsymbol{g} \right\} \tag{8.9}$$

$$Z = \left\{ \bar{\mathbf{z}} \mid \bar{\mathbf{z}} \in [H^1(\Omega)]^N, \bar{\mathbf{z}}|_{\Gamma^g} = 0 \right\}, \tag{8.10}$$

where Γ^g is the essential boundary where displacement is prescribed as $\mathbf{z} = \boldsymbol{g}$, and $H^1(\Omega)$ is the space of functions whose first-order derivatives are bounded in the energy norm. The variational equation (8.5) is called the *material description* or the *total Lagrangian formulation*, since the stress tensor $\boldsymbol{S}$ and the strain tensor $\boldsymbol{E}$ use the initial configuration as a reference. Note that $a(\mathbf{z}, \bar{\mathbf{z}})$ and $\ell(\bar{\mathbf{z}})$ are linear with respect to $\bar{\mathbf{z}}$ but are nonlinear with respect to displacement $\mathbf{z}$. Nonlinearity is due to the stress and strain tensors' implicit dependence on $\mathbf{z}$.

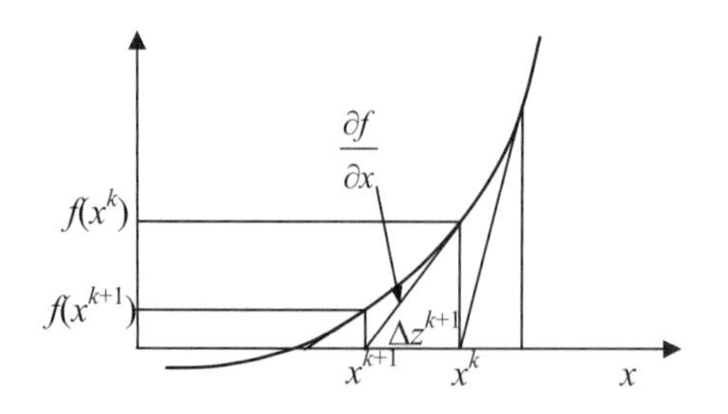

Figure 8.2. Newton-Raphson method for nonlinear equation $f = 0$.

Nonlinear variational equation (8.5) cannot be solved easily due to the nonlinearity involved in the constitutive relation. A general nonlinear functional can be solved using a Newton-Raphson iterative method through linearization. Let the linearization of a functional $f(\boldsymbol{x})$ in the direction $\Delta\boldsymbol{z}$ be denoted as

$$L[f] \equiv \frac{d}{d\omega} f(\boldsymbol{x} + \omega\Delta\boldsymbol{z})\bigg|_{\omega=0} = \frac{\partial f}{\partial \boldsymbol{x}}^T \Delta\boldsymbol{z}. \tag{8.11}$$

If right superscript k denotes the iteration counter, then the linear incremental solution procedure of the nonlinear equation $f(\boldsymbol{x}^{k+1}) = 0$ becomes

$$\frac{\partial f}{\partial \boldsymbol{x}^k}^T \Delta\boldsymbol{z}^{k+1} = -f(\boldsymbol{x}^k)$$

$$\boldsymbol{z}^{k+1} = \boldsymbol{z}^k + \Delta\boldsymbol{z}^{k+1} \tag{8.12}$$

$$\boldsymbol{x}^{k+1} = \boldsymbol{X} + \boldsymbol{z}^{k+1}.$$

Equation (8.12) is solved iteratively until the right side (or the residual term) vanishes. Figure 8.2 provides a one-dimensional example of the Newton-Raphson iterative method.

The nonlinear variational equation (8.5) can be linearized following the same procedure explained in (8.11). The linearization of the structural energy form in (8.6) can be written as

$$L[a(\boldsymbol{z}, \bar{\boldsymbol{z}})] = \int_{{}^0\Omega} [\Delta\boldsymbol{S} : \bar{\boldsymbol{E}} + \boldsymbol{S} : \Delta\bar{\boldsymbol{E}}]\, d\Omega, \tag{8.13}$$

where the incremental stress-strain relation is linear and can be written as

$$\Delta\boldsymbol{S} = \boldsymbol{C} : \Delta\boldsymbol{E}. \tag{8.14}$$

The constitutive relation for general isotropic, hyperelastic materials will be introduced in Section 8.1.3. To take a simple example using St. Vernant-Kirchhoff nonlinear elastic material [9], $\boldsymbol{S}$ is a linear function of $\boldsymbol{E}$, and $\boldsymbol{C}$ is constant with two independent parameters:

$$\boldsymbol{S} = \lambda\, tr(\boldsymbol{E})\boldsymbol{1} + 2\mu\boldsymbol{E} \tag{8.15}$$

$$C_{ijkl} = \lambda\delta_{ij}\delta_{kl} + \mu(\delta_{ik}\delta_{jl} + \delta_{il}\delta_{jk}), \tag{8.16}$$

where λ and μ are Lame's constants for isotropic material, and the Kronecker delta symbol is

$$\delta_{ij} = \begin{cases} 1 & if\ i = j \\ 0 & if\ i \neq j. \end{cases} \tag{8.17}$$

By noting that the deformation gradient increment is $\Delta \boldsymbol{F} = \nabla_0 \Delta \boldsymbol{z}$ from (8.2), the increments of the Green-Lagrange strain tensor and its variation can be obtained as

$$\Delta \boldsymbol{E} = sym(\nabla_0 \Delta \boldsymbol{z}^T \boldsymbol{F}) \tag{8.18}$$

$$\Delta \overline{\boldsymbol{E}} = sym(\nabla_0 \overline{\boldsymbol{z}}^T \nabla_0 \Delta \boldsymbol{z}). \tag{8.19}$$

Thus, the linearization of the structural energy form in (8.13) can be explicitly derived with respect to displacement and its variation as

$$\begin{aligned} L[a(\boldsymbol{z},\overline{\boldsymbol{z}})] &= \int_{{}^0\Omega} [\overline{\boldsymbol{E}} : \boldsymbol{C} : \Delta \boldsymbol{E} + \boldsymbol{S} : \Delta \overline{\boldsymbol{E}}]\, d\Omega \\ &\equiv a^*(\boldsymbol{z}; \Delta \boldsymbol{z}, \overline{\boldsymbol{z}}). \end{aligned} \tag{8.20}$$

The notation $a^*(\boldsymbol{z};\Delta\boldsymbol{z},\overline{\boldsymbol{z}})$ is used such that the configuration implicitly depends on total displacement $\boldsymbol{z}$, and $a^*(\boldsymbol{z};\Delta\boldsymbol{z},\overline{\boldsymbol{z}})$ is linear with respect to $\Delta\boldsymbol{z}$ and $\overline{\boldsymbol{z}}$. The first part of $a^*(\boldsymbol{z};\Delta\boldsymbol{z},\overline{\boldsymbol{z}})$ in (8.20) depends on the constitutive relation, while the second part is called the initial stiffness. Let the current time be t_n and let the current iteration counter be $k + 1$. Assuming that the external force is independent of displacement, the linearized incremental equation of (8.5) is obtained as

$$a^*({}^n\boldsymbol{z}^k; \Delta\boldsymbol{z}^{k+1}, \overline{\boldsymbol{z}}) = \ell(\overline{\boldsymbol{z}}) - a({}^n\boldsymbol{z}^k, \overline{\boldsymbol{z}}), \quad \forall \overline{\boldsymbol{z}} \in Z, \tag{8.21}$$

and the total displacement is updated using ${}^n\boldsymbol{z}^{k+1} = {}^n\boldsymbol{z}^k + \Delta\boldsymbol{z}^{k+1}$. Note that incremental equation (8.21) is in the form of ${}^n\boldsymbol{K}^k\Delta\boldsymbol{z}^{k+1} = {}^n\boldsymbol{F}^k$ after it has been discretized. Equation (8.21) is solved iteratively until the right side (the residual load) vanishes, which means that the original nonlinear equation (8.5) is satisfied.

Updated Lagrangian Formulation

A spatial description or an updated Lagrangian formulation uses stress and strain measures, such as the Cauchy stress tensor $\boldsymbol{\sigma}$ and the engineering strain tensor $\boldsymbol{\varepsilon}$, defined at the current configuration. The load form $\ell(\overline{\boldsymbol{z}})$ is the same as the total Lagrangian formulation in (8.7). The relation between material tensors ($\boldsymbol{S}$, $\boldsymbol{E}$) and spatial tensors ($\boldsymbol{\sigma}$, $\boldsymbol{\varepsilon}$) can be obtained through the transformation as

$$\boldsymbol{S} = J\boldsymbol{F}^{-1}\boldsymbol{\sigma}\boldsymbol{F}^{-T} \tag{8.22}$$

$$\overline{\boldsymbol{E}} = \boldsymbol{F}^T \boldsymbol{\varepsilon}(\overline{\boldsymbol{z}})\boldsymbol{F}, \tag{8.23}$$

where $\boldsymbol{\varepsilon}(\overline{\boldsymbol{z}}) = sym(\nabla_n \overline{\boldsymbol{z}})$ is the variation of the engineering strain tensor at the current configuration. Using (8.22) and (8.23), the structural energy form $a(\boldsymbol{z},\overline{\boldsymbol{z}})$ from (8.6) can be expressed in terms of the spatial description as

$$
\begin{aligned}
a(z,\bar{z}) &= \int_{{}^0\Omega} \boldsymbol{S} : \bar{\boldsymbol{E}}\, d\Omega \\
&= \int_{{}^0\Omega} (J\boldsymbol{F}^{-1}\boldsymbol{\sigma}\boldsymbol{F}^{-T}) : (\boldsymbol{F}^T \boldsymbol{\varepsilon}(\bar{z})\boldsymbol{F})\, d\Omega \\
&= \int_{{}^n\Omega} \boldsymbol{\sigma} : \boldsymbol{\varepsilon}(\bar{z})\, d\Omega.
\end{aligned} \tag{8.24}
$$

In (8.24), the property $d^n\Omega = Jd^0\Omega$ has been used. Using the updated Lagrangian formulation, the variational equation of a nonlinear elastic material has the same form as (8.5), namely,

$$
a(z,\bar{z}) = \ell(\bar{z}), \quad \forall \bar{z} \in Z. \tag{8.25}
$$

However, the expression of $a(z,\bar{z})$ is different from that of (8.6).

The linearization of (8.25) involves that of the Cauchy stress tensor, which is not easy to linearize because the Cauchy stress tensor is defined on the current configuration that is also changed during deformation. This difficulty can be overcome by transforming the material linearization, given in (8.20), to the current configuration. The integrands of (8.20) are transformed into the current configuration using the same method described in (8.24) as

$$
\begin{aligned}
(\boldsymbol{S} : \Delta\bar{\boldsymbol{E}}) &= J(\boldsymbol{F}^{-1}\boldsymbol{\sigma}\boldsymbol{F}^{-T}) : sym(\nabla_0 \bar{z}^T \nabla_0 \Delta z) \\
&= J\boldsymbol{\sigma} : sym(\nabla_n \bar{z}^T \nabla_n \Delta z) \\
&\equiv J\boldsymbol{\sigma} : \boldsymbol{\eta}(\Delta z, \bar{z})
\end{aligned} \tag{8.26}
$$

$$
\begin{aligned}
(\bar{\boldsymbol{E}} : \boldsymbol{C} : \Delta\boldsymbol{E}) &= (\boldsymbol{F}^T \boldsymbol{\varepsilon}(\bar{z})\boldsymbol{F}) : \boldsymbol{C} : (\boldsymbol{F}^T \boldsymbol{\varepsilon}(\Delta z)\boldsymbol{F}) \\
&= J\boldsymbol{\varepsilon}(\bar{z}) : \boldsymbol{c} : \boldsymbol{\varepsilon}(\Delta z),
\end{aligned} \tag{8.27}
$$

where the relation between the material and spatial constitutive tensors is given by

$$
c_{ijkl} = \frac{1}{J} F_{ir} F_{js} F_{km} F_{ln} C_{rsmn}. \tag{8.28}
$$

In the case of St. Vernant-Kirchhoff nonlinear elastic material from (8.16), by using the left Cauchy-Green deformation tensor $\boldsymbol{G} = \boldsymbol{F}\boldsymbol{F}^T$, the spatial constitutive tensor becomes

$$
c_{ijkl} = \frac{1}{J}\left[\lambda G_{ij} G_{kl} + \mu(G_{ik} G_{jl} + G_{il} G_{jk})\right]. \tag{8.29}
$$

Note that $\boldsymbol{c}$ is not constant for the updated Lagrangian formulation, and that $\boldsymbol{\sigma} \neq \boldsymbol{c} : \boldsymbol{\varepsilon}$. Linearization of the structural energy form can be obtained from (8.26) and (8.27) as

$$
\begin{aligned}
L[a(z,\bar{z})] &= \int_{{}^n\Omega} [\boldsymbol{\varepsilon}(\bar{z}) : \boldsymbol{c} : \boldsymbol{\varepsilon}(\Delta z) + \boldsymbol{\sigma} : \boldsymbol{\eta}(\Delta z, \bar{z})]\, d\Omega \\
&\equiv a^*(z; \Delta z, \bar{z}).
\end{aligned} \tag{8.30}
$$

The same notation $a^*(z;\Delta z,\bar{z})$ is used as the total Lagrangian formulation such that the configuration implicitly depends on total displacement z, and $a^*(z;\Delta z,\bar{z})$ is linear with respect to Δz and $\bar{z}$. If the current time is t_n, the current iteration counter is $k + 1$, and the external force is independent of displacement, then the linearized equation of (8.25) can be obtained as

$$
a^*({}^n z^k; \Delta z^{k+1}, \bar{z}) = \ell(\bar{z}) - a({}^n z^k, \bar{z}), \quad \forall \bar{z} \in Z \tag{8.31}
$$

Provided that appropriate constitutive relations are used, as in (8.28), the two linear formulations, (8.21) and (8.31), are theoretically equivalent but with different expressions. The choice of method should depend on how effective the numerical implementation is, and how convenient it is to generate the constitutive relation. For example, the strain measure of a total Lagrangian formulation is more complicated than that of an updated Lagrangian formulation. However, the constitutive relation in (8.15) can easily be used in the total Lagrangian formulation without transforming the constitutive relation into the current configuration, as in (8.28). In the case of elastoplasticity, the plastic evolution process always appears in the current configuration. Because it is difficult to express plastic evolution in terms of material stress measures, the updated Lagrangian formulation is a more attractive option for this case.

With a small deformation problem, it is possible to approximate $\boldsymbol{F}$ for $\boldsymbol{1}$. From this approximation and from their definitions, the two stress measures ($\boldsymbol{S}$ and $\boldsymbol{\sigma}$) become identical, and the same is true for the two strain measures ($\boldsymbol{E}$ and $\boldsymbol{\varepsilon}$).

The general formulations presented in this section will be applied to truss, beam, plane elastic solids, and plate components. The main purpose is to develop a structural energy form and its linearization to be used in later chapters when design sensitivity is discussed. For simplicity, examples are given for linear elastic material with large displacement, large rotation, and small strains.

Beam/Truss Components

Consider the beam/truss component in Fig. 8.3, with variable cross-sectional area $A(X_1)$. Let us consider the technical beam theory first. Using the undeformed configuration as a reference, the structural energy and load forms of this component [10] can be written as

$$a(z,\bar{z}) = \int_0^l [S_{11}(\bar{z}_{1,1} + \bar{z}_{1,1}z_{1,1})A + EIz_{2,11}\bar{z}_{2,11}]\, dX_1 \tag{8.32}$$

$$\ell(\bar{z}) = \int_0^l (\bar{z}_1 f_1 + \bar{z}_2 f_2)\, dX_1, \tag{8.33}$$

where E is the Young's modulus, $I(X_1)$ is the moment of inertia, $A(X)$ is the cross-sectional area, and l is the initial length of the neutral axis. In (8.32), $z_{1,1} = \partial z_1/\partial X_1$ and $z_{2,11} = \partial^2 z_2/\partial X_1^2$ are partial derivatives. For St. Vernant-Kirchhoff nonlinear elastic material, the second Piola-Kirchhoff stress, due to straining of the neutral axis, is defined as

$$S_{11} = E(z_{1,1} + \frac{1}{2}z_{1,1}z_{1,1}), \tag{8.34}$$

where the terms $z_{1,1} + \frac{1}{2}z_{1,1}z_{1,1}$ is the first component of Green-Lagrange strain E_{11}.

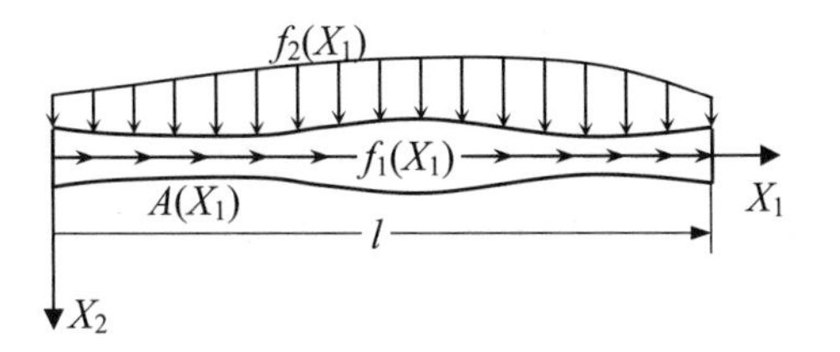

Figure 8.3. Beam/truss component.

The second Piola-Kirchhoff stress in (8.34) is nonlinear function of the displacement and, thus, the structural energy form in (8.32). In order to solve the nonlinear equation (8.5), the second Piola-Kirchhoff must be linearized. The incremental form of (8.34) becomes

$$\Delta S_{11} = E(\Delta z_{1,1} + z_{1,1}\Delta z_{1,1}). \tag{8.35}$$

In linear elasticity, the variation of the engineering strain ε is independent of the displacement. In nonlinear problem, however, the variation of Green-Lagrange strain is a linear function of displacement, whose linearization can be achieved by

$$\Delta \overline{E}_{11} = \overline{z}_{1,1}\Delta z_{1,1}. \tag{8.36}$$

By using (8.35) and (8.36), the linearization of the structural energy form can be obtained by

$$\begin{aligned} a^*(z;\Delta z,\overline{z}) = &\int_0^l [E(1+z_{1,1})^2\overline{z}_{1,1}\Delta z_{1,1} + S_{11}\overline{z}_{1,1}\Delta z_{1,1}]A\,dX_1 \\ &+ \int_0^l EI\overline{z}_{2,11}\Delta z_{2,11}\,dX_1. \end{aligned} \tag{8.37}$$

Note that the linearized structural energy form in (8.37) is bilinear with respect to Δz and $\overline{z}$.

As has been shown in Chapter 11, the technical beam theory has difficulty in configuration design sensitivity analysis due to its requirement in C^1-continuous design velocity fields. On the other hand, the Timoshenko beam theory only requires the C^0-continuous design velocity fields and, thus, it is a practical choice for the configuration design sensitivity analysis. The structural energy form for the Timoshenko beam/truss component can be written as

$$\begin{aligned} a(z,\overline{z}) = &\int_0^l S_{11}(\overline{z}_{1,1} + \overline{z}_{1,1}z_{1,1})A\,dX_1 \\ &+ \int_0^l [EI\overline{\theta}_{,1}\theta_{,1} + k\mu A(\overline{z}_{2,1} - \overline{\theta})(z_{2,1} - \theta)]dX_1, \end{aligned} \tag{8.38}$$

where k (=5/6) is the shear correction factor, μ is the shear modulus, θ is the rotation in X_3-coordinate, and $z = [z_1, z_2, \theta]^T$ is the state response. Note that the truss component is the same as (8.32) and, thus, its linearization. In the total Lagrangian formulation, the beam formulation in (8.38) is linear at the undeformed configuration. Thus, linearization of (8.38) can be can be carried out to obtain

$$\begin{aligned} a^*(z;\Delta z,\overline{z}) = &\int_0^l [E(1+z_{1,1})^2\overline{z}_{1,1}\Delta z_{1,1} + S_{11}\overline{z}_{1,1}\Delta z_{1,1}]A\,dX_1 \\ &+ \int_0^l [EI\overline{\theta}_{,1}\Delta\theta_{,1} + k\mu A(\overline{z}_{2,1} - \overline{\theta})(\Delta z_{2,1} - \Delta\theta)]dX_1. \end{aligned} \tag{8.39}$$

The load form for the for the Timoshenko beam/truss component includes additional moment. The load form can be defined as

$$\ell(\overline{z}) = \int_0^l \overline{z}^T f\,dX_1, \tag{8.40}$$

where $f = [f_1, f_2, m]^T$ and m is the distributed moment.

Plate/Plane Elastic Solids
Consider the thin plate/plane elastic solid component in Fig. 8.4 with thickness h. The structural energy form is

$$a(z,\bar{z}) = \iint_{^0\Omega} h\boldsymbol{S} : \bar{\boldsymbol{E}}\, d\Omega + \iint_{^0\Omega} \boldsymbol{\kappa}(\bar{z}_3)^T \boldsymbol{C}^b \boldsymbol{\kappa}(z_3)\, d\Omega, \tag{8.41}$$

where $z = [z_1, z_2, z_3]^T$ is the displacement of the neutral surface, $\boldsymbol{S}$ and $\boldsymbol{E}$ are stress and strain due to the straining of the neutral surface, $\boldsymbol{\kappa}(z)$ is the curvature vector, $\boldsymbol{C}^b$ is the bending stiffness matrix, and ${}^0\Omega$ is the undeformed neutral surface. The first term in the equation is associated with the stretching of the neutral surface, while the second term corresponds to the curvature of that stretched surface. In the thin plate formulation, only the vertical displacement is the state response, and the structural equaqtion (8.41) includes the second-order derivative of the vertical displacement. The load form of the component is

$$\ell(\bar{z}) = \iint_{^0\Omega} \bar{z}^T \boldsymbol{f}^B\, d\Omega + \int_{^0\Gamma^S} \bar{z}^T \boldsymbol{f}^S\, d\Gamma, \tag{8.42}$$

where $\boldsymbol{f}^B = [f_1^B, f_2^B, f_3^B]^T$ is the body force and $\boldsymbol{f}^S = [f_1^S, f_2^S, 0]^T$ is the lateral traction force, as shown in Fig. 8.4.

If a St. Vernant-Kirchhoff nonlinear elastic material is used for the constitutive model of the two-dimensional plane sold, the incremental stress-strain relation is the same as (8.14) with the stiffness tensor in (8.16). In this case, the indices of C_{ijkl} in (8.16) are i,j,k,l = 1, 2. The bending part of the structural energy form is basically the same with the linear problem. Thus, linearization of the structural energy form yields

$$\begin{aligned} a^*(z;\Delta z,\bar{z}) = &\iint_{^0\Omega} h(\bar{\boldsymbol{E}} : \boldsymbol{C} : \Delta\boldsymbol{E} + \boldsymbol{S} : \Delta\bar{\boldsymbol{E}})\, d^0\Omega \\ &+ \iint_{^0\Omega} \boldsymbol{\kappa}(\bar{z}_3)^T \boldsymbol{C}^b \boldsymbol{\kappa}(\Delta z_3)\, d^0\Omega, \end{aligned} \tag{8.43}$$

where the expressions of the incremental strain $\Delta\boldsymbol{E}$ and $\Delta\bar{\boldsymbol{E}}$ can be obtained from (8.18) and (8.19), respectively, by limiting the dimension of the matrix to 2 × 2.

In the case of the thick plate component, the structural energy form can be defined as

$$\begin{aligned} a(z,\bar{z}) = &\iint_{^0\Omega} h\boldsymbol{S} : \bar{\boldsymbol{E}}\, d^0\Omega \\ &+ \iint_{^0\Omega} [\boldsymbol{\kappa}(\bar{z})^T \boldsymbol{C}^b \boldsymbol{\kappa}(z) + \boldsymbol{\gamma}(\bar{z})^T \boldsymbol{C}^s \boldsymbol{\gamma}(z)]\, d^0\Omega, \end{aligned} \tag{8.44}$$

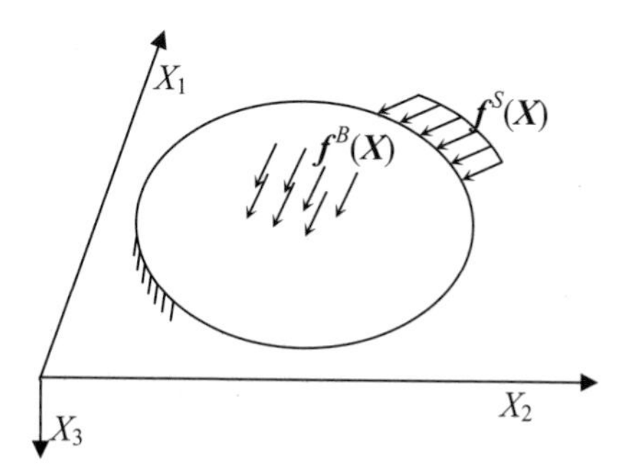

Figure 8.4. Plate/plane solid component.

where the state response $\boldsymbol{z} = [z_1, z_2, z_3, \theta_1, \theta_2]^T$, and θ_1 and θ_2 are rotations in X_1 and X_2 coordinates, respectively. In (8.44), the curvature and transverse shear vectors are defined as

$$\boldsymbol{\kappa}(\boldsymbol{z}) = \begin{Bmatrix} \theta_{1,1} \\ \theta_{2,2} \\ \theta_{1,2} + \theta_{2,1} \end{Bmatrix} \tag{8.45}$$

$$\boldsymbol{\gamma}(\boldsymbol{z}) = \begin{Bmatrix} z_{3,2} - \theta_2 \\ z_{3,1} - \theta_1 \end{Bmatrix}. \tag{8.46}$$

By using the fact that the plane stress part is the same as (8.41), the linearization of the structural energy form in (8.44) can be obtained as

$$\begin{aligned} a^*(\boldsymbol{z};\Delta\boldsymbol{z},\bar{\boldsymbol{z}}) &= \iint_{^0\Omega} h(\bar{\boldsymbol{E}}:\boldsymbol{C}:\Delta\boldsymbol{E} + \boldsymbol{S}:\Delta\bar{\boldsymbol{E}})\, d^0\Omega \\ &+ \iint_{^0\Omega} [\boldsymbol{\kappa}(\bar{\boldsymbol{z}})^T \boldsymbol{C}^b \boldsymbol{\kappa}(\Delta\boldsymbol{z}) + \boldsymbol{\gamma}(\bar{\boldsymbol{z}})^T \boldsymbol{C}^s \boldsymbol{\gamma}(\Delta\boldsymbol{z})] d^0\Omega. \end{aligned} \tag{8.47}$$

8.1.2 Critical Load

The critical load can be found by using the property that at least two adjacent configurations exist at that load [11]. The mathematical basis for such a situation immediately follows from the incremental equilibrium equations, that is, the left side of (8.21) or (8.31) vanishes at the critical limit point. To make it easier to follow the derivations, the linearized structural energy form in (8.20) can be divided into two parts

$$A(\boldsymbol{z};\Delta\boldsymbol{z},\bar{\boldsymbol{z}}) = \int_{^0\Omega} (\bar{\boldsymbol{E}}:\boldsymbol{C}:\Delta\boldsymbol{E})\, d\Omega \tag{8.48}$$

$$D(\boldsymbol{z};\Delta\boldsymbol{z},\bar{\boldsymbol{z}}) = -\int_{^0\Omega} (\boldsymbol{S}:\Delta\bar{\boldsymbol{E}})\, d\Omega. \tag{8.49}$$

With the critical displacement ${}^{cr}\boldsymbol{z}$ at the critical limit point $t_n = t_{cr}$, the stability equation becomes

$$a^*({}^{cr}\boldsymbol{z};\boldsymbol{y},\bar{\boldsymbol{y}}) \equiv A({}^{cr}\boldsymbol{z};\boldsymbol{y},\bar{\boldsymbol{y}}) - D({}^{cr}\boldsymbol{z};\boldsymbol{y},\bar{\boldsymbol{y}}) = 0, \qquad \forall \bar{\boldsymbol{y}} \in Z, \tag{8.50}$$

where the incremental displacement $\Delta\boldsymbol{z}$ in (8.48) and (8.49) is replaced with $\boldsymbol{y}$ to distinguish an eigenfunction from a real incremental displacement. The existence of nontrivial eigenfunction $\boldsymbol{y}$ of this nonlinear relation serves to identify a point of instability, that is, if the final equilibrium configuration is at the critical limit point, then solution $\boldsymbol{y}$ must be nontrivial.

The nonlinear incremental analysis is carried out until load ${}^n\boldsymbol{p} \le \boldsymbol{p}_{cr}$, corresponding to time t_n. To estimate critical load $\boldsymbol{p}_{cr}$, it is necessary to evaluate the left side of (8.50) at the critical limit point using the information that is available at a final prebuckling equilibrium configuration at time t_n. Linear extrapolation can be used to approximate the left side of (8.50) to form an eigenvalue problem. After solving the eigenvalue problem, the lowest eigenvalue is the critical load factor ${}^n\zeta$. Assuming a proportional conservative static loading, the estimated critical load vector ${}^n\boldsymbol{p}_{cr}$ can be expressed with the given load vector ${}^n\boldsymbol{p}$ and the critical load factor ${}^n\zeta$. Two commonly used approaches, one- and two-point linear eigenvalue approaches, are formulated in variational form, and expressions of the corresponding estimated critical load are presented.

One-Point Approach

Utilizing the information at the equilibrium configuration time t_n, (8.50) can be rewritten as an eigenvalue problem. By linearizing the nonlinear relationship between energy form D and the additional load increment at the critical limit point, D is approximated using the critical load factor ${}^n\zeta$ at time t_n: $D({}^{cr}z; y, \bar{y}) = {}^n\zeta D({}^nz; y, \bar{y})$. In addition, by neglecting the variations of energy form A due to the loading change, we can write $A({}^{cr}z; y, \bar{y})$ as $A({}^nz; y, \bar{y})$. Then, (8.50) becomes an eigenvalue problem, which can be called a one-point linear eigenvalue problem, in the form

$$A({}^nz; y, \bar{y}) - {}^n\zeta D({}^nz; y, \bar{y}) = 0, \quad \forall \bar{y} \in Z. \tag{8.51}$$

Solving this eigenvalue problem at a given load level ${}^n\boldsymbol{p}$ that is lower than the true critical load $\boldsymbol{p}_{cr}$ leads to the following estimated critical load:

$${}^n\boldsymbol{p}_{cr} = {}^n\zeta\, {}^n\boldsymbol{p}. \tag{8.52}$$

Two-Point Approach

Utilizing the information at two configurations, at time t_{n-1} and t_n where t_n is the final equilibrium configuration time, (8.50) can be rewritten as an eigenvalue problem. With the assumption that from time t_{n-1} onward the energy form $(A - D)$ changes linearly up to one additional load increment, but with the same ratio between the two configurations at time t_{n-1} and t_n, the energy form can be written as

$$A({}^{cr}z; y, \bar{y}) - D({}^{cr}z; y, \bar{y}) = B({}^{n-1}z; y, \bar{y}) + \zeta E({}^{n-1}z, {}^nz; y, \bar{y}), \tag{8.53}$$

where the energy forms B and E are defined as

$$B({}^{n-1}z; y, \bar{y}) \equiv A({}^{n-1}z; y, \bar{y}) - D({}^{n-1}z; y, \bar{y}) \tag{8.54}$$

$$E({}^{n-1}z, {}^nz; y, \bar{y}) \equiv B({}^nz; y, \bar{y}) - B({}^{n-1}z; y, \bar{y}), \tag{8.55}$$

and ζ is the critical load factor at time t_n. Consequently, (8.53) becomes a two-point linear eigenvalue problem, written as

$$B({}^{n-1}z; y, \bar{y}) + \zeta E({}^{n-1}z, {}^nz; y, \bar{y}) = 0, \quad \forall \bar{y} \in Z. \tag{8.56}$$

Solving this problem at a given load level ${}^n\boldsymbol{p}$ that is lower than the true critical load $\boldsymbol{p}_{cr}$ leads to an estimated critical load

$$\boldsymbol{p}_{cr} = {}^{n-1}\boldsymbol{p} + \zeta \Delta \boldsymbol{p}. \tag{8.57}$$

In (8.51) and (8.56), $\zeta \geq 1$ is the smallest eigenvalue, and $\boldsymbol{y}$ is the corresponding eigenfunction. Note that if the final equilibrium configuration time t_n is at the critical limit point, then $\zeta = 1$ in (8.51) and (8.56), and these two equations become identical to (8.50). The stability analysis of (8.51) or (8.56) can be applied to the any prebuckling configuration, and the estimated critical load becomes more accurate as the final equilibrium configuration approaches the critical limit point [5]. The estimated critical loads for both approaches are not conservative, that is, they are larger than the true critical load ${}^n\boldsymbol{p}_{cr} \geq \boldsymbol{p}_{cr}$. A stability analysis equation for linear structural systems can be obtained as a special case of the nonlinear stability equation, with the assumption of linearly elastic material and a small displacement.

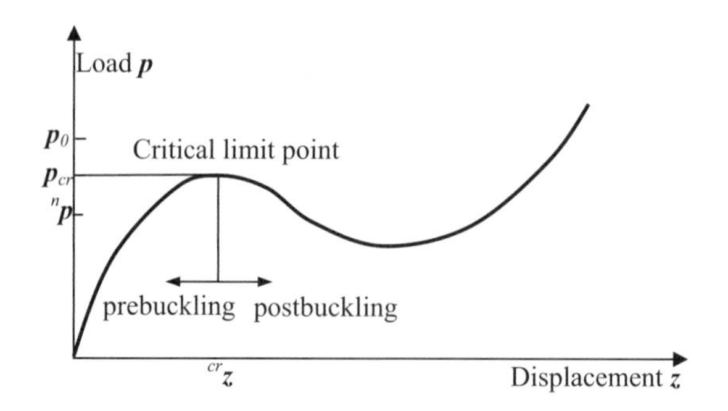

Figure 8.5. Equilibrium path of nonlinear structural system.

Stability Equation with Actual Critical Load Factor

Consider a structural system with the equilibrium path shown in Fig. 8.5. The critical limit point in Fig. 8.5 is a relative maximum point in the nonlinear load-displacement curve and defines the boundary between the prebuckling and the postbuckling equilibrium paths. Assume that the magnitude of the total applied load vector $\boldsymbol{p}_0$ is larger than the magnitude of the critical load vector $\boldsymbol{p}_{cr}$, and that, unlike the previous cases, the critical load vector $\boldsymbol{p}_{cr}$ occurs at the final prebuckling equilibrium configuration at time $t_n = t_{cr}$, i.e., $\boldsymbol{p}_{cr} = {}^n\boldsymbol{p}$. Note that the load vectors $\boldsymbol{p}_{cr}$, $\boldsymbol{p}_0$, and ${}^n\boldsymbol{p}$ have the same directions, since they are assumed to be proportional loadings. The magnitude of the critical load is unknown before the system is analyzed.

With the structural energy form $a(z,\bar{z})$ in the final prebuckling equilibrium configuration at time $t_n = t_{cr}$ and the load form $\ell(\bar{z})$ that is the virtual work done by the total applied load vector $\boldsymbol{p}_0$, the equilibrium equation of (8.5) can be rewritten as

$$a({}^n z,\bar{z}) = \beta \ell(\bar{z}), \qquad \forall \bar{z} \in Z, \tag{8.58}$$

where the actual critical load factor β is defined as the ratio of the magnitude of the critical load vector $\boldsymbol{p}_{cr} = {}^n\boldsymbol{p}$ to the magnitude of the total applied load vector $\boldsymbol{p}_0$, that is,

$$\boldsymbol{p}_{cr} = \beta \boldsymbol{p}_0. \tag{8.59}$$

The actual critical load factor β can be evaluated only after the critical load is known. Note that β is a function of design variable $\boldsymbol{u}$ and that $\beta \le 1.0$. When the total applied load vector $\boldsymbol{p}_0$ is equal to the critical load vector $\boldsymbol{p}_{cr} = {}^n\boldsymbol{p}$, $\beta = 1$ and the equilibrium equation of (8.58) is the same as the equilibrium equation of (8.5). Equation (8.58) serves as the stability equation for deriving the design sensitivity formulation for the critical load. Note that the actual critical load factor β does not depend on the configuration at t_n and is not related to the estimated critical load factor ${}^n\zeta$ which varies with configuration t_n.

8.1.3 Hyperelastic Material

In this section the response analysis of a hyperelastic problem is formulated based on a material description. In general, hyperelastic material exhibits the property of being incompressible during finite deformation, from which numerical difficulties associated with volumetric locking appear. The penalty method [12], the selective reduced integration method [13], and the mixed formulation [14] have been successfully used for incompressible and nearly incompressible media. To avoid volumetric locking for such material, Chen et al. [15] recently proposed a pressure projection method, which is a generalization of the $\bar{B}$ method [16] for linear problems. This method projects pressure by imposing a constraint condition between the hydrostatic pressure calculated from the displacement and the pressure obtained from the interpolation functions that are in a lower-order space than the displacement in a least-squares sense. If certain projection procedures are applied, this method degenerates into a form of the selective reduced integration method. When constant pressure is used for an element, the pressure projection method becomes equivalent to the perturbed Lagrangian formulation for nearly incompressible material.

If a strain energy density function exists, such that stress can be obtained from the derivative of the strain energy with respect to strain, the system is called path-independent. However, the nonlinear variational equation is solved by using a step-by-step incremental process with a number of load steps to finally reach the total applied load for computational purposes. The hyperelasticity problem contains both material nonlinearity from constitutive relations and geometric nonlinearity from kinematics.

Pressure Projection Method

A strain energy density function exists for hyperelastic material, and the stress-strain relation can be established using its derivative. Due to the assumption of isotropy, the strain energy density has to be a function of invariants. The near-incompressibility of hyperelastic material can be imposed by using the bulk modulus of a large magnitude, which relates the volumetric stress to the strain. Although a very simple material model is discussed here, the method can be extended to general hyperelastic material models. For the Mooney-Rivlin type of material model with a displacement-based single field formulation, the strain energy density function W in (8.3) is defined by

$$\begin{aligned} W(J_1,J_2,J_3) &= C_{10}(J_1-3)+C_{01}(J_2-3)+\frac{K}{2}(J_3-1)^2 \\ &= \qquad W_1(J_1,J_2) \qquad + \quad W_2(J_3), \end{aligned} \tag{8.60}$$

where C_{10}, C_{01} are the material constants, and K is the bulk modulus. For a small strain, $2(C_{10} + C_{01})$ represents the shear modulus, and $6(C_{10} + C_{01})$ represents the equivalent Young's modulus for a three-dimensional solid structure, and $8(C_{10} + C_{01})$ represents Young's modulus for a two-dimensional solid structure. In addition, J_1, J_2 and J_3 are the reduced invariants, defined by

$$J_1 = I_1 I_3^{-1/3}, \qquad J_2 = I_2 I_3^{-2/3}, \qquad J_3 = I_3^{1/2}, \tag{8.61}$$

where I_1, I_2, I_3 are the invariants of the Green deformation tensor $\boldsymbol{C} = \boldsymbol{F}^T\boldsymbol{F}$. The reason for introducing these reduced invariants is to separate dilation from distortion. J_1 and J_2 are constants in pure dilation; thus, they only contain distortional energy. Since W_1 in (8.60) is independent of dilation, volumetric locking is only related to W_2. The hydrostatic pressure is defined as the derivative of W_2 with respect to J_3, as

$$\tilde{p} \equiv \frac{\partial W(J_1,J_2,J_3)}{\partial J_3} = \frac{\partial W_2(J_3)}{\partial J_3} = K(J_3-1). \tag{8.62}$$

Since a large K is used to impose near-incompressibility, numerical instability can result in computing the pressure from the displacement. Chen et al. [15] proposed a pressure projection method, or volumetric locking, to relieve this instability. In this method, the pressure obtained from (8.62) is projected onto a lower-order space in the least-squares sense. In other words, $\boldsymbol{p} = [p_1, p_2, \ldots, p_n]^T$ is chosen to minimize

$$\Phi(\boldsymbol{p}) = \left\|\tilde{p} - \boldsymbol{Q}^T\boldsymbol{p}\right\|_{L_2}^2 = \int_{{}^0\Omega}\left(\tilde{p} - \boldsymbol{Q}^T\boldsymbol{p}\right)^2 d\Omega, \tag{8.63}$$

where $\boldsymbol{Q}(x) = [Q_1, Q_2, \ldots, Q_n]^T$ is a basis function for the projected space. The stationary condition of (8.63) becomes

$$\int_{{}^0\Omega}\overline{\boldsymbol{p}}^T\boldsymbol{Q}\left(\tilde{p} - \boldsymbol{Q}^T\boldsymbol{p}\right)d\Omega = 0. \tag{8.64}$$

The new variable $\boldsymbol{p}$ is added to the system, and (8.64) provides additional equations to determine $\boldsymbol{p}$.

The pressure projection method is a generalized approach that can be degenerated into another method by selecting an appropriate basis function $\boldsymbol{Q}$ and an integration method. For example, if reduced integration is used to evaluate (8.64), it is equivalent to the selective reduced integration method. If the full integration scheme is used, then the mixed displacement/pressure formulation for nearly incompressible material is obtained. A constant basis function is used for the basis function vector $\boldsymbol{Q}$, which implies constant pressure interpolation within a specific region. In this case, the basis function vector is a scalar, i.e., $Q = 1$, and accordingly the new pressure vector $\boldsymbol{p}$ is a scalar variable p. Although the choice of a basis function with a four-node finite element does not strictly satisfy the Babuska-Brezzi condition [17], it performs satisfactorily in practical situations and one can condense the constant pressure in each finite element. For this basis function choice, (8.64) becomes

$$\int_{{}^0\Omega}\overline{p}\left(J_3 - 1 - \frac{p}{K}\right)d\Omega = 0, \tag{8.65}$$

where constant $1/K$ is multiplied. Equation (8.65) will be added later to the structural variational form.

For hyperelastic material, the constitutive equation can be obtained by using strain energy density function W with an independently projected pressure term. The second Piola-Kirchhoff stress tensor can be obtained by

$$\boldsymbol{S} = W_{,\boldsymbol{E}} = C_{10}J_{1,\boldsymbol{E}} + C_{01}J_{2,\boldsymbol{E}} + pJ_{3,\boldsymbol{E}}. \tag{8.66}$$

Note that the projected pressure p is used in (8.66). The derivatives of the reduced invariants with respect to Green-Lagrange strain become

$$\begin{aligned}
J_{1,\boldsymbol{E}} &= I_{1,\boldsymbol{E}}(I_3)^{-1/3} - \frac{1}{3}I_1(I_3)^{-4/3}I_{3,\boldsymbol{E}} \\
J_{2,\boldsymbol{E}} &= I_{2,\boldsymbol{E}}(I_3)^{-2/3} - \frac{2}{3}I_2(I_3)^{-5/3}I_{3,\boldsymbol{E}} \\
J_{3,\boldsymbol{E}} &= \frac{1}{2}(I_3)^{-1/2}I_{3,\boldsymbol{E}},
\end{aligned}$$

where

$$\begin{aligned}
I_{1,\boldsymbol{E}} &= 2\boldsymbol{1} \\
I_{2,\boldsymbol{E}} &= 4\boldsymbol{1}(1 + tr(\boldsymbol{E})) - 4\boldsymbol{E} \\
I_{3,\boldsymbol{E}} &= 2\boldsymbol{1} + 4tr(\boldsymbol{E})\boldsymbol{1} - 4\boldsymbol{E} + \frac{9}{4}e_{imn}e_{jrs}E_{mr}E_{ns},
\end{aligned}$$

and $tr(\bullet)$ is a trace operator and e_{ijk} is a permutation symbol. The first variation of the strain energy density function in (8.60) can be written as

$$\overline{W} = W_{,E} : \overline{E} = S : \overline{E}. \tag{8.67}$$

The structural variational form is composed of the variation of the strain energy density function from (8.67) and the pressure projection term from (8.65). As discussed in Section 8.1.1, the variation of the total potential energy is defined by

$$a(\boldsymbol{r}, \overline{\boldsymbol{r}}) \equiv \int_{{}^0\Omega} \boldsymbol{S} : \overline{\boldsymbol{E}} \, d\Omega + \int_{{}^0\Omega} \overline{p} H \, d\Omega \tag{8.68}$$

with a new response variable $\boldsymbol{r} = [z^T, p]^T$, which contains all unknown variables in the analysis, and $H = J_3 - 1 - p/K$ corresponds to the volumetric strain. The structural energy form $a(\boldsymbol{r},\overline{\boldsymbol{r}})$ is nonlinear through the constitutive relation, strain tensor, and pressure projection. The linearization of the stress tensor can be expressed in terms of the displacement and pressure increment as

$$\begin{aligned} \Delta \boldsymbol{S} &= W_{,E,E} : \Delta \boldsymbol{E} + W_{,E,p} \Delta p \\ &= \boldsymbol{C} : \Delta \boldsymbol{E} + J_{3,E} \Delta p, \end{aligned} \tag{8.69}$$

where $\boldsymbol{C}$ is the fourth-order incremental stress-strain tensor at time t_n, referred to the configuration at time t_0, and $\Delta \boldsymbol{E}$ and Δp are the incremental strain and pressure. The linearization of structural energy form $a(\boldsymbol{r},\overline{\boldsymbol{r}})$ can be obtained using (8.18), (8.19), and (8.69) as

$$\begin{aligned} a^*(\boldsymbol{r}; \Delta \boldsymbol{r}, \overline{\boldsymbol{r}}) \equiv & \int_{{}^0\Omega} \left[\overline{\boldsymbol{E}} : (\boldsymbol{C} : \Delta \boldsymbol{E} + J_{3,E} \Delta p) + \boldsymbol{S} : \Delta \overline{\boldsymbol{E}} \right] d\Omega \\ & + \int_{{}^0\Omega} \overline{p} \left(J_{3,E} : \Delta \boldsymbol{E} - \frac{\Delta p}{K} \right) d\Omega. \end{aligned} \tag{8.70}$$

The pressure term can be condensed on the finite element level by directly solving the terms that contain the pressure variation. This can easily be done if constant pressure approximation is used, which can be done within a finite element.

8.2 Elastoplastic Problems

Elastoplastic material, combined with the hyperelastic material discussed in the previous section, represents material nonlinearity. Permanent material dislocation during plastic deformation is mathematically expressed as the evolution of internal plastic variables. Small material deformation is additively decomposed by elastic and plastic parts. A strain energy function is defined for elastic strain alone. Mathematically, elastoplasticity can be viewed as a projection of stress onto the elastic domain, which can be accomplished using a return-mapping algorithm. To guarantee the quadratic convergence of the response analysis, an algorithmic tangent operator that is consistent with the return-mapping algorithm can be obtained.

When the structure experiences a large deformation, the classical theory of elastoplasticity with the assumption of infinitesimal deformation is modified to consider rigid body motion. The objective rate plays an important role in elastoplasticity problems to systematically express the rigid body motion. The constitutive model, which represents a stress-strain relation, has to satisfy the principle of material frame indifference. Although a good deal of research has been performed on the objective rate, difficulties still remain concerning numerical integration methods that satisfy all physical requirements. The difficulty in obtaining an exact tangent stiffness operator is another drawback to this approach.

A new method for expressing the kinematics of finite deformation elastoplasticity using the hyperelastic constitutive relation, but not the hypo-elastic relation as mentioned above, is becoming a desirable approach to isotropic material. This method defines a stress-free intermediate configuration composed of a plastic deformation, and obtains the stress simply by taking a derivative of the strain energy density function with respect to the intermediate configuration. By using the constitutive relation between principal stress and logarithmic strain, better accuracy is obtained for a large elastic strain problem than with the classical elastoplasticity method. In addition, the same return-mapping algorithm from classical theory can be used in the principal stress space. The exact tangent stiffness operator especially guarantees quadratic convergence of the response analysis.

8.2.1 Small Deformation

A plastic deformation can be physically explained by atomic dislocation. An elastic deformation corresponds to the variation in the interatomic distance without causing atomic dislocation, while a plastic deformation implies relative sliding of the atomic layers and a permanent shape change without changing the structural volume. Plastic behavior can be efficiently described by the deviator of the tensors, which preserve the volumetric components. The deviatoric stress and strain tensors are defined as

$$\boldsymbol{s} \equiv \boldsymbol{\sigma} - \tfrac{1}{3}tr(\boldsymbol{\sigma})\boldsymbol{1} = \boldsymbol{I}_{dev} : \boldsymbol{\sigma} \tag{8.71}$$

$$\boldsymbol{e} \equiv \boldsymbol{\varepsilon} - \tfrac{1}{3}tr(\boldsymbol{\varepsilon})\boldsymbol{1} = \boldsymbol{I}_{dev} : \boldsymbol{\varepsilon}, \tag{8.72}$$

where $tr(\bullet)$ is a trace operator, $\boldsymbol{1}$ is the second-order unit tensor, $\boldsymbol{I}_{dev} = \boldsymbol{I} - \tfrac{1}{3}\boldsymbol{1}\otimes\boldsymbol{1}$ is the fourth-order unit deviatoric tensor, $\otimes$ is the standard tensor product, $I_{ijkl} = \tfrac{1}{2}(\delta_{ik}\delta_{jl} + \delta_{il}\delta_{jk})$ is the fourth-order unit symmetric tensor, and ":" is the contraction operator of tensors.

Assuming a small elastic strain, the stress and its rate can be additively decomposed into elastic and plastic parts as

$$\boldsymbol{\sigma} = \boldsymbol{\sigma}^e + \boldsymbol{\sigma}^p, \qquad \dot{\boldsymbol{\sigma}} = \dot{\boldsymbol{\sigma}}^e + \dot{\boldsymbol{\sigma}}^p, \tag{8.73}$$

where superscript e and p denote elastic and plastic parts, respectively. The superposed "dot" denotes the time rate of a quantity for temporal purposes. The notation in this section should not be confused with the symbol for a material derivative in sensitivity analysis. It is usually assumed that an elastic strain energy function exists for the plastic part, such that the stress can be determined by taking a derivative of the elastic energy function with respect to the elastic strain. The elastic part is assumed to be linear, and

$$W(\boldsymbol{\varepsilon}^e) = \tfrac{1}{2}\boldsymbol{\varepsilon}^e : \boldsymbol{C} : \boldsymbol{\varepsilon}^e = \tfrac{1}{2}(\boldsymbol{\varepsilon} - \boldsymbol{\varepsilon}^p) : \boldsymbol{C} : (\boldsymbol{\varepsilon} - \boldsymbol{\varepsilon}^p) \tag{8.74}$$

$$\boldsymbol{\sigma} = \frac{\partial W(\boldsymbol{\varepsilon}^e)}{\partial \boldsymbol{\varepsilon}^e} = \boldsymbol{C} : \boldsymbol{\varepsilon}^e = \boldsymbol{C} : (\boldsymbol{\varepsilon} - \boldsymbol{\varepsilon}^p) \tag{8.75}$$

$$\dot{\boldsymbol{\sigma}} = \boldsymbol{C} : (\dot{\boldsymbol{\varepsilon}} - \dot{\boldsymbol{\varepsilon}}^p), \tag{8.76}$$

where $\boldsymbol{C} = (\lambda + \tfrac{2}{3}\mu)\boldsymbol{1}\otimes\boldsymbol{1} + 2\mu\boldsymbol{I}_{dev}$ is the fourth-order isotropic constitutive tensor, and λ and μ are Lame's constants. From (8.76), the rate of stress can be decomposed into a volumetric (pressure) and a deviatoric component, as

$$\dot{p} = \tfrac{1}{3}tr(\dot{\boldsymbol{\sigma}}) = (\lambda + \tfrac{2}{3}\mu)tr(\dot{\boldsymbol{\varepsilon}}) \tag{8.77}$$

$$\dot{s} = 2\mu(\dot{e} - \dot{e}^p), \tag{8.78}$$

respectively. For rate independent plasticity, the von Mises pressure insensitive yield criterion with the associative flow rule is the commonly used method to describe material behavior after elastic deformation. Accordingly, the yield criterion or *yield function* is formulated as

$$f(\boldsymbol{\eta}, e_p) \equiv \|\boldsymbol{\eta}\| - \sqrt{\tfrac{2}{3}}\kappa(e_p) = 0, \tag{8.79}$$

where $\boldsymbol{\eta} = \boldsymbol{s}-\boldsymbol{\alpha}$, $\boldsymbol{\alpha}$ is the *back stress*, which is the center of the yield surface (the elastic domain), and is determined by the kinematic hardening law; $\kappa(e_p)$ is the radius of the elastic domain determined by the isotropic hardening rule; and e_p is an *effective plastic strain*. The combined isotropic/kinematic hardening law is used in (8.79). The elastic domain generated by the yield function in (8.79) forms a convex set as

$$E = \left\{(\boldsymbol{\eta}, e_p) \middle| f(\boldsymbol{\eta}, e_p) \le 0\right\}. \tag{8.80}$$

In mathematical terms, the plasticity can be thought of as a projection of the stress onto the yield surface. Since the yield surface is convex, the projection becomes a contraction mapping, which guarantees the existence of the projection.

It is assumed that a flow potential exists, and that plastic strain is proportional to the normal of the flow potential. If plastic flow is assumed to be associative, then the flow potential is the same as the yield function. Thus,

$$\dot{\boldsymbol{e}}^p = \gamma \frac{\partial f(\boldsymbol{\eta}, e_p)}{\partial \boldsymbol{\eta}} = \gamma \frac{\boldsymbol{\eta}}{\|\boldsymbol{\eta}\|} = \gamma \boldsymbol{N}, \tag{8.81}$$

where $\boldsymbol{N}$ is a unit deviatoric tensor normal to the yield surface, and γ is a *plastic consistency parameter*, which is nonnegative. If the material status is elastic, γ must be zero, but if it is plastic, then γ must be positive. The plastic strain rate is in the direction normal to the yield surface and has the magnitude of plastic consistency parameter γ. As the material deforms, the internal parameter (back stress, effective plastic strain, etc.) changes due to the hardening law. The rate of back stress can be determined by the kinematic hardening law as

$$\dot{\boldsymbol{\alpha}} = H_\alpha(e_p)\gamma \frac{\partial f(\boldsymbol{\eta}, e_p)}{\partial \boldsymbol{\eta}} = H_\alpha(e_p)\gamma \boldsymbol{N}, \tag{8.82}$$

where $H_\alpha(e_p)$ is the *plastic modulus for kinematic hardening*. The rate of effective strain can be expressed by

$$\dot{e}_p = \sqrt{\tfrac{2}{3}}\|\dot{\boldsymbol{e}}^p(t)\| = \sqrt{\tfrac{2}{3}}\gamma. \tag{8.83}$$

The Kuhn-Tucker loading/unloading condition becomes

$$\gamma \ge 0, \;\; f \le 0, \;\; \gamma f = 0. \tag{8.84}$$

It is possible to view the nonpositiveness of the yield function as a constraint, and plastic consistency parameter γ can be seen as the Lagrange multiplier corresponding to the inequality constraint.

Since all constitutive equations are in the form of rates in the elastoplastic model, the calculation of stresses requires integration. It is well known that the return-mapping algorithm, with the radial return method as a special case, is an effective and robust method for rate independent plasticity [18]. It is assumed that the solution and the status of material at time t_n is known. Since most solution procedures for elastoplasticity problems use the displacement-driven method, the configuration at time t_{n+1} is computed using the given displacement. In the return-mapping algorithm, a two-step method is often used. First, the elastic trial status is computed, and if the status of stress is outside the elastic domain, then the trial stress is projected onto the yield surface, which is a convex set. During the projection step, the yield surface itself is changed due to the evolution of internal variables.

For associative plasticity, it is well known that the backward Euler method produces the closest point projection. Since the displacement at time t_{n+1} is known in the displacement driven method, the incremental strain at time t_{n+1} can be computed from the definition of strain. The first step is called the elastic predictor and uses this incremental strain. The stress and hardening parameters are predicted elastically as

$$^{tr}\boldsymbol{s} = {}^{n}\boldsymbol{s} + 2\mu\Delta\boldsymbol{e} \tag{8.85}$$

$$^{tr}\boldsymbol{\alpha} = {}^{n}\boldsymbol{\alpha} \tag{8.86}$$

$$^{tr}e_p = {}^{n}e_p \tag{8.87}$$

$$^{tr}\boldsymbol{\eta} = {}^{tr}\boldsymbol{s} - {}^{tr}\boldsymbol{\alpha}, \tag{8.88}$$

where the left superscript n denotes the configuration time t_n, and "tr" denotes the trial status.

If the trial stress $^{tr}\boldsymbol{\eta}$ is within the elastic domain, then the stress is updated using the trial predictor, and elastic material status is declared. If the trial stress $^{tr}\boldsymbol{\eta}$ is outside the elastic domain, then the plastic corrector is carried out to find the plastic material status. The stress and hardening parameters are corrected by considering plastic deformation, given as

$$\begin{aligned} ^{n+1}\boldsymbol{s} &= {}^{tr}\boldsymbol{s} - 2\mu\Delta\boldsymbol{e}^{p} \\ &= {}^{tr}\boldsymbol{s} - 2\mu\hat{\gamma}\boldsymbol{N} \end{aligned} \tag{8.89}$$

$$^{n+1}\boldsymbol{\alpha} = {}^{tr}\boldsymbol{\alpha} + H_{\alpha}\hat{\gamma}\boldsymbol{N} \tag{8.90}$$

$$\begin{aligned} ^{n+1}\boldsymbol{\eta} &= {}^{n+1}\boldsymbol{s} - {}^{n+1}\boldsymbol{\alpha} \\ &= {}^{tr}\boldsymbol{\eta} - (2\mu + H_{\alpha})\hat{\gamma}\boldsymbol{N}, \end{aligned} \tag{8.91}$$

where $\boldsymbol{N} = {}^{n+1}\boldsymbol{\eta}/\|{}^{n+1}\boldsymbol{\eta}\|$, which is a unit deviatoric tensor and normal to the yield function at time t_{n+1}, and $\hat{\gamma} = \gamma\Delta t$ is computed from the yield condition at time t_{n+1}. An important feature of (8.91) is that the trial stress moves in the same direction as the final stress. Thus, the unit normal tensor to the yield surface can be computed from the trial stress by

$$\boldsymbol{N} = \frac{^{tr}\boldsymbol{\eta}}{\|{}^{tr}\boldsymbol{\eta}\|}, \tag{8.92}$$

which is known from the elastic predictor step.

At the return-mapped point the following yield condition is satisfied:

$$\begin{aligned} f({}^{n+1}\boldsymbol{\eta}, {}^{n+1}e_p) &= \left\| {}^{n+1}\boldsymbol{\eta} \right\| - \sqrt{\tfrac{2}{3}}\kappa({}^{n+1}e_p) \\ &= \left\| {}^{tr}\boldsymbol{\eta} \right\| - (2\mu + H_\alpha({}^{n+1}e_p))\hat{\gamma} - \sqrt{\tfrac{2}{3}}\kappa({}^{n+1}e_p) = 0, \end{aligned} \tag{8.93}$$

which is a nonlinear equation in terms of $\hat{\gamma}$. Equation (8.93) can be solved to compute $\hat{\gamma}$ using the local Newton method. If isotropic/kinematic hardening is a linear function of $\hat{\gamma}$, or of the effective plastic strain, then only one iteration is required to compute the return map point. After $\hat{\gamma}$ is found, the stress and hardening parameters can be updated at time t_{n+1} by

$${}^{n+1}\boldsymbol{s} = {}^{n}\boldsymbol{s} + 2\mu\Delta\boldsymbol{e} - 2\mu\hat{\gamma}\boldsymbol{N} \tag{8.94}$$

$${}^{n+1}\boldsymbol{\sigma} = {}^{n}\boldsymbol{\sigma} + \Delta\boldsymbol{\sigma} \tag{8.95}$$

$$\Delta\boldsymbol{\sigma} = \boldsymbol{C} : \Delta\varepsilon - 2\mu\hat{\gamma}\boldsymbol{N} \tag{8.96}$$

$${}^{n+1}\boldsymbol{\alpha} = {}^{n}\boldsymbol{\alpha} + H_\alpha\hat{\gamma}\boldsymbol{N} \tag{8.97}$$

$${}^{n+1}e_p = {}^{n}e_p + \sqrt{\tfrac{2}{3}}\hat{\gamma}. \tag{8.98}$$

Note that the stress and back-stress increments corresponding to the plastic correction component in (8.96) and (8.97) are in the same direction as $\boldsymbol{N}$, which is a radial direction of the yield surface, as shown in Fig. 8.6.

As discussed by Simo and Taylor [19], the tangent operator must be consistent with the time integration algorithm to achieve quadratic convergence of the Newton-Raphson method. Differentiation of the incremental stress tensor in (8.96) is taken with respect to the incremental strain tensor, which produces a consistent constitutive relation with the return-mapping algorithm as

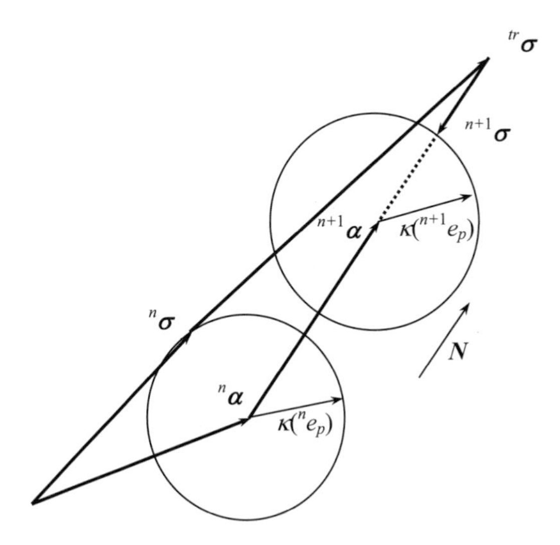

Figure 8.6. Return-mapping of isotropic elastoplasticity.

$$\begin{aligned}\boldsymbol{C}^{alg} &= \frac{\partial \Delta\boldsymbol{\sigma}}{\partial \Delta\boldsymbol{\varepsilon}} \\ &= \boldsymbol{C} - 2\mu\boldsymbol{N} \otimes \frac{\partial\hat{\gamma}}{\partial \Delta\boldsymbol{\varepsilon}} - 2\mu\hat{\gamma}\frac{\partial \boldsymbol{N}}{\partial \Delta\boldsymbol{\varepsilon}}.\end{aligned}$$

Since the consistency condition in (8.93) must satisfy for all strain status between time steps t_n and t_{n+1}, the differential of f with respect to $\delta\boldsymbol{\varepsilon}$ must vanish, from which the relation between $\hat{\gamma}$ and $\boldsymbol{\varepsilon}$ can be obtained as

$$\frac{\partial f}{\partial \Delta\boldsymbol{\varepsilon}} = 2\mu\boldsymbol{N} - \left(2\mu + H_\alpha + \sqrt{\tfrac{2}{3}}H_{\alpha,e_p}\hat{\gamma} + \tfrac{2}{3}\kappa_{,e_p}\right)\frac{\partial\hat{\gamma}}{\partial \Delta\boldsymbol{\varepsilon}} = 0.$$

The increment of unit normal to the yield function can also be expressed as

$$\begin{aligned}\frac{\partial \boldsymbol{N}}{\partial \Delta\boldsymbol{\varepsilon}} &= \frac{\partial \boldsymbol{N}}{\partial\, {}^{tr}\boldsymbol{\eta}}\frac{\partial\, {}^{tr}\boldsymbol{\eta}}{\partial \Delta\boldsymbol{\varepsilon}} \\ &= \left[\frac{\boldsymbol{I}}{\left\|{}^{tr}\boldsymbol{\eta}\right\|} - \frac{{}^{tr}\boldsymbol{\eta} \otimes {}^{tr}\boldsymbol{\eta}}{\left\|{}^{tr}\boldsymbol{\eta}\right\|^3}\right] : 2\mu\boldsymbol{I}_{dev} \\ &= \frac{2\mu}{\left\|{}^{tr}\boldsymbol{\eta}\right\|}\left[\boldsymbol{I}_{dev} - \boldsymbol{N} \otimes \boldsymbol{N}\right]\end{aligned}$$

Thus, from (8.96) the consistent or algorithmic tangent operator becomes

$$\boldsymbol{C}^{alg} = \frac{\partial \Delta\boldsymbol{\sigma}}{\partial \Delta\boldsymbol{\varepsilon}} = \boldsymbol{C} - 4\mu^2 A\boldsymbol{N} \otimes \boldsymbol{N} - \frac{4\mu^2\hat{\gamma}}{\left\|{}^{tr}\boldsymbol{\eta}\right\|}\left[\boldsymbol{I}_{dev} - \boldsymbol{N} \otimes \boldsymbol{N}\right], \tag{8.99}$$

where $A \equiv 1\Big/\left(2\mu + H_\alpha + \sqrt{\tfrac{2}{3}}H_{\alpha,e_p}\hat{\gamma} + \tfrac{2}{3}\kappa_{,e_p}\right)$. For notational convenience, the structural energy form and its linearization are defined as

$$a({}^{n+1}\boldsymbol{z}, \bar{\boldsymbol{z}}) \equiv \int_\Omega \boldsymbol{\varepsilon}(\bar{\boldsymbol{z}}) : {}^{n+1}\boldsymbol{\sigma}\, d\Omega \tag{8.100}$$

$$a^*({}^{n}\boldsymbol{z}; \Delta\boldsymbol{z}, \bar{\boldsymbol{z}}) \equiv \int_\Omega \boldsymbol{\varepsilon}(\bar{\boldsymbol{z}}) : \boldsymbol{C}^{alg} : \boldsymbol{\varepsilon}(\Delta\boldsymbol{z})\, d\Omega. \tag{8.101}$$

Note that unlike (8.30), the initial stiffness term does not appear, since only infinitesimal deformation is being considered. Total and updated Lagrangian formulations are identical for the infinitesimal deformation problem. The incremental iteration equation is the same as (8.21), with the structural energy form in (8.100) and its linearization in (8.101).

Combined Linear Isotropic/Kinematic Hardening Model

Although the plastic evolution models in (8.79) and (8.82) are general, many applications use a simple evolution model. A combined linear isotropic/kinematic hardening model is frequently used with $\kappa(e_p) = \sigma_0 + (1-\beta)He_p$ and $H_\alpha = \tfrac{2}{3}\beta H$. The yield function and hardening law can be defined as

$$f(\boldsymbol{\eta}, e_p) \equiv \|\boldsymbol{\eta}\| - \sqrt{\tfrac{2}{3}}[\sigma_0 + (1-\beta)He_p] = 0 \tag{8.102}$$

$$\dot{\boldsymbol{\alpha}} = \tfrac{2}{3}\beta H\gamma\boldsymbol{N}, \tag{8.103}$$

where H is the plastic modulus, which is constant for this model, and σ_0 is the initial yield stress. $\beta \in [0,1]$ is a parameter, used to consider the Baushinger effect. β equals one for kinematic hardening and zero for isotropic hardening. Note that the plastic consistency parameter $\hat{\gamma}$ in (8.93) can be explicitly solved without iteration. For example, when the status of a material is plastic, the $\hat{\gamma}$ is obtained by

$$\hat{\gamma} = A\left(\left\|{}^{tr}\boldsymbol{\eta}\right\| - \sqrt{\tfrac{2}{3}}\kappa({}^{n}e_p)\right)$$

$$\frac{1}{A} = 2\mu + \frac{2}{3}H.$$

8.2.2 Finite Rotation with Objective Integration

For a finite rotational problem, objective rate tensors have to be used to describe the motion of the structure and to obtain the material frame independent results. Although an extensive amount of research has been done pertaining to objective rates, only co-rotational Cauchy stress will be introduced here. Other types of objective rates can be used in a similar manner. Constitutive relations, however, use the rate-independent elastoplasticity introduced in Section 8.2.1. That approach is therefore valid for small elastic deformation, relatively large plastic deformation, and large rigid body rotation problems. Numerical difficulties associated with the objective rate involve the intricate transformation of a stress tensor into a rotation-free configuration, the unsymmetric properties of the tangent operator, and the difficulty in obtaining an exact tangent operator. Nevertheless, this model has been implemented in a good deal of application software.

It is well known that the rate of the Cauchy stress tensor is not objective, even if the Cauchy stress itself is objective. As a body rotates without deformation, the Cauchy stress tensor changes because the direction of the stress tensor has also changed. Consider a unit vector $\boldsymbol{e}_j$ in the jth direction of Cartesian coordinates under rigid body rotation. The increment of $\boldsymbol{e}_j$ can be computed by

$$\Delta\boldsymbol{e}_j = \boldsymbol{W}\,\boldsymbol{e}_j, \tag{8.104}$$

where $W_{i,j} = (\Delta z_{i,j} - \Delta z_{j,i})/2$ is a component of the spin tensor that represents a rigid body rotational motion. The relation between rotation tensor $\boldsymbol{R}$ and $\boldsymbol{W}$ is $\boldsymbol{W} = \Delta\boldsymbol{R}\,\boldsymbol{R}^T$ for a rigid body rotation, and the rate form of $\boldsymbol{e}_j = \boldsymbol{R}\boldsymbol{E}_j$ was used in (8.104), where $\boldsymbol{E}_j$ is a material unit vector. In Cartesian coordinates, the Cauchy stress tensor can be written as

$$\boldsymbol{\sigma} = \sigma_{ij}\boldsymbol{e}_i \otimes \boldsymbol{e}_j, \tag{8.105}$$

where $\boldsymbol{e}_i$ is a unit vector in the ith direction of Cartesian coordinates. The incremental form of the Cauchy stress tensor can be obtained by taking increments of (8.104) and (8.105) as

$$\begin{aligned}\Delta\boldsymbol{\sigma} &= \Delta\sigma_{ij}\,\boldsymbol{e}_i \otimes \boldsymbol{e}_j + \sigma_{ij}\,\Delta\boldsymbol{e}_i \otimes \boldsymbol{e}_j + \sigma_{ij}\,\boldsymbol{e}_i \otimes \Delta\boldsymbol{e}_j \\ &= \Delta\sigma_{ij}^J\,\boldsymbol{e}_i \otimes \boldsymbol{e}_j + \sigma_{ij}W_{ik}\,\boldsymbol{e}_k \otimes \boldsymbol{e}_j + \sigma_{ij}\,\boldsymbol{e}_i \otimes W_{jk}\boldsymbol{e}_k \\ &= (\Delta\sigma_{ij}^J + \sigma_{kj}W_{ik} - \sigma_{ik}W_{kj})\,\boldsymbol{e}_i \otimes \boldsymbol{e}_j,\end{aligned} \tag{8.106}$$

where $\Delta\sigma_{ij}^J$ is the Jaumann or co-rotational Cauchy stress increment, which is the objective rate because it takes an increment of the tensor with respect to the principal axis of the deformation rate tensor. Although the Jaumann stress rate is deficient with large shear strain problems because it produces an artificial oscillation for a simple shear problem, it is the one most frequently used. The constitutive equation in (8.99) must be written in the form of an objective rate

$$\Delta \boldsymbol{\sigma}^J = \boldsymbol{C}^{alg} : \Delta \boldsymbol{\varepsilon}. \tag{8.107}$$

The other two terms in (8.106) represent rigid body rotational effects. The incremental constitutive relation in (8.106) and (8.107) is only accurate for small, rigid body rotations. For finite rotation, the spin tensor $\boldsymbol{W}$ is not constant throughout the increment. Hughes and Winget [20] observed an excessive amount of rate form error and proposed an algorithm to preserve objectivity for large rotational increments. The strain increment and spin tensor are defined using a midpoint configuration as

$$\Delta \varepsilon_{ij} = \frac{1}{2}\left(\frac{\partial \Delta z_i}{\partial x_j^{n+\frac{1}{2}}} + \frac{\partial \Delta z_j}{\partial x_i^{n+\frac{1}{2}}} \right) \tag{8.108}$$

$$W_{ij} = \frac{1}{2}\left(\frac{\partial \Delta z_i}{\partial x_j^{n+\frac{1}{2}}} - \frac{\partial \Delta z_j}{\partial x_i^{n+\frac{1}{2}}} \right), \tag{8.109}$$

and stress at the previous iteration is updated to the rotation-free configuration by

$$\begin{aligned} {}^n\bar{\boldsymbol{\sigma}} &= \boldsymbol{Q}\, {}^n\boldsymbol{\sigma} \boldsymbol{Q}^T \\ {}^n\bar{\boldsymbol{\alpha}} &= \boldsymbol{Q}\, {}^n\boldsymbol{\alpha} \boldsymbol{Q}^T, \end{aligned} \tag{8.110}$$

where $\boldsymbol{Q} = (\boldsymbol{1} - ½\boldsymbol{W})^{-1}(\boldsymbol{1} + ½\boldsymbol{W}) = \boldsymbol{1} + (\boldsymbol{1} - ½\boldsymbol{W})^{-1}\boldsymbol{W}$, which is orthogonal and incrementally objective, and can be obtained by applying the generalized midpoint rule to $d\boldsymbol{Q}/dt = \boldsymbol{WQ}$. The usual return-mapping algorithm from Section 8.2.1 is applied to (8.110) after the transformation.

Variational Principle for Finite Rotation

It is inconvenient to express the variational principle of an elastoplastic problem in a total Lagrangian formulation, since the evolution of plastic variables is directly related to the Cauchy stress tensor. Thus, the updated Lagrangian formulation is a natural choice in terms of convenience. It is assumed that the solution and all configurations of the problem up to time t_{n-1} are known, and that the solution and configurations at current time t_n need to be computed. Variational equation (8.25) can be used for the spatial description, with the structural energy form defined as

$$a(\boldsymbol{z}, \bar{\boldsymbol{z}}) \equiv \int_{{}^n\Omega} \nabla_n \bar{\boldsymbol{z}} : \boldsymbol{\sigma}\, d\Omega, \tag{8.111}$$

where the symmetric property of the Cauchy stress tensor is used in (8.111), which is equivalent to (8.24). Equation (8.111) is a nonlinear function of displacement since the configuration and stress of the current time are unknown a priori. A linearization is required to solve variational equation (8.25) using the Newton-Raphson method iteratively. Let the external load be displacement independent, that is, conservative. Since the derivative and integration in (8.111) are carried out with respect to the current configuration, it is convenient to transform them to an undeformed configuration using the deformation gradient and the Jacobian. The structural energy form can be transformed into an undeformed configuration by

$$a(z,\bar{z}) = \int_{{}^n\Omega} \nabla_n \bar{z} : \boldsymbol{\sigma}\, d\Omega = \int_{{}^0\Omega} (\nabla_0 \bar{z} \boldsymbol{F}^{-1}) : \boldsymbol{\sigma} J\, d\Omega. \tag{8.112}$$

The integrand in this equation is the same as $\boldsymbol{T} : \bar{\boldsymbol{F}}$, where $\boldsymbol{T} = J\boldsymbol{F}^{-1}\boldsymbol{\sigma}$ is the first Piola-Kirchhoff stress tensor. Since the constitutive equation is given by the rate form of Cauchy stress, the linearization is carried out with respect to Cauchy stress in the undeformed configuration. To linearize (8.112), the linearized form of the deformation gradient and the Jacobian of deformation are introduced as

$$\Delta \boldsymbol{F} = \frac{\partial}{\partial \omega}\left[\frac{\partial (x_i + \omega \Delta z)}{\partial \boldsymbol{X}}\right]_{\omega=0} = \frac{\partial \Delta z}{\partial \boldsymbol{X}} = \nabla_0 \Delta z. \tag{8.113}$$

In addition, from the incremental relation of $\boldsymbol{F}\boldsymbol{F}^{-1} = \boldsymbol{1}$, the following is given:

$$\Delta \boldsymbol{F}^{-1} = -\boldsymbol{F}^{-1} \nabla_0 \Delta z \boldsymbol{F}^{-1} = -\boldsymbol{F}^{-1} \nabla_n \Delta z. \tag{8.114}$$

The incremental form of the Jacobian of the deformation gradient can be derived by direct linearization of the identity $|F_{mn}| = \frac{1}{6} e_{ijk} e_{rst} F_{ir} F_{js} F_{kt}$ with $e_{ijk} e_{ijr} = 2\delta_{kr}$

$$\Delta J = \Delta |\boldsymbol{F}| = J\, div(\Delta z). \tag{8.115}$$

By using (8.113), (8.114), and (8.115), the linearization of the structural energy form in (8.112) can be obtained as

$$\begin{aligned}
&\Delta \int_{{}^0\Omega} (\nabla_0 \bar{z} \boldsymbol{F}^{-1}) : \boldsymbol{\sigma} J d\Omega \\
&= \int_{{}^0\Omega} \left[(\nabla_0 \bar{z} \Delta \boldsymbol{F}^{-1}) : \boldsymbol{\sigma} J + (\nabla_0 \bar{z} \boldsymbol{F}^{-1}) : \Delta \boldsymbol{\sigma} J + (\nabla_0 \bar{z} \boldsymbol{F}^{-1}) : \boldsymbol{\sigma} \Delta J \right] d\Omega \\
&= \int_{{}^0\Omega} \left[-\nabla_0 \bar{z} \boldsymbol{F}^{-1} \nabla_n \Delta z : \boldsymbol{\sigma} + \nabla_0 \bar{z} \boldsymbol{F}^{-1} : \Delta \boldsymbol{\sigma} + \nabla_0 \bar{z} \boldsymbol{F}^{-1} : \boldsymbol{\sigma} div(\Delta z) \right] J d\Omega \\
&= \int_{{}^n\Omega} \left[-\nabla_n \bar{z} \nabla_n \Delta z : \boldsymbol{\sigma} + \nabla_n \bar{z} : \Delta \boldsymbol{\sigma} + \nabla_n \bar{z} : \boldsymbol{\sigma} div(\Delta z) \right] d\Omega \\
&= \int_{{}^n\Omega} \nabla_n \bar{z} : \left[\Delta \boldsymbol{\sigma} + \boldsymbol{\sigma} div(\Delta z) - \boldsymbol{\sigma} (\nabla_n \Delta z)^T \right] d\Omega \\
&= \int_{{}^n\Omega} \nabla_n \bar{z} : \left[\Delta \boldsymbol{\sigma}^J + \boldsymbol{W}\boldsymbol{\sigma} - \boldsymbol{\sigma}\boldsymbol{W} + \boldsymbol{\sigma} div(\Delta z) - \boldsymbol{\sigma} (\nabla_n \Delta z)^T \right] d\Omega \\
&= \int_{{}^n\Omega} \left[\nabla_n \bar{z} : (\boldsymbol{C}^{alg} - \boldsymbol{C}^*) : \nabla_n \Delta z + \boldsymbol{\sigma} : \boldsymbol{\eta}(\bar{z}, \Delta z) \right] d\Omega \\
&\equiv a^*(z; \Delta z, \bar{z}),
\end{aligned} \tag{8.116}$$

where the same notation in (8.30) is used for the linearized structural energy form $a^*(z;\Delta z,\bar{z})$ for the updated Lagrangian formulation, and where

$$C^*_{ijkl} = -\sigma_{ij}\delta_{kl} + \tfrac{1}{2}(\sigma_{il}\delta_{jk} + \sigma_{jl}\delta_{ik} + \sigma_{ik}\delta_{jl} + \sigma_{jk}\delta_{il}), \tag{8.117}$$

represents the rotational effect of the Cauchy stress tensor. With (8.116), the linearized variational equation using the Newton-Raphson method can be obtained. Since the reference configuration is the current, unknown one, the last iteration of current time is chosen as the reference configuration. Let the current time be t_n and let the current iteration counter be $k + 1$. Assuming that the external force is independent of displacement, the linearized incremental equation of (8.25) is obtained as

$$a^*({}^n z^k; \Delta z^{k+1}, \bar{z}) = \ell(\bar{z}) - a({}^n z^k, \bar{z}), \qquad \forall \bar{z} \in Z. \tag{8.118}$$

After the displacement increment is computed by solving linear systems of equations using (8.118), the return mapping is carried out to obtain the status of stress for each integration point, including internal plastic variables.

8.2.3 Finite Deformation with Hyperelasticity

Many finite deformation difficulties can be resolved by using a new plasticity model, where the constitutive relation is based on hyperelasticity. The multiplicative decomposition of an elastic-plastic deformation is converted into an additive decomposition by defining appropriate stress and strain measures. Even if the final variational equation is represented using the updated Lagrangian formulation, the reference for a constitutive relation is implicitly a stress-free intermediate configuration.

The theory of multiplicative plasticity proposed by Lee [21] is used to overcome the assumption of small elastic strain in the theory of classical infinitesimal plasticity, which uses an additive decomposition of the strain and its rate. The computational framework of this theory is proposed by Simo [22], which preserves the conventional return-mapping algorithm in the principal stress space. In this section, the constitutive relation and the return-mapping algorithm, both advocated by Simo, are summarized for a design sensitivity formulation that will later be developed. The deformation gradient $\boldsymbol{F}(\boldsymbol{X})$ of the motion at $\boldsymbol{X}$ in the undeformed configuration is assumed to take the following form of the local multiplicative decomposition:

$$\boldsymbol{F}(\boldsymbol{X}) = \boldsymbol{F}^{e}(\boldsymbol{X})\boldsymbol{F}^{p}(\boldsymbol{X}), \tag{8.119}$$

where $\boldsymbol{F}^{p}(\boldsymbol{X})$ denotes the deformation through the intermediate configuration, which is related to the internal variables, and $\boldsymbol{F}^{e^{-1}}(\boldsymbol{X})$ defines the local, stress-free, unloaded process. Figure 8.7 shows the decomposition of $\boldsymbol{F}(\boldsymbol{X})$ into the intermediate configuration followed by elastic deformation.

As in infinitesimal theory, the elastic domain is defined with the Kirchhoff stress tensor $\boldsymbol{\tau} = J\boldsymbol{\sigma}$, as

$$E \equiv \{(\boldsymbol{\tau},\boldsymbol{q}) \mid f(\boldsymbol{\tau},\boldsymbol{q}) \le 0\}, \tag{8.120}$$

where $\boldsymbol{q}$ is the vector of stresslike internal variables that characterize the hardening property of the material. The yield function $f(\boldsymbol{\tau},\boldsymbol{q})$ in (8.120) is an isotropic function of $\boldsymbol{\tau}$ due to the principle of objectivity, that is, the yield function does not depend on the orientation of the stress or on internal variables. It is assumed that the free energy

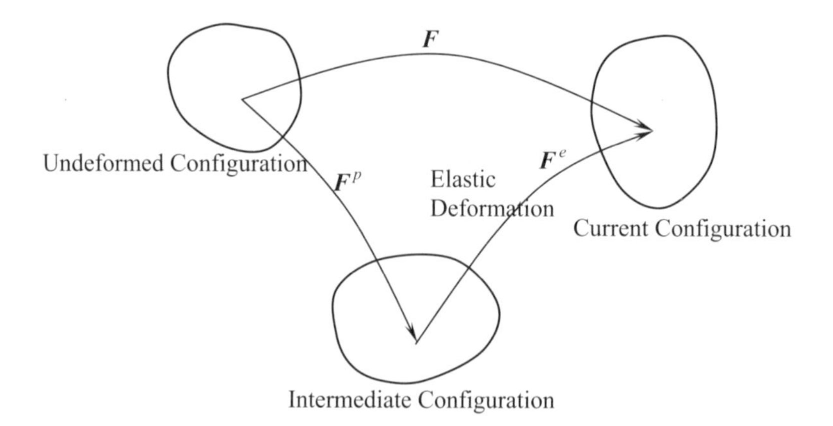

Figure 8.7. Multiplicative decomposition of deformation.

function locally depends on $\boldsymbol{F}^e(X)$ only, since the free energy represents stored energy through elastic deformation, and the free energy function is independent of the orientation, in the same context of the yield function as

$$\psi = \psi(\boldsymbol{b}^e, \xi), \tag{8.121}$$

where $\boldsymbol{b}^e \equiv \boldsymbol{F}^e \boldsymbol{F}^{eT}$ is the elastic left Cauchy-Green deformation tensor, and ξ is the vector of strainlike internal variables conjugate to $\boldsymbol{q}$ in the sense that $\boldsymbol{q} \equiv -\partial\psi/\partial\xi$.

The stress-strain relation can be obtained by defining a local dissipation function and using the principle of maximum dissipation. By ignoring thermoelastic parts, local dissipation function D is defined per unit reference volume as the difference between the rate of stress work and the rate of free energy change as

$$D \equiv \boldsymbol{\tau} : \boldsymbol{d} - \frac{d}{dt}\psi(\boldsymbol{b}^e, \xi) \geq 0, \tag{8.122}$$

where $\boldsymbol{d} = sym[\boldsymbol{L}]$ denotes the rate of deformation and $\boldsymbol{L} \equiv \frac{d}{dt}(\boldsymbol{F})\boldsymbol{F}^{-1}$ is the spatial velocity gradient. The time rate of the free energy function can be obtained using the chain rule and the time rate of $\boldsymbol{b}^e = \boldsymbol{F}\boldsymbol{C}^{p-1}\boldsymbol{F}^T$, with $\boldsymbol{C}^p = \boldsymbol{F}^{pT}\boldsymbol{F}^p$ as

$$\frac{d}{dt}(\boldsymbol{b}^e) = \boldsymbol{L}\boldsymbol{b}^e + \boldsymbol{b}^e\boldsymbol{L}^T + L_v\boldsymbol{b}^e, \tag{8.123}$$

where $\boldsymbol{C}^p$ is the plastic, right Cauchy-Green tensor, and $L_v\boldsymbol{b}^e$ is referred to as the Lie derivative of $\boldsymbol{b}^e$, which is obtained by pulling $\boldsymbol{b}^e$ back to the reference configuration and, after taking a time derivative, pushing $\boldsymbol{b}^e$ forward to the current configuration.

After expanding the time rate of the free energy function and using the fact that (8.122) holds for all admissible stresses and internal variables, the following constitutive relation and a reduced form of the dissipation inequality can be obtained, as detailed in Section 8.5.1:

$$\boldsymbol{\tau} = 2\frac{\partial\psi}{\partial\boldsymbol{b}^e}\boldsymbol{b}^e \tag{8.124}$$

$$D = \boldsymbol{\tau} : \left[-\tfrac{1}{2}(L_v\boldsymbol{b}^e)\boldsymbol{b}^{e-1}\right] + \boldsymbol{q}\tfrac{d}{dt}(\xi) \geq 0. \tag{8.125}$$

For the symmetric matrix, the property that $\boldsymbol{A}:\boldsymbol{BC} = \boldsymbol{AC}:\boldsymbol{B}$ is used to derive (8.125), and the skew-symmetric part of $\boldsymbol{L}$ (the spin) is canceled by multiplying it with the symmetric matrix. Using the principle of maximum dissipation and the classical results of the variational inequality, the dissipation inequality satisfies if and only if the coefficients of (8.125) lie within the normal cone of elastic domain E, defined in (8.120). The evolution equations can be obtained by using the normal property and plastic consistency parameter γ, as

$$-\tfrac{1}{2}L_v\boldsymbol{b}^e = \gamma\frac{\partial f(\boldsymbol{\tau},\boldsymbol{q})}{\partial\boldsymbol{\tau}}\boldsymbol{b}^e \tag{8.126a}$$

$$\tfrac{d}{dt}(\xi) = \gamma\frac{\partial f(\boldsymbol{\tau},\boldsymbol{q})}{\partial\boldsymbol{q}} \tag{8.126b}$$

$$\gamma \geq 0,\ f(\boldsymbol{\tau},\boldsymbol{q}) \leq 0,\ \gamma f(\boldsymbol{\tau},\boldsymbol{q}) = 0. \tag{8.126c}$$

Equation (8.126)c is the same as the Kuhn-Tucker condition of the classical elastoplasticity problem. It can be shown that the above evolution equations preserve the plastic volume change exactly for the pressure insensitive plasticity model. Note that the stress is computed by evaluating the free energy function at the current configuration.

For a displacement controlled problem, assume that the configuration at time t_n, $\{{}^n\boldsymbol{F}, {}^n\boldsymbol{b}^e, {}^n\xi\}$, and the incremental displacement $\Delta\boldsymbol{z}$ are known. Note that $\boldsymbol{b}^e$ is a primary variable instead of stress, as in the classical plasticity model. The relative deformation gradient is obtained from

$$\boldsymbol{f}(\boldsymbol{x}) = \boldsymbol{1} + \nabla_n \Delta\boldsymbol{z}, \tag{8.127}$$

which is a deformation gradient between time t_n and t_{n+1}. The total deformation gradient at time t_{n+1} is then ${}^{n+1}\boldsymbol{F}(\boldsymbol{X}) = \boldsymbol{f}(\boldsymbol{x})^n\boldsymbol{F}(\boldsymbol{X})$. The first-order system of evolution equations can be obtained by inserting (8.126) into (8.123) as

$$\tfrac{d}{dt}(\boldsymbol{f}) = \boldsymbol{L}\boldsymbol{f} \tag{8.128a}$$

$$\tfrac{d}{dt}(\boldsymbol{b}^e) = \left[\boldsymbol{L}\boldsymbol{b}^e + \boldsymbol{b}^e\boldsymbol{L}^T\right] - 2\gamma\frac{\partial f(\boldsymbol{\tau},\boldsymbol{q})}{\partial \boldsymbol{\tau}}\boldsymbol{b}^e \tag{8.128b}$$

$$\tfrac{d}{dt}(\xi) = \gamma\frac{\partial f(\boldsymbol{\tau},\boldsymbol{q})}{\partial \boldsymbol{q}} \tag{8.128c}$$

$$\gamma \ge 0,\ f(\boldsymbol{\tau},\boldsymbol{q}) \le 0,\ \gamma f(\boldsymbol{\tau},\boldsymbol{q}) = 0, \tag{8.128d}$$

with initial conditions $\{\boldsymbol{f}, \boldsymbol{b}^e, \xi\}\big|_{t=t_n} = \{\boldsymbol{1}, {}^n\boldsymbol{b}^e, {}^n\xi\}$. Kirchhoff stress and stresslike internal variables are dependent, and are defined in terms of primary variables by the following constitutive law:

$$\boldsymbol{\tau} = 2\frac{\partial \psi(\boldsymbol{b}^e,\xi)}{\partial \boldsymbol{b}^e}\boldsymbol{b}^e \tag{8.129}$$

$$\boldsymbol{q} = -\frac{\partial \psi(\boldsymbol{b}^e,\xi)}{\partial \xi}. \tag{8.130}$$

This process represents the major difference from the classical plasticity model, where an evolution equation is expressed in terms of stress. As in classical infinitesimal plasticity, the evolution equations in (8.128) can be split into a trial elastic state and a plastic return mapping.

The trial elastic state can be obtained by eliminating plastic flow and pushing the elastic, left Cauchy-Green deformation tensor forward to the current configuration using the relative deformation gradient as

$${}^{tr}\boldsymbol{f} = \boldsymbol{f},\ {}^{tr}\boldsymbol{b}^e = \boldsymbol{f}\,{}^n\boldsymbol{b}^e\boldsymbol{f}^T,\ {}^{tr}\xi = {}^n\xi. \tag{8.131}$$

If $\boldsymbol{\tau}$ and $\boldsymbol{q}$ in (8.129) and (8.130), which are evaluated using the trial state in (8.131), are within the elastic domain of (8.120), then the trial stress in (8.129) and the stress-like internal variable in (8.130) are exact. Otherwise, plastic return-mapping is carried out on the current, fixed configuration by integrating (8.128) between time t_n and t_{n+1} with constraints imposed on the stress through the yield function in (8.128)d. By integrating the differential part of (8.128) the following is obtained:

$$\boldsymbol{b}^e = {}^{tr}\boldsymbol{b}^e \exp\left[-2\hat{\gamma}\frac{\partial f(\boldsymbol{\tau}^e,\boldsymbol{q})}{\partial \boldsymbol{\tau}}\right] \tag{8.132a}$$

$$\xi = {}^{tr}\xi + \hat{\gamma}\frac{\partial f(\boldsymbol{\tau}^e,\boldsymbol{q})}{\partial \boldsymbol{q}} \tag{8.132b}$$

$$\hat{\gamma} \geq 0,\ f(\boldsymbol{\tau},\boldsymbol{q}) \leq 0,\ \hat{\gamma} f(\boldsymbol{\tau},\boldsymbol{q}) = 0. \tag{8.132c}$$

The differential equation $\dot{y} = Ay$ has $y = y_0\exp(At)$ as the solution to (8.132)a, and $\hat{\gamma} \equiv \gamma\Delta t$ is used. This integration algorithm has first-order accuracy and unconditional stability. Note that the return-mapping algorithms in (8.132) are carried out for the left Cauchy-Green tensor instead of for stress, as with the classical plasticity model.

The plastic evolution algorithm (8.132) can be implemented using the return-mapping algorithm of principal Kirchhoff stress for the isotropic material with a fixed current configuration. The principal direction of $\boldsymbol{\tau}$ is aligned with that of $\boldsymbol{b}^e$ by using the isotropic assumption, and their spectral decomposition becomes

$$\boldsymbol{b}^e = \sum_{i=1}^{3} \lambda_i^2\, \hat{\boldsymbol{n}}^i \otimes \hat{\boldsymbol{n}}^i \tag{8.133}$$

$$\boldsymbol{\tau} = \sum_{i=1}^{3} \tau_i^p\, \hat{\boldsymbol{n}}^i \otimes \hat{\boldsymbol{n}}^i, \tag{8.134}$$

where λ_i is the principal stretch, $\boldsymbol{\tau}^p$ is the principal Kirchhoff stress, and $\hat{\boldsymbol{n}}$ is the spatial eigenvector corresponding to the material eigenvector $\hat{\boldsymbol{N}}$. Since ${}^{tr}\boldsymbol{b}^e$ has the same principal direction as $\boldsymbol{b}^e$ in (8.132)a, the principal direction in (8.133) and (8.134) can be computed from the known principal direction of ${}^{tr}\boldsymbol{b}^e$.

For derivational simplicity, the following vector notations of the principal quantities are defined. A principal stress vector and a logarithmic elastic principal stretch vector are defined by $\boldsymbol{\tau}^p = [\tau_1^p, \tau_2^p, \tau_3^p]^T$ and $\boldsymbol{e} = [e_1, e_2, e_3]^T = [\log(\lambda_1), \log(\lambda_2), \log(\lambda_3)]^T$, respectively. In the J_2-flow theory, it can be assumed that the free energy function is in the following quadratic form:

$$\psi(\boldsymbol{e},\xi) = \tfrac{1}{2}\lambda[e_1 + e_2 + e_3]^2 + \mu[e_1^2 + e_2^2 + e_3^2] + \hat{K}(\xi), \tag{8.135}$$

where $\hat{K}(\xi)$ denotes energy from the isotropic hardening law. The constitutive relation in (8.129) can be reduced to principal space by $\boldsymbol{\tau}^p = \partial\psi/\partial\boldsymbol{e}$. The relation between principal stress and the logarithmic principal elastic stretch becomes

$$\boldsymbol{\tau}^p = \boldsymbol{c}^e\boldsymbol{e}, \tag{8.136}$$

where $\boldsymbol{c}^e = (\lambda + \frac{2}{3}\mu)\hat{\boldsymbol{1}} \otimes \hat{\boldsymbol{1}} + 2\mu\boldsymbol{1}_{dev}$ is the usual 3 × 3 constitutive tensor for an isotropic material, $\hat{\boldsymbol{1}} = [1, 1, 1]^T$ is the first-order vector, and $\boldsymbol{1}_{dev} = \boldsymbol{1} - \frac{1}{3}(\hat{\boldsymbol{1}} \otimes \hat{\boldsymbol{1}})$ is the second-order deviatoric tensor. These notations can be thought of as second-order extensions of the fourth-order notations given in (8.71). Taking the logarithm of (8.132)a and premultiplying $\boldsymbol{c}^e$ yields the following return-mapping algorithm forms in the principal stress space:

$${}^{tr}\boldsymbol{\tau}^p = \boldsymbol{c}^e\boldsymbol{e}^{tr} \tag{8.137a}$$

$$\boldsymbol{\tau}^p = {}^{tr}\boldsymbol{\tau}^p - \hat{\gamma}\,\boldsymbol{c}^e \frac{\partial \hat{f}(\boldsymbol{\tau}^p, \boldsymbol{q})}{\partial \boldsymbol{\tau}^p} \tag{8.137)b}$$

$$\xi = {}^n\xi + \hat{\gamma} \frac{\partial \hat{f}(\boldsymbol{\tau}^p, \boldsymbol{q})}{\partial \boldsymbol{q}} \tag{8.137)c}$$

$$\hat{\gamma} \ge 0,\ \hat{f}(\boldsymbol{\tau}^p, \boldsymbol{q}) \le 0,\ \hat{\gamma}\hat{f}(\boldsymbol{\tau}^p, \boldsymbol{q}) = 0, \tag{8.137)d}$$

where $f(\boldsymbol{\tau}, \boldsymbol{q}) = \hat{f}(\boldsymbol{\tau}^p, \boldsymbol{q})$ and $\partial f / \partial \boldsymbol{\tau} = \sum_{i=1}^{3} \partial \hat{f} / \partial \tau_i^p \hat{\boldsymbol{n}}^i \otimes \hat{\boldsymbol{n}}^i$ can be obtained by direct calculation. These return-mapping algorithms are the same as those from classical plasticity, with a difference that principal Kirchhoff stress and logarithmic strain are used instead of Cauchy stress and engineering strain.

Return-Mapping Algorithms

Since a plastic behavior can be efficiently described by the deviatoric part of a vector, which preserves the volume change, a deviatoric principal stress is defined by

$$\boldsymbol{s} = \boldsymbol{\tau}^p - \tfrac{1}{3}(\boldsymbol{\tau}^p \cdot \hat{\boldsymbol{1}})\hat{\boldsymbol{1}} = \boldsymbol{I}_{dev} \cdot \boldsymbol{\tau}^p. \tag{8.138}$$

For rate-independent plasticity, the von Mises pressure insensitive yield criterion and the associative flow rule are the commonly used methods to describe metal-like material behavior after elastic deformation. Accordingly, the yield criterion or yield function is formulated as

$$f(\boldsymbol{\eta}, e_p) = \|\boldsymbol{\eta}\| - \sqrt{\tfrac{2}{3}}\kappa(e_p) = 0, \tag{8.139}$$

where $\boldsymbol{\eta}$ is equal to $\boldsymbol{s} - \boldsymbol{\alpha}$ and $\boldsymbol{\alpha}$ is the back stress, which is the center of the yield surface and is determined by the kinematic hardening law. It can be shown that $\|dev(\boldsymbol{\tau})\| = \|\boldsymbol{s}\|$. $\kappa(e_p)$ is the radius of the yield surface and is determined by the isotropic hardening rule. For a general nonlinear hardening rule, we can express $\kappa(e_p)$ in the form

$$\kappa(e_p) = \sigma_0 + \hat{K}_{,e_p}, \tag{8.140}$$

where σ_0 is the initial yield stress of a uniaxial tension test, and $\hat{K}_{,e_p}(e_p)$ is the isotropic hardening law. e_p is an effective plastic strain and can be determined using the uniaxial tension test as

$$e_p = \int_0^t \sqrt{\tfrac{2}{3}} \|\dot{\boldsymbol{e}}^p(\tau)\| d\tau. \tag{8.141}$$

The internal variables are reduced to the effective plastic strain and back stress.

Now, the return-mapping algorithms in (8.137) can be implemented in the principal Kirchhoff stress space by

$${}^{n+1}\boldsymbol{\tau}^p = \boldsymbol{c}^e \boldsymbol{e}^{tr} \tag{8.142)a}$$

$${}^{n+1}\boldsymbol{\tau}^p = {}^{tr}\boldsymbol{\tau}^p - 2\mu\hat{\gamma}\boldsymbol{N} \tag{8.142)b}$$

$${}^{n+1}\boldsymbol{\alpha} = {}^n\boldsymbol{\alpha} + \hat{\gamma} H_\alpha \boldsymbol{N} \tag{8.142)c}$$

$$ {}^{n+1}e_p = {}^n e_p + \sqrt{\tfrac{2}{3}}\hat{\gamma}, \tag{8.142)d} $$

where $H_\alpha(e_p)$ is a plastic modulus for the kinematic hardening and

$$ \boldsymbol{N} \equiv \frac{\boldsymbol{\eta}}{\|\boldsymbol{\eta}\|} = \frac{{}^{tr}\boldsymbol{\eta}}{\|{}^{tr}\boldsymbol{\eta}\|} \tag{8.143} $$

is an outward unit normal vector on the yield surface of (8.139), and where $\hat{\gamma}$ is computed using a local Newton-Raphson method by imposing the consistency condition

$$ \begin{aligned} f(\boldsymbol{\eta}, e_p) &= \|\boldsymbol{\eta}\| - \sqrt{\tfrac{2}{3}}\kappa(e_p) \\ &= \|{}^{tr}\boldsymbol{\eta}\| - (2\mu + H_\alpha(e_p))\hat{\gamma} - \sqrt{\tfrac{2}{3}}\kappa(e_p) = 0, \end{aligned} \tag{8.144} $$

which is a nonlinear equation with respect to $\hat{\gamma}$. Equation (8.144) can be solved using a local Newton method to compute $\hat{\gamma}$. The convexity of the elastic domain in (8.120) guarantees the stability of this return-mapping method, which is a contraction mapping. If the isotropic/kinematic hardening is a linear function of $\hat{\gamma}$ or of the effective plastic strain, then only one iteration is required to compute the return-mapped point. The gradient of (8.144) can be evaluated by

$$ \frac{\partial f}{\partial \hat{\gamma}} = -\left(2\mu + H_\alpha + \sqrt{\tfrac{2}{3}}H_{\alpha,e_p}\hat{\gamma} + \tfrac{2}{3}\kappa_{,e_p}\right) \equiv -\frac{1}{A}. \tag{8.145} $$

The Kirchhoff stress tensor can be obtained from (8.134) using the principal stress and principal direction as

$$ \boldsymbol{\tau} = \sum_{i=1}^{3} \tau_i^p \boldsymbol{m}^i, \qquad \text{where } \boldsymbol{m}^i = \hat{\boldsymbol{n}}^i \otimes \hat{\boldsymbol{n}}^i. \tag{8.146} $$

The left Cauchy-Green deformation tensor is updated and stored by the formula in (8.133), which represents the intermediate configuration

$$ \boldsymbol{b}^e = \sum_{i=1}^{3} \exp[2e_i]\hat{\boldsymbol{n}}^i \otimes \hat{\boldsymbol{n}}^i, \tag{8.147} $$

where $\boldsymbol{e} = \boldsymbol{e}^{tr} - \hat{\gamma}\boldsymbol{N}$ is an elastic logarithmic principal strain. Equation (8.147) corresponds to the update of $\boldsymbol{C}^{p-1} = \boldsymbol{F}^{-1}\boldsymbol{b}^e\boldsymbol{F}^{-T}$.

Consistent Algorithmic Tangent Operator

The exact tangent operator can be obtained by taking the derivative of the Kirchhoff stress tensor in (8.146) with respect to the strain. This spatial tangent operator has the following relation to the material tangent operator:

$$ c_{ijkl} = 2F_{iI}F_{jJ}F_{kK}F_{lL}\frac{\partial S_{IJ}}{\partial C_{KL}}, \tag{8.148} $$

where C_{KL} is a component of the right Cauchy-Green deformation tensor, and S_{IJ} is a component of the second Piola-Kirchhoff stress defined by $\boldsymbol{S} = \boldsymbol{F}^{-1}\boldsymbol{\sigma}\boldsymbol{F}^{-T}$. Since stress is a function of elastic trial strain and since the intermediate configuration is held fixed in the elastic trial process, all material tensors are referred to the intermediate configuration and linearization is carried out with respect to that configuration. The return-mapping in the previous section gives the following relation (refer to Section 8.5.2 for detailed derivations):

$$\mathbf{c}^{alg} \equiv \frac{\partial \boldsymbol{\tau}^{p}}{\partial \mathbf{e}^{tr}} = \mathbf{c}^{e} - 4\mu^{2} A \mathbf{N} \otimes \mathbf{N} - \frac{4\mu^{2}\hat{\gamma}}{\left\| \boldsymbol{\eta}^{tr} \right\|} [\mathbf{I}_{dev} - \mathbf{N} \otimes \mathbf{N}], \tag{8.149}$$

which is a 3 × 3 symmetric matrix in the principal stress space. Equation (8.149) has the same form as the classical plasticity model in (8.99), except that principal stress and logarithmic stretch are now used. Using the property in (8.148), the stress in (8.146) is differentiated to yield

$$\frac{\partial \boldsymbol{\tau}}{\partial \boldsymbol{\varepsilon}} = \sum_{i=1}^{3} \left[\frac{\partial \tau_i^{p}}{\partial e_j^{tr}} \left(2\mathbf{F}^{e} \frac{\partial e_j^{tr}}{\partial \mathbf{C}^{e}} \mathbf{F}^{eT} \right) \otimes \mathbf{m}^{i} + 2\tau_i^{p} \left(\mathbf{F}^{e} \frac{\partial \mathbf{m}^{i}}{\partial \mathbf{C}^{e}} \mathbf{F}^{eT} \right) \right], \tag{8.150}$$

where e_j^{tr} is a function of total deformation and is independent of the plastic evolution law. The following relation is the differential version of the eigenvalue problem $\mathbf{C}^{e}\hat{\mathbf{N}} = \lambda^{2}\hat{\mathbf{N}}$, derived in Section 8.5.3:

$$2\mathbf{F}^{e} \frac{\partial e_j^{tr}}{\partial \mathbf{C}^{e}} \mathbf{F}^{eT} = \mathbf{m}^{j}. \tag{8.151}$$

The last term in (8.150) is independent of plastic flow because plastic evolution is carried out in the fixed, principal direction. In Section 8.5.4, it is explicitly shown from finite elasticity that

$$\begin{aligned} \mathbf{F}^{e} \frac{\partial \mathbf{m}^{i}}{\partial \mathbf{C}^{e}} \mathbf{F}^{eT} &= \frac{1}{d_i} \left\{ \mathbf{I}_{b^e} - \mathbf{b}^{e} \otimes \mathbf{b}^{e} - I_3 \lambda_i^{-2} \left[\mathbf{I} - (\mathbf{1} - \mathbf{m}^{i}) \otimes (\mathbf{1} - \mathbf{m}^{i}) \right] \right\} \\ &\quad + \frac{\lambda_i^{2}}{d_i} \left\{ \mathbf{b}^{e} \otimes \mathbf{m}^{i} + \mathbf{m}^{i} \otimes \mathbf{b}^{e} + (I_1 - 4\lambda_i^{2}) \mathbf{m}^{i} \otimes \mathbf{m}^{i} \right\} \\ &\equiv \hat{\mathbf{c}}^{i}, \end{aligned} \tag{8.152}$$

where $d_i = (\lambda_i^2 - \lambda_j^2)(\lambda_i^2 - \lambda_k^2)$ with an even permutation between *i-j-k*; I_1, and I_3 are the first and third invariant of $\mathbf{b}^{e}$; and where $I_{bijkl} = \frac{1}{2}(b_{ik}b_{jl} + b_{il}b_{jk})$ can be obtained by pushing $\mathbf{I}$ forward to the current configuration. The algorithmic tangent operator in (8.148) can thus be expressed as

$$\mathbf{c} = \sum_{i=1}^{3} \sum_{j=1}^{3} c_{ij}^{alg} \mathbf{m}^{i} \otimes \mathbf{m}^{j} + 2\sum_{i=1}^{3} \tau_i^{p} \hat{\mathbf{c}}^{i}, \tag{8.153}$$

which contains all symmetric properties between indices.

Variational Principles for Finite Deformation

The structural energy form at time t_n can be written using Kirchhoff stress and engineering strain at the current time as

$$a(\mathbf{z}, \bar{\mathbf{z}}) = \int_{{}^{0}\Omega} \boldsymbol{\tau} : \boldsymbol{\varepsilon}(\bar{\mathbf{z}}) \, d\Omega. \tag{8.154}$$

Since Kirchhoff stress is used, the integration domain is an undeformed configuration. It is assumed that the constitutive equation is given in the total form for elastic material as

$$C_{ijkl} = \frac{\partial S_{ij}}{\partial E_{kl}}, \qquad c_{ijkl} = F_{ir} F_{js} F_{kq} F_{lt} C_{rsqt}, \tag{8.155}$$

where C_{ijkl} and c_{ijkl} are fourth-order material and spatial constitutive tensors. This relation is slightly different from (8.28) since the Kirchhoff stress measure is used instead of the Cauchy stress tensor. The explicit form of c_{ijkl} can be obtained from (8.153). The updated Lagrangian formulation of the structural energy form derived in (8.30) has a specific expression as

$$a^*(z;\Delta z,\bar{z}) \equiv \int_{{}^0\Omega} [\bar{\boldsymbol{\varepsilon}} : \boldsymbol{c} : \boldsymbol{\varepsilon}(\Delta z) + \boldsymbol{\tau} : \boldsymbol{\eta}(\Delta z,\bar{z})]\, d\Omega. \tag{8.156}$$

If current time is t_n and the iteration counter is $k + 1$, then the linearized incremental equation becomes

$$a^*({}^n z^k;\Delta z^{k+1},\bar{z}) = \ell(\bar{z}) - a({}^n z^k,\bar{z}), \qquad \forall \bar{z} \in Z. \tag{8.157}$$

Equation (8.157) is solved iteratively until the right side (the residual force) vanishes. After convergence, time is increased and the same analysis procedure described above is repeated until a final configuration is reached. Note that integration of the internal energy term is carried out on the undeformed configuration because Kirchhoff stress is used.

8.3 Contact Problems

Contact problems are common and important aspects of mechanical systems. Metal formation, vehicle crashes, projectile penetration, various seal designs, and bushing and gear systems are only a few examples of possible contact problems. Recent developments in computational mechanics make it possible to solve frictional contact problems accurately and efficiently. But the design of the structural system, which has frictional contact along with other structures, has been of substantial interest for a decade without noticeable progress in real engineering applications. In this section, contact analysis is introduced for use in later chapters for design sensitivity analysis. A contact problem is classified as a boundary nonlinearity, in contrast to either a geometric nonlinearity, which emerges from finite deformation problems, or a material nonlinearity, which is a product of nonlinear constitutive relations.

Since a contact problem usually accompanies the other two nonlinearities, consistent linearization is critical to achieve convergence in the analysis, and accurate sensitivity results. The continuum-based variational form of frictional contact constraints is developed by taking the first variation of the penalty function, constructed from a normal gap function and tangential slip function. In this section, the contact condition of a flexible body–rigid wall is considered using the concept of a slave–master segment. However, this problem can easily be extended to flexible-flexible body and self-contact problems, as shown in Kim et al. [23]. It will be shown that the tangent stiffness matrix is symmetric in the frictionless contact problem, while it is unsymmetric in the frictional contact problem.

8.3.1 Contact Condition and Variational Inequality

Contact Condition with Rigid Surface

Contact conditions can be divided into normal impenetrability and tangential slip. The impenetrability condition prevents one body from penetrating into another, while the tangential slip represents the frictional behavior of a contact surface. Figure 8.8 illustrates a general contact condition with a rigid surface in R^2. Since the motion of the rigid surface is prescribed throughout the analysis, a natural coordinate ξ is used to represent the location on a rigid surface. The coordinates of contact point $\boldsymbol{x}_c$ can be represented using a natural coordinate at the contact point ξ_c by

$$\boldsymbol{x}_c = \boldsymbol{x}_c(\xi_c). \tag{8.158}$$

The impenetrability condition can be imposed on the structure by measuring the distance between a part of the structural boundary Γ_c and the rigid surface, as shown in Fig. 8.8. The impenetrability condition can be defined by using the normal gap function g_n, which measures the normal distance as

$$g_n \equiv (\boldsymbol{x} - \boldsymbol{x}_c(\xi_c))^T \boldsymbol{e}_n(\xi_c) \geq 0, \qquad \boldsymbol{x} \in \Gamma_c, \tag{8.159}$$

where $\boldsymbol{e}_n(\xi_c)$ is the unit outward normal vector of the rigid surface at the contact point. The contact point $\boldsymbol{x}_c$ that corresponds to body point $\boldsymbol{x} \in \Gamma_c$ is determined by solving the following nonlinear equation:

$$\varphi(\xi_c) = (\boldsymbol{x} - \boldsymbol{x}_c(\xi_c))^T \boldsymbol{e}_t(\xi_c) = 0, \tag{8.160}$$

where $\boldsymbol{e}_t = \boldsymbol{t}/\|\boldsymbol{t}\|$ is the unit tangential vector, and $\boldsymbol{t} = \boldsymbol{x}_{c,\xi}$ is the tangential vector at the contact point. Equation (8.160) is called the contact consistency condition, and $\boldsymbol{x}_c(\xi_c)$ is the closest projection point of $\boldsymbol{x} \in \Gamma_c$ onto the rigid surface that satisfies (8.160).

As the contact point moves along the rigid surface, a frictional force in a tangential direction to the rigid surface resists the tangential relative movement. Tangential slip function g_t is the measure of the relative movement of the contact point along the rigid surface as

$$g_t \equiv \|\boldsymbol{t}^0\|(\xi_c - \xi_c^0), \tag{8.161}$$

where both the tangential vector $\boldsymbol{t}^0$ and the natural coordinate ξ_c^0 are the values at the previously converged time step or load step. Right superscript 0 will denote the previous configuration time in the following derivations.

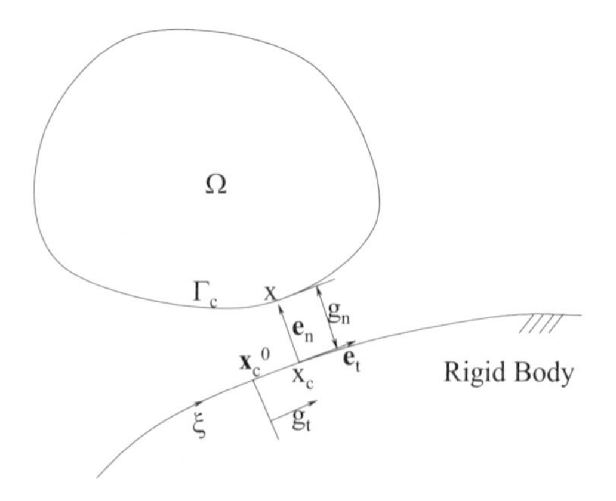

Figure 8.8. Contact condition in R^2.

Variational Inequality in the Contact Problem
Before deriving a contact variational formula, it is beneficial to discuss the fundamental properties of the contact problem. Although only a linear elastic problem will be considered for simplicity, due to the inequality constraint on the deformation field, the contact problem is nonlinear even in a linear elastic case. The differential equation of the contact problem can be written as follows:

Governing equation

$$\begin{aligned} &\sigma_{ij,j} + f_i^B = 0 && \boldsymbol{x} \in \Omega \\ &\boldsymbol{z}(\boldsymbol{x}) = 0 && \boldsymbol{x} \in \Gamma^g \\ &\sigma_{ij} n_j = f_i^S && \boldsymbol{x} \in \Gamma^S \end{aligned} \tag{8.162}$$

Contact conditions

$$\begin{aligned} &\boldsymbol{z}^T \boldsymbol{e}_n + g_n \geq 0 & \\ &\sigma_n \geq 0 & \boldsymbol{x} \in \Gamma_c. \\ &\sigma_n (\boldsymbol{z}^T \boldsymbol{e}_n + g_n) = 0 & \end{aligned} \tag{8.163}$$

The first inequality in (8.163) can be obtained from the incremental impenetrability condition in (8.159), since a small deformation linear problem is assumed. The inequality contact constraint in (8.163) can be considered by constructing a closed convex set K, defined as

$$K = \left\{ \boldsymbol{w} \in [H^1(\Omega)]^N \mid \boldsymbol{w}|_{\Gamma^g} = 0 \text{ and } \boldsymbol{w}^T \boldsymbol{e}_n + g_n \geq 0 \text{ on } \Gamma_c \right\}. \tag{8.164}$$

If $\boldsymbol{z}$ is the solution to (8.162) and (8.163), then $\boldsymbol{z} \in K$. The variational inequality can be derived from the weak formulation of the differential equation in (8.162), as well as by integrating by parts as

$$\begin{aligned} &\int_\Omega \sigma_{ij}(\boldsymbol{z}) \varepsilon_{ij}(\boldsymbol{w} - \boldsymbol{z}) \, d\Omega \\ &\quad = -\int_\Omega \sigma_{ij,j}(w_i - z_i) \, d\Omega + \int_{\Gamma^S \cup \Gamma_c} \sigma_{ij} n_j (w_i - z_i) \, d\Gamma \\ &\quad = \ell(\boldsymbol{w} - \boldsymbol{z}) + \int_{\Gamma_c} \sigma_{ij} n_j (w_i - z_i) \, d\Gamma, \end{aligned} \tag{8.165}$$

where the last term in (8.165), which is not known until the solution is obtained, is always nonnegative, as shown below:

$$\begin{aligned} &\int_{\Gamma_c} \sigma_{ij} n_j (w_i - z_i) \, d\Gamma \\ &\quad = \int_{\Gamma_c} \sigma_n (w_n - z_n) \, d\Gamma \\ &\quad = \int_{\Gamma_c} \sigma_n (w_n + g_n - z_n - g_n) \, d\Gamma \\ &\quad = \int_{\Gamma_c} \sigma_n (w_n + g_n) \, d\Gamma \geq 0, \forall \boldsymbol{w} \in K. \end{aligned}$$

Thus, variational equation (8.165) becomes a variational inequality as

$$a(\boldsymbol{z}, \boldsymbol{w} - \boldsymbol{z}) \geq \ell(\boldsymbol{w} - \boldsymbol{z}), \quad \forall \boldsymbol{w} \in K, \tag{8.166}$$

where $z \in K$ is a solution. The same variational inequality in (8.166) can be used for the nonlinear elastic contact problem with the appropriate structural energy form, as seen in (8.6) or (8.24). The constraint set of the large deformation problem contains the impenetrability condition in (8.159) as

$$K = \left\{ \mathbf{w} \in [H^1(\Omega)]^N \mid \mathbf{w}|_{\Gamma^g} = 0 \text{ and } (\mathbf{x} - \mathbf{x}_c(\xi_c))^T \mathbf{e}_n \geq 0 \text{ on } \Gamma_c \right\}. \tag{8.167}$$

The existence and uniqueness of the solution to the variational inequality has been extensively studied for linear elastic material [24] and [25]. The existence of a solution to (8.166) for the nonlinear elastic problem has been proved by Ciarlet [9] for a polyconvex strain energy function. However, from an engineering point of view, it is not convenient to solve the variational inequality directly without mentioning the construction of a test function on constraint set K. It is possible to show that the variational inequality is equivalent to the constrained optimization problem of the total potential energy. If the total potential energy is

$$\Pi(\mathbf{z}) = \tfrac{1}{2} a(\mathbf{z}, \mathbf{z}) - \ell(\mathbf{z}), \tag{8.168}$$

where $a(\mathbf{z},\mathbf{z})$ is positive-definite, then the directional derivative of $\Pi(\mathbf{z})$ in the direction of $\mathbf{v}$ is defined as

$$\langle D\Pi(\mathbf{z}), \mathbf{v} \rangle = a(\mathbf{z}, \mathbf{v}) - \ell(\mathbf{v}). \tag{8.169}$$

Then, the variational inequality $a(\mathbf{z},\mathbf{w}-\mathbf{z}) \geq \ell(\mathbf{w}-\mathbf{z})$ can be rewritten as

$$\langle D\Pi(\mathbf{z}), \mathbf{w} - \mathbf{z} \rangle \geq 0. \tag{8.170}$$

To show that (8.170) is equivalent to the constrained minimization problem, let us consider the following relation. For arbitrary $\mathbf{w} \in K$,

$$\Pi(\mathbf{w}) - \Pi(\mathbf{z}) = \langle D\Pi(\mathbf{z}), \mathbf{w} - \mathbf{z} \rangle + \tfrac{1}{2} a(\mathbf{w} - \mathbf{z}, \mathbf{w} - \mathbf{z}) \tag{8.171}$$

Since $a(\mathbf{z},\mathbf{z})$ is positive-definite, the last term in (8.171) is always nonnegative, thus

$$\Pi(\mathbf{w}) \geq \Pi(\mathbf{z}) + \langle D\Pi(\mathbf{z}), \mathbf{w} - \mathbf{z} \rangle, \quad \forall \mathbf{w} \in K, \tag{8.172}$$

which means

$$\Pi(\mathbf{z}) = \min_{\mathbf{w} \in K} \Pi(\mathbf{w}) = \min_{\mathbf{w} \in K} \left[\frac{1}{2} a(\mathbf{w}, \mathbf{w}) - \ell(\mathbf{w}) \right]. \tag{8.173}$$

If $\Pi(\mathbf{z})$ is convex, and set K is closed convex, then both the constrained minimization problem in (8.173) and the variational inequality have the unique solution $\mathbf{z}$. The variational inequality in (8.166) can be solved using the constrained minimization problem in (8.173). Many optimization theories can be used, including mathematical programming, sequential quadratic programming, and active set strategies.

Penalty Regularization

To maintain $\mathbf{w} \in K$ in (8.173), if a region Γ_c exists that violates the impenetrability condition in (8.159), then it is penalized using a penalty function. Similarly, the tangential movement of (8.161) can also be penalized under the stick condition. The contact penalty function must first be defined for the penetrated region by

$$P = \frac{1}{2}\omega_n \int_{\Gamma_C} g_n^{\,2}\, d\Gamma + \frac{1}{2}\omega_t \int_{\Gamma_C} g_t^{\,2}\, d\Gamma, \tag{8.174}$$

where ω_n and ω_t are the penalty parameters for normal contact and tangential slip. The penalty function defined in (8.174) leads to an exterior penalty method whereby the solution approaches from the infeasible region. This means that the impenetrability condition will be violated, but the amount of violation decreases as the penalty parameter is increased. The constrained minimization problem in (8.173) is converted to an unconstrained minimization problem by adding a penalty function to the total potential energy. Thus,

$$\Pi(z) = \min_{w \in K} \Pi(w) \approx \min_{w \in Z}\left[\Pi(w) + P(w)\right]. \tag{8.175}$$

Note that the solution space is changed to Z from K because of the penalty function. The variation of (8.175) contains two contributions that will be examined in this section: one from the structural potential and the other from the penalty function. The first variation of P yields the contact variational form, which is defined by

$$\begin{aligned} b(z,\bar{z}) &\equiv \omega_n \int_{\Gamma_C} g_n \bar{g}_n\, d\Gamma + \omega_t \int_{\Gamma_C} g_t \bar{g}_t\, d\Gamma \\ &= \quad b_N(z,\bar{z}) \quad + \quad b_T(z,\bar{z}), \end{aligned} \tag{8.176}$$

where $b_N(z,\bar{z})$ and $b_T(z,\bar{z})$ are the normal and tangential contact variational forms, respectively. $b_T(z,\bar{z})$ appears only when there is friction between contact surfaces. In (8.176), $\omega_n g_n$ corresponds to the compressive normal force, and $\omega_t g_t$ corresponds to the tangential traction force. The latter increases linearly with tangential slip g_t until it reaches a normal force multiplied by the friction coefficient. The contact variational form in (8.176) can be expressed in terms of the displacement variation. To make subsequent derivations easier to follow, it is necessary to define several scalar symbols, as follows:

$$\begin{aligned} \alpha \equiv \boldsymbol{e}_n^T \boldsymbol{x}_{c,\xi\xi}, \qquad \beta \equiv \boldsymbol{e}_t^T \boldsymbol{x}_{c,\xi\xi}, \qquad \gamma \equiv \boldsymbol{e}_n^T \boldsymbol{x}_{c,\xi\xi\xi} \\ c \equiv \|\boldsymbol{t}\|^2 - g_n \alpha, \qquad v \equiv \|\boldsymbol{t}\| \|\boldsymbol{t}^0\| / c. \end{aligned} \tag{8.177}$$

Note that if the rigid surface is approximated by a piecewise linear function, then $\alpha = \beta = \gamma = 0$ and $v = 1$.

8.3.2 Frictionless Contact Formulation

In the ideal case, there is no friction between contact surfaces. Computationally, the frictionless contact problem with elastic material is path-independent. All field variables are functions of the current configuration. The first step is to express the normal contact variational form in terms of displacement variation. By taking the first variation of the normal gap function in (8.159) and using the variation of the contact consistency condition in (8.160), the first variation of the normal gap function can be obtained as

$$\bar{g}_n(z;\bar{z}) = \bar{z}^T \boldsymbol{e}_n, \tag{8.178}$$

where the variation of the natural coordinate at the contact point is canceled by an orthogonal condition. The normal gap function can vary only in a normal direction to the rigid surface, which physically makes sense. By using (8.178), the normal contact form is expressed in terms of displacement variation as

$$b_N(z,\bar{z}) = \omega_n \int_{\Gamma_c} g_n \bar{z}^T e_n \, d\Gamma. \tag{8.179}$$

This contact form originates in the impenetrability condition, and the fact that the magnitude of the impenetrability force is proportional to the violation of the impenetrability condition. Note that $b_N(z,\bar{z})$ is linear with respect to $\bar{z}$ and implicit with respect to z through g_n and e_n. Since $b_N(z,\bar{z})$ is nonlinear in displacement, the same linearization procedure is required that was used for the structural energy form in Section 8.1. The increment of the normal gap function can be obtained in a similar procedure to (8.178) as

$$\Delta g_n(z;\Delta z) = e_n^T \Delta z. \tag{8.180}$$

To obtain the increment of the unit normal vector, it is necessary to compute the increment of natural coordinate ξ_c at the contact point using (8.160), since the normal vector changes along ξ_c. The increment of (8.160) solves $\Delta\xi_c$ in terms of Δz as

$$\begin{aligned}
&\Delta[(x - x_c)^T e_t] \\
&\quad = (\Delta z - t\Delta\xi_c)^T e_t + (x - x_c)^T \Delta e_t \\
&\quad = \Delta z^T e_t - \|t\|\Delta\xi_c + (x - x_c)^T e_n \left(\frac{1}{\|t\|} e_n^T x_{c,\xi\xi} \right) \Delta\xi_c = 0
\end{aligned}$$

$$\Delta\xi_c = \frac{\|t\|}{c} e_t^T \Delta z. \tag{8.181}$$

If e_3 is the fixed unit vector in the out-of-plane direction, then the increment of the unit normal vector can be obtained from the relation $e_n = e_3 \times e_t$ as

$$\begin{aligned}
\Delta e_n &= e_3 \times \Delta e_t \\
&= e_3 \times \Delta \left[\frac{t}{\|t\|} \right] \\
&= e_3 \times \frac{1}{\|t\|} \left[\Delta t - e_t (e_t^T \Delta t) \right] \\
&= -\frac{1}{\|t\|} \left[e_t (e_n^T \Delta t) \right] \\
&= -\frac{\alpha \Delta\xi_c}{\|t\|} e_t \\
&= -\frac{\alpha}{c} e_t (e_t^T \Delta z).
\end{aligned} \tag{8.182}$$

Thus, from (8.180) and (8.182) the linearization of the normal contact variational form is obtained as

$$b_N^*(z;\Delta z,\bar{z}) = \omega_n \int_{\Gamma_c} \bar{z}^T e_n e_n^T \Delta z \, d\Gamma - \omega_n \int_{\Gamma_c} \frac{\alpha g_n}{c} \bar{z}^T e_t e_t^T \Delta z \, d\Gamma. \tag{8.183}$$

Note that there is a component in the tangential direction because of curvature effects. The first term is the conventional contact tangent term for linear kinematics. The contribution of the second term is usually small, as the contact violation is reduced.

In the case of general nonlinear material with a frictionless contact problem, the variational principle for virtual work can be written as

$$a(z,\bar{z}) + b_N(z,\bar{z}) = \ell(\bar{z}), \quad \forall \bar{z} \in Z \tag{8.184}$$

that is obtained from the first variation of the penalized potential energy function in (8.175), which is equated to zero to satisfy the Kuhn-Tucker condition. Suppose the current time is t_n and the current iteration count is $k + 1$. Assuming that the external force is independent of displacement, the linearized incremental equation of (8.184) is obtained as

$$\begin{aligned} & a^*({}^n z^k; \Delta z^k, \bar{z}) + b_N^*({}^n z^k; \Delta z^{k+1}, \bar{z}) \\ & \quad = \ell(\bar{z}) - a({}^n z^k, \bar{z}) - b_N({}^n z^k, \bar{z}), \qquad \forall \bar{z} \in Z. \end{aligned} \tag{8.185}$$

Equation (8.185) is linear in incremental displacement for a given displacement variation. The linearized system of (8.185) is solved iteratively with respect to incremental displacement until the residual forces on the right side of the equation vanish at each time step.

8.3.3 Frictional Contact Formulation

When friction exists in the contact problem, a solution depends on the history of the load applied to the structure. The classical Coulomb friction law is commonly used in computational mechanics. As an alternative, the frictional interface law of Wriggers et al. [26] is employed here. This friction law is a regularized version of Coulomb law, such that the regularization parameter can be related to experimental observation.

The tangent slip variational form $b_T(z,\bar{z})$ in (8.176) can be expressed in terms of a displacement variation. The first variation of the tangential slip function, presented in (8.161), becomes

$$\bar{g}_t = \|\boldsymbol{t}^0\| \bar{\xi}_c = v \bar{z}^T \boldsymbol{e}_t, \tag{8.186}$$

where a procedure similar to (8.181) is used. Note that the first variations of $\|\boldsymbol{t}^0\|$ and ξ_c^0 are zero, since they are the solutions to the previous time step and fixed at the current time. By using (8.186), the tangential slip variational form in (8.176) can be rewritten in terms of the displacement variation as

$$b_T(z,\bar{z}) = \omega_t \int_{\Gamma_c} v g_t \bar{z}^T \boldsymbol{e}_t \, d\Gamma. \tag{8.187}$$

The frictional traction force works in the tangential direction, is proportional to the tangential slip, and is scaled by curvature through v. The frictional force is bounded above by a compressive normal force multiplied by the friction coefficient in the Coulomb friction law. But in the case of a small slip (microdisplacement), traction force is proportional to the tangential slip. The penalty parameter ω_t is a constant for this case. An exact stick condition represented by a step function in the classical Coulomb friction law is now regularized by a piecewise linear function, with the penalty parameter ω_t serving as a regularization parameter. As shown in Fig. 8.9, this regularized friction law is reduced to the classical law as $\omega_t \to \infty$. A regularized stick condition occurs when

$$\omega_t g_t \le |\mu \omega_n g_n|. \tag{8.188}$$

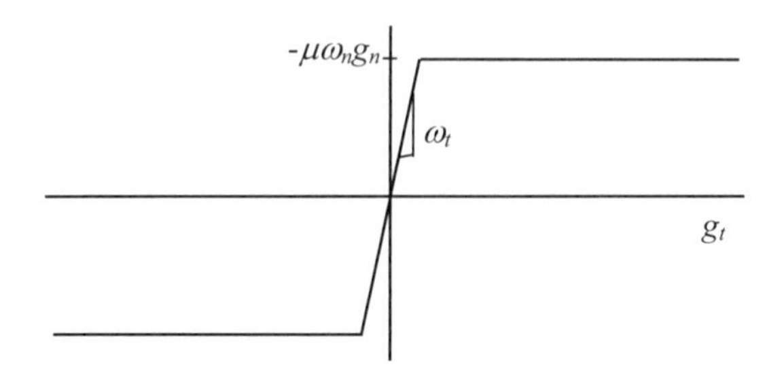

Figure 8.9. Frictional interface model.

Otherwise, it becomes a slip condition and $\omega_t g_t = -\mu\omega_n g_n$. In (8.188), μ is the Coulomb friction coefficient. In the case of a slip condition, the contact variational form has to be modified. Thus, (8.187) must be divided into two cases as

$$b_T(z,\bar{z}) = \begin{cases} \omega_t \int_{\Gamma_c} \nu g_t \bar{z}^T \boldsymbol{e}_t \, d\Gamma, & \text{if } |\omega_t g_t| \leq |\mu\omega_n g_n| \\ -\mu\omega_n \operatorname{sgn}(g_t) \int_{\Gamma_c} \nu g_n \bar{z}^T \boldsymbol{e}_t \, d\Gamma, & \text{otherwise.} \end{cases} \tag{8.189}$$

Thus, linearization of the tangential slip variational form has to be separated into stick and slip conditions.

Linearization of Stick Condition

The first equation in (8.189) implicitly depends on displacement through ν, g_t, and $\boldsymbol{e}_t$. The incremental form of g_t can be obtained using the relation in (8.181) as

$$\begin{aligned} \Delta g_t(z;\Delta z) &= \|\boldsymbol{t}^0\| \Delta\xi_c \\ &= \nu \boldsymbol{e}_t^T \Delta z. \end{aligned} \tag{8.190}$$

The incremental form of the unit tangential vector can be derived using a procedure similar to that used in (8.182) with $\boldsymbol{e}_t = -\boldsymbol{e}_3 \times \boldsymbol{e}_n$ as

$$\begin{aligned} \Delta\boldsymbol{e}_t &= -\boldsymbol{e}_3 \times \Delta\boldsymbol{e}_n \\ &= \frac{\alpha}{c} \boldsymbol{e}_n (\boldsymbol{e}_t^T \Delta z). \end{aligned} \tag{8.191}$$

Also, the increment of ν can be obtained from its definition in (8.177). After some algebraic calculation, the linearization of (8.187) leads to the tangential stick bilinear form as

$$\begin{aligned} b_T^*(z;\Delta z,\bar{z}) = {} & \omega_t \int_{\Gamma_c} \nu^2 \bar{z}^T \boldsymbol{e}_t \boldsymbol{e}_t^T \Delta z \, d\Gamma \\ & + \omega_t \int_{\Gamma_c} \frac{\alpha\nu g_t}{c} \bar{z}^T (\boldsymbol{e}_n \boldsymbol{e}_t^T + \boldsymbol{e}_t \boldsymbol{e}_n^T) \Delta z \, d\Gamma \\ & + \omega_t \int_{\Gamma_c} \frac{\nu g_t}{c^2} \left((\gamma \|\boldsymbol{t}\| - 2\alpha\beta) g_n - \beta \|\boldsymbol{t}\|^2 \right) \bar{z}^T \boldsymbol{e}_t \boldsymbol{e}_t^T \Delta z \, d\Gamma. \end{aligned} \tag{8.192}$$

The contact bilinear form is the sum of (8.183) and (8.192) as

$$b^*(z;\Delta z,\bar{z}) = b_N^*(z;\Delta z,\bar{z}) + b_T^*(z;\Delta z,\bar{z}). \tag{8.193}$$

In the case of a stick condition, the contact bilinear form in (8.193) is symmetric with respect to the incremental displacement and variation of displacement. It is noted that the elastic stick contact condition is a conservative system.

Linearization of Slip Condition

As the contact point is forced to move along the contact surface, leading to a violation of (8.188), the slip contact condition is applied and the second equation from (8.189) is used. In the case of a slip contact condition, the tangential penalty parameter ω_t is related to the impenetrability penalty parameter ω_n according to the relation

$$\omega_t = -\mu\omega_n \operatorname{sgn}(g_t). \tag{8.194}$$

The tangential slip form for the slip condition is

$$b_T(\mathbf{z},\bar{\mathbf{z}}) = \omega_t \int_{\Gamma_c} \nu\, g_n \bar{\mathbf{z}}^T \mathbf{e}_t \, d\Gamma. \tag{8.195}$$

The linearization of (8.195) leads to the tangential slip bilinear form as

$$\begin{aligned} b_T^*(\mathbf{z};\Delta\mathbf{z},\bar{\mathbf{z}}) &= \omega_t \int_{\Gamma_c} \nu \bar{\mathbf{z}}^T \mathbf{e}_t \mathbf{e}_n^T \Delta\mathbf{z}\, d\Gamma \\ &+ \omega_t \int_{\Gamma_c} \frac{\alpha \nu g_n}{c} \bar{\mathbf{z}}^T (\mathbf{e}_n \mathbf{e}_t^T + \mathbf{e}_t \mathbf{e}_n^T) \Delta\mathbf{z}\, d\Gamma \\ &+ \omega_t \int_{\Gamma_c} \frac{\nu g_n}{c^2} \left((\gamma \|\mathbf{t}\| - 2\alpha\beta) g_n - \beta \|\mathbf{t}\|^2 \right) \bar{\mathbf{z}}^T \mathbf{e}_t \mathbf{e}_t^T \Delta\mathbf{z}\, d\Gamma. \end{aligned} \tag{8.196}$$

In the case of a slip condition, the contact bilinear form in (8.196) is not symmetric with respect to the incremental displacement and variation of the displacement. The system is no longer conservative because frictional slip dissipates energy.

In the case of general nonlinear material with a frictional contact problem, the variational principle for virtual work can be written as

$$a(\mathbf{z},\bar{\mathbf{z}}) + b(\mathbf{z},\bar{\mathbf{z}}) = \ell(\bar{\mathbf{z}}), \quad \forall \bar{\mathbf{z}} \in Z. \tag{8.197}$$

The current time is t_n and the current iteration count is $k + 1$. Assuming that the external force is independent of displacement, the linearized incremental equation of (8.197) is obtained as

$$\begin{aligned} &a^*({}^n\mathbf{z}^k;\Delta\mathbf{z}^k,\bar{\mathbf{z}}) + b^*({}^n\mathbf{z}^k;\Delta\mathbf{z}^{k+1},\bar{\mathbf{z}}) \\ &\quad = \ell(\bar{\mathbf{z}}) - a({}^n\mathbf{z}^k,\bar{\mathbf{z}}) - b({}^n\mathbf{z}^k,\bar{\mathbf{z}}), \qquad \forall \bar{\mathbf{z}} \in Z. \end{aligned} \tag{8.198}$$

Equation (8.198) is linear in incremental displacement for a given displacement variation. This linearized equation is solved iteratively with respect to incremental displacement until the residual forces (the right side of the equation) vanish at each time step.

8.4 Nonlinear Dynamic Problems

When a time-dependent load and/or a boundary condition is applied to the structure, the transient response becomes important. Inertia effect and wave propagation are at times dominant throughout the structure compared with the quasi-static response. In a linear system, the mode superposition method is frequently used to solve the transient dynamic problem because it provides insight into the behavior of a structure, and is usually more cost effective than the direct integration method that is generally used for nonlinear problems. Since nonlinear responses of a structure are considered in this section, the direct integration method will be used.

Integration of the dynamic equation can be achieved using either an implicit or explicit method. The explicit integration method is usually used with a very small time-step size to solve a wave propagation problem associated with a relatively local structural response. This method is useful for impact and crashworthiness problems where such wave effects as focusing, reflection, and diffraction are important. In contrast, the implicit integration method is usually used to solve an inertial problem in which the overall dynamic response of a structure is sought. A structural problem is considered inertial when the response time is longer than the time required for the wave to traverse the structure. Since the implicit integration scheme is unconditionally stable when appropriate integration parameters are chosen, the time-step is usually one or two orders of magnitude larger than that of the explicit integration method. However, a reasonable time-step must be chosen to achieve accuracy of the solution. Explicit and implicit time integration methods are discussed in this section

8.4.1 Implicit Method

The response analysis of nonlinear, transient dynamics is reviewed in this section to be used in later sensitivity formulations. The implicit time integration method is discussed in conjunction with the acceleration/displacement formulation. When the structural response is nonlinear, the dynamic problem also solves for incremental acceleration/displacement through the linearization process. For more details, refer to Hughes [16] for the linear problem, and Bathe [1] for the nonlinear problem. In the weak formulation of a structural dynamic problem at time $t \in [0,T]$, the displacement function $\mathbf{z}(\mathbf{x},t) \in V$ satisfies the following equation:

$$d(\mathbf{z}_{,tt},\bar{\mathbf{z}}) + a(\mathbf{z},\bar{\mathbf{z}}) = \ell(\bar{\mathbf{z}}), \ \forall \bar{\mathbf{z}} \in Z, \tag{8.199}$$

where

$$V = \left\{ \mathbf{z}(\mathbf{x},t) \middle| \ \mathbf{z}(\mathbf{x},t) \in [H^1(\Omega)]^N \times R^+, \ \mathbf{z}(\mathbf{x},t)\big|_{\Gamma_g} = \mathbf{g}(\mathbf{x},t) \right\} \tag{8.200}$$

is the solution space, Z is the space of kinematically admissible displacements, $\mathbf{g}$ is the prescribed displacement vector, and $\bar{\mathbf{z}}$ is the displacement variation. If the problem contains structural impact, then the contact variational form $b(\mathbf{z},\bar{\mathbf{z}})$ from Section 8.3 needs to be added. In (8.199), the kinetic energy form is defined as

$$d(\mathbf{z}_{,tt},\bar{\mathbf{z}}) = \int_\Omega \rho \bar{\mathbf{z}}^T \mathbf{z}_{,tt} \, d\Omega, \tag{8.201}$$

where $\mathbf{z}_{,tt}$ is the second time derivative of displacement, i.e., acceleration, and ρ is the current density of a material. From the conservation of mass, the integration in (8.201) can be carried out at the undeformed configuration using the density at that configuration. In (8.199), $a(\mathbf{z},\bar{\mathbf{z}})$ and $\ell(\bar{\mathbf{z}})$ are the structural energy form and load form, respectively. The structural energy form depends on the constitutive model of the material. An appropriate structural energy form has to be used for the corresponding constitutive model, as discussed in Sections 8.1 and 8.2. For simplicity, only the case of a conservative load is considered for the load form. Note that the structural energy form is generally nonlinear, whereas kinetic energy and load forms are linear. Since (8.199) is an initial-boundary value problem (IBVP), the following initial conditions are necessary:

$$z(x,0) = z^0(x), \quad x \in \Omega \tag{8.202}$$

$$z_{,t}(x,0) = z_{,t}^0(x), \quad x \in \Omega, \tag{8.203}$$

where $z^0(x)$ and $z_{,t}^0(x)$ are the prescribed initial displacement and velocity vectors, respectively. To numerically solve IBVP in (8.199), the finite difference method is used to discretize the time domain. Time discretization is derived in the following section.

Acceleration Form

To solve the differential equation given in (8.199), with the initial conditions given in (8.202) and (8.203), time interval $[0,T]$ is discretized by $[t_1, t_2, \ldots, t_n, \ldots, t_N]$, and the equilibrium equation in (8.199) is imposed at each discrete time. The equation of motion in (8.199) is satisfied at time t_n, and the solution is given at time t_{n+1}. Newmark family time integration using the predictor-corrector method is used to integrate (8.199) as

$${}^{n+1}z_{,t} = {}^{pr}z_{,t} + \gamma\Delta t \, {}^{n+1}z_{,tt} \tag{8.204}$$

$${}^{n+1}z = {}^{pr}z + \beta\Delta t^2 \, {}^{n+1}z_{,tt}, \tag{8.205}$$

where

$${}^{pr}z_{,t} = {}^{n}z_{,t} + (1-\gamma)\Delta t \, {}^{n}z_{,tt} \tag{8.206}$$

$${}^{pr}z = {}^{n}z + \Delta t \, {}^{n}z_{,t} + (\tfrac{1}{2} - \beta)\Delta t^2 \, {}^{n}z_{,tt}. \tag{8.207}$$

Equation (8.206) is the velocity predictor and (8.207) is the displacement predictor, both constructed based on the configuration at previous time t_n. Here, β and γ are the integration parameters for the Newmark method and Δt is the time-step size. Since the right sides of (8.204) and (8.205) have responses at time t_{n+1}, this integration method is implicit, and iteration is required.

The stability and accuracy of the time integration method for a linear system is thoroughly examined by Hughes [16]. The unconditionally stable condition for the Newmark family integration method is given by $2\beta \geq \gamma \geq ½$, and second order accuracy is preserved only when $\gamma = ½$, which does not show any viscous damping effects. Choosing a different value for $\gamma\,(>½)$ shows first-order accuracy with viscous damping effects.

Since the structural energy form in (8.199) is nonlinear, linearization is necessary to solve the nonlinear equation iteratively using the Newton-Raphson method. The linearization of the structural variational form was discussed in Sections 8.1 and 8.2. The linearized form of $a(z,\bar{z})$ is denoted by $a^*(z;\Delta z,\bar{z})$, which is linear with respect to Δz and $\bar{z}$, and depends on the configuration at t_n. Let the current configuration time be t_{n+1}, and let the right superscript k+1 denote the current iteration counter. The linearized incremental equation of motion becomes

$$\begin{aligned} d(\Delta z_{,tt}^{k+1},\bar{z}) + a^*({}^{n+1}z^k;\Delta z^{k+1},\bar{z}) \\ = \ell(\bar{z}) - a({}^{n+1}z^k,\bar{z}) - d({}^{n+1}z_{,tt}^k,\bar{z}), \quad \forall \bar{z} \in Z. \end{aligned} \tag{8.208}$$

Since only the relation between incremental displacement and incremental acceleration is required in (8.208), the following incremental integration formulation of displacement is used

$$\Delta z^{k+1} = \beta \Delta t^2 \Delta z_{,tt}^{k+1} \tag{8.209}$$

$$^{n}z^{k+1} = {}^{n}z^{k} + \Delta z^{k+1}, \tag{8.210}$$

where ${}^{n+1}z^{0} = {}^{pr}z$ is the predictor at time t_{n+1}. The velocity term is updated after the solution is converged using (8.204). Note that since the kinetic energy form is already linear, its linearized form is the same as the original in (8.201). Equation (8.208) is solved iteratively until the residual terms on the right side vanish at configuration t_{n+1}. To solve (8.208), the kinematic relation in (8.209) is substituted into the structural energy form $a^{*}({}^{n+1}z^{k};\Delta z^{k+1},\bar{z})$ to express unknown Δz^{k+1} in terms of $\Delta z_{,tt}^{k+1}$ as

$$\begin{aligned} d(\Delta z_{,tt}^{k+1},\bar{z}) &+ \beta\Delta t^2 a^{*}({}^{n+1}z^{k};\Delta z_{,tt}^{k+1},\bar{z}) \\ &= \ell(\bar{z}) - a({}^{n+1}z^{k},\bar{z}) - d({}^{n+1}z_{,tt}^{k},\bar{z}), \quad \forall \bar{z} \in Z, \end{aligned} \tag{8.211}$$

and the linear system of equations is solved for the incremental acceleration $\Delta z_{,tt}^{k+1}$. After computing $\Delta z_{,tt}^{k+1}$, the displacement and velocity are updated using (8.210) and (8.204). Since acceleration is obtained from (8.211), it is called the acceleration form, as distinguished from the displacement form in which displacement is obtained by expressing acceleration in terms of displacement.

Displacement Form

In the displacement form, acceleration is expressed in terms of displacement and velocity as

$${}^{n+1}z_{,tt} = {}^{pr}z_{,tt} + \frac{1}{\beta\Delta t^2}\,{}^{n+1}z, \tag{8.212}$$

where ${}^{pr}z_{,tt}$ is the acceleration predictor, defined by

$${}^{pr}z_{,tt} = -\frac{1}{\beta\Delta t^2}\left({}^{n}z + \Delta t\,{}^{n}z_{,t} + (\tfrac{1}{2}-\beta)\Delta t^2\,{}^{n}z_{,tt}\right), \tag{8.213}$$

and the velocity integrator is the same as that of the acceleration form

$${}^{n+1}z_{,t} = {}^{pr}z_{,t} + \gamma\Delta t\,{}^{n+1}z_{,tt}. \tag{8.214}$$

Unlike the acceleration form, this method uses acceleration and velocity predictors. For an incremental analysis of a nonlinear problem, the incremental kinematic relation is

$$\Delta z_{,tt}^{k+1} = \frac{1}{\beta\Delta t^2}\Delta z^{k+1} \tag{8.215}$$

$${}^{n+1}z_{,tt}^{k+1} = {}^{n+1}z_{,tt}^{k} + \Delta z_{,tt}^{k+1}, \tag{8.216}$$

where ${}^{n+1}z_{,tt}^{0} = {}^{pr}z_{,tt}$ is the acceleration predictor at time t_{n+1}. By substituting (8.215) into (8.208), the linearized form of the equation of motion is obtained as

$$\begin{aligned} \frac{1}{\beta\Delta t^2} d(\Delta z^{k+1},\bar{z}) &+ a^{*}({}^{n+1}z^{k};\Delta z^{k+1},\bar{z}) \\ &= \ell(\bar{z}) - a({}^{n+1}z^{k},\bar{z}) - d({}^{n+1}z_{,tt}^{k},\bar{z}), \quad \forall \bar{z} \in Z, \end{aligned} \tag{8.217}$$

and the linear system of equations is solved for the incremental displacement Δz^{k+1}. After computing Δz^{k+1}, the acceleration and velocity are corrected using (8.216) and (8.214). Note that the displacement form uses the same structural energy form $a^*(z;\Delta z,\bar{z})$ as the quasi-static problem. Thus, it may be more convenient to simultaneously implement a quasi-static and dynamic problem in one analysis code when using this method.

8.4.2 Explicit Time Integration Method

The explicit integration method is very efficient with a diagonal mass matrix. With this combination, matrix equations become a system of algebraic equations. To solve the differential equation given in (8.199) with the initial conditions given in (8.202) and (8.203), the time interval $[0,T]$ is discretized by $[t_1, t_2, \ldots, t_n, \ldots, t_N]$, and the equilibrium equation in (8.199) is imposed at each discrete time. Belonging to the family of Newmark methods, the explicit integration method corresponds to the case in which $\beta = 0$. The displacement at time t_{n+1} is approximated using the configuration at time t_n as

$$ {}^{n+1}z = {}^{n}z + \Delta t\, {}^{n}z_{,t} + \tfrac{1}{2}\Delta t^2\, {}^{n}z_{,tt}. \tag{8.218} $$

Using this kinematic relation, the weak form of the equation of motion can be obtained as

$$ d({}^{n+1}z_{,tt},\bar{z}) = \ell(\bar{z}) - a({}^{n+1}z,\bar{z}), \qquad \forall \bar{z} \in Z \tag{8.219} $$

If the mass matrix is diagonal, then (8.219) is decoupled and can easily be solved without iteration. After computing ${}^{n+1}z_{,tt}$, the velocity at time t_{n+1} is updated by

$$ {}^{n+1}z_{,t} = {}^{n}z_{,t} + (1-\gamma)\Delta t\, {}^{n}z_{,tt} + \gamma\Delta t\, {}^{n+1}z_{,tt}. \tag{8.220} $$

Another explicit time integration method that is popular in applications is to use an intermediate time-step to integrate the velocity field. The velocity field at time $t_{n+1/2}$ and the displacement field at time t_{n+1} are approximated as

$$ {}^{n+\frac{1}{2}}z_{,t} = {}^{n-\frac{1}{2}}z_{,t} + \Delta t_n\, {}^{n}z_{,tt} \tag{8.221} $$

$$ {}^{n+1}z = {}^{n}z + \frac{1}{2}(\Delta t_n + \Delta t_{n+1})\, {}^{n+\frac{1}{2}}z_{,t}. \tag{8.222} $$

Equation (8.219) is solved with an extrapolated velocity and displacement in (8.221) and (8.222). The major advantage of the explicit time integration method is that no linearization is required for the structural energy form and only the mass matrix has to be inverted after discretization. If a lumped mass matrix is used for reasons of computational efficiency, the mass matrix is diagonal so that the inversion of a mass matrix is simple algebra. Among the many mass-lumping methods, the row-sum lumping method is used most frequently. However, this scheme is conditionally stable so that very small time-step sizes are required to achieve numerical stability. The time step size is governed by the length of the smallest element and by the material properties. Mathematically, the time step size is determined so that the next time step can be within the domain of influence in the hyperbolic system.

8.5 Mathematical Formulas for Finite Deformation Elastoplasticity

In this section, mathematical formulas used in the finite deformation elastoplasticity model are derived in detail.

8.5.1 Principle of Maximum Dissipation

The local dissipation function D per unit reference volume is defined as the difference between the rate of stress work and the rate of change of the free energy, as

$$\begin{aligned} D &\equiv \boldsymbol{\tau}:\boldsymbol{d} - \frac{d}{dt}\psi(\boldsymbol{b}^e,\xi) \\ &= \boldsymbol{\tau}:\boldsymbol{d} - \frac{\partial\psi}{\partial\boldsymbol{b}^e}\tfrac{d}{dt}(\boldsymbol{b}^e) - \frac{\partial\psi}{\partial\xi}\tfrac{d}{dt}(\xi) \\ &= \boldsymbol{\tau}:\boldsymbol{d} - \frac{\partial\psi}{\partial\boldsymbol{b}^e}:\left(\boldsymbol{L}\boldsymbol{b}^e + \boldsymbol{b}^e\boldsymbol{L}^T + L_v\boldsymbol{b}^e\right) + \boldsymbol{q}\tfrac{d}{dt}(\xi) \\ &= \boldsymbol{\tau}:\boldsymbol{d} - 2\frac{\partial\psi}{\partial\boldsymbol{b}^e}\boldsymbol{b}^e:\boldsymbol{L} + \left(2\frac{\partial\psi}{\partial\boldsymbol{b}^e}\boldsymbol{b}^e\right):\left[-\tfrac{1}{2}(L_v\boldsymbol{b}^e)\boldsymbol{b}^{e-1}\right] + \boldsymbol{q}\tfrac{d}{dt}(\xi) \\ &= \left(\boldsymbol{\tau} - 2\frac{\partial\psi}{\partial\boldsymbol{b}^e}\boldsymbol{b}^e\right):\boldsymbol{d} + \left(2\frac{\partial\psi}{\partial\boldsymbol{b}^e}\boldsymbol{b}^e\right):\left[-\tfrac{1}{2}(L_v\boldsymbol{b}^e)\boldsymbol{b}^{e-1}\right] + \boldsymbol{q}\tfrac{d}{dt}(\xi) \geq 0, \end{aligned} \tag{8.223}$$

where $\boldsymbol{d} = sym[\boldsymbol{L}]$ denotes the rate of deformation and $\boldsymbol{L} \equiv \tfrac{d}{dt}(\boldsymbol{F})\boldsymbol{F}^{-1}$ is the spatial velocity gradient. For the symmetric matrix the property $\boldsymbol{A}:\boldsymbol{BC} = \boldsymbol{AC}:\boldsymbol{B}$ is used, and the skew-symmetric component of $\boldsymbol{L}$ (the spin) vanishes by multiplying it with the symmetric matrix. Inequality in (8.223) holds for all admissible stresses and internal variables. When the material is in the elastic range, the rate of internal variables becomes zero, i.e., $L_v\boldsymbol{b}^e = \tfrac{d}{dt}(\xi) = 0$. Thus, the following constitutive relations and a reduced form of dissipation inequality can be obtained:

$$\boldsymbol{\tau} = 2\frac{\partial\psi}{\partial\boldsymbol{b}^e}\boldsymbol{b}^e \tag{8.224}$$

$$D = \boldsymbol{\tau}:\left[-\tfrac{1}{2}(L_v\boldsymbol{b}^e)\boldsymbol{b}^{e-1}\right] + \boldsymbol{q}\tfrac{d}{dt}(\xi) \geq 0. \tag{8.225}$$

For given rate $\left\{L_v\boldsymbol{b}^e, \tfrac{d}{dt}(\xi)\right\}$, state variables $\{\boldsymbol{\tau}, \boldsymbol{q}\}$ maximize the dissipation function D,

$$D = (\boldsymbol{\tau} - \boldsymbol{\tau}^*):\left[-\tfrac{1}{2}(L_v\boldsymbol{b}^e)\boldsymbol{b}^{e-1}\right] + (\boldsymbol{q} - \boldsymbol{q}^*)\tfrac{d}{dt}(\xi) \geq 0 \quad \forall\left\{\boldsymbol{\tau}^*, \boldsymbol{q}^*\right\} \in E. \tag{8.226}$$

A classical result of the variational inequality, the dissipation inequality satisfies if and only if the coefficients of (8.226) lie in the normal cone of elastic domain, defined in (8.120). One can obtain the evolution equations using the normal property and the plastic consistency parameter γ, as

$$-\tfrac{1}{2}L_v\boldsymbol{b}^e = \gamma\frac{\partial f(\boldsymbol{\tau},\boldsymbol{q})}{\partial\boldsymbol{\tau}}\boldsymbol{b}^e \tag{8.227}$$

$$\tfrac{d}{dt}(\xi) = \gamma\frac{\partial f(\boldsymbol{\tau},\boldsymbol{q})}{\partial\boldsymbol{q}}. \tag{8.228}$$

8.5.2 Algorithmic Tangent Operator for Principal Stress

The algorithmic tangent operator with respect to the principal stress and principal logarithmic stretch can be obtained by taking a derivative of the stress-updating algorithm. The principal stress is updated following (8.142)b as

$$\begin{aligned}\boldsymbol{\tau}^p &= {}^{tr}\boldsymbol{\tau}^p - 2\mu\hat{\gamma}\,\boldsymbol{N} \\ &= \boldsymbol{c}^e\boldsymbol{e}^{tr} - 2\mu\hat{\gamma}\,\boldsymbol{N},\end{aligned} \tag{8.229}$$

and its derivative with respect to the trial elastic logarithmic stretch becomes

$$\frac{\partial\boldsymbol{\tau}^p}{\partial\boldsymbol{e}^{tr}} = \boldsymbol{c}^e - 2\mu\frac{\partial\hat{\gamma}}{\partial\boldsymbol{e}^{tr}}\boldsymbol{N} - 2\mu\hat{\gamma}\frac{\partial\boldsymbol{N}}{\partial\boldsymbol{e}^{tr}}. \tag{8.230}$$

From this definition, the derivative of the unit normal vector to the yield surface can be easily obtained as

$$\frac{\partial\boldsymbol{N}}{\partial\boldsymbol{e}^{tr}} = \frac{\partial}{\partial\boldsymbol{e}^{tr}}\left[\frac{{}^{tr}\boldsymbol{\eta}}{\left\|{}^{tr}\boldsymbol{\eta}\right\|}\right] = \frac{2\mu}{\left\|{}^{tr}\boldsymbol{\eta}\right\|}\left[\boldsymbol{1}_{dev} - \boldsymbol{N}\otimes\boldsymbol{N}\right]. \tag{8.231}$$

The derivative of the plastic consistency parameter $\hat{\gamma}$ can be obtained from the derivative of the consistency condition in (8.144) as

$$\frac{\partial\hat{\gamma}}{\partial\boldsymbol{e}^{tr}} = -\frac{\partial f}{\partial\boldsymbol{e}^{tr}}\Big/\frac{\partial f}{\partial\hat{\gamma}} = 2\mu A\boldsymbol{N}, \tag{8.232}$$

where A is defined in (8.145). Thus, the algorithmic tangent operator with respect to the principal stress and principal logarithmic stretch becomes

$$\boldsymbol{c}^{alg} \equiv \frac{\partial\boldsymbol{\tau}^p}{\partial\boldsymbol{e}^{tr}} = \boldsymbol{c}^e - 4\mu^2 A\boldsymbol{N}\otimes\boldsymbol{N} - \frac{4\mu^2\hat{\gamma}}{\left\|{}^{tr}\boldsymbol{\eta}\right\|}\left[\boldsymbol{1}_{dev} - \boldsymbol{N}\otimes\boldsymbol{N}\right], \tag{8.233}$$

which is a symmetric second-order tensor and is in the same form as that of classical infinitesimal plasticity proposed by Simo and Taylor [19].

8.5.3 Linearization of Principal Logarithmic Stretches

Let the reference frame be the intermediate configuration for an elastoplastic problem, which is fixed in the trial state. The principal stretches λ_i are functions of the total deformation and independent of the plastic flow. For simplicity, all variables denote elastic trial status in this section without being given a specific notation. The right and left Cauchy-Green deformation tensors have the same principal values $\lambda_i 2$, and the eigenvalue problem is

$$\boldsymbol{C}\hat{\boldsymbol{N}}^i = \lambda_i^2\hat{\boldsymbol{N}}^i \tag{8.234}$$

$$\boldsymbol{b}\hat{\boldsymbol{n}}^i = \lambda_i^2\hat{\boldsymbol{n}}^i. \tag{8.235}$$

The relation between principal directions is

$$\boldsymbol{F}\hat{\boldsymbol{N}}^i = \lambda_i\hat{\boldsymbol{n}}^i. \tag{8.236}$$

By differentiating (8.234), the following relation can be obtained:

$$d\boldsymbol{C}\hat{\boldsymbol{N}}^i + \boldsymbol{C}d\hat{\boldsymbol{N}}^i = 2\lambda_i d\lambda_i \hat{\boldsymbol{N}}^i + \lambda_i^2 d\hat{\boldsymbol{N}} \qquad \text{no sum on } i. \tag{8.237}$$

Then, taking an inner product with $\hat{\boldsymbol{N}}$ and using the property that $\hat{\boldsymbol{N}} \cdot d\hat{\boldsymbol{N}} = 0$, the following can be obtained:

$$2\lambda_i d\lambda_i = \hat{\boldsymbol{N}}^i \cdot d\boldsymbol{C}\hat{\boldsymbol{N}}^i = tr\left[d\boldsymbol{C}(\hat{\boldsymbol{N}}^i \otimes \hat{\boldsymbol{N}}^i)\right] \tag{8.238}$$

or

$$\frac{\partial \lambda_i}{\partial \boldsymbol{C}} = \frac{1}{2\lambda_i}\hat{\boldsymbol{N}}^i \otimes \hat{\boldsymbol{N}}^i. \tag{8.239}$$

Since the logarithmic strain is defined by the principal stretch,

$$e_i = \log(\lambda_i) \tag{8.240}$$

and

$$2\frac{\partial e_i}{\partial \boldsymbol{C}} = 2\frac{\partial e_i}{\partial \lambda_i}\frac{\partial \lambda_i}{\partial \boldsymbol{C}} = \lambda_i^{-2}\hat{\boldsymbol{N}}^i \otimes \hat{\boldsymbol{N}}^i. \tag{8.241}$$

The push-forward of this result along with the relation in (8.236) yields

$$2\boldsymbol{F}\frac{\partial e_i}{\partial \boldsymbol{C}}\boldsymbol{F}^T = \hat{\boldsymbol{n}}^i \otimes \hat{\boldsymbol{n}}^i. \tag{8.242}$$

Since all the relations are transformed to the current configuration, the properties of the intermediate configuration are completely removed.

8.5.4 Linearization of the Eigenvector of the Elastic Trial Left Cauchy-Green Tensor

For simplicity, all superscripts of elastic trial status are ignored. Let $\hat{\boldsymbol{n}}^A$ be the principal direction of $\boldsymbol{b}$ corresponding to the principal value λ_A^2, and let $\hat{\boldsymbol{N}}^A$ be the principal direction of $\boldsymbol{C}$. The following relation is thus satisfied:

$$\boldsymbol{C} = \sum_{A=1}^{3}\lambda_A^2 \hat{\boldsymbol{N}}^A \otimes \hat{\boldsymbol{N}}^A, \qquad \boldsymbol{b} = \sum_{A=1}^{3}\lambda_A^2 \hat{\boldsymbol{n}}^A \otimes \hat{\boldsymbol{n}}^A, \tag{8.243}$$

and I_1, I_2, and I_3 are the principal invariants of $\boldsymbol{C}$. The relation of eigenvector bases between material and spatial description is

$$\boldsymbol{M}^A \equiv \lambda_A^{-2}\hat{\boldsymbol{N}}^A \otimes \hat{\boldsymbol{N}}^A = \boldsymbol{F}^{-1}(\hat{\boldsymbol{n}}^A \otimes \hat{\boldsymbol{n}}^A)\boldsymbol{F}^{-T}. \tag{8.244}$$

The explicit form of $\boldsymbol{M}^A$ can be computed by Serrin's representation theorem, namely,

$$\boldsymbol{M}^A = \frac{1}{d_A}\left[\boldsymbol{C} - (I_1 - \lambda_A^2)\boldsymbol{1} + I_3\lambda_A^{-2}\boldsymbol{C}^{-1}\right], \tag{8.245}$$

where $d_A = (\lambda_A^2 - \lambda_C^2)(\lambda_A^2 - \lambda_C^2) = 2\lambda_A^4 - I_1\lambda_A^2 + I_3\lambda_A^{-2}$ with A, B, and C having an even permutation. The following properties can be derived by using the chain rule of differentiation and direct computation:

$$\frac{\partial \boldsymbol{C}}{\partial \boldsymbol{C}} = \boldsymbol{I} = \tfrac{1}{2}(\delta_{ik}\delta_{jl} + \delta_{il}\delta_{jk}) \tag{8.246}$$

$$\frac{\partial \boldsymbol{C}^{-1}}{\partial \boldsymbol{C}} = -\boldsymbol{C}^{-1}\frac{\partial \boldsymbol{C}}{\partial \boldsymbol{C}}\boldsymbol{C}^{-1} = -\boldsymbol{I}_{C^{-1}} = -\tfrac{1}{2}(C_{ik}^{-1}C_{jl}^{-1} + C_{il}^{-1}C_{jk}^{-1}) \tag{8.247}$$

$$\frac{\partial \lambda_A^2}{\partial \boldsymbol{C}} = \lambda_A^2 \boldsymbol{M}^A \tag{8.248}$$

$$\frac{\partial I_1}{\partial \boldsymbol{C}} = \boldsymbol{1}, \qquad \frac{\partial I_3}{\partial \boldsymbol{C}} = I_3 \boldsymbol{C}^{-1} \tag{8.249}$$

$$\frac{\partial d_A}{\partial \boldsymbol{C}} = (4\lambda_A^4 - I_1\lambda_A^2 - I_3\lambda_A^{-2})\boldsymbol{M}^A - \lambda_A^2\boldsymbol{1} + I_3\lambda_A^{-2}\boldsymbol{C}^{-1}. \tag{8.250}$$

The derivative of (8.245) with respect to $\boldsymbol{C}$ becomes

$$\begin{aligned}\frac{\partial \boldsymbol{M}^A}{\partial \boldsymbol{C}} &= \frac{1}{d_A}\left[\frac{\partial \boldsymbol{C}}{\partial \boldsymbol{C}} - \boldsymbol{1}\otimes\left(\frac{\partial I_1}{\partial \boldsymbol{C}} - \frac{\partial \lambda_A^2}{\partial \boldsymbol{C}}\right)\right] \\ &+ \frac{1}{d_A}\left[\lambda_A^{-2}\boldsymbol{C}^{-1}\otimes\frac{\partial I_3}{\partial \boldsymbol{C}} - I_3\lambda_A^{-4}\boldsymbol{C}^{-1}\otimes\frac{\partial \lambda_A^2}{\partial \boldsymbol{C}} + I_3\lambda_A^2\frac{\partial \boldsymbol{C}^{-1}}{\partial \boldsymbol{C}} - \boldsymbol{M}^A\otimes\frac{\partial d_A}{\partial \boldsymbol{C}}\right].\end{aligned} \tag{8.251}$$

By using the property of (8.246) throuth (8.250), the following explicit form can be obtained:

$$\begin{aligned}\frac{\partial \boldsymbol{M}^A}{\partial \boldsymbol{C}} &= \frac{1}{d_A}\left[\boldsymbol{I} - \boldsymbol{1}\otimes\boldsymbol{1} - I_3\lambda_A^{-2}\left[\boldsymbol{I}_{C^{-1}} - (\boldsymbol{C}^{-1} - \boldsymbol{M}^A)\otimes(\boldsymbol{C}^{-1} - \boldsymbol{M}^A)\right]\right] \\ &+ \frac{\lambda_A^2}{d_A}\left[(\boldsymbol{1}\otimes\boldsymbol{M}^A + \boldsymbol{M}^A\otimes\boldsymbol{1}) + (I_1 - 4\lambda_A^{-4})\boldsymbol{M}^A\otimes\boldsymbol{M}^A\right].\end{aligned} \tag{8.252}$$

This relation was originally derived by Simo and Taylor [27]. The spatial version of (8.252) can be obtained through a transformation as

$$\begin{aligned}\frac{\partial \boldsymbol{m}^A}{\partial \boldsymbol{g}} &= \frac{1}{d_A}\left[\boldsymbol{I}_b - \boldsymbol{b}\otimes\boldsymbol{b} - I_3\lambda_A^{-2}\left[\boldsymbol{I} - (\boldsymbol{1} - \boldsymbol{m}^A)\otimes(\boldsymbol{1} - \boldsymbol{m}^A)\right]\right] \\ &+ \frac{\lambda_A^2}{d_A}\left[(\boldsymbol{b}\otimes\boldsymbol{m}^A + \boldsymbol{m}^A\otimes\boldsymbol{b}) + (I_1 - 4\lambda_A^{-4})\boldsymbol{m}^A\otimes\boldsymbol{m}^A\right].\end{aligned} \tag{8.253}$$

Equations (8.252) and (8.253) can be further simplified using $\boldsymbol{1} = \sum_{B=1}^{3}\lambda_B^2\boldsymbol{M}^B$ and $\boldsymbol{C}^{-1} = \sum_{B=1}^{3}\boldsymbol{M}^B$ as

$$\frac{\partial \boldsymbol{M}^A}{\partial \boldsymbol{C}} = \frac{1}{d_A}\left[\boldsymbol{I} - I_3\lambda_A^{-2}\boldsymbol{I}_{C^{-1}} + \sum_{B=1}^{3}(I_3\lambda_A^{-2} - \lambda_B^4)\boldsymbol{M}^B\otimes\boldsymbol{M}^B - \boldsymbol{M}^A\otimes\boldsymbol{M}^A\right] \tag{8.254}$$

and

$$\frac{\partial \boldsymbol{m}^A}{\partial \boldsymbol{g}} = \frac{1}{d_A}\left[\boldsymbol{I}_b - I_3\lambda_A^{-2}\boldsymbol{I} + \sum_{B=1}^{3}(I_3\lambda_A^{-2} - \lambda_B^4)\boldsymbol{m}^B\otimes\boldsymbol{m}^B - \boldsymbol{m}^A\otimes\boldsymbol{m}^A\right]. \tag{8.255}$$

9
Nonlinear Sizing Design Sensitivity Analysis

The nonlinear design sensitivity formulation of a structural system is substantially more complicated than that of a linear design examined in Part II. As discussed in Chapter 8, the energy form of a structural problem is not bilinear but nonlinear. Thus, the nonlinear state equation is solved iteratively using the Newton-Raphson method, which requires linearization of the energy form. Desirable features of the linear problem cannot be used for the nonlinear problem. However, the design sensitivity formulation in this section provides a significant amount of computational efficiency as compared with the nonlinear state problem. Such efficiency comes from the fact that the proposed design sensitivity formulation does not require any iteration. Based on the fact that first-order variation with respect to the design parameter includes linearization of the energy form, the linearized form that appears during design sensitivity analysis is the same as the linearized form that appears during the state problem. Thus, design sensitivity analysis uses the factorized stiffness matrix from the state problem. In addition, differentiation of the state equation is carried out at the converged configuration, and design sensitivity analysis does not follow the iterative procedure of the state problem. After the state problem is converged at the current load step, the design sensitivity equation is solved using the factorized stiffness matrix at the converged load step without iteration.

Among the many nonlinear properties of the structural problem, a nonlinear elastic problem is first described in Section 9.1 in which the structure experiences a large deformation and/or a large rotation. A desirable feature of nonlinear elastic material is that the potential function exists even if it is nonlinear function of displacement. This means that the solution to the state problem is independent of the loading path. The path-independent property of the state problem contributes significantly to the efficiency of design sensitivity analysis. More specifically, the nonlinear state problem is divided into several load steps to guarantee the convergence of the solution. For each load step, the nonlinear equation is solved iteratively using the linearization process. In design sensitivity analysis, however, the linearized design sensitivity equation is solved once at the final load step without any iteration. Thus, sensitivity calculation costs are fundamentally the same as those for the linear problem.

Unlike the nonlinear elastic problem, the elastoplastic problem in Section 9.2 includes the path-dependent, internal variable that represents plastic evolution, which is also related to energy dissipation during plastic deformation. These internal variables contribute to the design sensitivity formulation to generate so-called path-dependent terms. These path-dependent terms require the sensitivity results from the previous time step. Thus, the design sensitivity equation must be solved at each time step, which reduces the efficiency of the design sensitivity formulation as compared with nonlinear elastic problems.

In the final section of this chapter, the design sensitivity formulation for the transient dynamic problem with explicit time integration is introduced. As explained in Chapter 8, the explicit time integration method does not require any convergence iteration. Thus, for this example the benefits of an iteration-free design sensitivity formulation with the static problem, or with the implicit time integration method cannot be preserved. The cost of design sensitivity analysis is approximately the same as the cost using the finite difference method. However, the difficulties involved in choosing an appropriate perturbation size for the finite difference method does not appear in the proposed design sensitivity formulation.

9.1 Design Sensitivity Formulation for Nonlinear Elastic Problems

The first-order variation of the nonlinear equilibrium equation with respect to sizing design variables will be considered in this section, focusing on the cross-sectional geometry of the truss/beam, the thickness of the membrane/plate, and the material property of the continuum solid, obtained for elastic and hyperelastic materials. Since sizing design variables are independent of the reference configuration that is chosen, a distinction between total and updated Lagrangian formulations is not required to develop the design sensitivity formulation. However, for the shape design problem, which will be discussed in Chapter 10, such a distinction is necessary. Both the adjoint variable and direct differentiation methods are presented for static response. Using an adjoint variable method that parallels the method presented in Chapter 5 for linear structural systems, a linear adjoint equation is obtained for each performance measure in which the design sensitivity will be computed. If the finite element method is used for numerical evaluation, then the stiffness matrix of the adjoint system is the tangent stiffness matrix of the nonlinear system at the final equilibrium configuration. With the direct differentiation method, the first-order variation of the state variable is obtained by solving the linear equation produced by differentiating the nonlinear equation with respect to the design variable. As with the adjoint variable method, when the finite element method is used the stiffness matrix of the linear design sensitivity equation is the same as the tangent stiffness matrix from the linearized state problem at the final equilibrium configuration. The design sensitivity formulation includes the effects of a large displacement, large strain, and material nonlinearity, with appropriate kinematic and constitutive relations. The design sensitivity of the actual critical load factor can be obtained by taking the first-order variation of the stability equation with respect to the design variable.

9.1.1 Static Problems

It is assumed that the structural system is in the equilibrium configuration at time t_n for a given design $\boldsymbol{u}$. Here, design vector $\boldsymbol{u}$ can function as the cross-sectional geometry of the truss/beam, the thickness of the membrane/plate, and the material property of the continuum solid. Consider an arbitrary design variation $\delta\boldsymbol{u}$ and a small parameter $\tau > 0$. When the design is perturbed to $\boldsymbol{u} + \tau\delta\boldsymbol{u}$, the structural system reaches a new equilibrium configuration at time t_n. Using either (8.5) or (8.25), depending on the formulation, equilibrium equations can be written as

$$a_{u}(z,\bar{z}) = \ell_{u}(\bar{z}), \quad \forall \bar{z} \in Z \tag{9.1}$$

and

$$a_{u+\tau\delta u}(z,\bar{z}) = \ell_{u+\tau\delta u}(\bar{z}), \quad \forall \bar{z} \in Z. \tag{9.2}$$

The subscript $\boldsymbol{u}$ and $\boldsymbol{u} + \tau\delta\boldsymbol{u}$ are used to indicate the forms' dependence on the design. As design perturbation $\tau\delta\boldsymbol{u}$ decreases in size, the difference between the equilibrium state of the original and perturbed design also decreases. As with the linear problem in Chapter 5, the first-order variations of the nonlinear energy and load forms in (9.2) can be defined with respect to their explicit dependence on the design variable $\boldsymbol{u}$ as

$$a'_{\delta u}(z,\bar{z}) \equiv \frac{d}{d\tau} a_{u+\tau\delta u}(\tilde{z},\bar{z})\Big|_{\tau=0} \tag{9.3}$$

and

$$\ell'_{\delta u}(\bar{z}) \equiv \frac{d}{d\tau} \ell_{u+\tau\delta u}(\bar{z})\Big|_{\tau=0}, \tag{9.4}$$

where $\bar{z}$ is independent of τ (the design), and $\tilde{z}$ denotes state variable z with the dependence on τ suppressed (i.e., z remains constant during design variation). The explicitly dependent terms in (9.3) and (9.4) can be calculated using the result of state variable z and using the design variation $\delta\boldsymbol{u}$. Note that these two forms are linear with respect to $\delta\boldsymbol{u}$.

In addition to the explicitly dependent terms in (9.3) and (9.4), it is necessary to define the implicitly dependent terms. To this end, define the first-order variations of the state variable at $\tau = 0$ with respect to the design $\boldsymbol{u}$ as

$$z' \equiv \frac{d}{d\tau} z(\boldsymbol{u}+\tau\delta\boldsymbol{u})\Big|_{\tau=0} = \lim_{\tau\to 0} \frac{z(\boldsymbol{u}+\tau\delta\boldsymbol{u}) - z(\boldsymbol{u})}{\tau}. \tag{9.5}$$

Using the above definition, the partial derivative and the first-order variation are interchangeable, that is, $(\nabla_0 z)' = \nabla_0 z'$ where ∇_0 is the gradient operator at the undeformed configuration (see Section 8.1.1). When z' is linear in $\delta\boldsymbol{u}$, it is the Fréchet derivative of state variable z with respect to the design, evaluated in the direction of $\delta\boldsymbol{u}$. Proof of the validity of this result is not easily obtained, although it might seem intuitive that the state variable of a system should be smoothly dependent on the design.

By noting that the energy form $a_{u+\tau\delta u}$ in (9.3) is nonlinear in z, the chain rule of differentiation can be used to obtain

$$\frac{d}{d\tau} a_{u+\tau\delta u}[z(\boldsymbol{u}+\tau\delta\boldsymbol{u}),\bar{z}]\Big|_{\tau=0} = a'_{\delta u}(z,\bar{z}) + a^*_u(z;z',\bar{z}). \tag{9.6}$$

The detailed derivation of the forms in (9.6) will be provided for each design component in subsequent subsections. For the second term on the right side of (9.6), the linearization process described in (8.13) through (8.20) has been used. Note that the form $a^*_u(z;\bullet,\bullet)$ is symmetric and bilinear with respect to its arguments. By taking the first-order variation of both sides of (9.2), and using (9.4) and (9.6), we obtain

$$a^*_u(z;z',\bar{z}) = \ell'_{\delta u}(\bar{z}) - a'_{\delta u}(z,\bar{z}), \quad \forall \bar{z} \in Z, \tag{9.7}$$

where the explicit expression of $a^*_u(z;z',\bar{z})$ can be obtained by replacing Δz with z' from (8.20). Presuming that state variable z is the solution to (9.1), (9.7) is the linear variational equation for the first-order variation z', yielding the *direct differentiation method*. Note that the linearized (8.21) solves for the incremental displacement, whereas the linear sensitivity (9.7) solve for the variation of the total displacement. This provides a significant efficiency because the linear sensitivity equation needs to be solve once at the final configuration. As previously mentioned, since this formulation includes a large displacement, a large degree of rotation, a large strain, and material nonlinearity when appropriate kinematic and constitutive relations are used, (9.7) is valid for any and all of these cases.

In order to use the adjoint variable method, consider a structural performance measure that can be written in integral form over the final equilibrium configuration at time t_n, with the reference configuration time t_0 written as

$$\psi = \int_{\Omega_0} g(z, \nabla_0 z, \boldsymbol{u})\, d\Omega_0. \tag{9.8}$$

The undeformed domain Ω_0 is used to define the performance measure ψ regardless of whether the total or updated Lagrangian formulation is used. In fact, the main difference between the total and updated Lagrangian formulation is in the way stress and strain are defined. Thus, the use of an undeformed domain as the performance measure should not be a restriction. Function g is continuously differentiable with respect to its arguments. The functional in (9.8) represents a variety of structural performance measures. For example, the volume of a structural element can be written with g depending only on $\boldsymbol{u}$; the average stress over the subset of a plane elastic solid can be written in terms of $\boldsymbol{u}$ and $\nabla_0 z$; and displacement at a point on a beam or plate can be formally written by multiplying the Dirac delta measure by the displacement function in the integrand. Taking the variation of the functional in (9.8) with respect to the sizing design variables, we obtain

$$\psi' = \int_{\Omega_0} (g_{,z} z' + g_{,\nabla_0 z} : \nabla_0 z' + g_{,\boldsymbol{u}} \delta \boldsymbol{u})\, d\Omega_0, \tag{9.9}$$

where ":" is a contraction operator such that $\boldsymbol{a}{:}\boldsymbol{b} = a_{ij}b_{ij}$. Leibnitz's rule allows the derivative with respect to τ to be taken inside the integral, and the chain rule of differentiation has been used to calculate the integrand of (9.9). An explicit expression for ψ in terms of $\delta\boldsymbol{u}$ is desirable, which requires rewriting the first two terms of the integrand in (9.9) explicitly in terms of $\delta\boldsymbol{u}$.

It is possible to introduce an adjoint equation in the same manner as in Chapter 5 by replacing z' in (9.9) with virtual displacement $\bar{\lambda}$, and by equating the terms involving $\bar{\lambda}$ in (9.9) to the energy bilinear form $a^*_{\boldsymbol{u}}(z; \lambda, \bar{\lambda})$ defined in (8.20), yielding the following linear adjoint equation for adjoint variable λ:

$$a^*_{\boldsymbol{u}}(z; \lambda, \bar{\lambda}) = \int_{\Omega_0} (g_{,z} \bar{\lambda} + g_{,\nabla_0 z} : \nabla_0 \bar{\lambda})\, d\Omega_0, \quad \forall \bar{\lambda} \in Z, \tag{9.10}$$

where the adjoint solution λ is sought. Equation (9.10) defines the adjoint equation of the performance measure in (9.8). Since $z' \in Z$, evaluate (9.10) at $\bar{\lambda} = z'$ to obtain

$$a^*_{\boldsymbol{u}}(z; \lambda, z') = \int_{\Omega_0} (g_{,z} z' + g_{,\nabla_0 z} : \nabla_0 z')\, d\Omega_0. \tag{9.11}$$

Since both $\bar{z}$ and λ belong to the same space Z, evaluate (9.7) at $\bar{z} = \lambda$ to obtain

$$a^*_{\boldsymbol{u}}(z; z', \lambda) = \ell'_{\delta\boldsymbol{u}}(\lambda) - a'_{\delta\boldsymbol{u}}(z, \lambda). \tag{9.12}$$

Using the symmetric property of the energy bilinear form $a^*_{\boldsymbol{u}}(z; \bullet, \bullet)$ in its arguments, the following can be obtained from (9.11) and (9.12):

$$\int_{\Omega_0} (g_{,z} z' + g_{,\nabla_0 z} : \nabla_0 z')\, d\Omega_0 = \ell'_{\delta\boldsymbol{u}}(\lambda) - a'_{\delta\boldsymbol{u}}(z, \lambda), \tag{9.13}$$

where the right side is linear in $\delta\boldsymbol{u}$ and can be evaluated once it is determined that state z and adjoint variable λ are the solutions to (9.1) and (9.12), respectively. Substituting this result into (9.1) yields

$$\psi' = \int_{\Omega_0} g_u \delta \boldsymbol{u}\, d\Omega_0 + \ell'_{\delta \boldsymbol{u}}(\boldsymbol{\lambda}) - a'_{\delta \boldsymbol{u}}(\boldsymbol{z}, \boldsymbol{\lambda}), \tag{9.14}$$

which expresses ψ' explicitly in terms of the design variation $\delta \boldsymbol{u}$, and the form of the last two terms on the right depends on the design component. Explicit expressions for these two terms will be provided later in this chapter for some prototypical structural components. It is interesting to note that even though the original governing equation is nonlinear, sensitivity equation (9.7) and adjoint equation (9.10) are linear, with the same tangent stiffness matrix at the final equilibrium configuration. This means that the effort to compute design sensitivity will be the same for both linear and nonlinear structural systems.

Beam/Truss Component

The sensitivity formulation of the Bernoulli-Euler beam/truss component is developed in this section. The total Lagrangian formulation for the beam/truss component is presented as an analytical example. The design vector is $\boldsymbol{u} = [A, E]^T$, where A is the cross-sectional area and E is Young's modulus. By freezing the implicit dependence on the design, the explicit dependence of stress S_{11} can be obtained from the first-order variation of (8.34) as

$$\left(AS_{11}(\tilde{z})\right)' = E(z_{1,1} + \tfrac{1}{2} z_{1,1} z_{1,1})\delta A + A(z_{1,1} + \tfrac{1}{2} z_{1,1} z_{1,1})\delta E. \tag{9.15}$$

The explicitly dependent variation of the structural energy form can be obtained from its definition in (8.32) and (9.15) as

$$\begin{aligned} a'_{\delta \boldsymbol{u}}(\boldsymbol{z}, \bar{\boldsymbol{z}}) &\equiv \frac{d}{d\tau} a_{\boldsymbol{u}+\delta \boldsymbol{u}}(\tilde{\boldsymbol{z}}, \bar{\boldsymbol{z}})\bigg|_{\tau=0} \\ &= \int_0^l [E(z_{1,1} + \tfrac{1}{2} z_{1,1} z_{1,1})(\bar{z}_{1,1} + \bar{z}_{1,1} z_{1,1})]\delta A\, dX_1 \\ &\quad + \int_0^l [A(z_{1,1} + \tfrac{1}{2} z_{1,1} z_{1,1})(\bar{z}_{1,1} + \bar{z}_{1,1} z_{1,1})]\delta E\, dX_1 \\ &\quad + \int_0^l (EI_{,A} z_{2,11} \bar{z}_{2,11} \delta A + I z_{2,11} \bar{z}_{2,11} \delta E)\, dX_1, \end{aligned} \tag{9.16}$$

where I_{A} is the partial derivative of the moment of inertia with respect to the cross-sectional area. Although $a'_{\delta \boldsymbol{u}}(\boldsymbol{z}, \bar{\boldsymbol{z}})$ is linear with respect to $\delta \boldsymbol{u}$, it is nonlinear with respect to state variable $\boldsymbol{z}$.

If the applied load depends on the deformation, then the linearization of the state problem includes load stiffness terms, and the variation of the load form contains the displacement sensitivity $\boldsymbol{z}'$. Such a situation is not investigated in this chapter. In the case of a displacement independent load, the variation of the load form becomes

$$\ell'_{\delta \boldsymbol{u}}(\bar{\boldsymbol{z}}) = \int_0^l \bar{\boldsymbol{z}}^T \boldsymbol{f}_{,A} \delta A\, dX_1. \tag{9.17}$$

Using the explicit expression of (9.16) and (9.17), the sensitivity of various performance measures can be derived in the same way as with the linear problem. Let the response analysis be solved up to the last configuration time t_n, and let the final solution be $\boldsymbol{z}$ from (9.1). The direct differentiation method solves (9.7) with respect to displacement variation $\boldsymbol{z}'$, and by then using (9.9), the sensitivity expressions of the performance measures can be obtained using the chain rule of differentiation. Sensitivity expressions using the adjoint variable method are discussed below.

Using the adjoint variable method, several alternative forms may now be considered for structural response functionals. Consider a functional that defines the value of displacement at an isolated point $\hat{X}_1$, that is,

$$\psi_1 = z(\hat{X}_1) = \int_0^l \delta(X_1 - \hat{X}_1) z(X_1)\, dX_1, \tag{9.18}$$

where $\delta(X_1)$ is the Dirac delta measure at the origin. Taking the first variation of (9.18), we obtain

$$\psi_1' = \int_0^l \delta(X_1 - \hat{X}_1) z'(X_1)\, dX_1. \tag{9.19}$$

No explicitly dependent term exists in the displacement performance measure, as illustrated in the above equation. Using the direct differentiation method, (9.19) can easily be obtained with z'. For this functional, the adjoint equation of (9.10) is

$$a_u^*(z;\lambda,\bar{\lambda}) = \int_0^l \delta(X_1 - \hat{X}_1)\bar{\lambda}\, dX_1, \quad \forall \bar{\lambda} \in Z. \tag{9.20}$$

In the above equation, the linear adjoint system has a point load at point $\hat{X}_1$ and its stiffness matrix is the same as the tangent stiffness of the state problem at the final equilibrium configuration. Using (9.20), the design sensitivity expression of the functional in (9.18) is obtained as

$$\psi_1' = \ell'_{\delta u}(\lambda) - a'_{\delta u}(z,\lambda), \tag{9.21}$$

which can be calculated using the state response z and adjoint response λ.

Another important functional in the design of a truss component is stress, appearing as

$$\psi_2 = \int_0^l [EA(z_{1,1} + \tfrac{1}{2} z_{1,1} z_{1,1}) + EI z_{2,11}] m_p\, dX_1, \tag{9.22}$$

where $m_p(X_1)$ is a characteristic function that is only nonzero on a small subinterval and whose integral is one. Taking the first-order variation of (9.22), we obtain

$$\begin{aligned} \psi_2' = &\int_0^l [EA(z'_{1,1} + z_{1,1} z'_{1,1}) + EI z'_{2,11}] m_p\, dX_1 \\ &+ \int_0^l [E(z_{1,1} + \tfrac{1}{2} z_{1,1} z_{1,1})\delta A + A(z_{1,1} + \tfrac{1}{2} z_{1,1} z_{1,1})\delta E] m_p\, dX_1 \\ &+ \int_0^l [EI_{,A} z_{2,11}\delta A + I z_{2,11}\delta E] m_p\, dX_1. \end{aligned} \tag{9.23}$$

Since the first integral of (9.23) contains an unknown displacement variation, the following adjoint equation is defined for the stress functional:

$$a_u^*(z;\lambda,\bar{\lambda}) = \int_0^l \left[EA\left(\bar{\lambda}_{1,1} + z_{1,1}\bar{\lambda}_{1,1}\right) + EI\bar{\lambda}_{2,11} \right] m_p\, dX_1, \quad \forall \bar{\lambda} \in Z. \tag{9.24}$$

Using state response z, adjoint response λ, and the explicitly dependent terms from (9.23), the sensitivity of the stress functional can be obtained as

$$\begin{aligned} \psi_2' = &\ \ell'_{\delta u}(\lambda) - a'_{\delta u}(z,\lambda) \\ &+ \int_0^l [E(z_{1,1} + \tfrac{1}{2} z_{1,1} z_{1,1})\delta A + A(z_{1,1} + \tfrac{1}{2} z_{1,1} z_{1,1})\delta E] m_p\, dX_1 \\ &+ \int_0^l [EI_{,A} z_{2,11}\delta A + I z_{2,11}\delta E] m_p\, dX_1. \end{aligned} \tag{9.25}$$

An important functional that arises in the design of beam and truss components is rotation at an isolated point $\hat{X}_1$, that is,

$$\psi_3 = z_{2,1}(\hat{X}_1) = \int_0^l \delta(X_1 - \hat{X}_1) z_{2,1}(X_1)\, dX_1. \tag{9.26}$$

The first-order variation of the functional in (9.26) yields

$$\psi_3' = \int_0^l \delta(X_1 - \hat{X}_1) z_{i,1}'(X_1)\, dX_1. \tag{9.27}$$

For this functional, the adjoint equation is

$$a_\Omega^*(z; \boldsymbol{\lambda}, \bar{\boldsymbol{\lambda}}) = \int_0^l \delta(X_1 - \hat{X}_1) \bar{\lambda}_{2,1}\, dX_1, \qquad \forall \bar{\boldsymbol{\lambda}} \in Z. \tag{9.28}$$

Interpreting the Dirac delta measure as a unit load applied at a point $\hat{X}_1$, a physical interpretation of $\boldsymbol{\lambda}$ can be obtained as the displacement of the beam or truss from the final equilibrium configuration due to a unit moment at $\hat{X}_1$ in direction $z_{2,1}$. Using (9.14), the design sensitivity expression is given as

$$\psi_3' = \ell'_{\delta \boldsymbol{u}}(\boldsymbol{\lambda}) - a'_{\delta \boldsymbol{u}}(\boldsymbol{z}, \boldsymbol{\lambda}). \tag{9.29}$$

Plate/Plane Solid Component

Consider a plane stress elastic solid component using the total Lagrangian formulation. The design variable $u = h(\boldsymbol{x}_1, \boldsymbol{x}_2)$ is the thickness of the plane elastic solid. The explicit variation of the structural energy form in (8.41) and the load form in (8.42) can easily be taken by freezing the design dependence through displacement $\boldsymbol{z}$, as

$$a'_{\delta \boldsymbol{u}}(\boldsymbol{z}, \bar{\boldsymbol{z}}) = \iint_{^0\Omega} \boldsymbol{S} : \bar{\boldsymbol{E}} \delta h\, d\Omega + \iint_{^0\Omega} \boldsymbol{\kappa}(\bar{z}_3)^T \boldsymbol{C}^b \boldsymbol{\kappa}(z_3) \frac{3\delta h}{h} d\Omega \tag{9.30}$$

$$\ell'_{\delta \boldsymbol{u}}(\bar{\boldsymbol{z}}) = \iint_{^0\Omega} \bar{\boldsymbol{z}}^T \boldsymbol{f}_{,h}^B \delta h\, d\Omega. \tag{9.31}$$

The expression of displacement sensitivity is exactly the same as the expression of the beam/truss component in (9.21) by using (9.30) and (9.31).

9.1.2 Critical Load

Under the assumption that the eigenvalues of nonlinear systems are differentiable with respect to the design, a method of design sensitivity analysis of the estimated/actual critical loads is developed for a conservative system. The adjoint variable method presented in Section 9.1.1 for nonlinear structural systems is used to obtain design sensitivity expressions of the critical load factor in terms of design perturbation. Design sensitivity equations consider the effect of a large displacement, a large degree of rotation, a large strain, and material nonlinearity with appropriate kinematic and constitutive descriptions. The stability equations in (8.51) and (8.56), which are nonlinear with respect to design variables, are linearized to obtain the first variation.

The first two methods are good choices when the actual critical load factor is not available, and the estimated critical load factor is used to calculate design sensitivity information. However, sensitivity results may not be sufficiently accurate. Even if the estimated critical load may converge to the actual critical load, the sensitivity of the estimated critical load may not converge to that of actual critical load as shown in numerical examples in Section 9.1.4. Thus, it is strongly recommended to use the actual critical load factor.

One-Point Method

For one-point linear eigenvalue problems, the stability equations at the equilibrium configurations at time t_n with design $\boldsymbol{u}$ and design $\boldsymbol{u} + \tau\boldsymbol{\delta u}$ can be written using (8.51), as

$$A_u(z; y, \bar{y}) = \zeta D_u(z; y, \bar{y}), \quad \forall \bar{y} \in Z \tag{9.32}$$

$$A_{u+\tau\delta u}(z; y, \bar{y}) = \zeta_{u+\tau\delta u} D_{u+\tau\delta u}(z; y, \bar{y}), \quad \forall \bar{y} \in Z, \tag{9.33}$$

where z is the displacement at time t_n. The first variation of the nonlinear energy forms in (9.33) can be defined with respect to its explicit dependence on the design variable $\boldsymbol{u}$ as

$$A'_{\delta u}(z; y, \bar{y}) \equiv \frac{d}{d\tau} A_{u+\tau\delta u}(\tilde{z}; \tilde{y}, \bar{y})\Big|_{\tau=0} \tag{9.34}$$

$$D'_{\delta u}(z; y, \bar{y}) \equiv \frac{d}{d\tau} D_{u+\tau\delta u}(\tilde{z}; \tilde{y}, \bar{y})\Big|_{\tau=0}. \tag{9.35}$$

With an assumption of differentiability of the eigenvalue and eigenfunction for the linear eigenvalue problem in (9.33), their first variation with respect to design variable $\boldsymbol{u}$ can be defined as

$$\zeta' = \frac{d}{d\tau}\zeta(\boldsymbol{u} + \tau\boldsymbol{\delta u})\Big|_{\tau=0} = \lim_{\tau\to 0}\frac{\zeta(\boldsymbol{u}+\tau\boldsymbol{\delta u}) - \zeta(\boldsymbol{u})}{\tau} \tag{9.36}$$

$$y' = \frac{d}{d\tau} y(\boldsymbol{u} + \tau\boldsymbol{\delta u})\Big|_{\tau=0} = \lim_{\tau\to 0}\frac{y(\boldsymbol{u}+\tau\boldsymbol{\delta u}) - y(\boldsymbol{u})}{\tau}. \tag{9.37}$$

Using the above definition, it can be noted that the order in which the first-order variation and the partial derivative of the eigenfunction are taken can be interchanged.

In (9.33), the energy forms A_u and D_u are nonlinear in z and linear in y. Using these properties, the chain rule of differentiation can be used to yield

$$\frac{d}{d\tau}[A_{u+\tau\delta u}(z; y, \bar{y})]\Big|_{\tau=0} = A'_{\delta u}(z; y, \bar{y}) + A^*_u(z; z', y, \bar{y}) + A_u(z; y', \bar{y}) \tag{9.38}$$

$$\frac{d}{d\tau}[D_{u+\tau\delta u}(z; y, \bar{y})]\Big|_{\tau=0} = D'_{\delta u}(z; y, \bar{y}) + D^*_u(z; z', y, \bar{y}) + D_u(z; y', \bar{y}), \tag{9.39}$$

where A^*_u and D^*_u are the first-order variations of A_u and D_u, respectively, with respect to the design variable, and they implicitly depend on the design through the response z. If the material stiffness tensor C_{ijkl} in (8.16) is a constant, then A^*_u and D^*_u can be derived by differentiating (8.48) and (8.49), as

$$A^*_u(z, z'; y, \bar{y}) = \int_{{}^0\Omega}\left[\Delta\bar{\boldsymbol{E}}(z', \bar{y}) : \boldsymbol{C} : \Delta\boldsymbol{E}(z, y) + \bar{\boldsymbol{E}}(z, \bar{y}) : \boldsymbol{C} : \Delta\bar{\boldsymbol{E}}(z', y)\right] d\Omega \tag{9.40}$$

$$D_u^*(z,z';y,\overline{y}) = -\int_{{}^0\Omega}\left[\Delta\overline{E}(y,\overline{y}):C:\Delta E(z,z')\right]d\Omega, \tag{9.41}$$

where the following notations for strains are used from their definitions in (8.8), (8.18), and (8.19):

$$\begin{aligned}\overline{E}(z,\overline{y}) &= sym(\nabla_0\overline{y}^T F)\\ \Delta E(z,y) &= sym(\nabla_0 y^T F)\\ \Delta\overline{E}(z',\overline{y}) &= sym(\nabla_0\overline{y}^T\nabla_0 z').\end{aligned} \tag{9.42}$$

In addition, the deformation gradient F is a linear function of the displacement z based on its definition in (8.2),

Taking the first-order variation of both sides of (9.33), replacing $\overline{y}$ with $y \in Z$ since (9.33) holds for all $\overline{y} \in Z$, and using symmetry of the energy forms $A_u(z;\bullet,\bullet)$ and $D_u(z;\bullet,\bullet)$ in their arguments, we obtain

$$\begin{aligned}D_u(z;y,y)\zeta' = {}&[A_u^*(z;z',y,y) - \zeta D_u^*(z;z',y,y)]\\ &+[A_u(z;y,y') - \zeta D_u(z;y,y')] + [A'_{\delta u}(z;y,y) - \zeta D'_{\delta u}(z;y,y)].\end{aligned} \tag{9.43}$$

By noting that $y' \in Z$, it can be seen that the term in the second bracket on the right side of (9.43) is zero using (9.33). Thus,

$$\zeta' = \frac{[A_u^*(z;z',y,y) - \zeta D_u^*(z;z',y,y)] + [A'_{\delta u}(z;y,y) - \zeta D'_{\delta u}(z;y,y)]}{D_u(z;y,y)}. \tag{9.44}$$

Equation (9.44) constitutes a direct differentiation method for calculating the sensitivity of the critical load factor ζ. First, a nonlinear static design sensitivity analysis presented in Section 9.1.1 must be performed at time t_n to calculate the displacement sensitivity z'. After solving the eigenfunction y from (9.32), the sensitivity of the critical load factor ζ can be calculated from (9.44). Numerical integration is involved in evaluating ζ'.

Recall that the first-order variation z' depends on the direction of design change δu. The objective is to obtain an explicit expression for ζ' in terms of design change δu. The adjoint variable method in Section 9.1.1 can be used to obtain the terms that include z' in the numerators on the right side of (9.44) explicitly in terms of δu. The adjoint equations corresponding to z' are introduced by replacing z' with virtual displacement $\overline{\lambda}$, and equating terms involving $\overline{\lambda}$ to the energy form a_u^*.

In the critical load problem the adjoint variable method is used to remove the displacement sensitivity z' from (9.44). Accordingly, the adjoint problem corresponding to (9.44) can be defined in a similar way in Section 9.1.1, as

$$a_u^*\left(z;\lambda,\overline{\lambda}\right) = A_u^*\left(z;\overline{\lambda},y,y\right) - \zeta D_u^*\left(z;\overline{\lambda},y,y\right), \quad \forall\overline{y}\in Z, \tag{9.45}$$

where the adjoint solution $\lambda \in Z$ is desired. Following the same procedure employed with the adjoint variable method in Section 9.1.1, the terms that include z' in (9.44) can be rewritten in terms of design variation using (9.12), as

$$A_u^*(z;z',y,y) - \zeta D_u^*(z;z',y,y) = \ell'_{\delta u}(\lambda) - a'_{\delta u}(z,\lambda), \tag{9.46}$$

where the right side is linear in δu and can be evaluated once the state z and adjoint solution λ are determined. Substituting this result into (9.44), the explicit design sensitivity of ζ is

$$\zeta' = \frac{[\ell'_{\delta u}(\lambda) - a'_{\delta u}(z,\lambda)] + [A'_{\delta u}(z;y,y) - \zeta D'_{\delta u}(z;y,y)]}{D_u(z;y,y)}. \tag{9.47}$$

Two-Point Method

For two-point linear eigenvalue problems, the stability equations at the equilibrium configuration at time t_n with design $\boldsymbol{u}$ and design $\boldsymbol{u} + \tau\boldsymbol{\delta u}$ can be written using (8.56), as

$$B_u({}^{n-1}z;y,\bar{y}) + \zeta E_u({}^{n-1}z, {}^{n}z;y,\bar{y}) = 0, \quad \forall \bar{y} \in Z \tag{9.48}$$

$$B_{u+\tau\delta u}({}^{n-1}z;y,\bar{y}) + \zeta_{u+\tau\delta u} E_{u+\tau\delta u}({}^{n-1}z, {}^{n}z;y,\bar{y}) = 0, \quad \forall \bar{y} \in Z. \tag{9.49}$$

Since the two-point method is related to the two configurations t_n and t_{n-1}, the left superscript is used to distinguish them.

Define the first-order variation of the nonlinear energy form B_u in (9.49) with respect to its explicit dependence on the design variable $\boldsymbol{u}$ as

$$B'_{\delta u}({}^{n-1}z;y,\bar{y}) \equiv \frac{d}{d\tau} B_{u+\tau\delta u}({}^{n-1}\tilde{z};\tilde{y},\bar{y})\bigg|_{\tau=0}. \tag{9.50}$$

With the assumption of differentiability of the eigenvalue and eigenfunction for the linear eigenvalue problem in (9.49), their first variations with respect to explicit dependence on design variable $\boldsymbol{u}$ can be defined as in (9.36) and (9.37), respectively.

In (9.48), the energy forms B_u and E_u are nonlinear in ${}^{n}z$ and ${}^{n-1}z$, and linear in $\boldsymbol{y}$. Using these properties, the chain rule of differentiation can be used to obtain

$$\frac{d}{d\tau}[B_{u+\tau\delta u}({}^{n-1}z;y,\bar{y})]\bigg|_{\tau=0} = B'_{\delta u}({}^{n-1}z;y,\bar{y}) + B^*_u({}^{n-1}z;{}^{n-1}z',y,\bar{y}) + B_u({}^{n-1}z;y',\bar{y}) \tag{9.51}$$

$$\begin{aligned} \frac{d}{d\tau}[E_{u+\tau\delta u}({}^{n-1}z, {}^{n}z;y,\bar{y})]\bigg|_{\tau=0} &= B'_{\delta u}({}^{n}z;y,\bar{y}) + B^*_u({}^{n}z;{}^{n}z',y,\bar{y}) + B_u({}^{n}z;y',\bar{y}) \\ &\quad - B'_{\delta u}({}^{n-1}z;y,\bar{y}) - B^*_u({}^{n-1}z;{}^{n-1}z',y,\bar{y}) - B_u({}^{n-1}z;y',\bar{y}). \end{aligned} \tag{9.52}$$

In (9.51) and (9.52), B^*_u is the first-order variation of B_u with respect to the design variable and its dependence is implicit through the responses ${}^{n}z$,

$$B^*_u({}^{n}z,{}^{n}z';y,\bar{y}) = A^*_u({}^{n}z,{}^{n}z';y,\bar{y}) - D^*_u({}^{n}z,{}^{n}z';y,\bar{y}). \tag{9.53}$$

Taking the first-order variations of both sides of (9.49), replacing $\bar{y}$ with $\boldsymbol{y} \in Z$, since (9.49) holds for all $\bar{y} \in Z$, and using symmetry of the energy form $B_u({}^{n}z;\bullet,\bullet)$ in their arguments, we obtain

$$\begin{aligned} &\left\{B'_{\delta u}({}^{n-1}z;y,y) + \zeta[B'_{\delta u}({}^{n}z;y,y) - B'_{\delta u}({}^{n-1}z;y,y)]\right\} \\ &+ \left\{B_u({}^{n-1}z;y,y') + \zeta[B_u({}^{n}z;y,y') - B_u({}^{n-1}z;y,y')]\right\} \\ &+ \left\{B^*_u({}^{n-1}z;{}^{n-1}z',y,y) + \zeta[B^*_u({}^{n}z;{}^{n}z',y,y) - B^*_u({}^{n-1}z;{}^{n-1}z',y,y)]\right\} \\ &+ \zeta'[B_u({}^{n}z;y,y) - B_u({}^{n-1}z;y,y)] = 0. \end{aligned} \tag{9.54}$$

By noting that $\boldsymbol{y}' \in Z$, it can be seen that the term in the second set of brackets on the left side of (9.54) is zero using (9.48). Thus,

$$\zeta' = \frac{(1-\zeta)[B_u^*({}^{n-1}z;{}^{n-1}z',y,y) + B'_{\delta u}({}^{n-1}z;y,y)] + \zeta[B_u^*({}^{n}z;{}^{n}z',y,y) + B'_{\delta u}({}^{n}z;y,y)]}{B_u({}^{n-1}z;y,y) - B_u({}^{n}z;y,y)}. \tag{9.55}$$

Since the two-point method refers to the two configurations t_{n-1} and t_n, the adjoint variable method requires two adjoint problems. These two adjoint problems are given as

$$a_u^*\left({}^{n}z;{}^{n}\lambda,\bar{\lambda}\right) = B_u^*\left({}^{n}z;\bar{\lambda},y,y\right), \quad \forall\bar{\lambda}\in Z \tag{9.56}$$

$$a_u^*\left({}^{n-1}z;{}^{n-1}\lambda,\bar{\lambda}\right) = B_u^*\left({}^{n-1}z;\bar{\lambda},y,y\right), \quad \forall\bar{\lambda}\in Z, \tag{9.57}$$

where the solutions ${}^{n}\lambda$ and ${}^{n-1}\lambda$ are desired. In fact, (9.56) and (9.57) present the same adjoint problem at a different configuration, i.e., t_n and t_{n-1}. Following the same adjoint variable procedure from Section 9.1.1, the terms that include ${}^{n}z'$ and ${}^{n-1}z'$ in (9.55) can be rewritten in terms of design variation, as

$$\zeta' = \frac{(1-\zeta)[\ell'_{\delta u}({}^{n-1}\lambda) - a'_{\delta u}({}^{n-1}z,{}^{n-1}\lambda) + B'_{\delta u}({}^{n-1}z;y,y)] + \zeta[\ell'_{\delta u}({}^{n}\lambda) - a'_{\delta u}({}^{n}z,{}^{n}\lambda) + B'_{\delta u}({}^{n}z;y,y)]}{B_u({}^{n-1}z;y,y) - B_u({}^{n}z;y,y)}. \tag{9.58}$$

Variations of energy forms on the right sides of (9.47) and (9.58) depend on the structural component. These two equations will serve as the principal tools for sizing design sensitivity of the estimated critical load of nonlinear structural systems with large displacement, large rotation, small strain, and nonlinear elastic material behavior under conservative static loading. The design sensitivity analysis results in (9.47) are applicable to linear structural systems by simply dropping the expression $\ell'_{\delta u}(\lambda) - a'_{\delta u}(z,\lambda)$.

Sensitivity Analysis with Actual Critical Load Factor

Design sensitivity formulations for the estimated critical load factor in previous sections are good choices when the critical load factor is not available. However, sensitivity results may not be sufficiently accurate. Even if the estimated critical load may converge to the actual critical load, the sensitivity of the estimated critical load may not converge to that of actual critical load as shown in numerical examples in Section 9.1.4. This is a good example that the derivative of an approximate response does not necessarily converge to the derivative of the exact response even though the approximate response converges to the exact response. In this section, design sensitivity analysis is carried out for the actual critical load factor. Suppose that the design is perturbed by $\tau\boldsymbol{\delta u}$. The new stability equation of the structural system at the critical configuration and at time t_n for the perturbed design $\boldsymbol{u} + \tau\boldsymbol{\delta u}$ becomes

$$a_{u+\delta u}({}^{n}z,\bar{z}) = \beta_{u+\delta u}\ell_{u+\delta u}(\bar{z}), \quad \forall\bar{z}\in Z. \tag{9.59}$$

The first-order variation of both sides of (9.59) can be taken to obtain

$$a'_{\delta u}(z,\bar{z}) + a_u^*(z;z',\bar{z}) = \beta'\ell_u(\bar{z}) + \beta_u\ell'_{\delta u}(\bar{z}), \qquad \forall\bar{z}\in Z, \tag{9.60}$$

where $a_u^*(z;z',\bar{z})$ is the first-order variation of $a_u(z,\bar{z})$ with respect to design variables implicitly through total displacement z, and is defined in (8.20) for the total Lagrangian formulation. Since (9.60) holds for all $\bar{z}\in Z$, this equation may be evaluated at $\bar{z} = y$, which is the eigenfunction corresponding to the lowest eigenvalue of the linear eigenvalue problem in (8.51). Equation (9.60) then becomes

$$a'_{\delta u}(z,y) + a^*_u(z;z',y) = \beta' \ell_u(y) + \beta_u \ell'_{\delta u}(y). \tag{9.61}$$

Using (8.50) and the symmetry of energy forms $A_u(z;\bullet,\bullet)$ and $D_u(z;\bullet,\bullet)$ in their arguments, a^*_u in (9.61) can be rewritten as

$$a^*_u(z;z',y) = A_u(z;y,z') - D_u(z;y,z'). \tag{9.62}$$

At the critical limit point where $\zeta = 1$ and $z' \in Z$, the right side of (9.62) vanishes using (8.51). Thus, in (9.61), the second term on the left-hand side vanishes and, using (8.59), the sizing design sensitivity expression of the critical load becomes

$$\boldsymbol{p}'_{cr} = \boldsymbol{p}_0 \beta' = \frac{\boldsymbol{p}_0}{\ell_u(y)} [a'_{\delta u}(z,y) - \beta_u \ell'_{\delta u}(y)]. \tag{9.63}$$

In (9.63), z is the displacement in the final, prebuckling equilibrium configuration. Note that in order to evaluate the design sensitivity expression in (9.63) no adjoint system is required, unlike the design sensitivity analysis of the estimated critical load in previous sections.

9.1.3 Hyperelastic Material

The material properties of hyperelasticity are relatively easy to change during the manufacturing process. In this section, design sensitivity analysis of hyperelasticity will be examined with respect to material constants. For material parameter design sensitivity analysis, (9.7) can be used for the direct differentiation method, while (9.14) can be used for the specific performance measure with adjoint equation (9.10). Thus, in this section the explicit formulation of $a'_{\delta u}(\boldsymbol{r},\overline{\boldsymbol{r}})$ is derived for hyperelastic material. Three material properties can be chosen for a Mooney-Rivlin type hyperelastic material:

$$\begin{aligned} \boldsymbol{u} &= [C_{10},\ C_{01},\ K]^T \\ \delta \boldsymbol{u} &= [\delta C_{10},\ \delta C_{01},\ \delta K]^T. \end{aligned} \tag{9.64}$$

The structural energy form in (8.68) can be rewritten at the perturbed design $\boldsymbol{u} + \tau\delta\boldsymbol{u}$ as

$$a_{u+\tau\delta u}(\boldsymbol{r}(\boldsymbol{u}+\tau\delta\boldsymbol{u}),\overline{\boldsymbol{r}}) \equiv \int_\Omega \boldsymbol{S}_{u+\tau\delta u} : \overline{\boldsymbol{E}}\, d\Omega + \int_\Omega \overline{p} H_{u+\tau\delta u}\, d\Omega. \tag{9.65}$$

It is important to remember that $\boldsymbol{r} = [z,\ p]^T$ is the response variable for hyperelastic material using the pressure projection method. $a'_{\delta u}(\boldsymbol{r},\overline{\boldsymbol{r}})$ can be obtained by taking the variation of $a_{u+\tau\delta u}$ by freezing the implicit dependence through response variable $\boldsymbol{r}$ as

$$a'_{\delta u}(\boldsymbol{r},\overline{\boldsymbol{r}}) \equiv \frac{d}{d\tau} a_{u+\delta u}(\tilde{\boldsymbol{r}},\overline{\boldsymbol{r}})\bigg|_{\tau=0}. \tag{9.66}$$

The expression of $a'_{\delta u}(\boldsymbol{r},\overline{\boldsymbol{r}})$ can be obtained by taking the variation of the second Piola-Kirchhoff stress and volumetric stress tensors with respect to explicit dependence on design as

$$\delta \boldsymbol{S}(\tilde{z},\tilde{p}) = \delta C_{10} J_{1,E} + \delta C_{01} J_{2,E} \tag{9.67}$$

$$\delta H(\tilde{z},\tilde{p}) = \frac{p}{K^2}\delta K. \tag{9.68}$$

Thus,

$$\begin{aligned} a'_{\delta u}(\boldsymbol{r},\bar{\boldsymbol{r}}) = \delta C_{10} \int_{\Omega} J_{1,E} : \bar{\boldsymbol{E}}\, d\Omega \\ +\delta C_{01} \int_{\Omega} J_{2,E} : \bar{\boldsymbol{E}}\, d\Omega \\ +\delta K \int_{\Omega} \bar{p}p/K^2\, d\Omega. \end{aligned} \tag{9.69}$$

Note that the variations of the design are placed out of integrals since they are constant throughout the analysis.

9.1.4 Numerical Examples

In this section, the design sensitivity formulation developed in Sections 9.1.1 and 9.1.2 is applied to evaluate the design sensitivity coefficients using structural analysis results from ANSYS [28]. Geometrically nonlinear truss and beam structures with linear elastic material are considered as numerical examples. For all examples in this section, the total Lagrangian formulation is used for equilibrium, and one-point linearized eigenvalue analysis is used for stability.

Two-Bar Truss

Consider a shallow two-bar truss that is subjected to a vertical, concentrated load p at node 2, with two truss elements, three nodal points, and two degrees-of-freedom, as illustrated in Fig. 9.1. For analysis data, constant cross-sectional area A = 0.8 in^2 and Young's modulus $E = 1.0 \times 10^7$ psi are used. The design variables are two cross-sectional areas A_1 and A_2 of elements 1 and 2, respectively. Kamat et al. [29] developed an explicit expression of the actual critical load for the two-bar truss. According to their analysis data, the actual critical load is 60.14 lb and the design sensitivity coefficient of the actual critical load with respect to design variable A_1 is 60.14, which is the characteristic value of the given structure independent of the magnitude of applied loading. Wu and Arora [30] obtained the same design sensitivity coefficients of the actual critical load. Both studies addressed the critical load, rather than the critical load factor.

Structural analysis is performed using the two-dimensional truss element STIF1 of ANSYS. Using the incremental analysis method of ANSYS, the actual critical load is found to be around 60.060911 lb. At several load levels between p = 1 lb and p = 60 lb, the critical load factors are calculated, and design sensitivity analyses are performed for the two design variables A_1 and A_2. For linear structural systems, the design sensitivity coefficients of the linear critical load factor are evaluated using the eigenvalue design sensitivity formula in (9.47) without $a'_{\delta u}$ and $\ell'_{\delta u}$, and results are provided in Table 9.1. The design sensitivity coefficients of the nonlinear critical load factor are evaluated using (9.47) with $a'_{\delta u}$ and $\ell'_{\delta u}$, and the results are presented in Table 9.2. At each load level, the estimated critical load can be evaluated by multiplying the applied load in the first column of Table 9.1 or 9.2 by the critical load factor ζ in the second column for the one-point linear eigenvalue analysis of nonlinear systems.

In the case of a linear structural system in Table 9.1, the estimated critical load for all applied load levels is 312 lb, which is more than five times the actual critical load of 60.060911 lb. The third and fourth columns in Table 9.1 represent design sensitivity coefficients for the two cross-sectional areas, while the adjacent numbers in parentheses indicate the degree of agreement in percentages with sensitivity coefficients calculated using the finite difference method. Agreement is close to 100%, with a 1% design perturbation, which indicates good design sensitivity results from the proposed numerical method of eigenvalue design sensitivity analysis for linear structural systems.

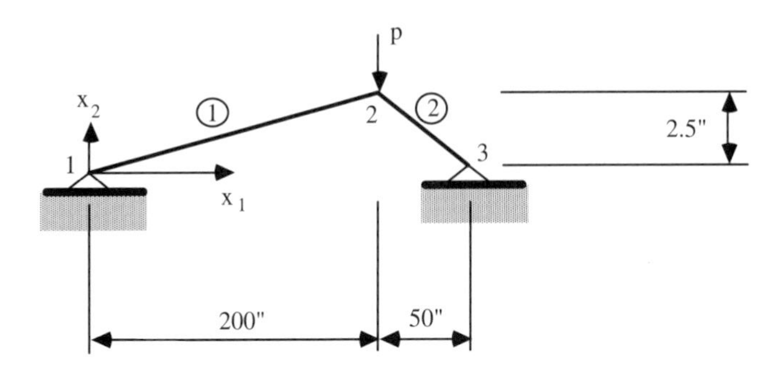

Figure 9.1. Two-bar truss.

Table 9.1. Design sensitivity of linear critical load factor for two-bar truss.

Load (lb)	ζ	$\partial\zeta/\partial A_1$	$\partial\zeta/\partial A_2$
1.00	0.31196E+03	311.5(100%)	75.93(100%)
10.0	0.31196E+02	30.71(100%)	5.922(100%)
20.0	0.15598E+02	15.11(100%)	2.154(100%)
30.0	0.10399E+02	9.907(100%)	0.984(100%)
40.0	0.77991E+01	7.308(100%)	0.462(100%)
50.0	0.62393E+01	5.750(100%)	0.201(100%)
60.0	0.51994E+01	4.712(100%)	0.069(100%)

Table 9.2. Design sensitivity of nonlinear critical load factor ζ for two-bar truss.

Load (lb)	ζ	$\partial\zeta/\partial A_1$	$\partial\zeta/\partial A_2$
1.00	0.30897E+03	312.01(100%)	77.895(101%)
10.0	0.28146E+02	31.257(100%)	7.8086(101%)
20.0	0.12485E+02	15.740(100%)	3.9303(101%)
30.0	0.72015E+01	10.683(100%)	2.6709(101%)
40.0	0.44886E+01	8.3460(100%)	2.0874(101%)
50.0	0.27510E+01	7.4445(101%)	1.8577(101%)
60.0	0.10808E+01	41.161(101%)	10.316(101%)

In the case of a nonlinear system in Table 9.2, the estimated critical load changes from 309.0 lb for the applied load $p = 1$ lb to 64.8 lb, which is 1.07 times the actual critical load (60.060911 lb), for the applied load $p = 60$ lb. If the eigenvalue problem is invoked at the critical limit point, the eigenvalue will be one, and the critical load is the same as the applied load. The third and fourth columns of Table 9.2 represent design sensitivity coefficients for the first and second design variables, respectively. The ratio of the design sensitivity coefficient of A_1 to the design sensitivity coefficient of A_2 is independent of the applied load level and is almost equal to 4. This means that the influence of the cross-sectional area A_1 is 4 times more than that of A_2 on the critical load factor. The numbers in parentheses represent agreement in percentages with the sensitivity coefficients calculated using the finite difference method. Agreement is almost uniformly 101.0%, which indicates good design sensitivity results from the proposed numerical method for nonlinear structural systems. With the finite difference method, a 1% design perturbation is used for loads up to 50 lb, while a 0.001% perturbation is used for a load level of 60 lb. The magnitude of the design sensitivity coefficient for design A_1 decreases to 7.4445 as the load increases to $p = 50$ lb. However, at the load level of 60 lb, design sensitivity coefficients grow rapidly to 41.16, and good agreement requires a very small perturbation for the finite difference; for example, an agreement of 101% requires a design perturbation of 0.001%.

Results for the design sensitivity analysis of displacement in the x_2-direction at node 2 using (9.21) without the bending term [31] and [32] are presented in Table 9.3. The design sensitivity coefficients in the second and third columns increase as the load increases. The increase in the size of the coefficient is especially rapid at a load level of 60 lb. Agreement between sensitivity coefficients and finite difference results are good, with a 1% design perturbation for all loads except at the load level of 60 lb where good agreement requires the very small design perturbation of 0.01%. These numerical results indicate that the design sensitivity of displacement may increase without bounds as the load approaches the critical value. Since the proposed critical load factor design sensitivity method is based on the design derivative of displacement, an increase in that derivative explains the rapid growth of design sensitivity of the critical load factor at a load level of 60 lb.

Table 9.3. Design sensitivity of the displacement z_2 at node 2 of two-bar truss.

Load (lb)	$\partial z_2/\partial A_1$	$\partial z_2/\partial A_2$
10.0	0.08887	0.022289
20.0	0.20183	0.050595
30.0	0.35625	0.089307
40.0	0.54597	0.14922
50.0	1.0852	0.27137
52.0	1.2745	0.31924
54.0	1.5425	0.38632
56.0	1.9791	0.49549
58.0	2.9291	0.73341
60.0	18.313	4.5837

Table 9.4. Limit study of design sensitivity of two-bar truss.

Load (lb)	ζ	Estimated Critical Load (lb)	$\partial\zeta/\partial A_1$
60.0	0.10808E+01	64.84800	41.161
60.05	0.10350E+01	62.15175	87.883
60.06	0.10098E+01	60.64859	306.86
60.0608	0.10049E+01	60.35510	550.08
60.0609	0.10021E+01	60.18703	1231.8

Design sensitivity analyses of the nonlinear critical load factor are performed as the applied load comes very close to a critical load of 60.060911 lb, and the results of four additional load levels are shown in Table 9.4. The design sensitivity coefficient in the fourth column increases drastically from 41.161 at load $p = 60.0$ lb, to 1231.8 at load $p = 60.0609$ lb. Good agreement is obtained with the finite difference results for all load levels with an extremely small step size. For example, a design perturbation of $10^{-6}\%$ is required to obtain an agreement of 105.2% for a load of 60.0609 lb. The critical load factor is very sensitive and highly nonlinear with design variations near the critical limit point. Even though the sensitivity of the critical load factor is correct, it does not approach the same sensitivity levels as the critical load 60.14. This indicates that the derivative of the estimated critical load does not converge to the derivative of the actual critical load even though the estimated critical load converges to the actual critical load.

Dome Structure

Consider a three-dimensional shallow dome with 30 truss elements, 19 nodal points, and 21 degrees of freedom, as shown in Fig. 9.2. Nodal point coordinates are given in Table 9.5 and element cross sections in Table 9.6. For material properties, Young's modulus E is equal to 1.0×10^7 psi, and Poisson's ratio is equal to 0.3. Loads of 2000 lb each are applied to nodes 1 and 7 in the vertical direction. Structural analysis is performed using the three-dimensional truss element STIF8 of ANSYS.

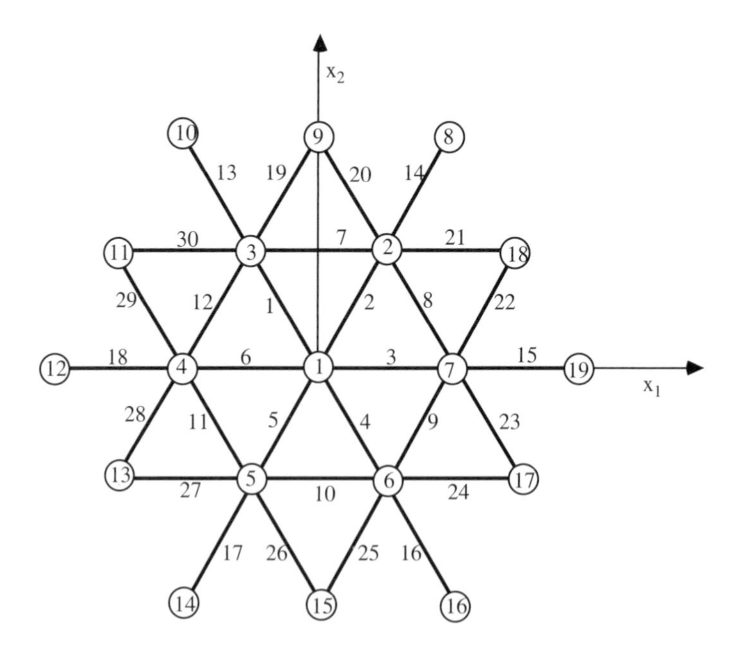

Figure 9.2. Plan of dome structure.

Table 9.5. Coordinates of nodal points of dome.

Node	X_1 (in)	X_2 (in)	X_3 (in)
1	0.0	0.000	85.912
2	180.0	311.769	64.662
7	360.0	0.000	64.662
8	360.0	623.538	0.000
9	0.0	623.538	21.709
18	540.0	311.769	21.709
19	720.0	0.000	0.000

Table 9.6. Element cross-sectional area of dome.

Element	Area (in^2)
1-6	1.6926
7-12	1.3754
13-18	.2693
19-30	.1

Sizing design sensitivity results are presented in Table 9.7 for displacement and in Table 9.8 for stress performance measures. In these tables, ψ^1 and ψ^2 are the values of the performance measures evaluated at the modified designs $u - \Delta u$ and $u + \Delta u$, respectively; $\Delta\psi = (\psi^2 - \psi^1)$ is twice the central finite difference; and ψ' is the predicted change by design sensitivity calculations, with δu indicating the variation in the design. For numerical calculation of the adjoint load in (9.10), the linear displacement shape function of element STIF8 is used. In Tables 9.7 and 9.8, a uniform design variation of $\Delta u = 0.01u$ has been used, where $u = [A_1, A_2, ..., A_{30}]^T$ and A_i the cross-sectional area of the ith element. For both displacement and stress performance measures, excellent agreement between design sensitivity predictions and finite differences is obtained.

Consider a design sensitivity analysis of critical loads in which a concentrated load is applied at node 1 in the vertical direction. At the initial design, cross-sectional areas of all elements are 1.528 in^2. Using the incremental analysis method of ANSYS, the actual critical load is found to be 1999.89172 lb. At several load levels between $p = 1500$ lb and $p = 1999.89$ lb, critical load factors are calculated, and design sensitivity analyses are performed with four design variables, A_1, A_2, A_3, and A_4. Element groups for each design variable are shown in Table 9.9.

The design sensitivity coefficients of the critical load factor are presented in Table 9.10 for seven loading cases. Using the applied load and the critical load factor in the first and second columns, respectively, the estimated critical loads are found to change from 5047.35 lb for applied load $p = 1500$ lb, to 2004.29 lb, which is 1.002 times the actual critical load (1999.89172 lb) for applied load $p = 1999.89$ lb. The next four columns in Table 9.10 represent design sensitivity coefficients for four design variables. Sensitivity results are verified as accurate using the finite difference method. To show the results of the proposed method, Table 9.11 compares design sensitivity coefficients with a uniform design at an applied load level of $p = 1999.89$ lb with the results from the finite

Table 9.7. Design sensitivity of dome (displacement).

Node & DOF	ψ^1	ψ^2	$\Delta\psi$	ψ'	$(\psi'/\Delta\psi \times 100)\%$
$3x$	0.1347	0.1325	–0.0022	–0.0022	100.0
$6x$	–0.1347	–0.1325	0.0022	0.0022	100.0
$2y$	–0.1167	–0.1148	0.0019	0.0019	100.0
$4y$	0.1167	0.1148	–0.0019	–0.0019	100.0
$1z$	6.9116	6.7872	–0.1243	–0.1243	100.0
$2z$	7.1085	6.9885	–0.1200	–0.1200	100.0

Table 9.8. Design sensitivity of dome (stress).

El. No.	ψ^1	ψ^2	$\Delta\psi$	ψ'	$(\psi'/\Delta\psi \times 100)\%$
1	3408.6340	3341.7957	–66.8383	–66.8320	100.0
7	3741.8592	3681.7637	–60.0955	–60.0911	100.0
13	32569.8438	31986.4356	–583.4082	–583.3583	100.0
19	23280.0493	22853.9780	–426.0713	–426.0336	100.0

Table 9.9. Design variable linking of dome.

Design	Element Linked	Number of Element
A1	1 – 6	6
A2	7 – 12	6
A3	13 – 18	6
A4	19 – 30	12

Table 9.10. Design sensitivity of nonlinear critical load factor ζ for dome.

Load p (lb)	ζ	$\partial\zeta/\partial A_1$	$\partial\zeta/\partial A_2$	$\partial\zeta/\partial A_3$	$\partial\zeta/\partial A_4$
1500.0	3.3649	2.51226	2.08182	0.050649	0.0091477
1750.0	2.2912	2.51124	2.02482	0.053239	0.0044556
1875.0	1.7856	1.79502	1.64736	0.041905	0.0036836
1938.0	1.5024	3.46200	2.73462	0.076044	0.0013737
1969.0	1.3337	4.40610	3.46842	0.097326	0.0007623
1999.0	1.0505	21.0024	16.4736	0.466344	-0.0009459
1999.89	1.0022	440.724	345.624	9.786600	-0.0226056

Table 9.11. Verification of design sensitivity of critical load factor ζ using finite difference method for dome (applied load = 1999.89 lb).

Perturbation (%)	Area (in^2)	ζ	$\Delta\zeta$	$\delta\zeta$	Agreement (%)
0	1.528	1.0022290			
1	1.54328	1.2595373	0.2573083	12.16459	4727.6
0.1	1.529528	1.0765525	0.0743235	1.216459	1636.7
0.01	1.5281528	1.0237544	0.0215254	0.1216459	565.1
0.001	1.52801528	1.0077491	0.0055201	0.01216459	220.4
0.0001	1.52800153	1.0032063	0.0009773	0.00121646	124.5
0.0000098	1.52800015	1.0023401	0.0001111	0.00011942	107.5
0.0000013	1.52800002	1.0022440	0.0000150	0.00001592	106.1

difference method for several design perturbations. In this table, $\Delta\zeta$ is the finite difference, and $\delta\zeta$ is the predicted change of the critical load factor by using design sensitivity analysis with the design perturbation. The agreement between $\delta\zeta$ and $\Delta\zeta$ is presented in the last column, and it ranges from 106.1% with a 1.3×10^{-6}% design perturbation to 4727.6% with a 1% design perturbation. This convergence indicates that the results of the finite difference method correspond to the results from this study, that is, the proposed design sensitivity analysis is correct. As with the two-bar truss, the critical load factor becomes very sensitive and highly nonlinear near the critical limit point with respect to the design.

Cap Model

Consider a shallow spherical cap with hinged rectangular boundaries, which is subjected to a concentrated load p at the crown [33], as shown in Fig. 9.3(a). Using symmetry of the geometric model, the quarter model is analyzed with 50 triangular plate finite elements as shown in Fig. 9.3(b), where two coordinate axes x_1 and x_2 are the axes of symmetry. For analysis data, a uniform thickness of h = 3.9154 cm, Young's modulus of $E = 1.0 \times 10^4$ kg/cm^2, and Poisson's ratio of $\nu = 0.3$ are used. Sizing design variables are the thicknesses of 50 plate elements. Structural analysis is performed using the triangular plate element STIF63 of ANSYS. Using the ANSYS incremental analysis method, the critical load for the quarter model is found to be between 2867.799 and 2867.800 kg.

At several load levels between p = 2865.0 and 2867.799 kg, design sensitivity coefficients of the critical load are evaluated for uniform thickness h as the design variable and presented in the last column of Table 9.12. This column is determined by multiplying the applied load in the first column with the design sensitivity of the actual critical load factor in the third column. Note that the design sensitivity of the actual critical load factor does not increase as the load level is getting near the critical load as previous examples. The second column represents the eigenvalue of the lowest

Table 9.12. Design sensitivity of critical load of cap.

Load	${}^{t}\zeta$	$\partial\beta/\partial h_{\text{uniform}}$	$\partial p_{cr}/\partial h_{\text{uniform}}$
2865.000	1.0500	0.56969	1632.16
2867.000	1.0284	0.57015	1634.61
2867.500	1.0213	0.57031	1635.37
2867.700	1.0181	0.57038	1635.68
2867.799	1.0163	0.57042	1635.84

Table 9.13. verification of design sensitivity of critical load using finite difference method for cap (applied load = 2867.799 kg).

Thickness	Perturb	p_{cr}	Δp_{cr}	p_{cr}'	$p_{cr}'/\Delta p_{cr}$ (%)
3.9154	—	2867.800	—	—	—
3.954554	1.0%	2934.800	67.0	64.0	95.6
3.9193154	0.1%	2874.450	6.65	6.40	96.3
3.91579154	0.01%	2868.446	0.646	0.64	99.0

eigenvalue problem in (8.51). In Table 9.13, the sensitivity results are verified as accurate. The design sensitivity coefficient with a uniform design at the applied load of 2867.799 kg is compared with the results using the finite difference method for three design perturbations: 1%, 0.1%, and 0.01%. In Table 9.13, Δp_{cr} is the finite difference of the critical load and p_{cr}' is the predicted change of the critical load by the proposed method for the corresponding design perturbation. In the last column, the agreements between Δp_{cr} and p_{cr}' approach 99.0% with a 0.01% design perturbation. The convergence to 1635.84 in the last column of Table 9.12 and good agreement with finite difference results in Table 9.13 indicate that the proposed method for design sensitivity analysis of the critical load the actual critical load factor is accurate.

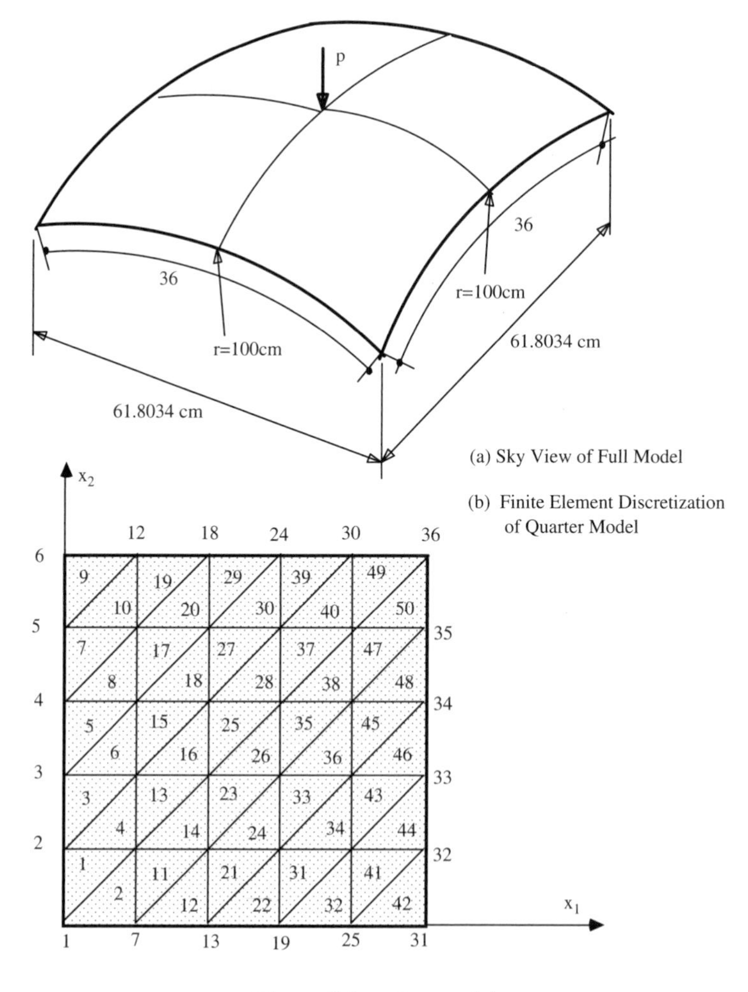

Figure 9.3. Cap model.

9.2 Design Sensitivity Analysis for Elastoplastic Problems

This section concentrates on the design sensitivity analysis of elastoplastic problems with respect to sizing design variables, including the cross-sectional area of the truss and beam, the thickness of plate and plane solids, and the material property of continuum solids. Rate-independent von Mises elastoplasticity is used with arbitrary isotropic and kinematic hardening rules. The major difference from a nonlinear elastic problem is that the design sensitivity equation is path-dependent. Computation of the design sensitivity equation at the current time requires sensitivity information from the previous time together with response analysis results at the current time. The design sensitivity equation has to be solved at each configuration time, followed by an update on variations of the stress tensor and internal evolution variables. The sensitivity computation process is divided into two parts: first computing the sensitivity of the incremental displacement, and then updating the stress and evolution variable sensitivities using the incremental displacement sensitivity. Design variation of the stress tensor can be obtained by satisfying the consistency condition at the perturbed design. No iteration is required to find the design variation at the return-mapped point, whereas the response analysis iterates in a return-mapping algorithm. Due to the path-dependent property, the direct differentiation method is preferred over the adjoint variable method. Inefficiencies associated with the adjoint variable method are briefly discussed below.

The Newton-Raphson iteration method is frequently used with a tangent operator for the response analysis of a nonlinear variational equation. If the tangent operator is exact, this method guarantees quadratic convergence when the initial estimate is close enough to the solution [19]. Even if the tangent operator is not exact, the response analysis may be converged by performing a large number of iterations. However, in design sensitivity analysis the inexact tangent operator produces erroneous sensitivity results. The accuracy of sensitivity coefficients using the consistent tangent operator and the rate form tangent operator are discussed by Vidal and Haber [34].

9.2.1 Static problems

Material Parameter Sensitivity Analysis of Elastoplastic Problems

For many applications, the property of a material is obtained through simple experimentation. When the constitutive relation contains many variables it is not easy to determine the material constants with a conventional approach. Accurate estimation of the material property is especially critical to the successful simulation and design of elastoplastic material. The problem of identifying material properties can be solved as an optimization problem to minimize the difference between an experimental and a computer simulation model. Design sensitivity information with respect to the material property can be efficiently used for the gradient-based optimization algorithm.

By assuming material nonlinearity alone through elastoplasticity, it is not necessary to distinguish between deformed and undeformed configurations, since all integration and differentiation is carried out in the undeformed configuration. The total and updated Lagrangian formulations are identical with engineering strain and Cauchy stress tensors. Using the structural energy form in (8.100), the variational equation can be written at initial and new design stages as

$$a_u({}^{n+1}z, \bar{z}) = \ell_u(\bar{z}), \quad \forall \bar{z} \in Z \tag{9.70}$$

$$a_{u+\tau\delta u}({}^{n+1}z_{u+\tau\delta u}, \bar{z}) = \ell_{u+\tau\delta u}(\bar{z}), \quad \forall \bar{z} \in Z, \tag{9.71}$$

where subscripts $\boldsymbol{u}$ and $\boldsymbol{u} + \tau\delta\boldsymbol{u}$ are used to denote dependence on the design. The structural energy form in (8.100) and load form in (8.7) are rewritten as

$$a_{\boldsymbol{u}}({}^{n+1}\boldsymbol{z}, \bar{\boldsymbol{z}}) = \int_{\Omega} {}^{n+1}\boldsymbol{\sigma} : \boldsymbol{\varepsilon}(\bar{\boldsymbol{z}})\, d\Omega \tag{9.72}$$

$$\ell_{\boldsymbol{u}}(\bar{\boldsymbol{z}}) = \int_{\Omega} \bar{\boldsymbol{z}}^T \boldsymbol{f}^B\, d\Omega + \int_{\Gamma^S} \bar{\boldsymbol{z}}^T \boldsymbol{f}^S\, d\Gamma. \tag{9.73}$$

From the infinitesimal assumption, the kinematics of variational equation (9.70) are linear, and nonlinearity is associated with the constitutive relation. Since a mathematical proof is not available, it is assumed that the structural energy form in (9.72) is differentiable with respect to the design. The variation of the structural energy form is directly related to the variation of the Cauchy stress tensor. From (8.85), the elastic trial of the stress tensor and its variation with respect to design can be obtained as

$${}^{tr}\boldsymbol{\sigma} = {}^{n}\boldsymbol{\sigma} + \boldsymbol{C} : \boldsymbol{\varepsilon}(\Delta\boldsymbol{z}) \tag{9.74}$$

$${}^{tr}\boldsymbol{\sigma}' = {}^{n}\boldsymbol{\sigma}' + \delta\boldsymbol{C} : \boldsymbol{\varepsilon}(\Delta\boldsymbol{z}) + \boldsymbol{C} : \boldsymbol{\varepsilon}(\Delta\boldsymbol{z}'), \tag{9.75}$$

where $\boldsymbol{\varepsilon}(\Delta\boldsymbol{z}')$ is as yet an unknown term, ${}^{n}\boldsymbol{\sigma}'$ is the variation of the stress tensor computed from the sensitivity analysis at the previous time, and $\delta\boldsymbol{C}$ denotes the explicit dependence of the elastic constitutive tensor $\boldsymbol{C} = (\lambda + \frac{2}{3}\mu)\boldsymbol{1} \otimes \boldsymbol{1} + 2\mu \boldsymbol{I}_{dev}$ on the design, as

$$\delta\boldsymbol{C} = \boldsymbol{1} \otimes \boldsymbol{1}\delta\lambda + 2\boldsymbol{I}\delta\mu. \tag{9.76}$$

If the material is in elastic domain E in (8.80), then the trial stress and its variation in (9.74) and (9.75) are exact. In the case of elastic status, by freezing the implicit dependence through $\boldsymbol{z}$ the variation of structural energy form in (9.72) can be obtained with respect to the design as

$$\begin{aligned} a'_{\delta\boldsymbol{u}}(\boldsymbol{z}, \bar{\boldsymbol{z}}) &\equiv \frac{d}{d\tau} a_{\boldsymbol{u}+\delta\boldsymbol{u}}(\tilde{\boldsymbol{z}}, \bar{\boldsymbol{z}})\bigg|_{\tau=0} \\ &= \int_{\Omega} \boldsymbol{\varepsilon}(\bar{\boldsymbol{z}}) : {}^{n}\boldsymbol{\sigma}'\, d\Omega \\ &\quad + \int_{\Omega} \boldsymbol{\varepsilon}(\bar{\boldsymbol{z}}) : \delta\boldsymbol{C} : \boldsymbol{\varepsilon}(\Delta\boldsymbol{z})\, d\Omega. \end{aligned} \tag{9.77}$$

The first integral in (9.77) represents the path-dependent component during the sensitivity analysis at previous time t_n, while the second integral denotes the explicitly dependent component through the variation of the constitutive tensor $\delta\boldsymbol{C}$ in (9.76). Note that $a'_{\delta\boldsymbol{u}}(\boldsymbol{z}, \bar{\boldsymbol{z}})$ is path-dependent, even if it has elastic material status, because of the incremental sensitivity formulation. The design sensitivity equation of the direct differentiation method is

$$a^*_{\boldsymbol{u}}(\boldsymbol{z}; \Delta\boldsymbol{z}', \bar{\boldsymbol{z}}) = \ell'_{\delta\boldsymbol{u}}(\bar{\boldsymbol{z}}) - a'_{\delta\boldsymbol{u}}(\boldsymbol{z}, \bar{\boldsymbol{z}}), \quad \forall \bar{\boldsymbol{z}} \in Z. \tag{9.78}$$

After computing the variation of the displacement vector, the variation of the stress tensor has to be updated for the computation at the next time, as in (9.75).

When the material is in the plastic status, the trial estimate of the stress tensor is outside elastic domain E. Response analysis carries out return mapping to project the trial stress onto the yield surface. This projection requires iteration with the general hardening model to find a solution. If the yield surface is convex, then this projection has a unique solution. Design sensitivity analysis, however, takes the variation of the projected stress tensor with respect to the design. Thus, no iteration is required to solve for the variation of plastic consistency parameter γ. From the definition of relative stress ${}^{tr}\boldsymbol{\eta} = {}^{tr}\boldsymbol{s} - {}^{tr}\boldsymbol{\alpha}$, the variation of ${}^{tr}\boldsymbol{\eta}$ is

$$ {}^{tr}\boldsymbol{\eta}' = {}^{tr}\boldsymbol{s}' - {}^{n}\boldsymbol{\alpha}' = {}^{n}\boldsymbol{\eta}' + 2\boldsymbol{I}_{dev} : \boldsymbol{\varepsilon}(\Delta \mathbf{z})\delta\mu + 2\mu \boldsymbol{I}_{dev} : \boldsymbol{\varepsilon}(\Delta \mathbf{z}'). \tag{9.79} $$

Note that only the last term of (9.79) is still unknown. The other terms can be easily computed from the sensitivity analysis of previous time and current response analysis results. All admissible stress status has to satisfy the Kuhn-Tucker loading/unloading condition in (8.84), which is equivalent to the consistency condition in (8.93) for plastic status. The variation of the yield function at the return-mapped point must satisfy

$$ \begin{aligned} f' = \left\| {}^{tr}\boldsymbol{\eta} \right\|' &- \left(2\mu + H_\alpha + \sqrt{\tfrac{2}{3}} H_{\alpha, e_p} \hat{\gamma} + \tfrac{2}{3} \kappa_{,e_p} \right) \hat{\gamma}' \\ &- (2\delta\mu + \delta H_\alpha)\hat{\gamma} - \left(H_{\alpha, e_p} \hat{\gamma} + \sqrt{\tfrac{2}{3}} \kappa_{,e_p} \right) {}^{n}e_p' - \sqrt{\tfrac{2}{3}} \delta\kappa = 0, \end{aligned} \tag{9.80} $$

where δH_α and $\delta\kappa$ represent the explicit dependence of these terms on design variables, and $\hat{\gamma}$ is the solution to the return-mapping algorithm in the response analysis in (8.84). By solving (9.80) with respect to $\hat{\gamma}'$ we have

$$ \hat{\gamma}' = A \left[\left\| {}^{tr}\boldsymbol{\eta} \right\|' - (2\delta\mu + \delta H_\alpha)\hat{\gamma} - \left(H_{\alpha, e^p} \hat{\gamma} + \sqrt{\tfrac{2}{3}} \kappa_{,e^p} \right) {}^{n}e_p' - \sqrt{\tfrac{2}{3}} \delta\kappa \right]. \tag{9.81} $$

As mentioned before, no iteration is required to compute (9.81). From (9.79), it can be shown that the variation of the trial relative stress norm can be

$$ \left\| {}^{tr}\boldsymbol{\eta} \right\|' = \boldsymbol{N} : {}^{n}\boldsymbol{\eta}' + 2\boldsymbol{N} : \boldsymbol{\varepsilon}(\Delta \mathbf{z})\delta\mu + 2\mu \boldsymbol{N} : \boldsymbol{\varepsilon}(\Delta \mathbf{z}'). \tag{9.82} $$

To obtain the variation of the stress updating algorithm, the variation of unit normal to the yield surface is required. By taking the variation of $\boldsymbol{N} = {}^{tr}\boldsymbol{\eta}/\|{}^{tr}\boldsymbol{\eta}\|$ and by using the relation in (9.82), we obtain

$$ \boldsymbol{N}' = \frac{1}{\left\| {}^{tr}\boldsymbol{\eta} \right\|} (\boldsymbol{I} - \boldsymbol{N} \otimes \boldsymbol{N}) : [{}^{n}\boldsymbol{\eta}' + 2\boldsymbol{I}_{dev} : \boldsymbol{\varepsilon}(\Delta \mathbf{z})\delta\mu + 2\mu \boldsymbol{I}_{dev} : \boldsymbol{\varepsilon}(\Delta \mathbf{z}')]. \tag{9.83} $$

Note that the direction of $\boldsymbol{N}'$ is tangential to the yield surface. Using the stress correction formula ${}^{n+1}\boldsymbol{\sigma} = {}^{tr}\boldsymbol{\sigma} - 2\mu\gamma \boldsymbol{N}$ and (9.75), (9.81), and (9.83), the variation of the stress correction formula can be obtained as

$$ \begin{aligned} {}^{n+1}\boldsymbol{\sigma}' &= {}^{tr}\boldsymbol{\sigma}' - 2\hat{\gamma}\boldsymbol{N}\delta\mu - 2\mu\hat{\gamma}'\boldsymbol{N} - 2\mu\hat{\gamma}\boldsymbol{N}' \\ &= \boldsymbol{C}^{alg} : \boldsymbol{\varepsilon}(\Delta \mathbf{z}') + \boldsymbol{\sigma}^{fic} + \boldsymbol{\sigma}_{\delta\boldsymbol{u}}, \end{aligned} \tag{9.84} $$

where $\boldsymbol{\sigma}^{fic}$ denotes path-dependent terms and $\boldsymbol{\sigma}_{\delta\boldsymbol{u}}$ represents explicitly dependent terms, written as

$$ \begin{aligned} \boldsymbol{\sigma}^{fic} = {}^{n}\boldsymbol{\sigma}' &- 2\mu A \boldsymbol{N} \left[\boldsymbol{N} : {}^{n}\boldsymbol{\eta}' - \left(H_{\alpha, e^p} \hat{\gamma} + \sqrt{\tfrac{2}{3}} \kappa_{,e^p} \right) {}^{n}e_p' \right] \\ &- \frac{2\mu\hat{\gamma}}{\left\| {}^{tr}\boldsymbol{\eta} \right\|} [{}^{n}\boldsymbol{\eta}' - (\boldsymbol{N} : {}^{n}\boldsymbol{\eta}')\boldsymbol{N}] \end{aligned} \tag{9.85} $$

$$\begin{aligned}\boldsymbol{\sigma}_{\delta \boldsymbol{u}} &= \delta \boldsymbol{C} : \boldsymbol{\varepsilon}(\Delta z) - 2\hat{\gamma}\boldsymbol{N}\delta\mu \\ &\quad -2\mu A\boldsymbol{N}\left[2\boldsymbol{N} : \boldsymbol{\varepsilon}(\Delta z)\delta\mu - 2\hat{\gamma}\delta\mu - \hat{\gamma}\delta H_{\alpha} - \sqrt{\tfrac{2}{3}}\delta\kappa\right] \\ &\quad -\frac{2\mu\hat{\gamma}}{\left\|{}^{tr}\boldsymbol{\eta}\right\|}(\boldsymbol{I} - \boldsymbol{N}\otimes\boldsymbol{N}) : (2\boldsymbol{I}_{dev} : \boldsymbol{\varepsilon}(\Delta z)\delta\mu),\end{aligned} \tag{9.86}$$

and $\boldsymbol{C}^{alg}$ is the algorithmic tangent stiffness of elastoplasticity derived in (8.99). When the material is in the plastic status, by freezing the implicit dependence through z the variation of the structural energy form in (9.71) with respect to the design is

$$\begin{aligned}a'_{\delta \boldsymbol{u}}(z,\bar{z}) &\equiv \frac{d}{d\tau} a_{\boldsymbol{u}+\delta\boldsymbol{u}}(\tilde{z},\bar{z})\bigg|_{\tau=0} \\ &= \int_{\Omega} \varepsilon(\bar{z}) : \boldsymbol{\sigma}^{fic}\, d\Omega + \int_{\Omega} \varepsilon(\bar{z}) : \boldsymbol{\sigma}_{\delta \boldsymbol{u}}\, d\Omega.\end{aligned} \tag{9.87}$$

The first integral in (9.87) represents the path-dependent component through the sensitivity analysis at previous time t_n, while the second integral denotes the explicitly dependent component on the design.

The direct differentiation method of the design sensitivity equation is

$$a^{*}_{\boldsymbol{u}}(z;\Delta z',\bar{z}) = a'_{\delta \boldsymbol{u}}(z,\bar{z}) - \ell'_{\delta \boldsymbol{u}}(\bar{z}), \quad \forall \bar{z} \in Z. \tag{9.88}$$

After solving $\Delta z'$ from (9.88), the total displacement sensitivity can be obtained as

$${}^{n+1}z' = {}^{n}z' + \Delta z', \tag{9.89}$$

and the variation of the stress tensor and the internal plastic variables are updated for computation of the next time as

$${}^{n+1}\boldsymbol{\sigma}' = \boldsymbol{C}^{alg} : \varepsilon(\Delta z') + \boldsymbol{\sigma}^{fic} + \boldsymbol{\sigma}_{\delta \boldsymbol{u}} \tag{9.90}$$

$${}^{n+1}\boldsymbol{\alpha}' = {}^{n}\boldsymbol{\alpha}' + \delta H_{\alpha}\hat{\gamma}\boldsymbol{N} + \left(\sqrt{\tfrac{2}{3}}H_{\alpha,e_p}\hat{\gamma} + H_{\alpha}\right)\hat{\gamma}'\boldsymbol{N} + H_{\alpha,e_p}\hat{\gamma}\,{}^{n}e'_p\boldsymbol{N} + H_{\alpha}\hat{\gamma}\boldsymbol{N}' \tag{9.91}$$

$${}^{n+1}e'_p = {}^{n}e'_p + \sqrt{\tfrac{2}{3}}\hat{\gamma}'. \tag{9.92}$$

Using (9.88) through (9.92), the sensitivity computation of the elastoplastic problem is a two-step process. The first step is to construct $a'_{\delta \boldsymbol{u}}(z,\bar{z})$ and $\ell'_{\delta \boldsymbol{u}}(\bar{z})$ using the sensitivity information from the previous time and using the response analysis results at the current time. If the finite element method is used with the Newton-Raphson iteration method, then the solution of linear equation (9.88) can easily be obtained using the factorized tangent stiffness matrix with a different right side. Only backward substitution is required to solve the sensitivity equation, which provides significant efficiency compared with an iterative response analysis. After solving the displacement sensitivity, the variations of path-dependent parameters are updated, as in (9.90) through (9.92). Since updating the response variable cannot be carried out before solving (9.88), design sensitivity analysis of the elastoplastic problem is less efficient than that of the elastic problem. In addition, because the elastoplasticity problem demands a step-by-step analysis procedure, it requires more computational effort.

The amount of path-dependent sensitivity information stored in the computer's memory can be calculated by assuming that the finite element method is used to discretize domain Ω with an NE number of finite elements. Let NG be the number of Gauss integration points within an element. Domain integration is carried out using the value at the Gauss integration point as well as the integration weight. For a three-dimensional solid element, the stress vector has six components, the same number as back stress $\boldsymbol{\alpha}$. By counting effective plastic strain as a scalar variable, 13 storage units are required for each Gauss integration point. Thus, the total storage required is $13 \times NG \times NE$ for path-dependent parameters. Design sensitivity formulation in (9.90) through (9.92) requires storing the variation of $\boldsymbol{\sigma}$, $\boldsymbol{\alpha}$, and e_p. If the size of the design vectors is NDV, then the amount of storage space required for the design sensitivity analysis is $13 \times NG \times NE \times NDV$. As NDV increases, the amount of storage space required for design sensitivity analysis also significantly increases.

Linear Isotropic/Kinematic Hardening Model

Isotropic hardening and kinematic hardening relations are given as a linear function of effective plastic strain e_p as

$$\kappa(e_p) = \sigma_Y + (1-\beta)He_p \tag{9.93}$$

$$^{n+1}\boldsymbol{\alpha} = {}^{n}\boldsymbol{\alpha} + \tfrac{2}{3}\beta H\gamma \boldsymbol{N}. \tag{9.94}$$

For this model, the design vector and its perturbation can be chosen generally as

$$\boldsymbol{u} = [\lambda, \mu, \beta, H, \sigma_Y]^T \tag{9.95}$$

$$\delta\boldsymbol{u} = [\delta\lambda, \delta\mu, \delta\beta, \delta H, \delta\sigma_Y]^T. \tag{9.96}$$

Then, $\boldsymbol{\sigma}_{\delta\boldsymbol{u}}$ in (9.86) can be denoted explicitly as a linear function of $\delta\boldsymbol{u}$ as

$$\boldsymbol{\sigma}_{\delta \boldsymbol{u}} = \boldsymbol{S}^T \delta\boldsymbol{u} \tag{9.97}$$

$$\boldsymbol{S} = \begin{bmatrix} tr(\boldsymbol{\varepsilon})\boldsymbol{1} \\ \boldsymbol{B} \\ 2\mu AH\boldsymbol{N}(\tfrac{2}{3}\hat{\gamma} - \sqrt{\tfrac{2}{3}}e_p) \\ 2\mu A\boldsymbol{N}(\tfrac{2}{3}\beta\hat{\gamma} + \sqrt{\tfrac{2}{3}}(1-\beta)e_p) \\ 2\sqrt{\tfrac{2}{3}}\mu A\boldsymbol{N} \end{bmatrix}, \tag{9.98}$$

where $\boldsymbol{B} = 2\boldsymbol{\varepsilon} - 2\hat{\gamma}\boldsymbol{N} - 4\left[\mu A\boldsymbol{N}(\boldsymbol{N}:\boldsymbol{\varepsilon} - \hat{\gamma}) - \mu\hat{\gamma}[\boldsymbol{e} - \boldsymbol{N}(\boldsymbol{N}:\boldsymbol{\varepsilon})]/\|{}^{tr}\boldsymbol{\eta}\|\right]$. A complicated expression is obtained with respect to shear modulus μ, since the plastic deformation is related to the deviatoric stress space.

Truss Component

An elastoplastic truss component is considered with the cross-sectional area taken as the design, i.e., $u = A(x)$. Since the material property is not part of the design, the $\boldsymbol{\sigma}_{\delta\boldsymbol{u}}$ term in (9.86), which represents explicit dependence on the material property design, vanishes. Instead, there is explicit dependence on the cross section of the design. The structural energy form can be written as

$$a_u(z, \bar{z}) = \int_0^l A(x)\sigma_{11}\varepsilon_{11}(x)\,dx. \tag{9.99}$$

Then, $a'_{\delta u}(z,\bar{z})$ in (9.87) contains the path-dependent term and the explicitly dependent term, as

$$a'_{\delta u}(z,\bar{z}) = \iint_{\Omega} \left[\sigma_{11}^{fic} \varepsilon_{11}(\bar{z}_1) + \sigma_{11}\varepsilon_{11}(\bar{z}_1)\delta A \right] d\Omega. \tag{9.100}$$

Difficulties in Adjoint Variable Method

A brief introduction is provided to the adjoint variable method in elastoplasticity, with its associated computational inefficiency problems. A structural performance measure can be written in integral form as

$${}^{n+1}\psi = \iint_{\Omega} g\left(z, \nabla z, {}^{n}\boldsymbol{\sigma}, {}^{n}\boldsymbol{\alpha}, {}^{n}e_p\right) d\Omega, \tag{9.101}$$

where g is continuously differentiable with respect to its arguments. Taking the variation of functional ψ gives

$$\begin{aligned} {}^{n+1}\psi' &= \iint_{\Omega} \left[g_{,\Delta z}^T \Delta z' + g_{,\nabla\Delta z} : \nabla\Delta z' \right] d\Omega \\ &+ \iint_{\Omega} \left[g_{,z}^T\, {}^{n}z' + g_{,\nabla z} : \nabla\, {}^{n}z' + g_{,\sigma} : {}^{n}\boldsymbol{\sigma}' + g_{,\alpha}^T\, {}^{n}\boldsymbol{\alpha}' + g_{,e_p}^T\, {}^{n}e'_p + g_{,u}^T \delta\boldsymbol{u} \right] d\Omega, \end{aligned} \tag{9.102}$$

where the second integration is assumed to be known at time t_{n+1}, provided that the sensitivity equation up to t_n is solved, and the first integration is explicitly expressed in terms of $\delta\boldsymbol{u}$.

An adjoint equation is introduced by replacing $\Delta z'$ in (9.102) with virtual displacement $\bar{\boldsymbol{\lambda}}$ and by equating the first integration of (9.102) with the incremental structural bilinear form $a_{\boldsymbol{u}}^*({}^{n+1}z;\boldsymbol{\lambda},\bar{\boldsymbol{\lambda}})$, yielding the following adjoint equation for adjoint variable $\boldsymbol{\lambda}$:

$$a_{\boldsymbol{u}}^*\left({}^{n+1}z;\boldsymbol{\lambda},\bar{\boldsymbol{\lambda}}\right) = \iint_{\Omega} \left[g_{,\Delta z}^T \bar{\boldsymbol{\lambda}} + g_{,\nabla\Delta z} : \nabla\bar{\boldsymbol{\lambda}} \right] d\Omega, \quad \forall \bar{\boldsymbol{\lambda}} \in Z, \tag{9.103}$$

which is associated with the performance measure in (9.101). Since (9.103) is satisfied for every $\bar{\boldsymbol{\lambda}} \in Z$ and $\Delta z' \in Z$, it can be evaluated at $\bar{\boldsymbol{\lambda}} = \Delta z'$ to obtain

$$a_{\boldsymbol{u}}^*({}^{n+1}z;\boldsymbol{\lambda},\Delta z') = \iint_{\Omega} \left[g_{,\Delta z}^T \Delta z' + g_{,\nabla\Delta z} : \nabla\Delta z' \right] d\Omega, \tag{9.104}$$

which is the first integration of (9.102) required to write it explicitly in terms of $\delta\boldsymbol{u}$. Similarly, the direct differentiation formulation in (9.88) may be evaluated at $\bar{z} = \boldsymbol{\lambda}$ since both are in Z, to obtain

$$a_{\boldsymbol{u}}^*({}^{n+1}z;\Delta z',\boldsymbol{\lambda}) = \ell'_{\delta u}(\boldsymbol{\lambda}) - a'_{\delta u}({}^{n+1}z,\boldsymbol{\lambda}). \tag{9.105}$$

Since the structural bilinear form is symmetric and bilinear in its arguments, that is, $a_{\boldsymbol{u}}^*(z;\Delta z',\boldsymbol{\lambda}) = a_{\boldsymbol{u}}^*(z;\boldsymbol{\lambda},\Delta z')$, the left side of (9.104) and (9.105) are equal, yielding the following desired result:

$$\iint_{\Omega} \left[g_{,\Delta z}^T \Delta z' + g_{,\nabla\Delta z} : \nabla\Delta z' \right] d\Omega = \ell'_{\delta u}(\boldsymbol{\lambda}) - a'_{\delta u}({}^{n+1}z,\boldsymbol{\lambda}), \tag{9.106}$$

where the right side is linear in $\delta\boldsymbol{u}$ and can be evaluated once adjoint variable $\boldsymbol{\lambda}$ and state variables z, $\boldsymbol{\sigma}$, $\boldsymbol{\alpha}$, and e_p are known. Substituting this result into (9.106), the explicit design sensitivity of performance ψ is

$$
\begin{aligned}
{}^{n+1}\psi' &= \ell'_{\delta u}(\boldsymbol{\lambda}) - a'_{\delta u}({}^{n+1}z, \boldsymbol{\lambda}) \\
&+ \int_{\Omega}\left[g_{,z}^{T}\,{}^{n}z' + g_{,\nabla z} : \nabla\,{}^{n}z' + g_{,\sigma} : {}^{n}\boldsymbol{\sigma}' + g_{,\alpha}^{T}\,{}^{n}\boldsymbol{\alpha}' + g_{,e_p}^{T}\,{}^{n}e'_p + g_{,u}^{T}\delta\boldsymbol{u} \right] d\Omega.
\end{aligned}
\tag{9.107}
$$

It is clear in (9.107) that the design sensitivity formulation of functional ψ using the adjoint variable method is path-dependent, since ψ' requires the variation of state variables at previous time t_n. For example, the displacement functional requires the $g_{,z}^{T}\,{}^{n}z'$ term from (9.107), while the stress functional requires $g_{,\sigma} : {}^{n}\boldsymbol{\sigma}'$ at the previous time. However, this approach suffers from a major inefficiency due to the $a'_{\delta u}({}^{n+1}z, \boldsymbol{\lambda})$ term. The computation of $a'_{\delta u}({}^{n+1}z, \boldsymbol{\lambda})$ requires the integration of the whole domain with the path-dependent parameters shown in (9.87). Thus, for any performance measure, variables z', $\boldsymbol{\sigma}'$, $\boldsymbol{\alpha}'$, and e_p' have to be computed in the entire domain, which is very inefficient. Consequently, there is no benefit in using the adjoint variable method for path-dependent problems.

9.2.2 Transient Problems with Explicit Time Integration

Mathematical theory for the differentiability of nonlinear structural responses is not yet well developed, whereas a reasonably complete theory of the differentiability of linear structural responses has been developed in Chapters 5 and 6. Thus, differentiability is assumed in deriving design sensitivity expressions for dynamic system. Equations of motion are considered that correspond to original design $\boldsymbol{u}$ and to perturbed design $\boldsymbol{u} + \tau\delta\boldsymbol{u}$ at the current configuration t_{n+1}, as

$$
d_u(z_{,tt}, \bar{z}) + a_u(z, \bar{z}) = \ell_u(\bar{z}), \qquad \forall \bar{z} \in Z \tag{9.108}
$$

and

$$
d_{u+\tau\delta u}((z_{,tt})_\tau, \bar{z}) + a_{u+\tau\delta u}(z_\tau, \bar{z}) = \ell_{u+\tau\delta u}(\bar{z}), \forall \bar{z} \in Z. \tag{9.109}
$$

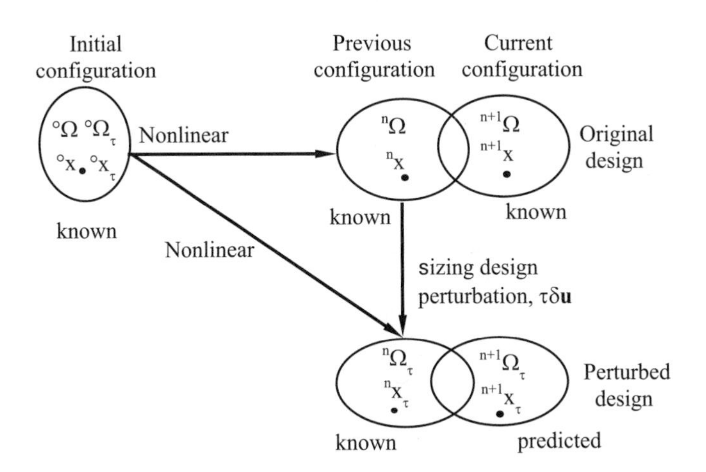

Figure 9.4. Deformation of body with sizing design perturbation.

Equations (9.108) and (9.109), along with Fig. 9.4, which provides a schema of the deformation for original and perturbed designs, indicate that the difference of response between the original and perturbed designs is small for a small design perturbation $\tau\delta\boldsymbol{u}$. Therefore, as design perturbation vanishes, so does the difference in the response.

The first-order variation of structural energy and load forms are developed in Section 9.2.1 for elastoplastic problems. The first-order variation of the kinetic energy form in (8.201), with respect to its explicit dependence on design variable $\boldsymbol{u}$, is defined as

$$d'_{\delta u}(z_{,tt},\bar{z}) \equiv \frac{d}{d\tau} d_{u+\tau\delta u}(\tilde{z}_{,tt},\bar{z})\Big|_{\tau=0}, \tag{9.110}$$

where the symbol "~" indicates that the dependence on the design variation is suppressed and that virtual displacement $\bar{z}$ is independent of τ. Note that the kinetic energy form $d_{\boldsymbol{u}}(z_{,tt},\bar{z})$ is linear with respect to its arguments. In addition to the first-order variation of response z in (9.5), the first-order variation of the solution $z_{,tt}$ in (9.108) with respect to the sizing design variable $\boldsymbol{u}$ is defined as

$$z'_{,tt} \equiv \frac{d}{d\tau} z_{,tt}(\boldsymbol{u}+\tau\delta\boldsymbol{u})\Big|_{\tau=0} = \lim_{\tau\to 0}\frac{z_{,tt}(\boldsymbol{u}+\tau\delta\boldsymbol{u}) - z_{,tt}(\boldsymbol{u})}{\tau}. \tag{9.111}$$

Using the chain rule of differentiation along with (9.110) and (9.111), the first-order variation of the kinetic energy form in (9.108) becomes

$$\frac{d}{d\tau} d_{u+\tau\delta u}(z_{,tt}(\boldsymbol{u}+\tau\delta\boldsymbol{u}),\bar{z})\Big|_{\tau=0} = d'_{\delta u}(z_{,tt},\bar{z}) + d_{u}(z'_{,tt},\bar{z}). \tag{9.112}$$

Using the relations in (9.4) and (9.6), The first-order variation of (9.108) yields

$$d'_{\delta u}(z_{,tt},\bar{z}) + d_{u}(z'_{,tt},\bar{z}) + a'_{\delta u}(z,\bar{z}) + a^{*}_{u}(z;z',\bar{z}) = \ell'_{\delta u}(\bar{z}), \quad \forall \bar{z}\in Z. \tag{9.113}$$

Note that the expressions of $a^{*}_{u}(z;z',\bar{z})$, $a'_{\delta u}(z,\bar{z})$, and $\ell'_{\delta u}(\bar{z})$ are exactly the same as the static problem in Section 9.2.1. The expression of $d_{\boldsymbol{u}}(z'_{,tt},\bar{z})$ is provided in (8.201) if $z_{,tt}$ is replaced by $z'_{,tt}$. In the explicit time integration method, the design sensitivity equation (9.113) is solved for $z'_{,tt}$. Let the left superscript denote the configuration time. Then,

$$d_{u}({}^{n+1}z'_{,tt},\bar{z}) = \ell'_{\delta u}(\bar{z}) - d'_{\delta u}({}^{n}z_{,tt},\bar{z}) - a'_{\delta u}({}^{n}z,\bar{z}) - a^{*}_{u}({}^{n}z;z',\bar{z}), \qquad \forall \bar{z}\in Z. \tag{9.114}$$

The right side of (9.114) is evaluated using the data at time t_n. If the lumped mass matrix is used in the discrete system, then (9.114) is uncoupled and ${}^{n+1}z'_{,tt}$ at time t_{n+1} can easily be calculated.

Design Sensitivity of Kinematic Responses

Once solution ${}^{n+1}z'_{,tt}$ is obtained from (9.114), the sensitivities of the other kinematic responses are found using the central difference method as

$${}^{n+\frac{1}{2}}z'_{,t} = {}^{n-\frac{1}{2}}z'_{,t} + {}^{n}\Delta t\,{}^{n}z'_{,tt} + {}^{n}\Delta t'\,{}^{n}z_{,tt} \tag{9.115}$$

and

$${}^{n+1}z' = {}^{n}z' + \tfrac{1}{2}({}^{n}\Delta t + {}^{n+1}\Delta t)\,{}^{n+\frac{1}{2}}z'_{,t} + \tfrac{1}{2}({}^{n}\Delta t + {}^{n+1}\Delta t)'\,{}^{n+\frac{1}{2}}z_{,t}. \tag{9.116}$$

Note that time step ${}^{n}\Delta t$ is determined according to its geometric and material properties.

To obtain kinematic responses in (9.115) and (9.116), ${}^{n+1}\Delta t$ should be less than the critical time step determined by the material property and element length, as illustrated by Hughes [35]:

$$
\begin{aligned}
{}^{n+1}\Delta t &= \kappa\sqrt{\frac{3\rho}{3K+4G}}\,{}^{n}\ell \\
&= \kappa\sqrt{\frac{3\rho}{3K+4G}\sum_{i=1}^{3}\left[\left({}^{0}x_i+{}^{n}z_i\right)_J-\left({}^{0}x_i+{}^{n}z_i\right)_I\right]^2},
\end{aligned}
\tag{9.117}
$$

where κ and ${}^{0}\boldsymbol{x}$ are the initial geometry and time step scale factor, respectively, and I and J represent the node numbers of the critical element. If the variable integration time step is used for nonlinear transient analysis, the design sensitivity of the time step can be obtained as

$$
\begin{aligned}
({}^{n+1}\Delta t)'_{\delta u} &= \kappa\sqrt{\frac{3\rho}{3K+4G}}({}^{n}\ell)'_{\delta u} \\
&= \kappa\sqrt{\frac{3\rho}{3K+4G}}\frac{\sum_{i=1}^{3}\left[\left({}^{0}x_i+{}^{n}z_i\right)_J-\left({}^{0}x_i+{}^{n}z_i\right)_I\right]\left[\{({}^{n}z_i)'_{\delta u}\}_J-\{({}^{n}z_i)'_{\delta u}\}_I\right]}{{}^{n}\ell}.
\end{aligned}
\tag{9.118}
$$

The design sensitivity of the time step will not predict the exact perturbation. Thus, there is a slight difference between predicted ($t\,'$) and actual time steps (t_τ) for the perturbed design, as shown in Fig. 9.5. Even though the difference is small, the error in the time step will increase the inaccuracy of the structural response design sensitivity, especially when using the explicit method. To avoid this complication, a small fixed time step, that is less than the critical time step used for analysis, should be used. In this case, the design sensitivity of those time steps given in (9.118) is not necessary. This case will be examined later in detail using a numerical example.

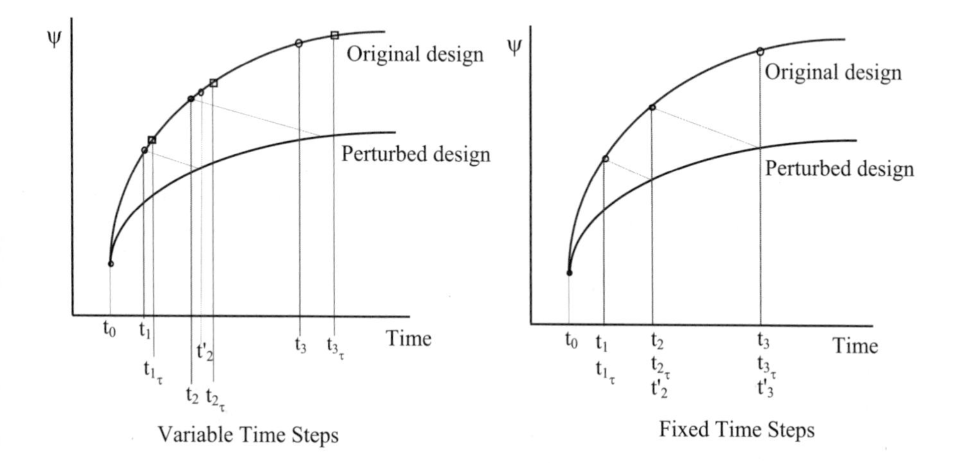

Figure 9.5. Performance measures with variable and fixed time steps.

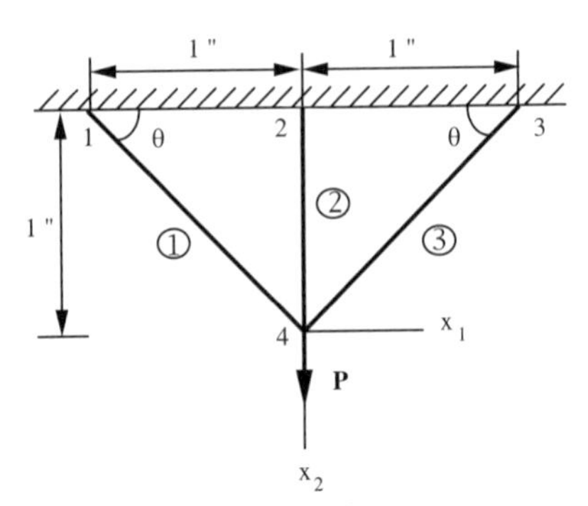

Figure 9.6. Three-member truss structure.

Table 9.14. Accuracy of design sensitivity results of three-member truss model.

Load (lb)	$z_2(b)$	$z_2(b+\Delta b)$	Δz_2	z_2'	$z_2'/\Delta z_2$
5	2.929E–4	2.923E–4	–6.000E–7	–6.059E–7	100.99%
10	6.835E–4	6.805E–4	–3.000E–6	–3.030E–6	100.99%
15	2.594E–3	2.568E–3	–2.600E–5	–2.625E–5	100.98%
20	5.816E–3	5.784E–3	–3.200E–5	–3.232E–5	101.01%
30	1.226E–2	1.221E–2	–5.000E–5	–5.005E–5	101.00%

9.2.3 Numerical Examples

Three-Bar Truss

In this example, the displacement design sensitivity of elastoplastic truss is calculated under a static load with isotropic hardening model. A three-member truss structure that is subjected to a conservative force at node 4 is considered, as shown in Figure 9.6. For analysis data, a constant cross-sectional area $A_1 = A_2 = A_3 = 0.1$ in^2, plastic modulus $H_1 = H_2 = H_3 = 1.0 \times 10^4$ psi, and Young's modulus $E_1 = E_2 = E_3 = 1.0 \times 10^5$ psi are used. The yield stress of each member is 50 psi.

The problem is to find the sizing design sensitivity of the displacement z_2 at node 4 with respect to the cross-sectional area of the truss. Finite element results show that element 2 is in the plastic range at a load level of 8.55 lb, and elements 1 and 3 are in the plastic range at a load level of 12.54 lb. In Table 9.14, analysis results with 1% forward perturbation in the cross-sectional area are listed in the third column. The fourth column is the finite difference $\Delta z_2 = z_2(b + \Delta b) - z_2(b)$, and the fifth column is the predicted change by the proposed sensitivity calculation method. The last column is the agreement between Δz_2 and z_2'. Agreement between Δz_2 and z_2' is excellent for all load steps.

One-Bar Truss

The one-bar truss example is designed to demonstrate the effects of the variable integration time step on the accuracy of the design sensitivity. When the integration time step varies according to the characteristic length, it is necessary to carry out a design sensitivity analysis of the integration time steps. The dimensions of the original and perturbed designs, and boundary conditions are shown in Fig. 9.7. Numerical integration

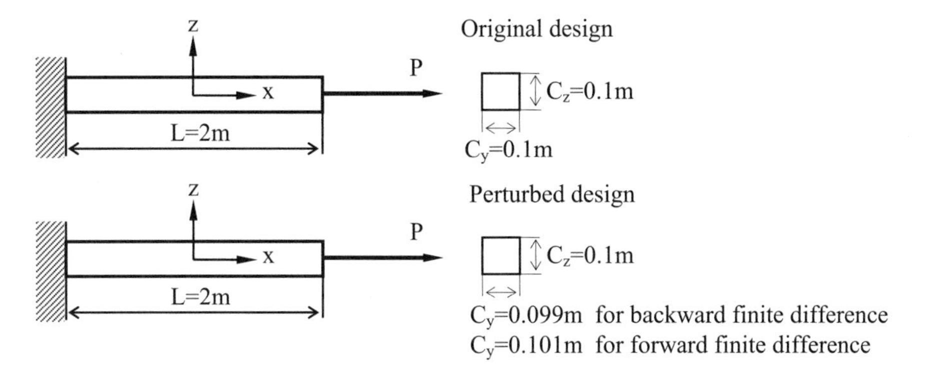

Figure 9.7. Original and perturbed designs of a truss.

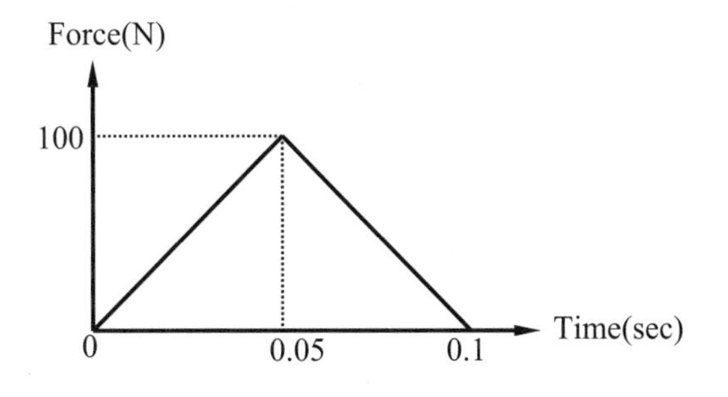

Figure 9.8. Loading history.

of the variational equation is carried out using one-point integration along the axis and one-point integration in the cross section. The elastoplastic material model with the linear isotropic hardening assumption is used. The loading history is shown in Fig. 9.8.

Design sensitivity results are compared with central finite difference results. To make the time step size large enough to be within the significant digits of single precision (7–8 digits), a fictitious material is assumed. Young's modulus, density, and Poisson's ratio are 20000, 50000, and 0.3, respectively. In Table 9.15, the time steps and their sensitivities are presented at initial and the final (70th) time steps. Note that the final time step is noticeably different from the initial one and the sensitivities of the time steps are not small.

In the explicit code, the time step is determined at each load step based on the critical time step in order to establish of numerical stability. In this case, design sensitivity analysis has to be carried out with a variable time step. However, if the fixed time step is employed in nonlinear finite element analysis, then design sensitivity analysis of the time step is not necessary. The results in Table 9.16 indicate that the variable time step affects the accuracy of design sensitivity results, even though design sensitivity analysis of the time step is used to obtain design sensitivity of the stress.

The reason for inaccurate design sensitivity results when the variable integration time step is used is that design sensitivity analysis of the time step is required. Design sensitivity of the time step will not predict the exact perturbation. Thus, there is a slight difference between predicted and the actual time steps in the perturbed design, as shown in Fig. 9.5. Even though the difference is small, the error in the time step will increase the inaccuracy of the design sensitivity of structural responses. The loading steps will also be different because of the change in the time step.

The same example is tested using a different material: $E = 2.0 \times 10^{11}$ [N/m^2], $\rho = 7.8 \times 10^3$ [kg/m^2], and $\nu = 0.3$. In Table 9.17, the time steps and their sensitivities are presented at the initial and final (500th) time steps. Note that the difference between initial and the final time steps is very small and the change in sensitivities of these time steps are insignificant.

The results in Table 9.18 indicate that the effects of variable time steps are not significant because the time steps do not significantly vary during the analysis. However, better sensitivity results are obtained when double precision is used. The results of this example suggest that it is best to use a combination of double precision with a fixed time step for both the analysis and design sensitivity analysis.

Table 9.15. Comparison of time steps and their sensitivities.

	Initial	Final
Δt	0.136277E+01	0.143058E+01
$\Delta t'$	−0.344898E−02	−0.693570E+00

Table 9.16. Comparison of sensitivity accuracy of stress performance measure.

Time	Single Precision		Double Precision	
Steps	Variable	Fixed	Variable	Fixed
10	100.47	99.96	100.46	99.97
20	100.94	99.95	100.93	99.94
30	101.47	99.92	101.47	99.91
40	102.15	99.88	102.16	99.87
50	103.15	99.88	103.16	99.82
60	104.76	99.73	104.89	99.76
70	108.76	99.54	109.15	99.64

Table 9.17. Comparison of time steps and their sensitivities.

	Initial	Final
Δt	0.170210E−03	0.170210E−03
$\Delta t'$	−0.538410E−13	−0.720552E−10

Table 9.18. Comparison of sensitivity accuracy of stress performance measure.

Time Steps	Single Precision		Double Precision	
	Variable	Fixed	Variable	Fixed
10	99.99	99.99	99.99	99.99
100	99.96	99.96	99.99	99.99
200	99.99	99.99	99.99	99.99
300	95.59	95.59	99.99	99.99
400	100.12	100.12	99.99	99.99
500	100.56	100.56	99.99	99.99

Table 9.19. Perturbed components for sizing DSA of vehicle frame.

Description of component	Element	Pert. of C_y	Pert. of C_z
Longitudinal-middle	4,5,6,7,8,9	1.000 %	0.000 %
Longitudinal-front	10,11,12,13,14,15	0.000 %	1.000 %
Second cross component	37,38,39,40,41,42	1.000 %	1.000 %
Third cross component	43,44,45,46,47,48	1.000 %	1.000 %

Vehicle Frame

Consider the vehicle frame structure shown in Fig. 9.9, where large strain and large rotation effects are considered. In this example, steel is chosen as a material. The layout of the original design and boundary conditions are shown in Fig. 9.9, where each component consists of six finite elements, with the exception of two longitudinal rear components, which each consist of three finite elements. The kinematic responses at selected nodes of the longitudinal components are selected as performance measures. Numerical integration of the variational equation is carried out using one-point integration along the axis and four-point integration in the cross section. The finite element model consists of 48 beam elements. To verify the accuracy of the design sensitivity, some longitudinal and cross components are perturbed by 1%, as shown in Table 9.19, where C_y and C_z are the cross-sectional dimensions of each rectangular beam.

Analysis with 118 load steps is performed for the impact loading given in Fig. 9.10 for duration of 2000 μsec. The deformed frame is shown in Fig. 9.11 at selected time steps.

The transient history of the design sensitivities of acceleration is shown in Table 9.20 for selected nodes and time steps. In this table, δa_x represents the predicted variation of acceleration in the x-direction using the proposed sensitivity calculation method, Δa_x represents the finite difference result, and $\delta a_x/\Delta a_x$ indicates agreement between the predicted variation and finite difference results. In Table 9.20, the sensitivity of y-acceleration at node 54 yields a 90.41% agreement at load step 40, but improves to 98.46% at time step 70, which implies that analytical sensitivity provides better sensitivity results than finite difference for a highly nonlinear problem. If the analytical sensitivities were inaccurate, the error would have increased at later load steps. In Table 9.21, the sensitivities of the shear stress at element 1 yields an agreement of approximately 110%, but improves at later load steps, which also indicates that the analytical sensitivity calculation method yields better sensitivity results.

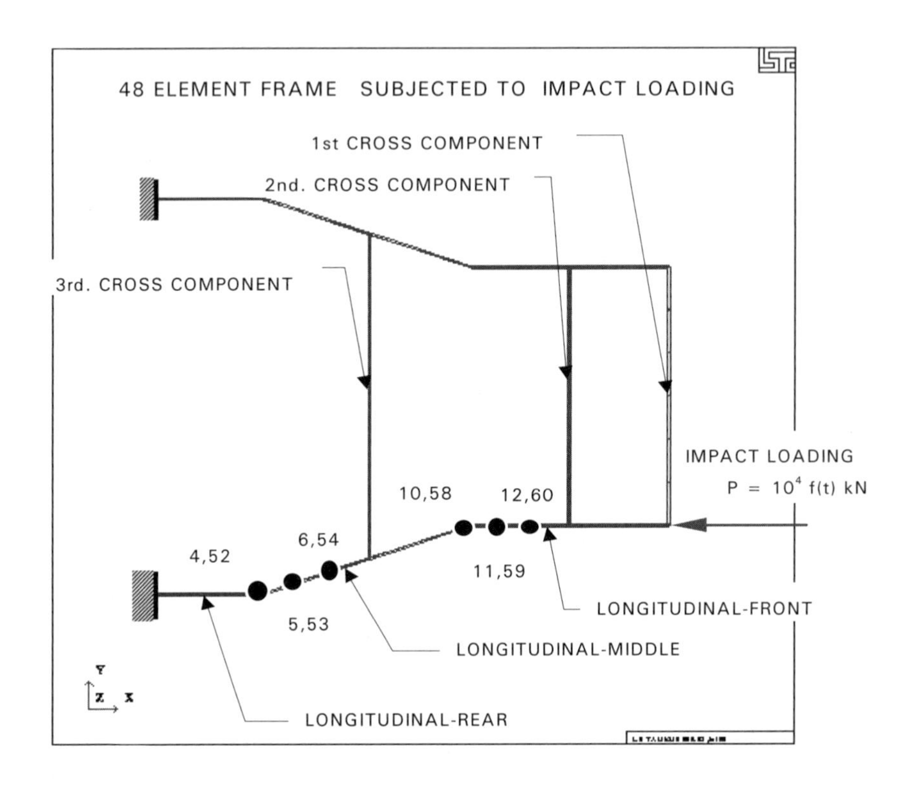

Figure 9.9. Original design of vehicle frame.

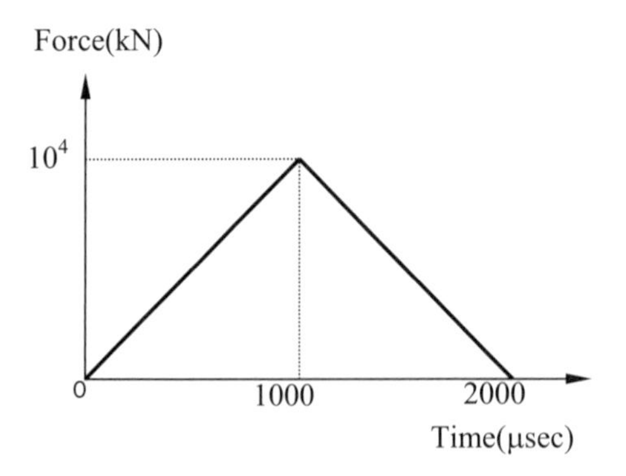

Figure 9.10. Impact loading.

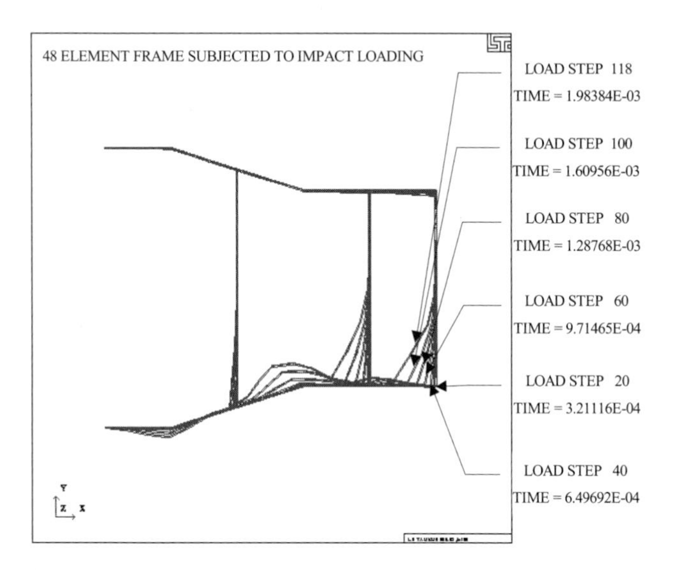

Figure 9.11. Deformed shape of vehicle frame.

Table 9.20. History of acceleration variation for vehicle frame.

Load	Node	δa_x	δa_y	δa_z	Δa_x	Δa_y	Δa_z	$\delta a_x/\Delta a_x$	$\delta a_y/\Delta a_y$	$\delta a_z/\Delta a_z$
40	4	.352196E3	.146095E3	-.146806E3	.352089E3	.146059E3	-.146770E3	100.03	100.02	100.02
	6	.125376E4	.153969E3	-.380134E3	.125495E4	.154053E3	-.380279E3	99.90	99.95	99.96
	52	-.100478E1	.188312E3	.184234E3	-.100762E1	.188535E3	.184434E3	99.72	99.88	99.89
	54	-.452294E3	-.863794E1	-.151563E4	-.452383E3	-.955376E1	-.151683E4	99.98	**90.41**	99.92
70	4	-.890262E3	.286738E4	-.252918E4	-.882564E3	.287105E4	-.253262E4	100.87	99.87	99.86
	6	-.706282E3	-.122787E3	-.462041E3	-.701525E3	-.125900E3	-.457392E3	100.68	97.53	101.02
	52	.757630E3	.306891E4	.558203E4	.757918E3	.306516E4	.557929E4	99.96	100.12	100.05
	54	.500331E3	-.274139E4	-.225288E4	.501913E3	-.278436E4	-.227601E4	99.68	**98.46**	98.98
110	4	.641833E4	-.935337E4	.923883E4	.631074E4	-.923569E4	.913752E4	101.70	101.27	101.11
	6	.130885E5	.187790E4	-.592426E4	.135265E5	.198851E4	-.605878E4	96.76	94.44	97.78
	52	.603457E4	-.176948E5	-.218121E5	.603234E4	-.176301E5	-.214210E5	100.04	100.37	101.83
	54	-.109009E5	.611409E4	-.223709E5	-.107913E5	.532070E4	-.229336E5	101.02	114.91	97.55

Table 9.21. History of global stress variation for vehicle frame.

Load	Elem	$\delta\sigma_{11}$	$\delta\sigma_{12}$	$\delta\sigma_{13}$	$\Delta\sigma_{11}$	$\Delta\sigma_{12}$	$\Delta\sigma_{13}$	$\delta\sigma_{11}/\Delta\sigma_{11}$	$\delta\sigma_{12}/\Delta\sigma_{12}$	$\delta\sigma_{13}/\Delta\sigma_{13}$
40	**1**	.193666E5	.443090E2	-.400608E2	.191829E5	.401785E2	-.356526E2	100.96	**110.28**	**112.36**
	4	.108491E7	.309987E6	-.311716E6	.108526E7	.310061E6	-.311793E6	99.97	99.98	99.98
	17	-.835844E5	-.589407E6	.902907E1	-.836059E5	-.589548E6	.884195E1	99.97	99.98	102.12
	18	-.217869E4	-.275631E4	.300073E2	-.217909E4	-.277040E4	.300097E2	99.98	99.49	99.99
70	**1**	.229054E8	-.137605E6	.141141E6	.229594E8	-.137932E6	.141474E6	99.76	**99.76**	**99.76**
	4	.334719E8	.785888E7	-.726770E7	.335118E8	.786713E7	-.727539E7	99.88	99.90	99.89
	17	-.137913E7	-.630847E7	.638194E5	-.138449E7	-.632129E7	.640026E5	99.61	99.80	99.71
	18	-.896337E5	-.123101E7	.416486E4	-.890443E5	-.122558E7	.402503E4	100.66	100.44	103.47
110	1	-.604567E8	.985217E7	-.100275E8	-.604288E8	.100179E8	-.101572E8	100.05	98.35	98.72
	4	-.637014E8	-.249843E8	.209632E8	-.637561E8	-.249072E8	.209241E8	99.91	100.31	100.19
	17	-.179760E8	-.259022E8	.159163E7	-.178243E8	-.256925E8	.158829E7	100.85	100.82	100.21
	18	-.214412E7	-.102335E8	.127174E6	-.213039E7	-.101675E8	.127006E6	100.64	100.65	100.13

10 Nonlinear Shape Design Sensitivity Analysis

Design sensitivity analysis of nonlinear problems when structural shape is considered part of the design is presented in this chapter. Since nonlinear analysis undergoes different configurations during the deformation process, the shape design sensitivity process is not as simple as defining the design at an initial configuration. When the initial configuration is used as a reference frame, which is known as the total Lagrangian formulation, it is relatively easy to formulate a sensitivity analysis with the response analysis using the same reference frame. When the updated Lagrangian formulation is used, which uses the current configuration as the reference point, the shape design sensitivity formulation has to be transformed to the initial configuration where design perturbation is defined [36]. After obtaining explicit design dependence, the sensitivity equation can be transformed back to the current configuration in order to use the response analysis results.

Two alternative methods are presented to numerically evaluate nonlinear elastic problem sensitivity expressions: the adjoint variable method and the direct differentiation method [37]. The advantages and disadvantages of each method are discussed. Using the domain formulation of shape design sensitivity analysis, as well as adjoint variable and direct differentiation methods, design sensitivity expressions are derived in a continuum setting in terms of shape design variations.

10.1 Nonlinear Elastic Problems

In nonlinear elastic design sensitivity analysis an adjoint variable method can be employed that is not available in the elastoplasticity model discussed in the following section. Using the adjoint variable method, a linear adjoint equation is generated for each constraint. In a nonlinear analysis, like a linear analysis, the linear adjoint equations are the same for sizing and shape design variables. Explicit design sensitivity expressions are obtained in terms of design velocity fields. Even though the response analysis is carried out incrementally; the adjoint system uses the same tangent stiffness operator of the original nonlinear system at the final equilibrium configuration [38]. Computation of design sensitivity expressions can be performed once structural responses are obtained for the original nonlinear and linear adjoint systems.

In the direct differentiation method, the first variation of the state variable with respect to the shape design is solved using a linear system of equations obtained by differentiating the original nonlinear equation with respect to the shape design variable. As with the adjoint variable method, if the finite element method is used for numerical analysis, the stiffness operator of the linear equation is the tangent stiffness operator of the original system at the final equilibrium configuration. Once the variation of the state variable is available, computation of design sensitivity expressions for performance measures can be carried out using the sensitivity information of the state variable.

10.1.1 Static Problems

As previously mentioned, two types of formulations are usually used in the shape sensitivity analysis of elastic material: the total and updated Lagrangian formulations. The former refers to the undeformed configuration as the reference point, while the latter refers to the current configuration. As shown in Chapter 8, this distinction is only made for convenience in the analysis process, since these two methods are mathematically equivalent. But from a shape design sensitivity standpoint, equality between the two formulations is not clearly established. Since design perturbation is always defined on the initial (undeformed) configuration, the total Lagrangian formulation is dominant, where the undeformed configuration is referred to as the frame of reference. Cho and Choi [39] discuss updating the design velocity fields at each time step to incorporate shape sensitivity analysis with respect to the updated Lagrangian formulation for elastoplastic material. Design velocity fields are computed at the current time by using the displacement sensitivity and design velocity from the previous time. Thus, the sensitivity equation must be solved at each time step even if the material model is elastic or hyperelastic. Such inefficiency in the updated Lagrangian formulation is not due to physics. For example, in a similar updated Lagrangian formulation for sizing design sensitivity analysis (Section 9.1.1), the sensitivity equation is only solved once, at the final converged time step. By using the fact that the total and updated Lagrangian formulations are equivalent, the same equivalence can be shown for the shape design sensitivity formulation without updating the design velocity field. An appropriate transformation between configurations can be used.

In shape design, the shape of the domain occupied by a structural component is treated as the design variable. Thus, it is convenient to think of the domain as a continuous medium and to utilize the notion of material derivatives from continuum mechanics. All material derivative results presented in Section 6.1 are valid for nonlinear sensitivity analysis in the undeformed configuration.

Consider a structural system in an equilibrium configuration at the final analysis time t_n, corresponding to the initial design ${}^0\Omega$, and a new equilibrium configuration corresponding to the perturbed design ${}^0\Omega_\tau$, as shown in Fig. 10.1. Using (8.5), the equilibrium equations can be written as

$$a_\Omega(z,\bar{z}) = \ell_\Omega(\bar{z}), \qquad \forall \bar{z} \in Z \tag{10.1}$$

and

$$a_{\Omega_\tau}(z_\tau,\bar{z}_\tau) = \ell_{\Omega_\tau}(\bar{z}_\tau), \qquad \forall \bar{z}_\tau \in Z_\tau. \tag{10.2}$$

Subscripts Ω and Ω_τ are used to indicate the dependence of these terms on the shape, and Z_τ is the space of kinematically admissible displacements at perturbed domain ${}^0\Omega_\tau$. As with the linear problem, it is assumed that only one parameter τ defines shape perturbation in the direction of design velocity field $\boldsymbol{V}(\boldsymbol{X})$.

Solution z_τ in (10.2), which refers to initial coordinates $\boldsymbol{X}_\tau$, is assumed to be a differentiable function of the shape design variables. The mapping $z_\tau(\boldsymbol{X}_\tau) \equiv z_\tau(\boldsymbol{X} + \tau\boldsymbol{V}(\boldsymbol{X}))$ is then defined on ${}^0\Omega$, and $z_\tau(\boldsymbol{X}_\tau)$ depends on τ in two ways. First, $z_\tau(\boldsymbol{X}_\tau)$ is the solution to the equilibrium equation on ${}^0\Omega_\tau$ [equation (10.2)]. Secondly, it is evaluated at a point $\boldsymbol{X}_\tau$ that moves with τ. If a pointwise material derivative of z_τ at $\boldsymbol{X} \in {}^0\Omega$ exists, it is defined as

$$\dot{z} \equiv \frac{d}{d\tau} z_\tau\left(\boldsymbol{X} + \tau\boldsymbol{V}(\boldsymbol{X})\right)\Big|_{\tau=0} = \lim_{\tau\to 0} \frac{z_\tau\left(\boldsymbol{X} + \tau\boldsymbol{V}(\boldsymbol{X})\right) - z(\boldsymbol{X})}{\tau}. \tag{10.3}$$

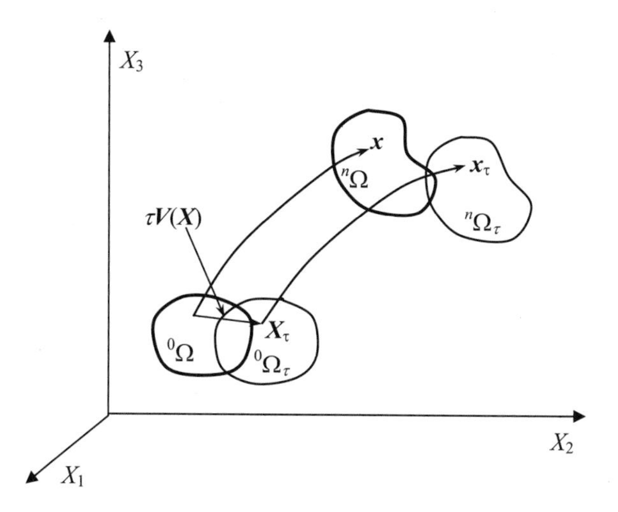

Figure 10.1. Variation of a structural domain.

If z_τ has a regular extension in the neighborhood of ${}^0\bar{\Omega}_\tau$, then

$$\dot{z}(X) = z'(X) + \nabla_0 z V(X), \tag{10.4}$$

where

$$z' \equiv \lim_{\tau \to 0} \frac{z_\tau(X) - z(X)}{\tau} \tag{10.5}$$

is the partial derivative of z_τ. As in the linear problem, using the above definition, it can be noted that the partial derivative commutes with the derivative with respect to undeformed coordinate X, that is,

$$(\nabla_0 z)' = \nabla_0 z'. \tag{10.6}$$

Since a gradient operator accompanies many structural problems, the following relation can be derived to reduce the complexities of partial derivatives:

$$\begin{aligned} \frac{d}{d\tau}(\nabla_0 z_\tau)\bigg|_{\tau=0} &= (\nabla_0 z)' + \nabla_0(\nabla_0 z)V \\ &= \nabla_0 z' + \nabla_0(\nabla_0 z)V \\ &= \nabla_0(\dot{z} - \nabla_0 z V) + \nabla_0(\nabla_0 z)V \\ &= \nabla_0 \dot{z} - \nabla_0 z \nabla_0 V. \end{aligned} \tag{10.7}$$

Note that the same order of differentiation is involved in displacement and design velocity fields. For design sensitivity analysis, variational equation (10.2) is differentiated with respect to parameter τ by using the above relations. Before deriving a detailed sensitivity formulation, consider the sensitivity of the displacement variation. Since virtual displacement $\bar{z}_\tau$ is arbitrary, it can be selected as $\bar{z}(X + \tau V(X)) = \bar{z}(X)$, that is, $\bar{z}$ can be chosen as constant along $X_\tau = X + \tau V(X)$. For this set of kinematically admissible virtual displacements from (10.4), it follows that

$$\dot{\bar{z}}(\boldsymbol{X}) = \bar{z}'(\boldsymbol{X}) + \nabla_0 \bar{z} \boldsymbol{V}(\boldsymbol{X}) = 0. \tag{10.8}$$

However, even if it is assumed that $\dot{\bar{z}} \neq \boldsymbol{0}$, since $\dot{\bar{z}}$ is in the space of kinematically admissible virtual displacements, the relation

$$a_\Omega(z, \dot{\bar{z}}) = \ell_\Omega(\dot{\bar{z}}), \quad \forall \dot{\bar{z}} \in Z \tag{10.9}$$

is valid. Thus, all terms containing $\dot{\bar{z}}$ will be ignored in the following derivations.

Taking the material derivative of both sides of (10.2), and using the relations in Section 6.1 along with the above relations, we have

$$[a_\Omega(z, \bar{z})]' = a^*_\Omega(z; \dot{z}, \bar{z}) + a'_V(z, \bar{z}) = \ell'_V(\bar{z}), \quad \forall \bar{z} \in Z. \tag{10.10}$$

In (10.10), $a'_V(z, \bar{z})$ is called the structural fictitious load and contains all terms that explicitly depend on the design through the structural energy form. $\ell'_V(\bar{z})$ is called the external fictitious load and contains all terms that explicitly depend on the design though the load form. By replacing Δz with $\dot{z}$, $a^*_\Omega(z; \dot{z}, \bar{z})$ is the same as $a^*(z; \Delta z, \bar{z})$ from (8.20). Note that the structural energy form and fictitious load are different for each formulation and constitutive relation. Equation (10.10) can be rewritten to yield the following result:

$$a^*_\Omega(z; \dot{z}, \bar{z}) = \ell'_V(\bar{z}) - a'_V(z, \bar{z}), \quad \forall \bar{z} \in Z, \tag{10.11}$$

which yields the direct differentiation method. Presuming that state variable z is known as the solution to (8.21) at final configuration t_n, (10.11) is the linear variational equation of the material derivative $\dot{z}$.

In order to utilize the adjoint variable method, consider a structural performance measure that can be written in integral form over the final equilibrium configuration ${}^n\Omega_\tau$ at time t_n, with the reference configuration ${}^0\Omega_\tau$ at time t_0, as

$$\psi_\tau = \int_{{}^0\Omega_\tau} g(z_\tau, \nabla_0 z_\tau)\, d\Omega, \tag{10.12}$$

where function g is continuously differentiable with respect to its arguments. The functional ψ_τ depends on ${}^0\Omega_\tau$ in two ways: (1) through the domain of integration, and (2) through the solution z_τ, which corresponds to the new shape ${}^0\Omega_\tau$. Carrying out the variation of the functional in (10.12) using the material derivative formula from (6.37) and (10.7) yields

$$\psi' = \int_{{}^0\Omega} (g^T_{,z}\dot{z} + g_{,\nabla_0 z} : \nabla_0 \dot{z} - g_{,\nabla_0 z} : (\nabla_0 z \nabla_0 \boldsymbol{V}) + g\, div \boldsymbol{V})\, d\Omega. \tag{10.13}$$

Both $\dot{z}$ and $\nabla_0\dot{z}$ implicitly depend on the design velocity field $\boldsymbol{V}$ through the variational equation (10.2), and are as yet unknown. In order to obtain an explicit expression for ψ' in terms of the design velocity field $\boldsymbol{V}$, the first two terms on the right of (10.13) must be expressed explicitly in terms of $\boldsymbol{V}$.

It is now necessary to use the same adjoint equation used with the sizing design problem from (9.10), written as

$$a^*_\Omega(z; \lambda, \bar{\lambda}) = \int_{{}^0\Omega} (g^T_{,z}\bar{\lambda} + g_{,\nabla_0 z} : \nabla_0 \bar{\lambda})\, d\Omega, \quad \forall \bar{\lambda} \in Z, \tag{10.14}$$

where the adjoint solution λ is sought by solving the linear system. Equation (10.14) defines the adjoint equation of the performance measure in (10.12). Since $\dot{z} \in Z$, evaluate (10.14) at $\bar{\lambda} = \dot{z}$ to obtain

$$a^*_\Omega(z;\lambda,\dot{z}) = \int_{^0\Omega} (g^T_{,z}\dot{z} + g_{,\nabla_0 z} : \nabla_0 \dot{z})\, d\Omega. \tag{10.15}$$

Since $\bar{z}$ and λ belong to the same Z, evaluate (10.11) at $\bar{z} = \lambda$ to obtain

$$a^*_\Omega(z;\dot{z},\lambda) = \ell'_V(\lambda) - a'_V(z,\lambda). \tag{10.16}$$

Using symmetry of the linearized energy form $a^*_\Omega(z;\bullet,\bullet)$ in its arguments, we obtain the following from (10.15) and (10.16):

$$\int_{^0\Omega} (g^T_{,z}\dot{z} + g_{,\nabla_0 z} : \nabla_0 \dot{z})\, d\Omega = \ell'_V(\lambda) - a'_V(z,\lambda). \tag{10.17}$$

Thus, the implicitly dependent terms in (10.13) can be obtained with the response result z and the adjoint result λ from (10.17). Substituting this result into (10.13) yields

$$\psi' = \ell'_V(\lambda) - a'_V(z,\lambda) + \int_{^0\Omega} [-g_{,\nabla_0 z} : (\nabla_0 z \nabla_0 V) + g\, divV]\, d\Omega, \tag{10.18}$$

where the right side can be evaluated once it is determined that the state z is the solution to (10.1), and the adjoint variable λ is the solution to (10.14). Note that adjoint equation (10.14) is the same as the sizing sensitivity problem in (9.10). This is advantageous when both sizing and shape design sensitivity information are simultaneously required. The sensitivity equation (10.11) and adjoint equation (10.14) are linear, even though the original governing equation (10.1) is nonlinear. Equations (10.11) and (10.14) can be solved with the same tangent stiffness matrix at the final equilibrium configuration, which is already in factorized form. The specific form of $a'_V(z,\bar{z})$ and $\ell'_V(\bar{z})$ depend on the constitutive model and load type, and will be derived in the following.

Total Lagrangian Formulation

Shape design sensitivity analysis using the total Lagrangian formulation begins with the undeformed configuration where perturbation of the structure is defined using the design velocity field V. The design derivative of the structural energy form becomes

$$[a_\Omega(z,\bar{z})]' = \left[\int_{^0\Omega_\tau} S : \bar{E}\, d\Omega\right]' = \int_{^0\Omega}\left[\dot{S} : \bar{E} + S : \dot{\bar{E}} + S : \bar{E} divV\right] d\Omega, \tag{10.19}$$

where $(\dot{\bullet})$ is the material derivative and V is the design velocity field, representing the direction and magnitude of shape perturbation at the initial geometry. The last term $divV$ comes from the domain perturbation effect. The first part of the right side of (10.19) can be expressed in terms of the displacement sensitivity and design velocity by using the constitutive relation. In the case of an elastic material, the material derivative of stress can be expressed in terms of the material derivative of strain using $S = \partial W / \partial E = W_{,E}$, as

$$\begin{aligned} \dot{S} &= W_{,E,E} : \dot{E} \\ &= C : \dot{E}, \end{aligned} \tag{10.20}$$

where $C \equiv W_{,E,E}$ is a fourth-order material constitutive tensor, defined in (8.16) for Saint Vernant-Kirchhoff elastic material. Since the Green-Lagrange strain tensor is defined by the deformation gradient, the design derivative form of F is first derived using (10.7) as

$$\dot{F} = (1 + \nabla_0 z)^{\cdot} = \nabla_0 \dot{z} - \nabla_0 z \nabla_0 V. \tag{10.21}$$

Using (10.21) and the definition of (8.4), the material derivative of the Green-Lagrange strain tensor can be expressed as

$$
\begin{aligned}
\dot{\boldsymbol{E}} &= \tfrac{1}{2}\Big[\left(\nabla_0\dot{\boldsymbol{z}} - \nabla_0\boldsymbol{z}\nabla_0\boldsymbol{V}\right)^T \boldsymbol{F} + \boldsymbol{F}^T\left(\nabla_0\dot{\boldsymbol{z}} - \nabla_0\boldsymbol{z}\nabla_0\boldsymbol{V}\right)\Big] \\
&\equiv \Delta\boldsymbol{E}(\dot{\boldsymbol{z}}) + \boldsymbol{E}_V(\boldsymbol{z}),
\end{aligned}
\tag{10.22}
$$

where, by substituting $\Delta\boldsymbol{z}$ into $\dot{\boldsymbol{z}}$, $\Delta\boldsymbol{E}(\dot{\boldsymbol{z}})$ represents the same form as $\Delta\boldsymbol{E}$ in (8.18), and where $\boldsymbol{E}_V(\boldsymbol{z})$ represents other terms that depend on the response analysis result and design velocity field. If the response analysis result is known, then $\boldsymbol{E}_V(\boldsymbol{z})$ can be obtained using

$$
\boldsymbol{E}_V(\boldsymbol{z}) = -sym[(\nabla_0\boldsymbol{z}\nabla_0\boldsymbol{V})^T \boldsymbol{F}]. \tag{10.23}
$$

The material derivative of the Lagrangian strain variation can be derived from the definition in (8.8) using (10.7) and (10.8) as

$$
\begin{aligned}
\dot{\overline{\boldsymbol{E}}} &= \tfrac{1}{2}\Big[\nabla_0\overline{\boldsymbol{z}}^T\boldsymbol{F} + \boldsymbol{F}^T\nabla_0\overline{\boldsymbol{z}}\Big]^{\cdot} \\
&= \tfrac{1}{2}\Big[-(\nabla_0\overline{\boldsymbol{z}}\nabla_0\boldsymbol{V})^T\boldsymbol{F} - \boldsymbol{F}^T(\nabla_0\overline{\boldsymbol{z}}\nabla_0\boldsymbol{V})\Big] \\
&\quad + \tfrac{1}{2}\Big[\nabla_0\overline{\boldsymbol{z}}^T\left(\nabla_0\dot{\boldsymbol{z}} - \nabla_0\boldsymbol{z}\nabla_0\boldsymbol{V}\right) + \left(\nabla_0\dot{\boldsymbol{z}} - \nabla_0\boldsymbol{z}\nabla_0\boldsymbol{V}\right)^T\nabla_0\overline{\boldsymbol{z}}\Big] \\
&\equiv \Delta\overline{\boldsymbol{E}}(\dot{\boldsymbol{z}},\overline{\boldsymbol{z}}) + \overline{\boldsymbol{E}}_V(\boldsymbol{z},\overline{\boldsymbol{z}}),
\end{aligned}
\tag{10.24}
$$

where, by substituting $\Delta\boldsymbol{z}$ into $\dot{\boldsymbol{z}}$, $\Delta\overline{\boldsymbol{E}}(\dot{\boldsymbol{z}},\overline{\boldsymbol{z}})$ is in the same form as $\Delta\overline{\boldsymbol{E}}$ in (8.19), and $\overline{\boldsymbol{E}}_V(\boldsymbol{z},\overline{\boldsymbol{z}})$ represents other terms that explicitly depend on the response analysis result and design velocity field. If the response analysis result is known, then $\overline{\boldsymbol{E}}_V(\boldsymbol{z},\overline{\boldsymbol{z}})$ can be obtained using

$$
\overline{\boldsymbol{E}}_V(\boldsymbol{z},\overline{\boldsymbol{z}}) = -sym[(\nabla_0\overline{\boldsymbol{z}}\nabla_0\boldsymbol{V})^T\boldsymbol{F}] - sym[\nabla_0\overline{\boldsymbol{z}}^T\left(\nabla_0\boldsymbol{z}\nabla_0\boldsymbol{V}\right)]. \tag{10.25}
$$

Thus, using (10.20), (10.22), and (10.24), (10.19) can be expressed as

$$
\begin{aligned}
[a_\Omega(\boldsymbol{z},\overline{\boldsymbol{z}})]' &= \int_{{}^0\Omega}\left\{\overline{\boldsymbol{E}}:\boldsymbol{C}:\Delta\boldsymbol{E}(\dot{\boldsymbol{z}}) + \boldsymbol{S}:\Delta\overline{\boldsymbol{E}}(\dot{\boldsymbol{z}},\overline{\boldsymbol{z}})\right\}d\Omega \\
&\quad + \int_{{}^0\Omega}\Big[\overline{\boldsymbol{E}}:\boldsymbol{C}:\boldsymbol{E}_V(\boldsymbol{z}) + \boldsymbol{S}:\overline{\boldsymbol{E}}_V(\boldsymbol{z},\overline{\boldsymbol{z}}) + \boldsymbol{S}:\overline{\boldsymbol{E}}\,div\boldsymbol{V}\Big]d\Omega \\
&\equiv a^*_\Omega(\boldsymbol{z};\dot{\boldsymbol{z}},\overline{\boldsymbol{z}}) + a'_V(\boldsymbol{z},\overline{\boldsymbol{z}}),
\end{aligned}
\tag{10.26}
$$

where

$$
a'_V(\boldsymbol{z},\overline{\boldsymbol{z}}) = \int_{{}^0\Omega}\Big[\overline{\boldsymbol{E}}:\boldsymbol{C}:\boldsymbol{E}_V(\boldsymbol{z}) + \boldsymbol{S}:\overline{\boldsymbol{E}}_V(\boldsymbol{z},\overline{\boldsymbol{z}}) + \boldsymbol{S}:\overline{\boldsymbol{E}}\,div\boldsymbol{V}\Big]d\Omega \tag{10.27}
$$

is the structural fictitious load form, which explicitly depends on the design velocity fields and the solution at time t_n. If a converged solution is obtained at the final configuration time t_n, then (10.27) can be computed using the given design velocity field $\boldsymbol{V}$. If the material derivative $\dot{\boldsymbol{z}}$ is substituted by the incremental displacement $\Delta\boldsymbol{z}$, then the expression $a^*_\Omega(\boldsymbol{z};\dot{\boldsymbol{z}},\overline{\boldsymbol{z}})$ in (10.26) is in the same form as in (8.20).

The load form defined in (8.7) can be perturbed in the same way as the structural energy form. At perturbed domain ${}^0\Omega_\tau$ the load form is

$$
\ell_{\Omega_\tau}(\overline{\boldsymbol{z}}_\tau) = \int_{{}^0\Omega_\tau}\overline{\boldsymbol{z}}_\tau^T\boldsymbol{f}_\tau^B\,d\Omega + \int_{{}^0\Gamma_\tau^S}\overline{\boldsymbol{z}}_\tau^T\boldsymbol{f}_\tau^S\,d\Gamma \tag{10.28}
$$

and the material derivative is

$$\ell'_V(\bar{z}) = \int_{^0\Omega}\left[\bar{z}^T(\nabla_0 f^B V) + \bar{z}^T f^B \, divV\right]d\Omega \\ + \int_{^0\Gamma^S}\left[\bar{z}^T(\nabla_0 f^S V) + \kappa\bar{z}^T f^S (V^T n)\right]d\Gamma. \tag{10.29}$$

It is assumed that the external force is independent of the design change, that is, $f^{B\prime} = f^{S\prime} = 0$. Equation (10.29) defines the external fictitious load.

Updated Lagrangian Formulation

In this section, a shape design sensitivity formulation is presented for the elastic material using the updated Lagrangian approach, which is as efficient as the total Lagrangian formulation. Given the fact that since these two formulations are equivalent, the design derivative is taken at the undeformed configuration, and all variables are then transformed to the current configuration using a procedure similar to that used in (8.26) and (8.27). As a result, the sensitivity equation at the current configuration is nothing but transformed equation from the undeformed configuration. The two formulations therefore yield the same fictitious load form, but different representations.

The material derivative of the Green-Lagrange strain tensor is first transformed to the current configuration using the relation in (8.27), as

$$F^{-T}\dot{E}F^{-1} = \tfrac{1}{2}F^{-T}\left[\left(\nabla_0\dot{z} - \nabla_0 z\nabla_0 V\right)^T F + F^T\left(\nabla_0\dot{z} - \nabla_0 z\nabla_0 V\right)\right]F^{-1} \\ \equiv \varepsilon(\dot{z}) + \varepsilon_V(z), \tag{10.30}$$

where

$$\varepsilon_V(z) = -sym(\nabla_0 z\nabla_n V) \tag{10.31}$$

denotes explicit dependence on the design velocity field and the response. Note that $\varepsilon_V(z)$ contains two different gradient operators, ∇_0 and ∇_n. Since the design velocity is defined at $^0\Omega$, the derivative of V is always taken with respect to the undeformed configuration. Thus, a spatial description of the first part of (10.19) can be expressed as

$$\dot{S}:\bar{E} = \bar{E}:C:\dot{E} \\ = J(F^{-T}\bar{E}F^{-1}):c:(F^{-T}\dot{E}F^{-1}) \\ = J\bar{\varepsilon}:c:\varepsilon(\dot{z}) + J\bar{\varepsilon}:c:\varepsilon_V(z). \tag{10.32}$$

Equation (10.32) splits stress sensitivity into known and unknown parts. The known part $J\bar{\varepsilon}:c:\varepsilon_V(z)$ can be computed from the design velocity and the current response. By replacing $\dot{z}$ with Δz, the unknown part $J\bar{\varepsilon}:c:\varepsilon(\dot{z})$ has the same form as the first part of $a^*_\Omega(z;\Delta z,\bar{z})$.

In the second part of (10.19), the design derivative of the strain variation in (10.24) is transformed to the current configuration using the same procedure in (10.30), as

$$F^{-T}\dot{\bar{E}}F^{-1} \equiv J\eta(\dot{z},\bar{z}) + J\eta_V(z,\bar{z}), \tag{10.33}$$

where

$$\eta_V(z,\bar{z}) = -sym\left[\nabla_n\bar{z}^T\left(\nabla_0 z\nabla_n V\right)\right] - sym\left[\nabla_0\bar{z}\nabla_n V\right]. \tag{10.34}$$

Thus, the second term in (10.19) can be expressed in terms of the current configuration as

$$S:\dot{\bar{E}} = J\sigma:\eta(\dot{z},\bar{z}) + J\sigma:\eta_V(z,\bar{z}). \tag{10.35}$$

By substituting (10.32) and (10.35) into (10.19), the material derivative of the structural energy form at the current configuration can be obtained as

$$\left[a(z,\bar{z})\right]' \equiv a^{*}_{\Omega}(z;\dot{z},\bar{z}) + a'_{V}(z,\bar{z}), \tag{10.36}$$

where

$$a'_{V}(z,\bar{z}) = \int_{{}^{n}\Omega} \left(\bar{\varepsilon} : \boldsymbol{c} : \varepsilon_{V}(z) + \boldsymbol{\sigma} : \boldsymbol{\eta}_{V}(z,\bar{z}) + \boldsymbol{\sigma} : \bar{\varepsilon}\, div\boldsymbol{V}\right) d\Omega \tag{10.37}$$

is the structural fictitious load form, and is linear in $\boldsymbol{V}$ and z. Since the results z and $\boldsymbol{\sigma}$ from the response analysis are given at the current time, $a'_{V}(z,\bar{z})$ can be explicitly computed from the previous configuration without any sensitivity information. Note that the spatial fictitious load form $a'_{V}(z,\bar{z})$ in (10.37) is a transformed version of the material fictitious load form given by Santos and Choi [38]. Because these two formulations have the same fictitious load vectors, the same result $\dot{z}$ is expected from both of them. Therefore, the sensitivity equation of the updated Lagrangian formulation is equivalent to the sensitivity equation for the total Lagrangian formulation.

Beam/Truss Component

Consider the beam/truss component in Fig. 8.3, with variable cross-sectional area $A(X_1)$. For simplicity, only the total Lagrangian formulation will be discussed in this section. Since only the X_1 direction is considered the structural domain, the design velocity field is given in one-dimension as $\boldsymbol{V} = [V(X_1)]$.

The material derivative of the structural energy form in (8.32) yields two forms: $a^{*}_{\Omega}(z;\dot{z},\bar{z})$ and $a'_{V}(z,\bar{z})$. The first form is the same as the linearized structural energy form in (8.37) by replacing $\dot{z}$ with Δz, while the second form represents the explicitly dependent terms on the design. From its expression in (10.27), the explicitly dependent term can be defined as

$$\begin{aligned} a'_{V}(z,\bar{z}) = &-\int_0^l \left[EA(\bar{z}_{1,1} + \bar{z}_{1,1}z_{1,1})(z_{1,1} + z_{1,1}z_{1,1})V_{,1}\right] dX_1 \\ &-\int_0^l \left[S_{11}\bar{z}_{1,1}z_{1,1}V_{,1} - A_{,1}S_{11}(\bar{z}_{1,1} + \bar{z}_{1,1}z_{1,1})V / A\right] dX_1 \\ &-\int_0^l \left[3EIz_{2,11}\bar{z}_{2,11}V_{,1} + EI(z_{2,11}\bar{z}_{2,1} + z_{2,1}\bar{z}_{2,11})V_{,11} - EI_{,1}z_{2,11}\bar{z}_{2,11}V\right] dX_1. \end{aligned} \tag{10.38}$$

Assuming that the partial derivative of the load vanishes, the external fictitious load becomes

$$\ell'_{V}(\bar{z}) = \int_0^l \left[f_{\alpha,1}\bar{z}_{\alpha}V + f_{\alpha}\bar{z}_{\alpha}V_{,1}\right] dX_1, \tag{10.39}$$

where the summation rule is used for repeated indices $\alpha = 1, 2$. The direct differentiation method can be used to solve (10.11), using (10.38) and (10.39) to find the material derivative of the displacement. The performance measure sensitivity can be computed using the chain rule of differentiation, as explained in the following discussion of the adjoint variable method.

Several alternative forms may now be considered as structural response functionals in the adjoint variable method. Consider a functional that defines the value of displacement at isolated point $\hat{X}_1$, that is,

$$\psi_1 = z(\hat{X}_1) = \int_0^l \delta(X_1 - \hat{X}_1) z(X_1)\, dX_1, \tag{10.40}$$

where $\delta(X_1)$ is the Dirac delta measure at the origin. By taking the first variation of (10.40) using the material derivative, we obtain

$$\psi_1' = \int_0^l \delta(X_1 - \hat{X}_1)\dot{z}(X_1)\, dX_1, \tag{10.41}$$

where no explicitly dependent term exists for the displacement. Using the direct differentiation method, (10.41) can easily be evaluated using $\dot{z}$ obtained from (10.11). For this functional, the adjoint equation of (10.14) is

$$a_\Omega^*(z;\lambda,\bar{\lambda}) = \int_0^l \delta(X_1 - \hat{X}_1)\bar{\lambda}\, dX_1, \quad \forall \bar{\lambda} \in Z. \tag{10.42}$$

Using (10.18), the design sensitivity expression is

$$\psi_1' = \ell_V'(\lambda) - a_V'(z,\lambda). \tag{10.43}$$

Another important functional that arises in the design of truss components is stress, given as

$$\begin{aligned}\psi_2 &= \int_0^l S_{11} m_p\, dX_1 \\ &= \int_0^l EA(z_{1,1} + \frac{1}{2} z_{i,1} z_{i,1}) m_p\, dX_1,\end{aligned} \tag{10.44}$$

where $m_p(X_1)$ is a characteristic function that is nonzero only on a small subinterval and whose integral is one. Taking the first variation of (10.44), we obtain

$$\begin{aligned}\psi_2' = &\int_0^l EA(\dot{z}_{1,1} + z_{\alpha,1}\dot{z}_{\alpha,1}) m_p\, dX_1 \\ &- \int_0^l EA(z_{1,1} + z_{\alpha,1} z_{\alpha,1}) V_{,1} m_p\, dX_1.\end{aligned} \tag{10.45}$$

Since the first integral of (10.45) contains an unknown displacement variation, the following adjoint equation is defined for the stress functional:

$$a_u^*(z;\lambda,\bar{\lambda}) = \int_0^l EA(\bar{\lambda}_{1,1} + z_{\alpha,1}\bar{\lambda}_{\alpha,1}) m_p\, dX_1, \quad \forall \bar{\lambda} \in Z. \tag{10.46}$$

Using the response analysis result z, the adjoint analysis result λ, and the explicitly dependent term in (10.45), the sensitivity of the stress functional can be obtained as

$$\psi_2' = \ell_{\delta u}'(\lambda) - a_{\delta u}'(z,\lambda) - \int_0^l EA(z_{1,1} + z_{\alpha,1} z_{\alpha,1}) V_{,1} m_p\, dX_1. \tag{10.47}$$

Another functional that arises in the design of beam components is rotation at an isolated point $\hat{X}_1$, that is,

$$\psi_3 = z_{2,1}(\hat{X}_1) = \int_0^l \delta(X_1 - \hat{X}_1) z_{2,1}(X_1)\, dX_1. \tag{10.48}$$

Computing the first variation of the functional in (10.48) by using the material derivative, we obtain

$$\psi_3' = \int_0^l \delta(X_1 - \hat{X}_1)\dot{z}_{2,1}(X_1)\, dX_1 - \int_0^l \delta(X_1 - \hat{X}_1) z_{2,1}(X_1) V_{,1}\, dX_1. \tag{10.49}$$

For this functional, the adjoint equation is

$$a^{*}_{\Omega}(z;\boldsymbol{\lambda},\bar{\boldsymbol{\lambda}}) = \int_0^l \delta(X_1 - \hat{X}_1)\bar{\lambda}_{2,1}\, dX_1, \quad \forall \bar{\boldsymbol{\lambda}} \in Z. \tag{10.50}$$

In (10.50), $\boldsymbol{\lambda}$ can be physically interpreted as the displacement of the beam/truss by applying a unit moment at $\hat{X}_1$ in the direction $z_{i,1}$ at the final equilibrium configuration. Using (10.18), the design sensitivity expression is

$$\psi'_3 = \ell'_V(\boldsymbol{\lambda}) - a'_V(z,\boldsymbol{\lambda}) - \int_0^l \delta(X_1 - \hat{X}_1) z_{2,1}(X_1) V_{,1}\, dX_1. \tag{10.51}$$

Plate/Plane Elastic Solid Component

Consider the plate/plane stress component in Fig. 8.4 with thickness h. Again, only the total Lagrangian formulation is discussed in this example. Since the component-fixed local coordinate system is defined on the X_1-X_2 plane, the design velocity field is given on the same two-dimensional plane, i.e., $\boldsymbol{V}(\boldsymbol{X}) = [V_1, V_2]^T$. Design perturbation in the X_3-direction will be categorized as a configuration design, and will be taken into account in Chapter 11.

The material derivative of the structural energy form in (8.41) yields two forms: $a^{*}_{\Omega}(z;\dot{z},\bar{z})$ and $a'_V(z,\bar{z})$. The first form is the same as the linearized structural energy form in (8.43) by replacing $\dot{z}$ with Δz, while the second form represents the explicitly dependent terms on the design. From its expression in (10.27), the explicitly dependent term can be defined as

$$\begin{aligned} a'_V(z,\bar{z}) = &\iint_{^0\Omega} h\left[\bar{\boldsymbol{E}} : \boldsymbol{C} : \boldsymbol{E}_V(z) + \boldsymbol{S} : \bar{\boldsymbol{E}}_V(z,\bar{z}) + \boldsymbol{S} : \bar{\boldsymbol{E}}\, div\boldsymbol{V}\right] d\Omega \\ &- \iint_{^0\Omega}\left[\boldsymbol{\kappa}(\bar{z}_3)^T \boldsymbol{C}^b \boldsymbol{\kappa}(\nabla z_3{}^T \boldsymbol{V}) + \boldsymbol{\kappa}(\nabla \bar{z}_3{}^T \boldsymbol{V})^T \boldsymbol{C}^b \boldsymbol{\kappa}(z) - div\left(\boldsymbol{\kappa}(\bar{z}_3)^T \boldsymbol{C}^b \boldsymbol{\kappa}(z_3)\boldsymbol{V}\right)\right] d\Omega, \end{aligned} \tag{10.52}$$

where the first integral is the reduced form of (10.27) on a two-dimensional domain, the curvature vector $\boldsymbol{\kappa}$ in the second integral is defined in (3.39), and the bending stiffness matrix $\boldsymbol{C}^b$ is defined in (3.40). Assuming that the partial derivative of the load vanishes, the external fictitious load becomes

$$\begin{aligned} \ell'_V(\bar{z}) = &\iint_{^0\Omega}\left[\bar{z}^T(\nabla_0 \boldsymbol{f}^B \boldsymbol{V}) + \bar{z}^T \boldsymbol{f}^B\, div\boldsymbol{V}\right] d\Omega \\ &+ \int_{^0\Gamma^S}\left[\bar{z}^T(\nabla_0 \boldsymbol{f}^S \boldsymbol{V}) + \kappa \bar{z}^T \boldsymbol{f}^S (\boldsymbol{V}^T \boldsymbol{n})\right] d\Gamma. \end{aligned} \tag{10.53}$$

Note that the curvature vector $\boldsymbol{\kappa}$ in (10.52) is generated during deformation, while the scalar curvature κ in (10.53) is generated from shape design perturbation.

10.1.2 Critical Load

Suppose that initial shape ${}^0\Omega$ is perturbed to a new shape ${}^0\Omega_\tau$ by the design velocity $\boldsymbol{V}$. In the perturbed shape ${}^0\Omega_\tau$, the new stability equation at its critical limit point becomes

$$a_{\Omega_\tau}(z_\tau,\bar{z}_\tau) = \beta_{\Omega_\tau}\, \ell_{\Omega_\tau}(\bar{z}_\tau), \quad \forall \bar{z}_\tau \in Z. \tag{10.54}$$

Assuming differentiability of the critical load factor with respect to the shape design variable, the first variation of (10.54) is

$$\begin{aligned} [a_{\Omega}(z,\bar{z})]' &\equiv a^{*}_{\Omega}(z;\dot{z},\bar{z}) + a'_V(z,\bar{z}) \\ &= \beta' \ell_{\Omega}(\bar{z}) + \beta_{\Omega} \ell'_V(\bar{z}). \end{aligned} \tag{10.55}$$

Since (10.55) holds for all $\bar{z} \in Z$, this equation may be evaluated with $\bar{z} = \boldsymbol{y}$, which is the eigenvector associated with the lowest eigenvalue of the following linearized eigenvalue problem:

$$A_\Omega(z; \boldsymbol{y}, \bar{\boldsymbol{y}}) = \zeta D_\Omega(z; \boldsymbol{y}, \bar{\boldsymbol{y}}), \qquad \forall \bar{\boldsymbol{y}} \in Z. \tag{10.56}$$

Thus,

$$a^*_\Omega(z; \dot{z}, \boldsymbol{y}) + a'_V(z, \boldsymbol{y}) = \beta' \ell_\Omega(\boldsymbol{y}) + \beta_\Omega \ell'_V(\boldsymbol{y}). \tag{10.57}$$

Using the definition of the energy linear form a^*_Ω given in (10.10), and the symmetry of energy bilinear forms $A_\Omega(z;\bullet,\bullet)$ and $D_\Omega(z;\bullet,\bullet)$ in their arguments, a^*_Ω can be written as

$$a^*_\Omega(z; \dot{z}, \boldsymbol{y}) = A_\Omega(z; \dot{z}, \boldsymbol{y}) - D_\Omega(z; \dot{z}, \boldsymbol{y}). \tag{10.58}$$

At the critical limit point, the right side of (10.58) vanishes, that is,

$$a^*_\Omega(z; \dot{z}, \boldsymbol{y}) = 0. \tag{10.59}$$

Using (10.57) and (10.59), the design sensitivity of the critical load factor β_Ω at the final prebuckling equilibrium configuration is

$$\beta' = \frac{1}{\ell_\Omega(\boldsymbol{y})} \{a'_V(z, \boldsymbol{y}) - \beta_\Omega \ell'_V(\boldsymbol{y})\}, \tag{10.60}$$

where z is the displacement at the final prebuckling equilibrium configuration. Note that critical factor β_Ω is defined as the ratio of the magnitude of critical load vector $\boldsymbol{p}_{cr}$ to the magnitude of total applied load vector $\boldsymbol{p}_0$, as shown in (8.52). Therefore, the design sensitivity expression of the critical load is

$$\begin{aligned} \boldsymbol{p}'_{cr} &= \boldsymbol{p}_0 \beta' \\ &= \frac{\boldsymbol{p}_0}{\ell_\Omega(\boldsymbol{y})} [a'_V(z, \boldsymbol{y}) - \beta_\Omega \ell'_V(\boldsymbol{y})] \end{aligned} \tag{10.61}$$

Note that no adjoint equation is necessary for the design sensitivity expression of the critical load. The eigenvector $\boldsymbol{y}$ in (10.56), the critical load factor β_Ω, and the explicit expression of fictitious load terms are used to obtain the critical load sensitivity using (10.61).

10.1.3 Hyperelastic Material

Using (8.5) and (8.68), the variational equation of the perturbed domain Ω_τ for hyperelastic material at time t_n is

$$a_{\Omega_\tau}(\boldsymbol{r}_\tau, \bar{\boldsymbol{r}}_\tau) = \ell_{\Omega_\tau}(\bar{\boldsymbol{r}}_\tau), \quad \forall \bar{\boldsymbol{r}}_\tau = Z_\tau, \tag{10.62}$$

where the load form is the same as in (10.28). Consider the structural energy form in (8.68) at Ω_τ:

$$a_{\Omega_\tau}(\boldsymbol{r}_\tau, \bar{\boldsymbol{r}}_\tau) \equiv \int_{\Omega_\tau} (\boldsymbol{S}_\tau : \bar{\boldsymbol{E}}_\tau + \bar{p}_\tau H_\tau) d\Omega. \tag{10.63}$$

Using a procedure similar to that used in (10.19) to take the material derivative of the structural energy form, the following can be obtained:

$$\left[a_{\Omega_\tau}(\boldsymbol{r}_\tau,\overline{\boldsymbol{r}}_\tau)\right]' = \int_\Omega (\dot{\boldsymbol{S}}:\overline{\boldsymbol{E}} + \boldsymbol{S}:\dot{\overline{\boldsymbol{E}}} + \boldsymbol{S}:\overline{\boldsymbol{E}}div\boldsymbol{V})\,d\Omega + \int_\Omega (\overline{p}\dot{H} + \overline{p}Hdiv\boldsymbol{V})\,d\Omega. \tag{10.64}$$

As with (10.8), $\dot{\overline{\boldsymbol{r}}}$ is ignored in the following derivations. The first integral on the right of (10.64) is evaluated as follows. For the mixed formulation, the second Piola-Kirchhoff stress sensitivity can be obtained by taking the derivative of (8.66) with respect to displacement and pressure, as

$$\begin{aligned}\dot{\boldsymbol{S}} &= W_{,\boldsymbol{E},\boldsymbol{E}}:\dot{\boldsymbol{E}} + W_{,\boldsymbol{E},p}\dot{p}\\ &= \boldsymbol{C}:\dot{\boldsymbol{E}} + J_{3,\boldsymbol{E}}\dot{p}.\end{aligned} \tag{10.65}$$

Thus, using (10.22) the first integral of (10.64) becomes

$$\begin{aligned}&\int_\Omega \left\{\dot{\boldsymbol{S}}:\overline{\boldsymbol{E}} + \boldsymbol{S}:\dot{\overline{\boldsymbol{E}}} + \boldsymbol{S}:\overline{\boldsymbol{E}}\,div\boldsymbol{V}\right\}d\Omega\\ &= \int_\Omega \left\{\overline{\boldsymbol{E}}:[\boldsymbol{C}:\Delta\boldsymbol{E}(\dot{\boldsymbol{z}}) + J_{3,\boldsymbol{E}}\dot{p}] + \boldsymbol{S}:\Delta\overline{\boldsymbol{E}}(\dot{\boldsymbol{z}},\overline{\boldsymbol{z}})\right\}d\Omega\\ &+ \int_\Omega \left[\overline{\boldsymbol{E}}:\boldsymbol{C}:\boldsymbol{E}_V(\boldsymbol{z}) + \boldsymbol{S}:\overline{\boldsymbol{E}}_V(\boldsymbol{z},\overline{\boldsymbol{z}}) + \boldsymbol{S}:\overline{\boldsymbol{E}}\,div\boldsymbol{V}\right]d\Omega.\end{aligned} \tag{10.66}$$

Evaluation of the second integral in (10.64) is straightforward. By inserting (10.66) into (10.64), the material derivative of the structural energy form becomes

$$\begin{aligned}&[a_{\Omega_\tau}(\boldsymbol{r}_\tau,\overline{\boldsymbol{r}}_\tau)]'\\ &= \int_\Omega \left\{\overline{\boldsymbol{E}}:[\boldsymbol{C}:\Delta\boldsymbol{E}(\dot{\boldsymbol{z}}) + J_{3,\boldsymbol{E}}\dot{p}] + \boldsymbol{S}:\Delta\overline{\boldsymbol{E}}(\dot{\boldsymbol{z}},\overline{\boldsymbol{z}})\right\}d\Omega\\ &+ \int_\Omega \overline{p}\left(J_{3,\boldsymbol{E}}:\Delta\boldsymbol{E}(\dot{\boldsymbol{z}}) - \frac{\dot{p}}{K}\right)d\Omega\\ &+ \int_\Omega \left[\overline{\boldsymbol{E}}:\boldsymbol{C}:\boldsymbol{E}_V(\boldsymbol{z}) + \boldsymbol{S}:\overline{\boldsymbol{E}}_V(\boldsymbol{z},\overline{\boldsymbol{z}}) + \boldsymbol{S}:\overline{\boldsymbol{E}}div\boldsymbol{V}\right]d\Omega\\ &+ \int_\Omega \left[\overline{p}J_{3,\boldsymbol{E}}:\boldsymbol{E}_V(\boldsymbol{z}) + \overline{p}Hdiv\boldsymbol{V}\right]d\Omega\\ &\equiv a^*_\Omega(\boldsymbol{r};\dot{\boldsymbol{r}},\overline{\boldsymbol{r}}) + a'_V(\boldsymbol{r},\overline{\boldsymbol{r}}),\end{aligned} \tag{10.67}$$

where

$$\begin{aligned}a'_V(\boldsymbol{r},\overline{\boldsymbol{r}}) &= \int_\Omega \left[\overline{\boldsymbol{E}}:\boldsymbol{C}:\boldsymbol{E}_V(\boldsymbol{z}) + \boldsymbol{S}:\overline{\boldsymbol{E}}_V(\boldsymbol{z},\overline{\boldsymbol{z}}) + \boldsymbol{S}:\overline{\boldsymbol{E}}div\boldsymbol{V}\right]d\Omega\\ &+ \int_\Omega \left[\overline{p}J_{3,\boldsymbol{E}}:\boldsymbol{E}_V(\boldsymbol{z}) + \overline{p}Hdiv\boldsymbol{V}\right]d\Omega\end{aligned} \tag{10.68}$$

is the structural fictitious load form, which explicitly depends on the design velocity fields and on the solution at time t_n. If a converged solution is obtained at t_n, then (10.68) can be computed using the given design velocity field $\boldsymbol{V}$. If material derivative $\dot{\boldsymbol{r}}$ is substituted by incremental response $\Delta\boldsymbol{r}$, then the expression $a^*_\Omega(\boldsymbol{r};\dot{\boldsymbol{r}},\overline{\boldsymbol{r}})$ in (10.67) will be in the same form as in (8.70).

10.1.4 Numerical Examples

In this section, three numerical examples are employed to verify the accuracy and efficiency of the presented shape design sensitivity analysis methods. The beam finite element STIF3 in ANSYS [28]; and the isoparametric plane strain element CPE4RH and the three-dimensional solid element C3D8RH in ABAQUS [40] are used for these examples. For shape design sensitivity analysis, the velocity computation method presented by Choi and Chang [41] is used. Both boundary displacement and isoparametric methods are employed. The geometric modeling tool PATRAN [42] is used for generating the geometric and finite element models. The tangent stiffness matrix at the final equilibrium configuration of the original structural analysis is stored for the adjoint variable and direct differentiation methods in order to obtain the adjoint response and the first-order variation of the structural response, respectively.

Cantilever Beam

Consider the cantilever beam shown in Fig. 10.2. In Chapter 9, the same example was used to demonstrate the application of design sensitivity analysis to a geometrically nonlinear structure with sizing design variables. It is modeled as a 60 in long beam, composed of 20 finite elements, 21 nodal points and 60 degrees-of-freedom. The cross section has constant width $b(X_1)$ = 0.25 in, constant height $h(X_1)$ = 0.5 in, Young's modulus $E = 30.0 \times 10^6$ psi, and Poisson's ratio $\nu = 0.3$. The beam is subjected to two concentrated forces of 100 lb each, one transverse and the other axial, and a concentrated moment of 100 lb-in parallel to the X_3-axis at the tip. Structural analysis is performed using ANSYS finite element beam STIF3 [28]. Beam length is considered the shape design parameter and a linear velocity field

$$V(X_1) = \frac{X_1}{l}\delta l, \quad X_1 \in [0,60]$$

is used for the purposes of perturbing the shape design parameter.

Displacement and stress sensitivity results of the linear and nonlinear beam models are shown in Table 10.1 and 10.2, respectively, for a design variable $\delta l = 0.01l$.

In the tables, ψ_p^1 and ψ_p^2 are the values of the performance functional evaluated at the modified designs $u - \Delta u$ and $u + \Delta u$, respectively; $\Delta\psi_p = (\psi_p^2 - \psi_p^1)$ is twice the central finite difference; and ψ_p' is the amount of change predicted by design sensitivity calculations, with Δu as the variation in design. As shown in Table 10.1, design sensitivity predictions are in close agreement with finite difference results for the displacement performance functional, for both linear and nonlinear models. However, in Table 10.2 the agreement of stress design sensitivity predictions with finite difference results is not as good as that achieved with displacement. The stress sensitivity at node 1 for the nonlinear model is not good since the finite difference is very small so that it falls beyond the digits of significant numbers.

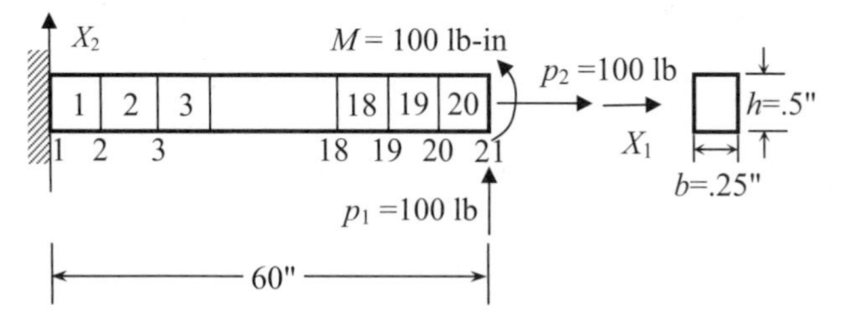

Figure 10.2. Cantilever beam finite element model.

Table 10.1 . design sensitivity of cantilever beam (vertical displacement).

(a) Linear Design Sensitivity Analysis

Node No.	ψ_p^1	ψ_p^2	$\Delta\psi_p$	ψ_p'	$(\psi_p'/\Delta\psi_p)\times 100$
2	0.3354	0.3560	0.0206	0.0206	100.0
4	2.9179	3.0973	0.1794	0.1794	100.0
6	7.8259	8.3069	0.4810	0.4810	100.0
8	14.7911	15.6999	0.9088	0.9088	100.0
10	23.5451	24.9915	1.4464	1.4463	100.0
12	33.8198	35.8968	2.0770	2.0769	100.0
14	45.3469	48.1310	2.7841	2.7840	100.0
16	57.8580	61.4092	3.5512	3.5510	100.0
18	71.0850	75.4465	4.3615	4.3614	100.0
20	84.7596	89.9852	5.1986	5.1984	100.0
21	91.6809	97.3028	5.6219	5.6218	100.0

(b) Nonlinear Design Sensitivity Analysis

Node No.	ψ_p^1	ψ_p^2	$\Delta\psi_p$	ψ_p'	$(\psi_p'/\Delta\psi_p)\times 100$
2	0.1371	0.1428	0.0056	0.0056	98.9
4	1.1380	1.1829	0.0449	0.0445	99.0
6	2.9138	3.0242	0.1104	0.1095	99.2
8	5.2720	5.4672	0.1921	0.1911	99.5
10	8.0637	8.3472	0.2836	0.2828	99.7
12	11.1762	11.5568	0.3807	0.3806	100.0
14	14.5248	15.0058	0.4810	0.4819	100.2
16	18.0451	18.6283	0.5832	0.5853	100.4
18	21.6876	22.3742	0.6867	0.6903	100.5
20	25.4127	26.2038	0.7911	0.7964	100.7
21	27.2958	28.1395	0.8437	0.8499	100.7

Table 10.2. Design sensitivity of cantilever beam (stress).

(a) Linear Design Sensitivity Analysis

Node No.	ψ_p^1	ψ_p^2	$\Delta\psi_p$	ψ_p'	$(\psi_p'/\Delta\psi_p)\times 100$
1	566383.9276	577615.9262	11231.9986	11231.9942	100.0
3	509359.9349	519439.9336	10079.9987	10079.9948	100.0
5	452335.9422	461263.9411	8927.9989	8972.9954	100.0
7	395311.9495	403087.9485	7775.9990	7775.9959	100.0
9	338287.9568	344911.9560	6623.9992	6623.9965	100.0
11	281263.9641	286735.9634	5471.9993	5471.9971	100.0
13	224239.9787	228559.9708	4319.9994	4319.9977	100.0
15	167215.9783	170383.9783	3167.9996	3167.9983	100.0
17	110191.9860	112207.9857	2015.9997	2015.9989	100.0
19	53167.9969	54031.9932	863.9999	863.9995	100.0
20	24655.9969	24943.9969	288.0000	287.9998	100.0

(b) Nonlinear Design Sensitivity Analysis

Node No.	ψ_p^1	ψ_p^2	$\Delta\psi_p$	ψ_p'	$(\psi_p'/\Delta\psi_p)\times 100$
1	229609.2018	229590.5986	–18.6032	–122.1261	656.5
3	180596.0289	179750.8418	–845.1808	–952.8020	112.7
5	141114.0089	139809.0420	–1304.9669	–1417.3788	108.6
7	109656.9489	108165.3232	–1491.6257	–1606.2522	107.7
9	84707.2047	83218.7339	–1488.4708	–1600.2993	107.5
11	64904.4799	63545.3321	–1359.1478	–1462.2258	107.6
13	49092.6459	47944.4639	–1148.1820	–1263.7196	107.7
15	36311.2677	35426.7429	–884.5248	–953.4425	107.8
17	25765.6530	25180.3031	–585.3499	–630.4023	107.7
19	16790.9896	16531.7666	–259.2229	–276.8910	106.8
20	12711.7527	12624.7701	–86.9826	–89.8080	103.2

Engine Mount

The engine mount shown in Fig. 10.3 is used as the first numerical example for design sensitivity of the hyperelastic material. Due to symmetry, only one-half of it is modeled. The engine mount is treated as a plane strain problem. The outer boundary of the engine mount is attached to a metal frame of the vehicle body, and thus fixed. In this model, the shaded area is metal with Young's modulus $E = 2.068 \times 10^7$ N/cm^2, while the material in the other area is rubber with material constants $C_{10} = 19.31$ N/cm^2 and $C_{01} = 8.27$ N/cm^2. Three concentration forces, $F_1 = 88.97$ N, $F_2 = 177.94$ N, and $F_3 = 88.97$ N, act at nodes 94, 107, and 121, respectively, as shown in Fig. 10.3(b). The mount is meshed using 154 ABAQUS four-node plain strain element CPE4RH, with a total of 207 nodes and 325 degrees-of-freedom. The displacement, stress, and hydrostatic pressure are chosen as the performance measures.

The material constants C_{10} and C_{01} are chosen as material property design variables. Table 10.3 shows material property design sensitivity results that are compared with central finite difference results where $\psi(u - \Delta u)$ and $\psi(u + \Delta u)$ are values of the performance measure at perturbed designs $u - \Delta u$ and $u + \Delta u$, respectively; $\Delta\psi$ is the central finite difference result; and ψ' is the predicted perturbation obtained from the presented sensitivity calculation method. For this example, a design perturbation of 1% (i.e., $\Delta u = 0.01u$) is used to compute $\Delta\psi$. As can be seen from the table, design sensitivity results predict central finite difference results very accurately.

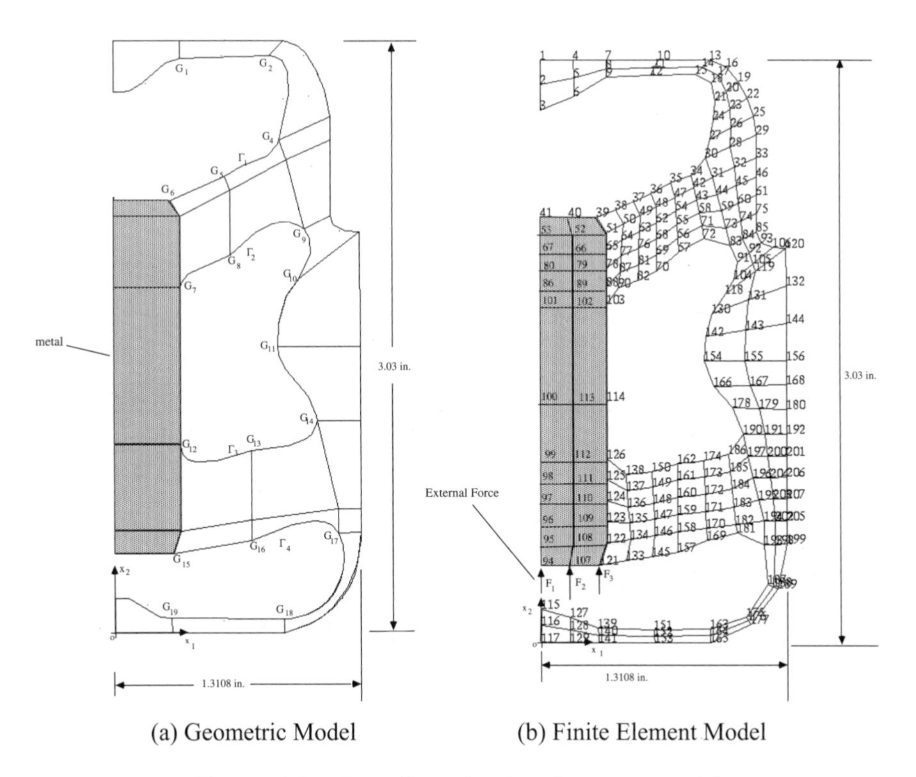

(a) Geometric Model (b) Finite Element Model

Figure 10.3. Two-dimensional engine mount model.

For shape design sensitivity analysis, the geometry of the engine mount is shown in Fig. 10.3(a). The boundary curves Γ_1 though Γ_4 are selected as the shape design boundaries and the locations of grid points G_1 through G_{19} are selected as shape design parameters. To avoid stress concentration, the design boundaries Γ_1 through Γ_4 are modeled using cubic geometric curves and C^1-continuity is maintained at the joints. In this example, the isoparametric mapping method is used to obtain the design velocity on the boundary, and the boundary displacement method [41] is used to compute the domain design velocity field. In Tables 10.4 through 10.6, grid points G_4, G_5, G_9, and G_{13} are perturbed in the y-direction. These design sensitivity results predict central finite difference results very accurately.

Table 10.3. Displacement sensitivity with respect to material property design variable C_{10}.

Node ID	Function	$\psi(u-\Delta u)$	$\psi(u+\Delta u)$	$\Delta\psi$	ψ'	$\psi'/\Delta\psi\%$
34	$z_1(x)$	.10460E+0	.10452E+0	−.40023E−4	−.40021E−4	100.48
34	$z_2(x)$	.34028E+0	.34015E+0	−.67819E−4	−.67839E−4	100.03
58	$z_1(x)$	.18232E−1	.18230E−1	−.52766E−6	−.48702E−6	92.30
58	$z_2(x)$	.21078E+0	.21009E+0	−.45448E−4	−.45418E−4	99.93
62	$z_1(x)$	.15433E−1	.15429E−1	−.19279E−4	−.19832E−4	102.87
62	$z_2(x)$	.74858E+0	.74813E+0	−.21928E−3	−.21924E−3	99.98
76	$z_1(x)$	−.33971E−1	−.33948E−1	.11479E−4	.11421E−4	99.49
76	$z_2(x)$	.97432E+0	.97374E+0	−.29648E−3	−.29642E−3	99.98
135	$z_1(x)$	−.13172E−1	−.13177E−1	−.21328E−4	−.21438E−4	100.52
135	$z_2(x)$	.11756E+1	.11749E+1	−.35042E−3	−.35032E−3	99.98

Table 10.4. Displacement sensitivity with respect to locations of grid points G_4 and G_5 in y-direction.

Node ID	Function	$\psi(x-\Delta x)$	$\psi(x+\Delta x)$	$\Delta\psi$	ψ'	$\psi'/\Delta\psi\%$
26	$z_1(x)$	−.25217E−1	−.18194E−1	.35111E−2	.34754E−2	98.98
26	$z_2(x)$	.85632E−1	.96006E−1	.51850E−2	.51029E−2	98.42
38	$z_1(x)$	.11592E+0	.10214E+0	−.68903E−2	−.70839E−2	102.81
38	$z_2(x)$	.12597E+1	.12077E+1	−.25970E−1	−.25188E−1	96.99
56	$z_1(x)$	−.56575E−1	−.36721E−1	.99271E−2	.98651E−2	99.38
56	$z_2(x)$	.55896E+0	.50555E+0	−.26707E−1	−.28257E−1	105.80
124	$z_1(x)$	.13566E−5	.17610E−5	.20216E−6	.21723E−6	107.45
124	$z_2(x)$	.13783E+1	.13110E+1	−.33655E−1	−.34142E−1	101.45
184	$z_1(x)$	.97032E−2	.11079E−1	.68800E−3	.66645E−3	96.87
184	$z_2(x)$	.16834E+0	.13368E+0	−.17326E−1	−.16698E−1	96.38

Table 10.5. Stress sensitivity with respect to location of grid point G_{13} in y-direction.

Element ID	Function	$\psi(x-\Delta x)$	$\psi(x+\Delta x)$	$\Delta\psi$	ψ'	$\psi'/\Delta\psi$ %
21	$S_{11}(x)$	.43700E+2	.43936E+2	.11827E+0	.11807E+0	99.83
21	$S_{12}(x)$	–.38917E+1	–.40364E+1	–.72382E–1	–.72372E–1	99.98
21	$S_{22}(x)$	–.55881E+0	–.73272E+0	–.86955E–1	–.86959E–1	100.00
44	$S_{11}(x)$	–.77567E+2	–.77109E+2	.22996E+0	.22929E+0	99.71
44	$S_{12}(x)$	–.66236E+2	–.66667E+2	–.21421E+0	–.21486E+0	100.30
44	$S_{22}(x)$	–.54213E+2	–.54057E+2	.78692E–1	.78480E–1	99.73
102	$S_{11}(x)$	.43879E+2	.43620E+2	–.12936E+0	–.13087E+0	101.66
102	$S_{22}(x)$	.29872E+2	.29534E+2	–.17107E+0	–.16996E+0	99.18
117	$S_{12}(x)$	–.41429E+2	–.41511E+2	–.41027E–1	–.42710E–1	104.10
117	$S_{22}(x)$	–.24404E+2	–.24244E+2	.80613E–1	.75457E–1	93.60

Table 10.6. Hydrostatic pressure sensitivity with respect to location of grid point G_9 in y-direction.

Element ID	Function	$\psi(x-\Delta x)$	$\psi(x+\Delta x)$	$\Delta\psi$	ψ'	$\psi'/\Delta\psi$ %
24	$p(x)$	.22460E+2	.22076E+2	–.19169E+0	–.19241E+0	100.38
39	$p(x)$	.16327E+2	.16545E+2	.10870E+0	.11088E+0	102.00
44	$p(x)$	.30002E+2	.30119E+2	.58995E–1	.57544E–1	97.54
52	$p(x)$	.21579E+2	.22375E+2	.39774E+0	.39818E+0	100.11
56	$p(x)$	.16378E+2	.17579E+2	.60058E+0	.60009E+0	99.92
99	$p(x)$	–.25940E+2	–.26230E+2	–.14447E+0	–.14498E+0	100.35
106	$p(x)$	.40072E+0	.35791E+0	–.21409E–1	–.21423E–1	100.06
110	$p(x)$	.60769E+1	.60863E+1	.47011E–2	.46289E–2	98.46
118	$p(x)$	.39136E+1	.39120E+1	–.82409E–3	–.80016E–3	97.10
139	$p(x)$	–.20402E+2	–.20556E+2	–.77555E–1	–.77534E–1	99.98

Bushing

The bushing shown in Fig. 10.4 is used as the second numerical example for design sensitivity of the hyperelastic material. Only half of the bushing is modeled due to symmetry. The material constants are $C_{10} = 80$ N/cm^2 and $C_{01} = 100$ N/cm^2. The outer surface of the bushing is formed by revolving boundaries Γ_1 through Γ_4. The design boundaries Γ_1 through Γ_4 are modeled using C^1-continuous lines and are parameterized by cubic geometric curves. The inner surface of the bushing is formed by revolving boundaries Γ_5 through Γ_8. The radii of the outer surface at grid points G_1 through G_5 are 3.0 cm and the radius of the inner surface is 1.0 cm. The length of the bushing is 5.0 cm.

To simulate the assembly process in which the bushing is squeezed into a cylindrical metal pipe, a prescribed displacement constraint is first imposed on the outer surface of the bushing, with the outer radius pressed to 2.7 cm (i.e., prescribed displacement). In the second step, two forces with magnitudes of 1000 N are applied at node 1 in axial and radial directions, as shown in Fig. 10.4(b). The bushing is modeled using 1024 ABAQUS 8-node solid elements C3D8RH with a total of 1445 nodes and 4164 degrees-of-freedom.

The boundary curves Γ_1 through Γ_4 are chosen as shape design boundaries, and the locations of grid points G_1 through G_5 in the radial direction are selected as shape design parameters. To make the design symmetric, three independent design parameters G_1 through G_3 are selected. The grid point G_4 is linked to G_2, and grid point G_5 is linked to

G_1. Thus, the curve Γ_3 is linked to Γ_2, and Γ_4 is linked to Γ_1. Curves Γ_1 and Γ_2 are chosen as independent shape design boundaries. For this example, the isoparametric mapping method is used to obtain the design velocity field.

The displacement and hydrostatic pressure are selected as performance measures. In order to verify the design sensitivity results, three shape design parameters are perturbed, as shown in Tables 10.7 through 10.9. The results in Tables 10.7 and 10.8 are for loading case 1 (prescribed displacement at the outer surface), while the results in Table 10.9 are for loading case 2 (loading case 1 plus 1000 N at node 1 in axial and radial directions). In Table 10.7, the location of grid point G_1 is perturbed in the radial direction and other design parameters are fixed. In Table 10.8, the grid point G_3 is perturbed in the radial direction and other parameters are fixed. In Table 10.9, the grid point G_2 is perturbed in the radial direction and the other design parameters are fixed. Accurate design sensitivity predictions are obtained in all tables using the sensitivity calculation method.

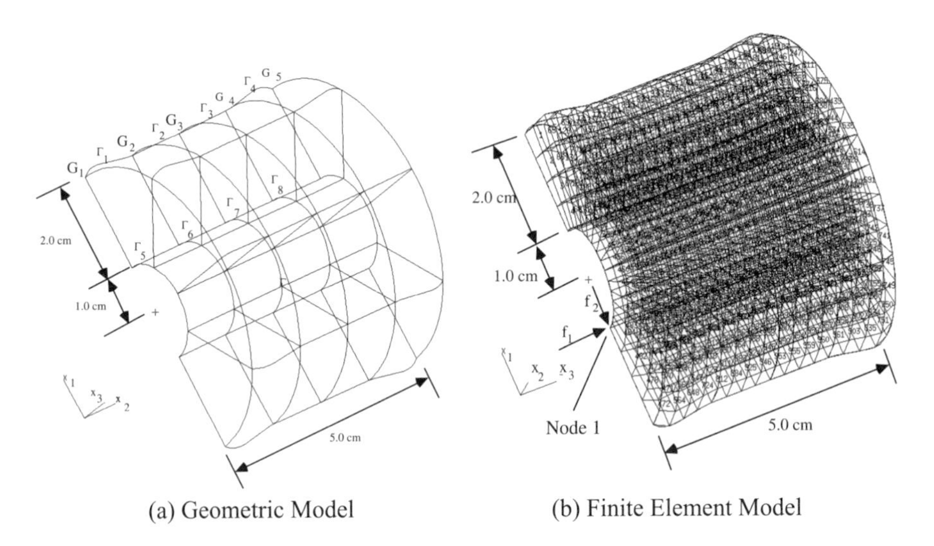

(a) Geometric Model (b) Finite Element Model

Figure 10.4. Three-dimensional bushing model.

Table 10.7. Displacement sensitivity with respect to location of grid point G_1 in radial direction (loading case 1).

Node ID	Function	$\psi(x-\Delta x)$	$\psi(x+\Delta x)$	$\Delta\psi$	ψ'	$\psi'/\Delta\psi$%
18	$z_1(x)$	.14687E+0	.14491E+0	–.97871E–3	–.98127E–3	100.26
97	$z_2(x)$	.41321E–2	.43055E–2	.86724E–4	.81724E–4	94.23
159	$z_1(x)$	.20836E+0	.20321E+0	–.25747E–2	–.26247E–2	101.94
330	$z_2(x)$	–.24633E+0	–.24106E+0	.26368E–2	.25868E–2	98.10
472	$z_3(x)$	.37932E+0	.37386E+0	–.27281E–2	–.26775E–2	98.14
518	$z_1(x)$	–.76902E–1	–.75748E–1	.57714E–3	.60214E–3	104.33
653	$z_2(x)$	–.20124E–1	–.25025E–1	–.24506E–2	–.23505E–2	95.92
857	$z_1(x)$	.11932E–1	.12405E–1	.23649E–3	.23149E–3	97.89
964	$z_2(x)$	–.25085E+0	–.24709E+0	.18807E–2	.19309E–2	102.66
1021	$z_3(x)$	.29931E+0	.29680E+0	–.12549E–2	–.13049E–2	103.98

Table 10.8. Hydrostatic pressure sensitivity with respect to location of grid point G_3 in radial direction (loading case 1).

Element ID	Function	$\psi(x-\Delta x)$	$\psi(x+\Delta x)$	$\Delta\psi$	ψ'	$\psi'/\Delta\psi\%$
58	$p(x)$	.23076E+3	.23021E+3	–.27514E+0	–.25117E+0	91.29
134	$p(x)$	.16589E+3	.16577E+3	–.59841E–1	–.60075E–1	100.39
275	$p(x)$	.59266E+3	.59285E+3	.94326E–1	.96002E–1	101.78
462	$p(x)$	–.18890E+3	–.18877E+3	.65184E–1	.63129E–1	96.85
521	$p(x)$	.70882E+3	.71367E+3	.24251E+1	.24199E+1	99.79
687	$p(x)$	.59102E+3	.59291E+3	.94506E+0	.98253E+0	103.96
743	$p(x)$	–.11460E+3	–.11575E+3	–.57524E+0	–.56658E+0	98.49
816	$p(x)$	.56446E+3	.56389E+3	–.28501E+0	–.28269E+0	99.19
983	$p(x)$	.60772E+3	.60765E+3	–.36248E–1	–.36956E–1	101.95
1021	$p(x)$	.61936E+3	.62090E+3	.77152E+0	.77160E+0	100.01

Table 10.9. Hydrostatic pressure sensitivity with respect to location of grid point G_2 in radial direction (loading case 2).

Element ID	Function	$\psi(x-\Delta x)$	$\psi(x+\Delta x)$	$\Delta\psi$	ψ'	$\psi'/\Delta\psi\%$
47	$p(x)$	.62921E+3	.62237E+3	–.34227E+1	–.34726E+1	101.45
126	$p(x)$	.14731E+3	.14602E+3	–.64883E+0	–.66884E+0	103.08
257	$p(x)$	.25136E+3	.24887E+3	–.12422E+1	–.12372E+1	99.60
381	$p(x)$	.59043E+2	.58534E+2	–.25448E+0	–.25149E+0	98.82
439	$p(x)$	.25182E+3	.24963E+3	–.10952E+1	–.11453E+1	104.57
562	$p(x)$	.30912E+3	.31199E+3	.14334E+1	.14635E+1	102.10
636	$p(x)$	.32628E+3	.32410E+3	–.10918E+1	–.10721E+1	98.20
784	$p(x)$	.75927E+3	.74586E+3	–.67098E+1	–.66999E+1	99.85
938	$p(x)$	.64487E+3	.63535E+3	–.47605E+1	–.47655E+1	100.11
1012	$p(x)$	.58867E+3	.58122E+3	–.37265E+1	–.37216E+1	99.84

10.2 Elastoplastic Problems

In this section, a shape design sensitivity analysis is presented for an infinitesimal elastoplasticity problem. Rate-independent plasticity is considered using a return-mapping algorithm and a von Mises yield criterion. The direct differentiation method is used to compute displacement sensitivity, and the sensitivities of various performance measures can be computed from the displacement sensitivity. Path dependency of the sensitivity equation due to constitutive relations is discussed. The sensitivity of the stress and the internal plastic variables are updated to construct a linear sensitivity equation at the next time step. It is shown that no iteration is required to solve the sensitivity equation. Difficulties in formulating the sensitivity for finite deformation are discussed.

As discussed in Section 8.2, in a nonlinear response analysis, the projection of elastic trial stress, known as return mapping, is carried out to satisfy the variational inequality through iteration in the stress space [24]. Design sensitivity analysis, on the other hand, computes the rate of change of the projected response in the tangential direction of the constraint set without iteration. It is important to note that sensitivity analysis is linear and can be computed without iteration, even though response analysis is nonlinear [34].

Unlike the nonlinear elastic problem, the sensitivity equation of the plastic problem requires stress sensitivity and internal evolution variables at the previous time step. Thus, the direct differentiation method is used to compute the displacement sensitivity. The sensitivity equation is solved at each time step without iteration, and the sensitivity of the stress and evolution variables are stored for the sensitivity equation at the next time step. For this reason, the sensitivity equation computes the material derivative at each time step, and the total displacement sensitivity can be obtained by adding together all material derivatives of these incremental displacements. Sensitivity computation is divided into two parts: first, computing the sensitivity of the incremental displacement, and second, updating the sensitivity of the stress and evolution variables using the incremental displacement sensitivity.

The Newton-Raphson iteration method is frequently used for response analysis of a nonlinear variational equation using a tangent operator. If the tangent operator is exact, then this method guarantees quadratic convergence when the initial estimate is close to the solution [19]. Even when the tangent operator is not exact, response analysis may be converged by running a large number of iterations. However, in design sensitivity analysis the inexact tangent operator produces an error in the sensitivity results. Vidal and Haber [34] discuss the accuracy of sensitivity coefficients using the consistent tangent operator and the rate form tangent operator.

10.2.1 Small Deformation

The material derivative formula presented in Section 6.1 is modified to account for incremental sensitivity relations. Because this approach is concerned with infinitesimal deformation, it is not necessary to distinguish between the initial and deformed configuration. If the response analyses up to time t_{n+1} and the sensitivity analyses up to time t_n are completed, then the sensitivity equation at time t_{n+1} can be solved to obtain the incremental displacement sensitivity. The incremental displacement sensitivity is defined as

$$\frac{d}{d\tau}(\Delta z_\tau)\bigg|_{\tau=0} \equiv \Delta\dot{z} = \lim_{\tau\to 0}\frac{1}{\tau}\big[\Delta z_\tau(\boldsymbol{X}+\tau\boldsymbol{V}(\boldsymbol{X})) - \Delta z(\boldsymbol{X})\big], \tag{10.69}$$

and the material derivative of total displacement at time t_{n+1} is obtained by

$$^{n+1}\dot{z} = {}^{n}\dot{z} + \Delta\dot{z}. \tag{10.70}$$

The decomposition of $^{n+1}\dot{z}$ is necessary, and will be discussed in the incremental constitutive relation as follows.

The constitutive relation of elastoplastic material is given by the following incremental form:

$$^{n+1}\boldsymbol{\sigma} = {}^{n}\boldsymbol{\sigma} + \boldsymbol{C} : (\Delta\varepsilon - \Delta\varepsilon^{p}). \tag{10.71}$$

The material derivative of the stress tensor includes that of incremental strain. Due to the small deformation assumption, using (10.7), it can be shown that

$$\begin{aligned}(\Delta\boldsymbol{\varepsilon})^{\cdot} &= \tfrac{1}{2}\big[\nabla_0(\Delta\dot{z}) + \nabla_0(\Delta\dot{z})^T\big] - \tfrac{1}{2}\big[\nabla_0(\Delta z)\nabla_0\boldsymbol{V} + \nabla_0\boldsymbol{V}^T\nabla_0(\Delta z)^T\big] \\ &\equiv \boldsymbol{\varepsilon}(\Delta\dot{z}) + \boldsymbol{\varepsilon}_V(\Delta z)\end{aligned} \tag{10.72}$$

and that

$$\boldsymbol{\varepsilon}(\bar{z})^{\cdot} = -\tfrac{1}{2}(\nabla_0\bar{z}\nabla_0\boldsymbol{V} + \nabla_0\boldsymbol{V}^T\nabla_0\bar{z}^T) = \boldsymbol{\varepsilon}_V(\bar{z}). \tag{10.73}$$

The same notations for ε and ε_V are used from (10.30), since no distinction need be made between the total and updated Lagrangian formulation. In (10.73), the terms that contain $\dot{\bar{z}}$ are ignored, as they were in (10.8). It is crucial that $\Delta\dot{z}$ is substituted into Δz so that the material derivative of the incremental strain has the same structure as the total strain, because the kinematic relation is linear even if the constitutive relation is nonlinear. Such is not the case for a nonlinear kinematic relation, where the sensitivity of the incremental strain has a different structure from the total strain.

The sensitivity of the stress tensor can be obtained by consistently taking the derivative of the return-mapping algorithm along with the response analysis. The sensitivities of the internal evolution variables are computed in the same way. First, when the material is in an elastic state, the Cauchy stress is increased elastically, and the internal variables remain constant throughout the deformation process, as

$$\begin{aligned} {}^{n+1}\dot{\boldsymbol{\sigma}} &= {}^{n}\dot{\boldsymbol{\sigma}} + \boldsymbol{C} : (\Delta\boldsymbol{\varepsilon})^{\boldsymbol{\cdot}} \\ &= {}^{n}\dot{\boldsymbol{\sigma}} + \boldsymbol{C} : \boldsymbol{\varepsilon}(\Delta\dot{\boldsymbol{z}}) + \boldsymbol{C} : \boldsymbol{\varepsilon}_V(\Delta\boldsymbol{z}) \end{aligned} \tag{10.74}$$

$${}^{n+1}\dot{\boldsymbol{\alpha}} = {}^{n}\dot{\boldsymbol{\alpha}} \tag{10.75}$$

$${}^{n+1}\dot{e}_p = {}^{n}\dot{e}_p. \tag{10.76}$$

However, when material is in a plastic state, the material derivative of the stress and the internal variables follow the return-mapping algorithm. The material derivative formulas for the elastic trial status are

$${}^{tr}\dot{\boldsymbol{s}} = {}^{n}\dot{\boldsymbol{s}} + 2\mu(\Delta\boldsymbol{e})^{\boldsymbol{\cdot}} \tag{10.77}$$

$${}^{tr}\dot{\boldsymbol{\alpha}} = {}^{n}\dot{\boldsymbol{\alpha}} \tag{10.78}$$

$${}^{tr}\dot{e}_p = {}^{n}\dot{e}_p \tag{10.79}$$

$$\begin{aligned} {}^{tr}\dot{\boldsymbol{\eta}} &= {}^{tr}\dot{\boldsymbol{s}} - {}^{tr}\dot{\boldsymbol{\alpha}} \\ &= {}^{n}\dot{\boldsymbol{\eta}} + 2\mu(\Delta\boldsymbol{e})^{\boldsymbol{\cdot}}. \end{aligned} \tag{10.80}$$

If the von Mises yield criterion is used with an associative plasticity assumption, the return-mapping direction is radial, and the normal direction of trial stress is the same as that of final stress. The material derivative of the normal tensor in (8.92) becomes

$$\dot{\boldsymbol{N}} = \frac{2\mu}{\|{}^{tr}\boldsymbol{\eta}\|}\left[\boldsymbol{I} - \boldsymbol{N}\otimes\boldsymbol{N}\right] : (\Delta\boldsymbol{e})^{\boldsymbol{\cdot}} + \frac{1}{\|{}^{tr}\boldsymbol{\eta}\|}\left[\boldsymbol{I} - \boldsymbol{N}\otimes\boldsymbol{N}\right] : {}^{n}\dot{\boldsymbol{\eta}}. \tag{10.81}$$

The radial return-mapping algorithm computes plastic consistency parameter γ through the plastic consistency condition. The material derivative of the plastic consistency condition in (8.93) is

$$\dot{f} = \|{}^{tr}\boldsymbol{\eta}\|^{\boldsymbol{\cdot}} - \left[(2\mu + H_\alpha)\hat{\gamma} + \sqrt{\tfrac{2}{3}}\kappa({}^{n+1}e_p)\right]^{\boldsymbol{\cdot}} = 0. \tag{10.82}$$

By solving this equation in terms of $\dot{\hat{\gamma}}$, it can be shown that

$$\dot{\hat{\gamma}} = 2\mu A\boldsymbol{N} : (\Delta\boldsymbol{e})^{\boldsymbol{\cdot}} + A\boldsymbol{N} : {}^{n}\dot{\boldsymbol{\eta}} - A(H_{\alpha,e_p}\hat{\gamma} + \sqrt{\tfrac{2}{3}}\kappa_{,e_p})\,{}^{n}\dot{e}_p. \tag{10.83}$$

Note that there is no iteration to compute $\dot{\hat{\gamma}}$ in (10.83), whereas response analysis is carried out iteratively to compute $\hat{\gamma}$ using the local Newton-Raphson method. By taking a derivative of the stress-updating algorithm in (8.95) and (8.96), the material derivative of Cauchy stress can be obtained as

$$\begin{aligned} {}^{n+1}\dot{\boldsymbol{\sigma}} &= {}^{n}\dot{\boldsymbol{\sigma}} + \boldsymbol{C} : (\Delta\boldsymbol{\varepsilon})^{\cdot} - 2\mu \boldsymbol{N}\dot{\hat{\gamma}} - 2\mu\hat{\gamma}\dot{\boldsymbol{N}} \\ &= \boldsymbol{C}^{alg} : \boldsymbol{\varepsilon}(\Delta\dot{\boldsymbol{z}}) + \boldsymbol{C}^{alg} : \boldsymbol{\varepsilon}_V(\Delta\boldsymbol{z}) + {}^{n+1}\boldsymbol{\sigma}^{fic}, \end{aligned} \tag{10.84}$$

where

$$\begin{aligned} {}^{n+1}\boldsymbol{\sigma}^{fic} &= {}^{n}\dot{\boldsymbol{\sigma}} - 2\mu A\boldsymbol{N}\left[\boldsymbol{N} : {}^{n}\dot{\boldsymbol{\eta}} - (H_{\alpha,e_p}\hat{\gamma} + \sqrt{\tfrac{2}{3}}\kappa_{,e_p})\,{}^{n}\dot{e}_p\right] \\ &\quad - \frac{2\mu\hat{\gamma}}{\left\|{}^{tr}\boldsymbol{\eta}\right\|}(\boldsymbol{I} - \boldsymbol{N}\otimes\boldsymbol{N})\,{}^{n}\dot{\boldsymbol{\eta}} \end{aligned} \tag{10.85}$$

can be computed from the information at time t_n and from the trial status. $\boldsymbol{C}^{alg}:\boldsymbol{\varepsilon}_V(\Delta\boldsymbol{z})$ is computed using the given design velocity field $\boldsymbol{V}$ and response $\Delta\boldsymbol{z}$.

The governing variational equation of plasticity for the perturbed design is

$$a_{\Omega_\tau}({}^{n+1}\boldsymbol{z}_\tau, \bar{\boldsymbol{z}}_\tau) = \ell_{\Omega_\tau}(\bar{\boldsymbol{z}}_\tau), \quad \forall \bar{\boldsymbol{z}}_\tau \in Z_\tau. \tag{10.86}$$

By using the stress sensitivity from (10.84), the derivative of the structural energy form in (10.86) can be obtained as

$$[a_\Omega({}^{n+1}\boldsymbol{z}, \bar{\boldsymbol{z}})]' = a^{*}_\Omega({}^{n+1}\boldsymbol{z}; \Delta\dot{\boldsymbol{z}}, \bar{\boldsymbol{z}}) + a'_V({}^{n+1}\boldsymbol{z}, \bar{\boldsymbol{z}}), \tag{10.87}$$

where

$$a'_V({}^{n+1}\boldsymbol{z}, \bar{\boldsymbol{z}}) = \int_\Omega \left[\boldsymbol{\varepsilon}_V(\bar{\boldsymbol{z}}) : {}^{n+1}\boldsymbol{\sigma} + \boldsymbol{\varepsilon}(\bar{\boldsymbol{z}}) : \boldsymbol{C}^{alg} : \boldsymbol{\varepsilon}_V(\Delta\boldsymbol{z}) + \boldsymbol{\varepsilon}(\bar{\boldsymbol{z}}) : {}^{n+1}\boldsymbol{\sigma}^{fic} + \boldsymbol{\varepsilon}(\bar{\boldsymbol{z}}) : {}^{n+1}\boldsymbol{\sigma}(div\boldsymbol{V})\right] d\Omega \tag{10.88}$$

is the structural fictitious load form, and can be obtained using the design velocity field $\boldsymbol{V}(\boldsymbol{x})$ and response ${}^{n+1}\boldsymbol{z}$.

Shape Design Sensitivity Equation and Updating Sensitivity Formula

The shape design sensitivity equation at time t_{n+1} can be obtained by taking the derivative of perturbed variational equation (10.86). When the applied load is independent of the deformation, the sensitivity equation is

$$a^{*}_\Omega({}^{n+1}\boldsymbol{z}; \Delta\dot{\boldsymbol{z}}, \bar{\boldsymbol{z}}) = \ell'_V(\bar{\boldsymbol{z}}) - a'_V({}^{n+1}\boldsymbol{z}, \bar{\boldsymbol{z}}). \tag{10.89}$$

By substituting $\Delta\boldsymbol{z}$ into $\Delta\dot{\boldsymbol{z}}$, the left side of (10.89) is the same as that of (8.101). Thus, the design sensitivity equation is solved with the same tangent stiffness matrix as the response analysis at the converged load step. Equation (10.89) is solved for each design parameter because the right-hand side (the fictitious load) changes accordingly. Thus, factorization of the stiffness matrix is important for numerical efficiency. Note that the solution to (10.89) is the incremental displacement sensitivity $\Delta\dot{\boldsymbol{z}}$. The total displacement sensitivity is computed by

$${}^{n+1}\dot{\boldsymbol{z}} = {}^{n}\dot{\boldsymbol{z}} + \Delta\dot{\boldsymbol{z}}. \tag{10.90}$$

After computing the incremental displacement sensitivity by solving (10.89), the Cauchy stress and internal variables are updated. The updating formula for the former is the same as (10.84), while the formula for the latter is given by

$$ {}^{n+1}\dot{\boldsymbol{\alpha}} = {}^{n}\dot{\boldsymbol{\alpha}} + \boldsymbol{N}\left(H_{\alpha} + \sqrt{\tfrac{2}{3}}H_{\alpha,e_p}\hat{\gamma}\right)\dot{\hat{\gamma}} + H_{\alpha}\hat{\gamma}\dot{\boldsymbol{N}} \tag{10.91} $$

$$ {}^{n+1}\dot{e}_p = {}^{n}\dot{e}_p + \sqrt{\tfrac{2}{3}}\dot{\hat{\gamma}}. \tag{10.92} $$

Note that the cost of sensitivity computation for the elastoplastic material is relatively expensive compared with the elastic material, because the material derivatives of stress and internal variables are updated at each integration point.

Difficulties in Design Sensitivity Analysis of a Finite Deformation Problem

For the geometric nonlinear problem, it is convenient to choose the current configuration as the frame of reference because plastic evolution appears in the current configuration. The constitutive relation and the return-mapping algorithm are given with respect to the current configuration. However, since the design is perturbed at the undeformed structure, from a design sensitivity analysis viewpoint, the material derivative must be taken at the undeformed configuration with a given design velocity field. It is necessary to transform the spatial strain tensor into the undeformed configuration using the deformation gradient. If classical plasticity theory is used for analysis, the material derivative of the incremental strain must be taken during the design sensitivity procedure as

$$ (\Delta\varepsilon)^{\boldsymbol{\cdot}} = \frac{1}{2}\left(\nabla_0(\Delta z)\boldsymbol{F}^{-1} + \boldsymbol{F}^{-T}\nabla_0(\Delta z)^T\right)^{\boldsymbol{\cdot}}. \tag{10.93} $$

Since the inverse of the deformation gradient exists in (10.93), additional terms that contain the total displacement sensitivity also exist, and yield a different stiffness matrix for analysis.

As previously discussed, however, if infinitesimal deformation is assumed, then the reference configuration is always undeformed and there is no need to update it. Thus, most of the research cited at the end of this chapter succeeds based on the infinitesimal deformation assumption. Kleiber [43] tried to resolve this problem by establishing a reference configuration using the previously converged time without numerical examples. However, the constitutive relation must be expressed in terms of the second Piola-Kirchhoff stress, which is inconvenient for the plasticity model. Since all configurations at the previous time are known, the sensitivity formulation is similar to that of the total Lagrangian formulation. Zhang et al. [44] proposed a large deformation sensitivity problem using the boundary element method. Displacement sensitivity was computed using iteration, which may significantly increase the computation costs.

Dutta [45] discussed a similar approach in which the sensitivity equation has a different stiffness matrix than the analysis stiffness matrix. He proposed an iterative method to compute sensitivity using the same stiffness matrix as for analysis. All of these difficulties stem from the existence of either a deformation gradient or its inverse in the strain measure. However, theoretically speaking, if the stiffness matrix of analysis is the exact tangent operator, then the sensitivity equation must use the same tangent operator. Either erroneous data or a slow convergence will result in sensitivity or response analyses if the nonexact tangent operator is used.

Consider the following incremental elastic constitutive relation:

$$ {}^{n+1}\boldsymbol{S} = {}^{n}\boldsymbol{S} + \boldsymbol{C}:\Delta\boldsymbol{E}, \tag{10.94} $$

where C_{ijkl} could be constant, or it could be a function of deformation. Is the sensitivity of this material model path-dependent? Obviously, the analysis itself is path-independent, which means that the same result will be obtained even if a different path is chosen. However, if the material derivative of (10.94) is taken, then the sensitivity equation becomes path-dependent because it needs the sensitivity information from the previous time. The sensitivity equation cannot be solved at the final converged time alone. If the constitutive relation in (10.94) is differentiated for the purposes of design sensitivity analysis, then the tangent operator will be different from the stiffness matrix of the response analysis, for the reason explained in (10.93). In the following section, a new approach is provided for the design sensitivity analysis of a finite deformation problem with elastoplastic material.

10.2.2 Finite Deformation

In this section, a shape design formulation for finite deformation elastoplasticity is presented. When the material is in the plastic range, the intermediate configuration contributes to the sensitivity formulation. The frame of reference for the response analysis, which is the intermediate configuration, is different from the frame of reference for design perturbation, which is an undeformed configuration. Figure 10.5 shows a procedure for response analysis and design perturbation. The transformation between the undeformed and intermediate configuration is ignored in response analysis linearization because the configuration is fixed in the elastic trial process. Since the intermediate configuration is changed as the design is perturbed in the undeformed configuration, this transformation is not fixed from the sensitivity viewpoint. Path-dependency of the sensitivity equation, as well as dependency of the plastic evolution variables, result from this transformation.

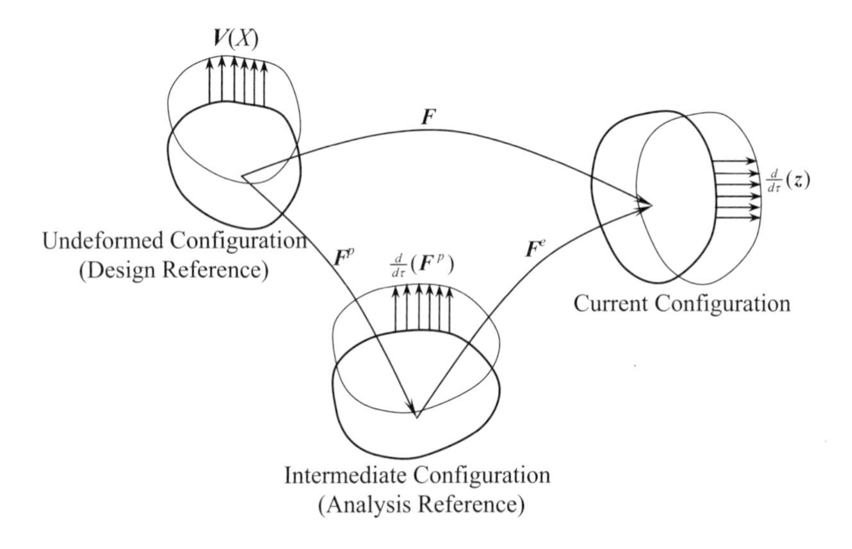

Figure 10.5. Illustration of analysis and design perturbation procedure.

Since the response analysis is formulated with respect to the intermediate configuration, the design derivative of the elastic trial Lagrangian strain tensor in terms of the intermediate configuration $\boldsymbol{E}^e = \frac{1}{2}[\boldsymbol{F}^{eT}\boldsymbol{F}^e - \boldsymbol{1}]$ can be expressed as

$$\dot{\boldsymbol{E}}^e = \tfrac{1}{2}(\dot{\boldsymbol{F}}^{e^T}\boldsymbol{F}^e + \boldsymbol{F}^{e^T}\dot{\boldsymbol{F}}^e), \tag{10.95}$$

and the transformation, like (10.30), into the current configuration leads to

$$\boldsymbol{F}^{e^{-T}}\dot{\boldsymbol{E}}^e\boldsymbol{F}^{e^{-1}} = \tfrac{1}{2}(\boldsymbol{F}^{e^{-T}}\dot{\boldsymbol{F}}^{e^T} + \dot{\boldsymbol{F}}^e\boldsymbol{F}^{e^{-1}}). \tag{10.96}$$

Note that the push-forward transformation is between the intermediate and current configuration. Since $\dot{\boldsymbol{F}}^e$ refers to the intermediate configuration, it should be expressed in terms of the undeformed configuration, where the design velocity is explicitly given by taking the material derivative of multiplicative decomposition in (8.119). After using the property that $\dot{\boldsymbol{F}}^e = \dot{\boldsymbol{F}}\boldsymbol{F}^{p^{-1}} - \boldsymbol{F}^e\dot{\boldsymbol{F}}^p\boldsymbol{F}^{p^{-1}}$ and defining a path-dependent tensor $\boldsymbol{G}$ by

$$\boldsymbol{G} = \boldsymbol{F}^e\dot{\boldsymbol{F}}^p\boldsymbol{F}^{-1}, \tag{10.97}$$

which is the transformation of $\dot{\boldsymbol{F}}^p$ into the current configuration, (10.96) can be rearranged as

$$\begin{aligned}\boldsymbol{F}^{e^{-T}}\dot{\boldsymbol{E}}^e\boldsymbol{F}^{e^{-1}} &= \frac{1}{2}(\boldsymbol{F}^{-T}\dot{\boldsymbol{F}}^T + \dot{\boldsymbol{F}}\boldsymbol{F}^{-1}) - \frac{1}{2}(\boldsymbol{G} + \boldsymbol{G}^T) \\ &\equiv \boldsymbol{\varepsilon}(\dot{\boldsymbol{z}}) + \boldsymbol{\varepsilon}_V(\boldsymbol{z}) + \boldsymbol{\varepsilon}_p(\boldsymbol{z}),\end{aligned} \tag{10.98}$$

where $\boldsymbol{\varepsilon}(\dot{\boldsymbol{z}})$ and $\boldsymbol{\varepsilon}_V(\boldsymbol{z})$ are in the same form as in the finite elasticity sensitivity formulation in (10.30), and $\boldsymbol{\varepsilon}_P(\boldsymbol{z})$ is the contribution from the elastic trial intermediate configuration, where path-dependency comes from

$$\boldsymbol{\varepsilon}_P(\boldsymbol{z}) = -\tfrac{1}{2}(\boldsymbol{G} + \boldsymbol{G}^T). \tag{10.99}$$

Equation (10.98) is equivalent to pull-back $\boldsymbol{E}^e$ in the undeformed configuration and to push-forward $\boldsymbol{E}^e$ in the current configuration, after the material derivative is taken. Trial elastic deformation gradient $\boldsymbol{F}^e$ must be extracted from the response analysis, and the design derivative of $\boldsymbol{F}^p$ must be stored from the previous sensitivity procedure. It is interesting to note that path-dependency in the rate form plasticity comes from the derivative of the stress tensor at the previous time step, while path-dependency in the multiplicative plasticity comes from the transformation between the intermediate and current configurations.

The same procedure must be applied to the variation of the Lagrangian strain tensor $\overline{\boldsymbol{E}}^e = \frac{1}{2}(\overline{\boldsymbol{F}}^{e^T}\boldsymbol{F}^e + \boldsymbol{F}^{e^T}\overline{\boldsymbol{F}}^e)$, with respect to the intermediate configuration, as

$$\boldsymbol{F}^{e^{-T}}\dot{\overline{\boldsymbol{E}}}^e\boldsymbol{F}^{e^{-1}} \equiv \boldsymbol{\eta}(\dot{\boldsymbol{z}}, \overline{\boldsymbol{z}}) + \boldsymbol{\eta}_V(\boldsymbol{z}, \overline{\boldsymbol{z}}) + \boldsymbol{\eta}_P(\boldsymbol{z}, \overline{\boldsymbol{z}}), \tag{10.100}$$

where, again using a procedure similar to that used for $\boldsymbol{\varepsilon}_P(\boldsymbol{z})$,

$$\boldsymbol{\eta}_P(\boldsymbol{z}, \overline{\boldsymbol{z}}) = -\tfrac{1}{2}(\nabla_n\overline{\boldsymbol{z}}^T\boldsymbol{G} + \boldsymbol{G}^T\nabla_n\overline{\boldsymbol{z}}) \tag{10.101}$$

is the contribution of the elastic trial intermediate configuration through the geometric strain terms.

To obtain the variation of the structural energy form, the following design derivative of Kirchhoff stress in (8.146) is required:

$$\dot{\boldsymbol{\tau}} = \sum_{i=1}^{3} (\dot{\tau}_i^P \boldsymbol{m}^i + \tau_i^P \dot{\boldsymbol{m}}^i). \tag{10.102}$$

The design derivative of principal stress is a function of principal logarithmic strain. The following relation can be derived by the chain rule and push-forward operation:

$$\begin{aligned} \dot{\tau}_i^P &= \sum_{j=1}^{3} \frac{\partial \tau_i^P}{\partial e_j^{tr}} \left(2\boldsymbol{F}^e \frac{\partial e_j^{tr}}{\partial \boldsymbol{C}^e} \boldsymbol{F}^{e^T} \right) : \left(\boldsymbol{F}^{e^{-T}} \dot{\boldsymbol{E}}^e \boldsymbol{F}^{e^{-1}} \right) + \frac{\partial \tau_i^P}{\partial e_p} {}^n\dot{e}_p + \frac{\partial \tau_i^P}{\partial \boldsymbol{\alpha}} {}^n\dot{\boldsymbol{\alpha}} \\ &= \sum_{j=1}^{3} c_{ij}^{alg} \boldsymbol{m}^j : [\varepsilon(\dot{\boldsymbol{z}}) + \varepsilon_V(\boldsymbol{z}) + \varepsilon_P(\boldsymbol{z})] + \frac{\partial \tau_i^P}{\partial e^p} {}^n\dot{e}_p + \frac{\partial \tau_i^P}{\partial \boldsymbol{\alpha}} {}^n\dot{\boldsymbol{\alpha}}, \end{aligned} \tag{10.103}$$

where

$$\frac{\partial \boldsymbol{\tau}^P}{\partial \boldsymbol{\alpha}} = 2\mu A \boldsymbol{N} \otimes \boldsymbol{N} + \frac{2\mu\hat{\gamma}}{\left\| {}^{tr}\boldsymbol{\eta} \right\|} (\boldsymbol{I}_{dev} - \boldsymbol{N} \otimes \boldsymbol{N}) \tag{10.104}$$

and

$$\frac{\partial \boldsymbol{\tau}^P}{\partial e_p} = 2\mu A \kappa_{,e_p} \boldsymbol{N} \tag{10.105}$$

are in the same form as the sensitivity formulation for classical infinitesimal plasticity. The path dependency of the design derivative of Kirchhoff stress comes from $\varepsilon_P(\boldsymbol{z})$, $\boldsymbol{\eta}_P(\boldsymbol{z}, \bar{\boldsymbol{z}})$, ${}^n\dot{\boldsymbol{\alpha}}$, and ${}^n\dot{e}_p$. Since $\boldsymbol{m}^i$ is related to the elastic trial status, it is independent of the plastic evolution and its design derivative can be obtained from the derivative of the elastic trial strain, as

$$\begin{aligned} \dot{\boldsymbol{m}}^i &= \left(2\boldsymbol{F}^e \frac{\partial \boldsymbol{m}^i}{\partial \boldsymbol{C}^e} \boldsymbol{F}^{e^T} \right) : \left(\boldsymbol{F}^{e^{-T}} \dot{\boldsymbol{E}}^e \boldsymbol{F}^{e^{-1}} \right) \\ &= 2\hat{\boldsymbol{c}}^i : [\varepsilon(\dot{\boldsymbol{z}}) + \varepsilon_V(\boldsymbol{z}) + \varepsilon_P(\boldsymbol{z})]. \end{aligned} \tag{10.106}$$

Thus, the design derivative of the Kirchhoff stress tensor can be expressed in terms of $\dot{\boldsymbol{z}}$, the configuration of the response analysis, and the sensitivity results of the previous time step, as

$$\begin{aligned} \dot{\boldsymbol{\tau}} &= \sum_{i=1}^{3} (\dot{\tau}_i^P \boldsymbol{m}^i + \tau_i^P \dot{\boldsymbol{m}}^i) \\ &= \sum_{i=1}^{3} \sum_{j=1}^{3} \left(c_{ij}^{alg} \boldsymbol{m}^i \otimes \boldsymbol{m}^j + 2\tau_i^P \hat{\boldsymbol{c}}^i \right) : \left(\varepsilon(\dot{\boldsymbol{z}}) + \varepsilon_V(\boldsymbol{z}) + \varepsilon_P(\boldsymbol{z}) \right) + \boldsymbol{\tau}^{fic} \\ &= \boldsymbol{c} : \left(\varepsilon(\dot{\boldsymbol{z}}) + \varepsilon_V(\boldsymbol{z}) + \varepsilon_P(\boldsymbol{z}) \right) + \boldsymbol{\tau}^{fic}, \end{aligned} \tag{10.107}$$

where $\boldsymbol{c}$ is the fourth-order consistent tangent stiffness tensor at the current configuration and

$$\boldsymbol{\tau}^{fic} = \sum_{i=1}^{3} \left[\frac{\partial \tau_i^P}{\partial \boldsymbol{\alpha}} {}^n\dot{\boldsymbol{\alpha}} + \frac{\partial \tau_i^P}{\partial e^p} {}^n\dot{e}_p \right] \boldsymbol{m}^i \tag{10.108}$$

is the path-dependent term from the plastic evolution procedure. $\boldsymbol{\tau}^{fic}$ must be included when the material is in the plastic range. It is clear from (10.108) that the sensitivity information ${}^n\dot{\boldsymbol{\alpha}}$ and ${}^n\dot{e}_p$ at the previous time step needs to be stored for displacement sensitivity computation at the current time step.

From the design derivative formula of Kirchhoff stress in (10.102) and from the transformation of the design derivative of the Lagrangian strain into the current configuration in (10.98) and (10.100), the design variation of the structural energy form can be obtained as

$$[a(z,\bar{z})]' = \left[\int_{\Omega} \boldsymbol{\tau} : \bar{\varepsilon}\, d\Omega\right]' \equiv a^{*}(z;\dot{z},\bar{z}) + a_V'(z,\bar{z}), \tag{10.109}$$

where

$$\begin{aligned} a_V'(z,\bar{z}) = & \int_{\Omega} \left(\bar{\varepsilon} : \boldsymbol{c} : \varepsilon_V(z) + \bar{\varepsilon} : \boldsymbol{c} : \varepsilon_P(z) + \boldsymbol{\tau}^{fic} : \bar{\varepsilon}\right) d\Omega \\ & + \int_{\Omega} \left(\boldsymbol{\tau} : \boldsymbol{\eta}_V(z,\bar{z}) + \boldsymbol{\tau} : \boldsymbol{\eta}_P(z,\bar{z}) + \boldsymbol{\tau} : \bar{\varepsilon} \operatorname{div} \boldsymbol{V}\right) d\Omega \end{aligned} \tag{10.110}$$

is the structural fictitious load form for finite plasticity. This form can be computed from the response analysis and sensitivity equation results at the previous time step for a given design velocity field. Using the same procedure as finite elasticity, the sensitivity equation is obtained in terms of the current configuration by

$$a_{\Omega}^{*}(z;\dot{z},\bar{z}) = \ell_V'(\bar{z}) - a_V'(z,\bar{z}). \tag{10.111}$$

The linear system of this equation can be solved at each time step to compute displacement sensitivity $\dot{z}$. Note that the sensitivity equation of classical rate-form plasticity in Section 8.2.1 solves the incremental displacement sensitivity, whereas sensitivity (10.111) solves the total displacement sensitivity, even if (10.111) is solved at each time step.

Updating Evolution Variables

After computing displacement sensitivity $\dot{z}$, the design derivatives of other path-dependent variables can be updated. The design derivative of the logarithmic principal stretch can be obtained from the definition of strain and principal direction, as

$$\begin{aligned} \dot{e}_j^{tr} &= \left(2\boldsymbol{F}^e \frac{\partial e_j^{tr}}{\partial \boldsymbol{C}^e} \boldsymbol{F}^{e^T}\right) : \left(\boldsymbol{F}^{e^{-T}} \dot{\boldsymbol{E}}^e \boldsymbol{F}^{e^{-1}}\right) \\ &= \boldsymbol{m}^j : \left[\varepsilon(\dot{z}) + \varepsilon_V(z) + \varepsilon_P(z)\right]. \end{aligned} \tag{10.112}$$

The design derivative of the unit normal vector to the yield surface in (8.143), and the design derivative of the plastic consistency parameter $\hat{\gamma}$ in (8.144) can be obtained by differentiating these equations as

$$\dot{\boldsymbol{N}} = \frac{1}{\left\|{}^{tr}\boldsymbol{\eta}\right\|}\left[\boldsymbol{I}_{dev} - \boldsymbol{N} \otimes \boldsymbol{N}\right]\left[2\mu\dot{\boldsymbol{e}}^{tr} - {}^{n}\dot{\boldsymbol{\alpha}}\right] \tag{10.113}$$

$$\dot{\hat{\gamma}} = A\boldsymbol{N}^T(2\mu\dot{\boldsymbol{e}}^{tr} - {}^{n}\dot{\boldsymbol{\alpha}}) - A\kappa_{,e_p}\, {}^{n}\dot{e}_p. \tag{10.114}$$

Note that local Newton's method is used to compute the plastic consistency parameter $\hat{\gamma}$ in (8.144), whereas no iteration is required to compute $\dot{\hat{\gamma}}$ in (10.114). By using (10.113) and (10.114), ${}^{n}\dot{\boldsymbol{\alpha}}$ and ${}^{n}\dot{e}_p$ are updated using the same procedure employed in the response analysis, namely,

$$^{n+1}\dot{\boldsymbol{\alpha}} = {}^{n}\dot{\boldsymbol{\alpha}} + \left(H_\alpha + \sqrt{\tfrac{2}{3}}H_{\alpha,e_p}\hat{\gamma}\right)\dot{\hat{\gamma}} + H_\alpha \hat{\gamma}\dot{\boldsymbol{N}} \tag{10.115}$$

$$^{n+1}\dot{e}_p = {}^{n}\dot{e}_p + \sqrt{\tfrac{2}{3}}\dot{\hat{\gamma}}. \tag{10.116}$$

All that remains is to evaluate the design derivative of the deformation gradient at the intermediate configuration, denoted as $\boldsymbol{G}$ in (10.97),

$$\boldsymbol{G} = \boldsymbol{F}^e \dot{\boldsymbol{F}}^p \boldsymbol{F}^{-1}. \tag{10.117}$$

Since the response analysis updates the symmetric left Cauchy-Green deformation tensor $\boldsymbol{b}^e = \boldsymbol{F}^e \boldsymbol{F}^{e^T}$ using (8.147), it is difficult to extract any information about $\boldsymbol{F}^e$ or $\boldsymbol{F}^p$ separately. It is also difficult to express $\boldsymbol{G}$ in (10.117) in terms of $\boldsymbol{b}^e$ and $\dot{\boldsymbol{b}}^e$. Based on the spatial formulation and assumption of isotropic material, specification of the intermediate configuration is not necessary in the response analysis procedure. The intermediate configuration has an ambiguity up to the order of rigid body rotation.

Since $\dot{\boldsymbol{F}}^p$ takes the role of design velocity field at the intermediate configuration, for design sensitivity purposes, it is necessary to specify the intermediate configuration. For example, the intermediate configuration can be defined, without any loss of generality, as the unrotated de-stressing process [21]. In general, the polar decomposition of the elastic deformation gradient can be written as $\boldsymbol{F}^e = \boldsymbol{V}^e\boldsymbol{R}^e$, where $\boldsymbol{V}^e$ is the principal stretch and $\boldsymbol{R}^e$ is the rigid-body rotation. By removing the rotational component from the polar decomposition of $\boldsymbol{F}^e$, the elastic deformation gradient can be redefined as

$$\boldsymbol{F}^e \equiv \boldsymbol{V}^e = \sqrt{\boldsymbol{b}^e}. \tag{10.118}$$

In this approach, the intermediate configuration defined by $\boldsymbol{F}^p$ contains rigid body rotation as well as local destressing. Since plastic evolution occurs on the principal logarithmic stretches with the current fixed configuration, the principal direction of the elastic deformation gradient remains constant, that is, $\boldsymbol{m}^i$ remains constant. Thus, from (8.147), $\boldsymbol{F}^e$ and its material derivative can be obtained as

$$\boldsymbol{F}^e = \sum_{i=1}^{3} \exp\left(e_i^{tr} - \hat{\gamma}N_i\right)\boldsymbol{m}^i \tag{10.119}$$

$$\begin{aligned}\dot{\boldsymbol{F}}^e &= \sum_{i=1}^{3} \exp\left(e_i^{tr} - \hat{\gamma}N_i\right)\left(\dot{e}_i^{tr} - \dot{\hat{\gamma}}N_i - \hat{\gamma}\dot{N}_i\right)\boldsymbol{m}^i \\ &+ \sum_{i=1}^{3} \exp\left(e_i^{tr} - \hat{\gamma}N_i\right)\boldsymbol{c}^i : \left(\boldsymbol{\varepsilon}(\dot{\boldsymbol{z}}) + \boldsymbol{\varepsilon}_V(\dot{\boldsymbol{z}}) + \boldsymbol{\varepsilon}_p(\boldsymbol{z})\right).\end{aligned} \tag{10.120}$$

Finally, the design derivative of the intermediate configuration can be obtained from

$$\dot{\boldsymbol{F}}^p = \boldsymbol{F}^{e^{-1}}\dot{\boldsymbol{F}} - \boldsymbol{F}^{e^{-1}}\dot{\boldsymbol{F}}^e\boldsymbol{F}^p. \tag{10.121}$$

An interesting result can be obtained if a material's elastic status is assumed. If $\boldsymbol{F}^e = \boldsymbol{F}$ and $\boldsymbol{F}^p = \boldsymbol{1}$, then $\dot{\boldsymbol{F}}^p = \boldsymbol{0}$ in (10.121) and the same formulation of the structural fictitious load form is recovered as the finite elasticity form in (10.37). However, from the assumption of $\boldsymbol{F}^e$ in (10.119), $\boldsymbol{F}^e = \boldsymbol{V}^e \neq \boldsymbol{F}$, $\boldsymbol{F}^p = \boldsymbol{R}^e$, and $\dot{\boldsymbol{F}}^p \neq \boldsymbol{0}$. The structural fictitious load form is different from that of finite elasticity. This situation occurs because the intermediate configuration is different from the undeformed one, even when the material is in an elastic state. Thus, from a design sensitivity viewpoint this type of decomposition is inappropriate even if the response analysis provides an equivalent result, as discussed by Simo [46]. One reason for this inconsistency is that the isotropic property of the material is imposed too early, whereas the design perturbation is generally not isotropic.

Miehe et al. [47] note that among rigid body rotations, only plastic spin is undetermined within the associative plastic theory. Furthermore, plastic spin can be represented by $\boldsymbol{W}^p = \frac{1}{2}(\boldsymbol{L}^p + \boldsymbol{L}^{p^T}) = 0$ within the isotropic assumption. To resolve the previous inconsistency, consider the stress return-mapping algorithm in (8.142)b in a strain space. The logarithmic elastic principal stretch vector can be updated, as in (8.14), by

$$\boldsymbol{e} = \boldsymbol{e}^{tr} - \hat{\gamma}\boldsymbol{N}, \qquad \lambda_j^e = \lambda_j^{etr}\exp(-\hat{\gamma}N_j), \tag{10.122}$$

where $\exp(-\hat{\gamma}N_j)$ is the principal value of incremental plastic deformation gradient $\boldsymbol{f}^p$ in the current fixed principal direction. Thus, the incremental plastic deformation gradient is defined as

$$\boldsymbol{f}^p = \sum_{j=1}^{3}\exp(-\hat{\gamma}N_j)\boldsymbol{m}^j, \tag{10.123}$$

and the updated elastic deformation gradient is

$${}^{n+1}\boldsymbol{F}^e = \boldsymbol{f}^p\,{}^{tr}\boldsymbol{F}^e. \tag{10.124}$$

Note that the incremental plastic deformation gradient $\boldsymbol{f}^p$ in (10.123) is a symmetric tensor, which means that incremental plastic spin vanishes. From the relation in (8.119), the plastic deformation gradient is updated by

$${}^{n+1}\boldsymbol{F}^p = {}^{n+1}\boldsymbol{F}^{e-1}\,{}^{n+1}\boldsymbol{F}, \tag{10.125}$$

and its material derivative is updated by

$${}^{n+1}\dot{\boldsymbol{F}}^p = \left({}^{n+1}\boldsymbol{F}^{e^{-1}}\right)^{\boldsymbol{\cdot}}\,{}^{n+1}\boldsymbol{F} + {}^{n+1}\boldsymbol{F}^{e^{-1}}\,{}^{n+1}\dot{\boldsymbol{F}}, \tag{10.126}$$

where ${}^{n+1}\dot{\boldsymbol{F}}$ is given in (10.21) and $\left({}^{n+1}\boldsymbol{F}^{e^{-1}}\right)^{\boldsymbol{\cdot}} = -{}^{n+1}\boldsymbol{F}^{e^{-1}}\,{}^{n+1}\dot{\boldsymbol{F}}^e\,{}^{n+1}\boldsymbol{F}^{e^{-1}}$ with

$${}^{n+1}\dot{\boldsymbol{F}}^e = \dot{\boldsymbol{f}}^p\,{}^{tr}\boldsymbol{F}^e + \boldsymbol{f}^p\,{}^{tr}\dot{\boldsymbol{F}}^e. \tag{10.127}$$

To be consistent with the finite elastic state presented in the previous section, consider the elastic state when $\hat{\gamma} = 0$, such that $\boldsymbol{f}^p = {}^{n+1}\boldsymbol{F}^p = \boldsymbol{1}$ and ${}^{n+1}\dot{\boldsymbol{F}}^p = \boldsymbol{0}$. Thus, the formulation fully recovers finite elasticity. It is necessary that intermediate configuration ${}^{n+1}\boldsymbol{F}^p$ be specified for sensitivity purposes, and that it be updated by removing the incremental plastic spin.

10.2.3 Numerical Examples

Thick-Walled Hollow Cylinder

A thick-walled hollow cylinder that is subject to an internal pressure, as shown in Fig. 10.6, is considered as the first example. Analysis data include the internal radius $r_i = 100$ mm, the external radius $r_o = 200$ mm, Young's Modulus $E = 2.1 \times 10^5$ N/mm^2, Poisson's ratio $\nu = 0.3$, and the strain hardening parameter $H = 2.1 \times 10^4$ N/mm^2. The yield stress is $\sigma_Y = 240.0$ N/mm^2. Plane strain conditions are assumed in the axial direction. For loading condition, the internal pressure is increased from 0 N/mm^2 to 2630.0 N/mm^2. Plastic deformation occurs at pressure 1640 N/mm^2.

The problem is to find the shape design sensitivities of displacements z_1 and z_2 on the interior surface with respect to changes in internal radius r_i. The design sensitivities of displacements z_1 and z_2 are shown in the third and sixth columns of Tables 10.10 through 10.12, respectively, for three different load levels. Note that sensitivity results are updated at each pressure level (load step). The central finite difference result $\Delta z = [z(x + \Delta x) - z(x - \Delta x)]/2$ is shown in the second and fifth columns for a 1% perturbation of r_i. The fourth column is the agreement between Δz_1 and z_1', and the seventh column is the agreement between Δz_2 and z_2'. The results obtained using the presented method agree very well with those obtained using the central finite difference method throughout the load steps.

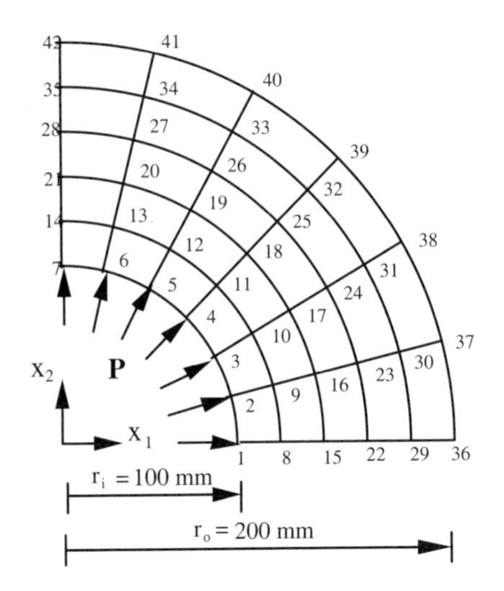

Figure 10.6. Thick-walled hollow cylinder under internal pressure.

Table 10.10. Shape design sensitivity results of displacement of the interior surface at the pressure level 1400 N/mm^2.

Node	Δz_1	z_1'	$z_1'/\Delta z_1$	Δz_2	z_2'	$z_2'/\Delta z_2$
1	8.480E–4	8.480E–4	100.0%	0	0	—
2	8.191E–4	8.190E–4	99.9%	2.195E–4	2.194E–4	99.9%
3	7.344E–4	7.343E–4	99.9%	4.240E–4	4.240E–4	100.0%
4	5.997E–4	5.995E–4	99.9%	5.996E–4	5.995E–4	99.9%
5	4.240E–4	4.240E–4	100.0%	7.344E–4	7.343E–4	99.9%
6	2.195E–4	2.194E–4	99.9%	8.191E–4	8.190E–4	99.9%
7	0	0	—	8.480E–4	8.479E–4	99.9%

Table 10.11. Shape design sensitivity results of displacement of the interior surface at the pressure level 1760 N/mm^2.

Node	Δz_1	z_1'	$z_1'/\Delta z_1$	Δz_2	z_2'	$z_2'/\Delta z_2$
1	1.032E–3	1.030E–3	99.8%	0	0	—
2	9.960E–3	9.950E–3	99.9%	2.670E–4	2.666E–4	99.9%
3	8.930E–3	8.922E–3	99.9%	5.158E–4	5.151E–4	99.9%
4	7.294E–3	7.284E–3	99.9%	7.294E–4	7.284E–4	99.9%
5	5.158E–4	5.151E–4	99.9%	8.930E–4	8.921E–4	99.9%
6	2.670E–4	2.666E–4	99.9%	9.965E–4	9.951E–4	99.9%
7	0	0	—	1.032E–3	1.030E–3	99.9%

Table 10.12. Shape design sensitivity results of displacement of the interior surface at the pressure level 2630 N/mm^2.

Node	Δz_1	z_1'	$z_1'/\Delta z_1$	Δz_2	z_2'	$z_2'/\Delta z_2$
1	4.766E–3	4.756E–3	99.8%	0	0	—
2	4.604E–3	4.594E–3	99.8%	1.234E–3	1.231E–3	99.8%
3	4.128E–3	4.118E–3	99.8%	2.384E–3	2.378E–3	99.7%
4	3.370E–3	3.363E–3	99.8%	3.370E–3	3.363E–3	99.8%
5	2.384E–3	2.378E–3	99.7%	4.128E–3	4.118E–3	99.8%
6	1.234E–3	1.231E–3	99.8%	4.604E–3	4.594E–3	99.8%
7	0	0	—	4.766E–3	4.756E–3	99.8%

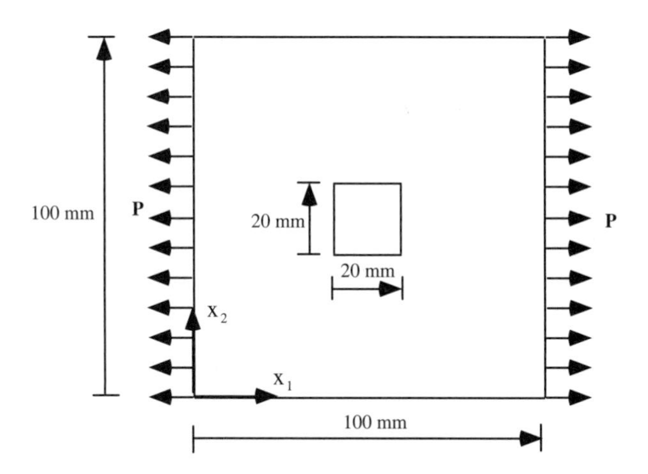

Figure 10.7. Membrane with a hole.

Membrane with a Hole

A membrane with a hole that is subject to a pressure load, as shown in Fig. 10.7, is considered as the second example. The following specifications are used for analysis data: Young's Modulus $E = 2.1 \times 10^5$ N/mm^2, Poisson's ratio $\nu = 0.3$, and strain hardening parameter $H = 2.1 \times 10^4$ N/mm^2. The yield stress is $\sigma_Y = 240.0$ N/mm^2. Plane stress conditions are assumed. The pressure load P is increased from 0 N/mm^2 to 1500.0 N/mm^2. Plastic deformation occurs at pressure level 1220 N/mm^2.

The problem is to find shape sensitivities of the displacement z_1 of nodes 6 through 35 (right edge of Fig. 10.8) with respect to the design change in a. The design sensitivities of the displacement z_1 are shown in the third columns of Tables 10.13 through 10.15 for different load levels. The central finite difference result $\Delta z = [z(x + \Delta x) - z(x - \Delta x)]/2$ is shown in the second column for a 1% perturbation of horizontal dimension a. The fourth column is the agreement between Δz_1 and z_1'. The results obtained with the presented method agree very well with those obtained by using the central finite difference method through each load step.

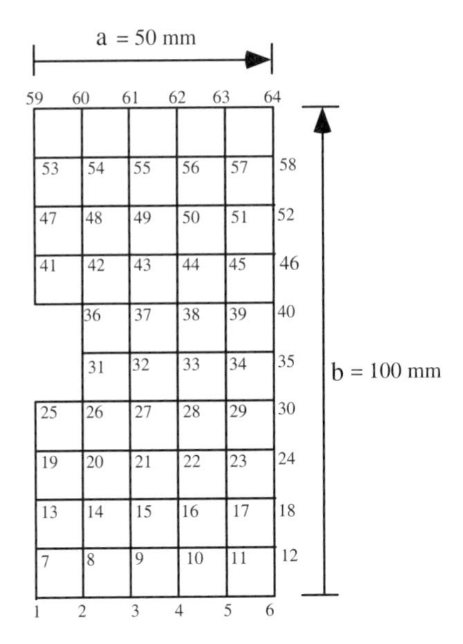

Figure 10.8. Finite elements of membrane with a hole.

Table 10.13. Shape design sensitivity results of displacement at the pressure level 1000 N/mm^2.

Node	Δz_1	z_1'	$z_1'/\Delta z_1$
6	2.642E–4	2.642E–4	100.0%
12	2.611E–4	2.611E–4	100.0%
18	2.578E–4	2.579E–4	100.0%
24	2.505E–4	2.505E–4	100.0%
30	2.404E–4	2.405E–4	100.0%
35	2.355E–4	2.356E–4	100.0%

Table 10.14. Shape design sensitivity results of displacement at the pressure level 1260 N/mm^2.

Node	Δz_1	z_1'	$z_1'/\Delta z_1$
6	3.329E–4	3.325E–4	99.9%
12	3.286E–4	3.276E–4	99.7%
18	3.240E–4	3.224E–4	99.5%
24	3.142E–4	3.120E–4	99.3%
30	3.012E–4	2.985E–4	99.1%
35	2.948E–4	2.919E–4	99.0%

Table 10.15. Shape design sensitivity results of displacement at the pressure level 1500 N/mm^2.

Node	Δz_1	z_1'	$z_1'/\Delta z_1$
6	4.004E–4	3.997E–4	99.8%
12	3.932E–4	3.920E–4	99.7%
18	3.856E–4	3.838E–4	99.5%
24	3.715E–4	3.691E–4	99.4%
30	3.539E–4	3.511E–4	99.2%
35	3.454E–4	3.432E–4	99.1%

10.3 Contact Problems

The shape design sensitivity formulation of the contact problem has been extensively developed using the linear variational inequality. The linear operator theory is not applicable to a nonlinear analysis, and the nonconvex property of the constraint set makes it difficult to prove the existence of the directional derivative of the projection. Proving the existence of the solution when frictional effects exist is an additional challenge, without mentioning the difficulty involved in proving the differentiability of the solution. Despite such mathematical uncertainty, the shape design sensitivity formulation for the contact problem is derived in a general continuum setting. As a result of the regularizing property of the penalty method, it is assumed that the solution continuously depends on shape design. Since the penalty method approximates the original variational inequality, taking a derivative of the approximated function may result in inaccuracy. However, it can be shown that the penalty-approximated sensitivity result approaches the approximated sensitivity from the variational inequality, which is obtained by taking its first order variation. As has been well established in the literature, differentiability fails in the region where contact status changes. One good feature of the penalty method is that the contact region is established using a violated region, thus avoiding a nondifferentiable region.

In this section, a continuum-based shape design sensitivity formulation is derived with respect to the perturbation of the original undeformed structure. For simplicity, a contact constraint between a flexible-rigid body is considered. A shape design sensitivity formulation for a multibody contact problem can be found in Kim et al. [23]. It is shown by Kim that the design sensitivity analysis of a frictionless contact problem is path independent, whereas that of a frictional contact problem is path dependent and requires information from the previous time step to compute sensitivity at the current time. By controlling die and punch shape, the die shape design sensitivity analysis presented in Section 10.3.3 is useful in the design of the manufacturing process.

10.3.1 Frictionless Contact

Contact surfaces generate resistant force against relative tangential motion. When surfaces are either well lubricated, or no relative movement is expected, frictional effects can be ignored and a very simple mathematical modeling can be obtained. In this section, the design sensitivity formulation of a frictionless contact problem is derived with respect to the structural shape design. Before taking the material derivative of the normal contact form, consider the fundamental properties of differentiation as related to the contact. Let the right subscript τ denote the shape perturbation parameter. Using $\boldsymbol{x}_\tau = \boldsymbol{X}_\tau + \boldsymbol{z}_\tau$, the material derivative of the structural point at time t_n is

$$\frac{d}{d\tau}(\boldsymbol{x}_\tau)\Big|_{\tau=0} \equiv \boldsymbol{V}(\boldsymbol{x}) = \boldsymbol{V}(\boldsymbol{X}) + \dot{\boldsymbol{z}}(\boldsymbol{X}), \quad \boldsymbol{X} \in \Omega, \tag{10.128}$$

where $\boldsymbol{V}(\boldsymbol{X})$ is the design velocity vector at point $\boldsymbol{X}$. The first term on the right of (10.128) explicitly depends on the design, while the second term implicitly depends on the design through the displacement. In addition to perturbation $\boldsymbol{x}_\tau$, the contact point $\boldsymbol{x}_c$ also changes. As shown in Fig. 10.9, new contact point $\boldsymbol{x}_{c\tau}$ can be obtained by projecting $\boldsymbol{x}_\tau$ onto the rigid surface in order to satisfy the contact consistency condition in (8.160). As $\tau \to 0$, $\boldsymbol{x}_{c\tau}$ approaches $\boldsymbol{x}_c$. Thus the material derivative of $\boldsymbol{x}_c$ is in the tangential direction of the rigid surface, and magnitude is determined by the variation of the natural coordinate, as

$$\frac{d}{d\tau}(\boldsymbol{x}_{c\tau})\Big|_{\tau=0} = \boldsymbol{x}_{c,\xi}\dot{\xi}_c = \boldsymbol{t}\dot{\xi}_c, \quad \boldsymbol{x}_c \in \Gamma_c^2. \tag{10.129}$$

In (10.129), $\dot{\xi}_c$ is computed from a variation of the contact consistency condition as follows. From the definition of unit tangential vector $\boldsymbol{e}_t = \boldsymbol{t}/\|\boldsymbol{t}\|$, the material derivative of $\boldsymbol{e}_t$ can be obtained, with $\boldsymbol{t} = \boldsymbol{x}_{c,\xi}$, as

$$\dot{\boldsymbol{e}}_t = \frac{\alpha\dot{\xi}_c}{\|\boldsymbol{t}\|}\boldsymbol{e}_n, \tag{10.130}$$

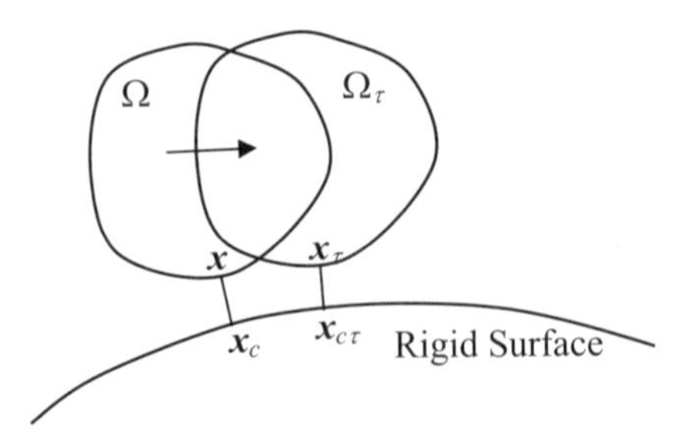

Figure 10.9. Perturbation of contact point.

where α is given in (8.177). Equations (10.129) and (10.130) need to be expressed in terms of the material derivative of displacement. Since the initial and perturbed configuration satisfy the orthogonal projection condition at $\boldsymbol{x}_c$ and $\boldsymbol{x}_{c\tau}$, respectively, the material derivative of natural coordinate ξ_c can be computed by taking a derivative of the consistency condition in (8.160), and solving for $\dot{\xi}_c$, as

$$\begin{aligned}\frac{d}{d\tau}(\varphi_\tau)\Big|_{\tau=0} &= \frac{d}{d\tau}\left[(\boldsymbol{x}_\tau - \boldsymbol{x}_{c_\tau})^T \boldsymbol{e}_{t_\tau}\right]\Big|_{\tau=0} \\ &= \boldsymbol{e}_t^T(\boldsymbol{V} + \dot{\boldsymbol{z}} - \boldsymbol{t}\dot{\xi}_c) + (g_n/\|\boldsymbol{t}\|)\alpha\dot{\xi}_c \\ &= 0\end{aligned}$$

$$\dot{\xi}_c = \left(\|\boldsymbol{t}\|/c\right)\boldsymbol{e}_t^T(\boldsymbol{V} + \dot{\boldsymbol{z}}). \tag{10.131}$$

As shown in (10.131), $\dot{\xi}_c$ is determined by the tangential component of the shape perturbation because ξ_c changes as the contact point moves along the master surface. Since the $\boldsymbol{x}_3$ directional coordinate is fixed along the shape perturbation, the material derivative of the unit tangential vector and normal vector can be expressed in terms of the design velocity and material derivative of the displacement, using the relation $\boldsymbol{e}_n = \boldsymbol{e}_3 \times \boldsymbol{e}_t$ as

$$\dot{\boldsymbol{e}}_t = \frac{\alpha}{c}\left[\boldsymbol{e}_t^T(\boldsymbol{V} + \dot{\boldsymbol{z}})\right]\boldsymbol{e}_n \tag{10.132}$$

$$\dot{\boldsymbol{e}}_n = -\frac{\alpha}{c}\left[\boldsymbol{e}_t^T(\boldsymbol{V} + \dot{\boldsymbol{z}})\right]\boldsymbol{e}_t. \tag{10.133}$$

Using the penalty method, the contact variational equation is composed of the normal contact form and the structural energy and load form, as in (8.184). By assuming an appropriate constitutive relation and applied load, the contact formulation can be used independently for an arbitrary material model. In this section, the material derivative of the normal contact form is discussed with respect to shape design parameters. The normal contact form in (8.179) at the perturbed design is

$$b_{N_\tau}(\boldsymbol{z}_\tau, \bar{\boldsymbol{z}}_\tau) = \omega_n \int_{\Gamma_{c_\tau}} g_{n_\tau} \bar{\boldsymbol{z}}_\tau^T \boldsymbol{e}_{n_\tau}\, d\Gamma. \tag{10.134}$$

Note that the contact region implicitly changes due to structural shape perturbation. By ignoring the $\dot{\bar{\boldsymbol{z}}}$ term, as discussed in (10.8), the material derivative of b_N contains that of g_n and $\boldsymbol{e}_n$. The material derivative of the normal gap function can be found by taking the derivative of (8.159) as

$$\begin{aligned}\dot{g}_n &= \frac{d}{d\tau}\left[(\boldsymbol{x}_\tau - \boldsymbol{x}_{c\tau})^T \boldsymbol{e}_{n\tau}\right]\Big|_{\tau=0} \\ &= (\boldsymbol{V} + \dot{\boldsymbol{z}} - \boldsymbol{t}\dot{\xi}_c)^T \boldsymbol{e}_n + (\boldsymbol{x} - \boldsymbol{x}_c)^T \dot{\boldsymbol{e}}_n \\ &= (\boldsymbol{V} + \dot{\boldsymbol{z}})^T \boldsymbol{e}_n.\end{aligned} \tag{10.135}$$

Equation (10.135) implies that, for an arbitrary perturbation of the structure, only the normal components of that perturbation will contribute to the material derivative of the normal gap function. In shape design sensitivity, the tangential component of the boundary perturbation does not contribute to the first-order shape change; only the normal component of the boundary perturbation affects structural performance. This is not true for the contact problem, since the material derivative of $\boldsymbol{e}_n$ in (10.133) has a tangential component. Even if the boundary perturbation is given in a tangential direction, the structural performance will change, since the contact point changes through the contact consistency condition in (8.130).

In the nonlinear sensitivity formulation, it is sufficient to differentiate the linearized equation rather than the nonlinear one, since differentiation includes the linearization process. This is true when the variational equation only contains the displacement, which is a relative measurement of the deformation. When the variational equation is a function of material point x, as in the case of the contact form, then the domain perturbation is explicitly embedded and the linearization process eliminates this dependency. Thus, it is necessary to take the derivative of the contact form before linearization. To do so, it is necessary to take the material derivative of the normal contact form in (10.134), using (10.133) and (10.135), to obtain

$$\begin{aligned}
[b_{N_\tau}(z_\tau, \bar{z}_\tau)]' &= \omega_n \int_{\Gamma_c} \bar{z}^T \boldsymbol{e}_n \boldsymbol{e}_n^T (\boldsymbol{V} + \dot{z})\, d\Gamma \\
&\quad - \omega_n \int_{\Gamma_c} \frac{\alpha g_n}{c} \bar{z}^T \boldsymbol{e}_t \boldsymbol{e}_t^T (\boldsymbol{V} + \dot{z})\, d\Gamma \\
&\quad + \omega_n \int_{\Gamma_c} \kappa g_n \bar{z}^T \boldsymbol{e}_n (\boldsymbol{V}^T \boldsymbol{n})\, d\Gamma \\
&\equiv b_N^*(z; \dot{z}, \bar{z}) + b'_N(z, \bar{z}),
\end{aligned} \tag{10.136}$$

where all terms containing $\boldsymbol{V}$ represent the explicit dependence on the shape design and terms with $\dot{z}$ denote the implicit dependence. The implicitly dependent term $b_N^*(z;\dot{z},\bar{z})$ is available in (8.183) by substituting the material derivative of the displacement into the incremental displacement. In (10.136), explicitly dependent term $b'_N(z,\bar{z})$ is defined as the normal contact fictitious load form and can be obtained by collecting all terms in (10.136) that have explicit dependency on the design velocity field, as

$$\begin{aligned}
b'_N(z, \bar{z}) &= \omega_n \int_{\Gamma_c} \bar{z}^T \boldsymbol{e}_n \boldsymbol{e}_n^T \boldsymbol{V}\, d\Gamma \\
&\quad - \omega_n \int_{\Gamma_c} \frac{\alpha g_n}{c} \bar{z}^T \boldsymbol{e}_t \boldsymbol{e}_t^T \boldsymbol{V}\, d\Gamma \\
&\quad + \omega_n \int_{\Gamma_c} \kappa g_n \bar{z}^T \boldsymbol{e}_n (\boldsymbol{V}^T \boldsymbol{n})\, d\Gamma.
\end{aligned} \tag{10.137}$$

Since the material derivative of the normal contact form in (10.137) depends on displacement at time t_n, as well as on the design velocity fields of the undeformed configuration, it is path-independent. However, the same is not true for the slip form, discussed in the next section. Thus, it is very efficient to compute the design sensitivity of a frictionless contact problem with elastic materials. The design sensitivity equation is solved only once for each design parameter at the last converged configuration with the same tangent stiffness matrix from the response analysis.

10.3.2 Frictional Contact

Friction is a mechanism that is related to two different configurations. In response analysis, the amount of slip along the contact surface is computed between current and previous time steps. By assuming a constant frictional coefficient and by using the friction model defined in (8.189), a design sensitivity formulation can be derived for the frictional contact problem. From the response analysis results, the status of the tangential movement can be determined as either stick or slip. Design sensitivity formulations for each status are derived as follows. Since friction only exists for those regions that violate the impenetrability condition given in (8.159), the normal contact condition in Section 10.3.1 needs to be considered in addition to the slip condition.

Stick Form

The slip function in (8.161) is expressed at the perturbed domain as

$$g_{t_\tau} \equiv \left\| t_\tau^0 \right\| (\xi_{c_\tau} - \xi_{c_\tau}^0). \tag{10.138}$$

Since two different configurations are involved, for simplicity, the right superscript "0" denotes the previous configuration time $t_{n-1.}$ That is, both t^0 and ξ_c^0 are evaluated at the contact point of the previously converged time. Before taking the material derivative of the slip function in (10.138), the material derivatives of $\left\| t^0 \right\|$ and ξ_c^0 are computed using the relation in (10.131) at time step t_{n-1} as

$$\frac{d}{d\tau} \left\| t^0 \right\| = \frac{\beta^0 \left\| t^0 \right\|}{c^0} e_t^{0T} (V + \dot{z}^0) \tag{10.139}$$

$$\dot{\xi}_c^0 = \frac{\left\| t^0 \right\|}{c^0} e_t^{0T} (V + \dot{z}^0), \tag{10.140}$$

where $\dot{z}^0$ is the material derivative of the displacement at time t_{n-1}. Even if all quantities are evaluated at time t_{n-1}, the design velocity is still evaluated at the undeformed configuration because the perturbation occurs at ${}^0\Omega$.

Using (10.131), (10.139), and (10.140), the material derivative of the slip function in (10.138) is

$$\begin{aligned} \dot{g}_t &= \frac{d}{d\tau} \left[\left\| t_\tau^0 \right\| (\xi_{c_\tau} - \xi_{c_\tau}^0) \right] \Big|_{\tau=0} \\ &= v e_t^T (V + \dot{z}) + \frac{\beta^0 g_t - \left\| t^0 \right\|^2}{c^0} e_t^{0T} (V + \dot{z}^0). \end{aligned} \tag{10.141}$$

Thus, the material derivative of the slip function at time t_n depends on quantities at configuration time t_{n-1}, which makes the problem path dependent. Before taking the derivative of the stick form, the following derivative needs to be computed by differentiating (8.177):

$$\begin{aligned} \dot{v} &= \frac{v}{c^2} \left[\left(\gamma \left\| t \right\| - 2\alpha\beta \right) g_n - \beta \left\| t \right\|^2 \right] e_t^T (V + \dot{z}) \\ &+ \frac{\alpha v}{c} e_n^T (V + \dot{z}) + \frac{\beta^0 v}{c^0} e_t^{0T} (V + \dot{z}^0). \end{aligned} \tag{10.142}$$

The following relations are used to evaluate the derivative in (10.142):

$$\dot{\alpha} = \frac{\gamma \left\| t \right\| - \alpha\beta}{c} e_t^T (V + \dot{z}) \tag{10.143}$$

$$\begin{aligned} \dot{c} &= \frac{1}{c} \left[2 \left\| t \right\|^2 \beta + (\alpha\beta - \gamma \left\| t \right\|) g_n \right] e_t^T (V + \dot{z}) \\ &- \alpha e_n^T (V + \dot{z}). \end{aligned} \tag{10.144}$$

The variation of the stick condition can now be derived as follows. The stick form in (8.189) at the perturbed design is

$$b_{T_\tau}(z_\tau, \bar{z}_\tau) = \omega_t \int_{\Gamma_{c_\tau}} v_\tau g_{t_\tau} \bar{z}_\tau^T e_{t_\tau} \, d\Gamma. \tag{10.145}$$

By taking the material derivative of (10.145), using (10.141) and (10.142), we obtain

$$\begin{aligned} \tfrac{d}{d\tau}\left[b_{T_\tau}(z_\tau, \bar{z}_\tau) \right] &= \omega_t \int_{\Gamma_c} [\dot{v} g_t \bar{z}^T e_t + v \dot{g}_t \bar{z}^T e_t + v g_t \bar{z}^T \dot{e}_t] d\Gamma \\ &+ \omega_t \int_{\Gamma_c} \kappa v g_t \bar{z}^T e_t (V^T n) d\Gamma \\ &\equiv b_T^*(z; \dot{z}, \bar{z}) + b_T'(z, \bar{z}). \end{aligned} \tag{10.146}$$

If the material derivative of the displacement is substituted for the incremental displacement, then $b_T^*(z;\dot{z},\bar{z})$ is the same as the linearized slip form in (8.192). In addition, $b_T'(z,\bar{z})$ is the stick fictitious load form, obtained from those terms explicitly dependent on the design in (10.146), as

$$\begin{aligned} b_T'(z,\bar{z}) &= b_T^*(z; V, \bar{z}) \\ &+ \omega_t \int_{\Gamma_c} [v(2\beta^0 g_t - \|t^0\|^2] \bar{z}^T e_t e_t^{0T} (V + \dot{z}^0) d\Gamma \\ &+ \omega_t \int_{\Gamma_c} [\kappa v g_t \bar{z}^T e_t](V^T n) d\Gamma. \end{aligned} \tag{10.147}$$

Since $b_T^*(z;V,\bar{z})$ is computed at the response analysis stage, it is very convenient to evaluate (10.147) with the sensitivity results from the previous time t_{n-1}. Even if the stick condition is elastic and $b_T^*(z;\dot{z},\bar{z})$ is symmetric in its arguments, the stick fictitious load in (10.147) depends on the sensitivity analysis from a previous time step. Thus, the sensitivity problem is path-dependent even when the response analysis only has a stick condition that is elastic.

Slip Form

The slip form in (8.189) at the perturbed design is

$$b_{T_\tau}(z_\tau, \bar{z}_\tau) = \omega_t \int_{\Gamma_{c_\tau}} v_\tau g_{n_\tau} \bar{z}_\tau^T e_{t_\tau} \, d\Gamma, \tag{10.148}$$

with $\omega_t = -\mu\omega_n \text{sgn}(g_t)$ for the slip condition. The following expression is obtained by taking the material derivative of (10.148) using a procedure similar to that of the stick condition, but with a normal gap function:

$$\begin{aligned} \tfrac{d}{d\tau}\left[b_{T_\tau}(z_\tau, \bar{z}_\tau) \right] &= \omega_t \int_{\Gamma_c} [\dot{v} g_n \bar{z}^T e_t + v \dot{g}_n \bar{z}^T e_t + v g_n \bar{z}^T \dot{e}_t] d\Gamma \\ &+ \omega_t \int_{\Gamma_c} \kappa v g_n \bar{z}^T e_t (V^T n) d\Gamma \\ &\equiv b_T^*(z; \dot{z}, \bar{z}) + b_T'(z, \bar{z}), \end{aligned} \tag{10.149}$$

where $b_T^*(z;\dot{z},\bar{z})$ is the same as the linearized slip form in (8.196), which is not symmetric if the incremental displacement is substituted for the material derivative of the displacement. Additionally, $b_T'(z,\bar{z})$ is the slip fictitious load form and is obtained from those terms that are explicitly dependent on the design from (10.149), as

$$\begin{aligned} b'_T(z,\bar{z}) &\equiv b^*_T(z;V,\bar{z}) \\ &+\omega_t \int_{\Gamma_c} \left(\nu\beta^0 g_n / c^0\right) \bar{z}^T e_t e_t^{0T} (V + \dot{z}^0)\, d\Gamma \\ &+\omega_t \int_{\Gamma_c} \kappa[\nu g_n \bar{z}^T e_t](V^T n)\, d\Gamma. \end{aligned} \quad (10.150)$$

Since $b^*_T(z;V,\bar{z})$ is computed during the response analysis, it is very convenient to evaluate (10.150) with the sensitivity results from previous time t_{n-1}.

By combining (10.136) and (10.146) for the stick condition, or (10.136) and (10.149) for the slip condition, the material derivative of the contact form can be obtained as

$$\frac{d}{d\tau}\left[b_{\Gamma_\tau}(z_\tau, \bar{z}_\tau) \right] = b^*_\Gamma(z;\dot{z},\bar{z}) + b'_V(z,\bar{z}), \quad (10.151)$$

where

$$b^*_\Gamma(z;\dot{z},\bar{z}) = b^*_N(z;\dot{z},\bar{z}) + b^*_T(z;\dot{z},\bar{z}) \quad (10.152)$$

$$b'_V(z,\bar{z}) = b'_N(z,\bar{z}) + b'_T(z,\bar{z}). \quad (10.153)$$

Since the equilibrium configuration at time t_n and the design velocity fields are known, the contact fictitious load form in (10.153) can be computed for a given displacement and its variation. If incremental displacement is substituted for the material derivative of the displacement, then the contact bilinear form in (10.152) is the same as the linearized contact form in (8.193).

Shape Sensitivity Equation of Frictional Contact Problem

The principle of virtual work at the perturbed design can be written as

$$a_{\Omega_\tau}(z_\tau, \bar{z}_\tau) + b_{\Gamma_\tau}(z_\tau, \bar{z}_\tau) = \ell_{\Omega_\tau}(\bar{z}_\tau), \quad \forall \bar{z}_\tau \in Z_\tau, \quad (10.154)$$

where $a_{\Omega_\tau}(z_\tau, \bar{z}_\tau)$ represents the arbitrary constitutive model discussed in Sections 8.1 and 8.2, and where $\ell_{\Omega_\tau}(\bar{z}_\tau)$ is the deformation-independent load form. After taking the material derivative of (10.154), the design sensitivity equation for the direct differentiation method is obtained as

$$a^*_\Omega(z;\dot{z},\bar{z}) + b^*_\Gamma(z;\dot{z},\bar{z}) = \ell'_V(\bar{z}) - a'_V(z,\bar{z}) - b'_V(z,\bar{z}), \quad \forall \bar{z} \in Z. \quad (10.155)$$

For the frictionless contact problem with an elastic constitutive model presented in Section 8.1, the energy of the system does not dissipate through the deformation process. The material derivative formulation in (10.155) only depends on the current configuration and the design velocity field. The linear system of equations only needs to be solved once at the final converged configuration, even though analysis is carried out using iteration. As compared with nonlinear response analysis, this property provides great efficiency in the sensitivity computation process.

Since the tangent stiffness operator in (10.155) is not symmetric for the frictional contact problem, the adjoint variable method requires more computational effort than the direct differentiation method. Since the slip fictitious load depends on the material derivative of the previous converged configuration, the linear system of equations needs to be solved at each load step. Thus, the design sensitivity is path dependent.

10.3.3 Die Shape Design

A die shape design is a special term used in the metal forming process. Figure 10.10 provides a simple example of the metal forming process. Complicated geometry can be created from flat, blank sheet metal through the process of permanent deformation. From the reciprocal action between punch and die, the final geometry of the workpiece is controlled. Through this stamping process, the geometry of the punch and die are as critical as the stages in the deformation process. The desired workpiece shape can be obtained by changing the shape of the punch and die. The difficulty associated with die shape design is related to elastoplasticity. When the punch is removed after the deformation process, the metal is usually affected by elastic springback, which is extremely difficult to estimate beforehand. A conventional trial-and-error approach requires the design of a cycle period, which is expensive and requires a tremendous amount of time. In this section, the sensitivity expression of such structural performance measures as displacement and plastic strain are obtained with respect to the shape change of the die geometry. Since the frictional contact condition is imposed between the workpiece and die, the shape change of the die is affected by structural performance through the contact form.

Let $\mathbf{x}_c(\xi_c)$ represent the coordinate of the master surface and let the geometry of the rigid surface be perturbed in the direction $\mathbf{V}_c(\mathbf{x}_c(\xi_c))$, as shown in Fig. 10.11. The contact point $\mathbf{x}_c$, normal vector $\mathbf{e}_n$, and corresponding natural coordinate ξ_c are changed along with the die shape perturbation. The variation of the contact point and the tangential vector on the master surface can be obtained by perturbing the master surface in the direction $\mathbf{V}_c(\mathbf{x}_c)$, and changing the natural coordinate to correspond to the new contact point in the tangential direction, from (10.129) as

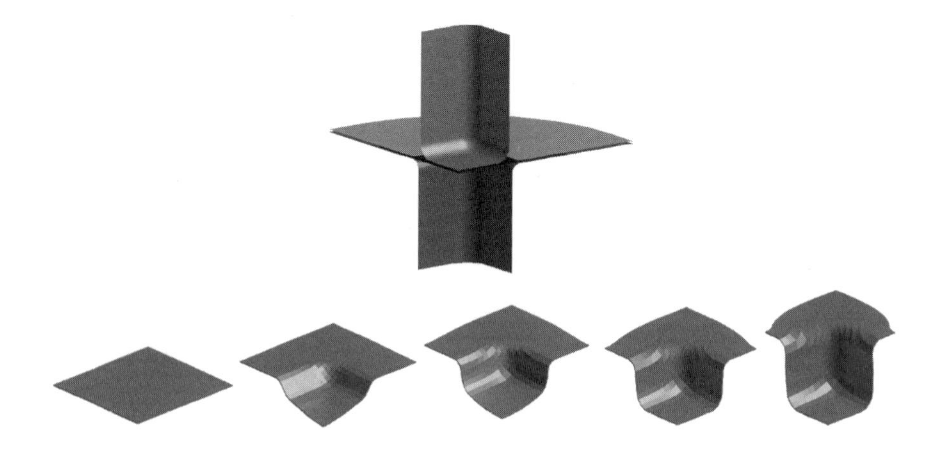

Figure 10.10. Sheet metal stamping process.

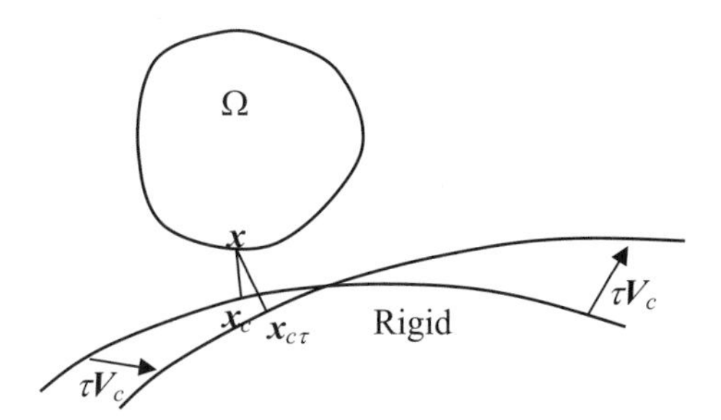

Figure 10.11. Perturbation of rigid surface.

$$\frac{d}{d\tau}(\boldsymbol{x}_{c_\tau})\Big|_{\tau=0} = \boldsymbol{V}_c(\boldsymbol{x}_c) + \boldsymbol{t}\dot{\xi}_c\,, \quad \boldsymbol{x}_c \in \Gamma_c^2 \tag{10.156}$$

$$\dot{\boldsymbol{t}} \equiv \frac{d}{d\tau}(\boldsymbol{x}_{c,\xi})\Big|_{\tau=0} = \boldsymbol{V}_{c,\xi} + \boldsymbol{x}_{c,\xi\xi}\dot{\xi}_c, \tag{10.157}$$

where $\boldsymbol{t} = \boldsymbol{x}_{c,\xi}$ is used. The first term in (10.156) represents the explicit dependence on the design perturbation at $\boldsymbol{x}_c$, and the second term represents the contact point change along the master surface due to a change in the natural coordinates. In die shape design, the structural domain Ω is fixed from a design point of view. Thus, there is no convective term for the material derivative formula in (10.4), and only a partial derivative term remains. In a strict sense, a die shape design sensitivity analysis should be treated as a size design problem where the structural integration domain is fixed. However, since the perturbation of the die shape changes the boundary integration domain in the contact form, die shape design is treated here as a shape design problem. In this way, the material derivative in (10.156) can be interpreted as a partial derivative.

In (10.156), $\dot{\xi}_c$ can be obtained from the variation of the contact consistency condition, as in (10.131). The only difference from (10.131) is that the variation of the tangential vector has the explicitly dependent term $V_{c,\xi}$, as in (10.157). The variation of the natural coordinate at the contact point can be obtained by following a similar procedure to (10.131) as

$$\dot{\xi}_c = \frac{\|\boldsymbol{t}\|}{c}\boldsymbol{e}_t^T\left(\dot{\boldsymbol{z}} - \boldsymbol{V}_c\right) + \frac{g_n}{c}\boldsymbol{e}_n^T\boldsymbol{V}_{c,\xi}. \tag{10.158}$$

Note that the unknown term that includes $\dot{\boldsymbol{z}}$ in (10.158) is the same as in (10.131), and the known term has an explicit expression with respect to die shape design velocity $\boldsymbol{V}_c$ and its derivative with respect to ξ. Using (10.158), the variation of the unit normal and tangential vector with respect to die shape design can be obtained by following similar procedures to (10.132) and (10.133) as

$$\tfrac{d}{d\tau}(\boldsymbol{e}_t) = \left(\frac{\alpha}{c}\boldsymbol{e}_t^T(\dot{\boldsymbol{z}} - \boldsymbol{V}_c) + \frac{\|\boldsymbol{t}\|}{c}\boldsymbol{e}_n^T\boldsymbol{V}_{c,\xi}\right)\boldsymbol{e}_n \tag{10.159}$$

$$\frac{d}{d\tau}(\boldsymbol{e}_n) = -\left(\frac{\alpha}{c}\boldsymbol{e}_t^T(\dot{\boldsymbol{z}} - \boldsymbol{V}_c) + \frac{\|\boldsymbol{t}\|}{c}\boldsymbol{e}_n^T\boldsymbol{V}_{c,\xi}\right)\boldsymbol{e}_t. \tag{10.160}$$

To obtain the design sensitivity expression of the normal contact form in (10.134), the variation of the normal gap function is obtained by following a similar procedure to (10.135) as

$$\dot{g}_n = (\dot{\boldsymbol{z}} - \boldsymbol{V}_c)^T\boldsymbol{e}_n. \tag{10.161}$$

Note that the design velocity fields of the die geometry negatively affect the variation of the normal gap function compared with the design velocity field of the structural domain in (10.135). For example, if $\boldsymbol{V}$ (the structural velocity) and $\boldsymbol{V}_c$ (the master surface velocity) move in the same direction, as suggested by Fig. 10.12, then the normal component of structural design velocity $\boldsymbol{V}$ will increase the value of $\dot{g}_n$, whereas the normal component of die shape velocity $\boldsymbol{V}_c$ will decrease that value.

Using (10.160) and (10.161), the variation of the normal contact form in (10.134) can be taken with respect to the die shape design, following the same procedure in (10.136). By substituting the material derivative of the displacement into the incremental displacement, the term that is implicitly dependent on the design, $b_N^*(\boldsymbol{z};\dot{\boldsymbol{z}},\bar{\boldsymbol{z}})$, is the same as the term in (8.183). The term that is explicitly dependent on the design, $b'_N(\boldsymbol{z},\bar{\boldsymbol{z}})$, is defined as the normal contact fictitious load form and can be obtained as

$$\begin{aligned} b'_N(\boldsymbol{z},\bar{\boldsymbol{z}}) &= -b_N^*(\boldsymbol{z};\boldsymbol{V}_c,\bar{\boldsymbol{z}}) \\ &\quad -\omega_n \int_{\Gamma_c} \frac{g_n\|\boldsymbol{t}\|}{c}\bar{\boldsymbol{z}}^T\boldsymbol{e}_t\boldsymbol{e}_n^T\boldsymbol{V}_{c,\xi}\,d\Gamma. \end{aligned} \tag{10.162}$$

Note that no curvature term exists since the undeformed structural domain is fixed for the design.

The die shape design sensitivity formulation of friction can be derived in a similar way as the description in Section 10.3.2. Since most derivations have previously been discussed, only final expressions are presented here. The variation of the slip function can be written as

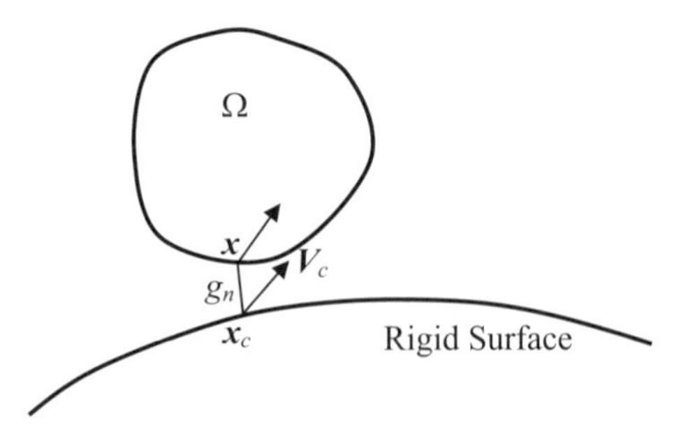

Figure 10.12. Effect of design velocity to the normal gap function.

$$\begin{aligned}\dot{g}_t &= v\boldsymbol{e}_t^T(\dot{\boldsymbol{z}} - \boldsymbol{V}_c) + (g_n\|\boldsymbol{t}^0\|/c)\boldsymbol{e}_n^T\boldsymbol{V}_{c,\xi}\\ &+((\beta^0 g_t - \|\boldsymbol{t}^0\|^2)/c^0)\boldsymbol{e}_t^{0T}(\dot{\boldsymbol{z}}^0 - \boldsymbol{V}_c)\\ &+(g_n^0\left(\beta^0(\xi_c - \xi_c^0) - \|\boldsymbol{t}^0\|\right)/c^0)\boldsymbol{e}_n^{0T}\boldsymbol{V}_{c,\xi}\\ &+(\xi_c - \xi_c^0)\boldsymbol{e}_t^{0T}\boldsymbol{V}_{c,\xi},\end{aligned} \tag{10.163}$$

where the term including $\boldsymbol{V}_{c,\xi}$ is new compared with (10.141). As in the response analysis, the tangential movement is divided into stick and slip conditions. In the case of a stick condition,

$$\begin{aligned}b'_T(\boldsymbol{z},\bar{\boldsymbol{z}}) &= -b_T^*(\boldsymbol{z};\boldsymbol{V}_c,\bar{\boldsymbol{z}})\\ &+\omega_t\int_{\Gamma_c}(2g_t\|\boldsymbol{t}\|/c)\bar{\boldsymbol{z}}^T\boldsymbol{e}_t\boldsymbol{e}_t^{0T}\boldsymbol{V}_{c,\xi}\,d\Gamma\\ &+\omega_t\int_{\Gamma_c}v(2\beta^0 g_t - \|\boldsymbol{t}^0\|^2)\bar{\boldsymbol{z}}^T\boldsymbol{e}_t\boldsymbol{e}_t^{0T}(\dot{\boldsymbol{z}}^0 - \boldsymbol{V}_c)\,d\Gamma\\ &+\omega_t\int_{\Gamma_c}(\beta^0 g_n g_t(\|\boldsymbol{t}^0\| + \|\boldsymbol{t}\|)/cc^0)\bar{\boldsymbol{z}}^T\boldsymbol{e}_t\boldsymbol{e}_n^{0T}\boldsymbol{V}_{c,\xi}\,d\Gamma\\ &-\omega_t\int_{\Gamma_c}(g_n\|\boldsymbol{t}\|\|\boldsymbol{t}^0\|^2/cc^0)\,\bar{\boldsymbol{z}}^T\boldsymbol{e}_t\boldsymbol{e}_n^{0T}\boldsymbol{V}_{c,\xi}\,d\Gamma,\quad \text{if } |\omega_t g_t| \le |\mu\omega_n g_n|,\end{aligned} \tag{10.164}$$

while for the slip condition,

$$\begin{aligned}b'_T(\boldsymbol{z},\bar{\boldsymbol{z}}) &= -b_T^*(\boldsymbol{z};\boldsymbol{V}_c,\bar{\boldsymbol{z}})\\ &+\omega_t\int_{\Gamma_c}(g_n\|\boldsymbol{t}\|/c)\bar{\boldsymbol{z}}^T\boldsymbol{e}_t\boldsymbol{e}_t^{0T}\boldsymbol{V}_{c,\xi}\,d\Gamma\\ &+\omega_t\int_{\Gamma_c}(v\beta^0 g_n/c)\bar{\boldsymbol{z}}^T\boldsymbol{e}_t\boldsymbol{e}_t^{0T}(\dot{\boldsymbol{z}}^0 - \boldsymbol{V}_c)\,d\Gamma\\ &+\omega_t\int_{\Gamma_c}(\beta^0 g_n g_n^0\|\boldsymbol{t}^0\|/cc^0)\bar{\boldsymbol{z}}^T\boldsymbol{e}_t\boldsymbol{e}_n^{0T}\boldsymbol{V}_{c,\xi}\,d\Gamma,\quad \text{if } |\omega_t g_t| > |\mu\omega_n g_n|\end{aligned} \tag{10.165}$$

are the fictitious load forms corresponding to friction. The same design sensitivity equation in (10.155) can be used to compute the displacement variation using the direct differentiation method.

10.3.4 Numerical Examples

Seal Problem

Automotive or refrigerator door seals are commonly used for noise isolation and sealing purposes. The performance of a door seal is evaluated based on the contact pressure distribution and size of the contact area. The door is modeled by a rigid body, based on the assumption that its stiffness is much greater than the rubber sealing material. Figure 10.13 shows the geometry of the door seal and a portion of the door with discrete particles of the meshfree method [48]. The geometry of the seal is approximated by a circular shape with a constant thickness except for the mounting component. For simplicity, only the contact region of the door is modeled, using a quarter circular section with a 4-mm radius. The geometry of the seal is discretized by 174 particles, and the rigid surface is modeled by 32 piecewise linear master segments. Mooney-Rivlin type hyperelastic material is used with a pressure projection formulation for nearly incompressibility constraints. Material constants C_{10} = 80 kPa, C_{01} = 20 kPa, and bulk modulus K = 80 MPa are used. The rigid door is pushed down 10 mm from the top of the rubber material. The bottom surface of the door seal corresponds to the interface of the installation and remains completely fixed. A flexible-rigid body contact condition is

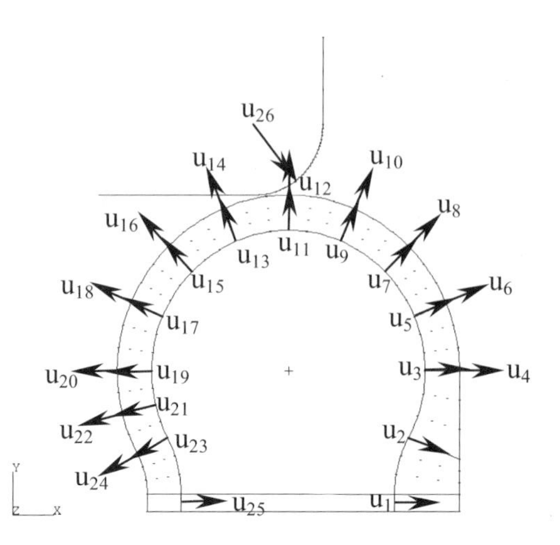

Figure 10.13. Design parameters of door seal problem.

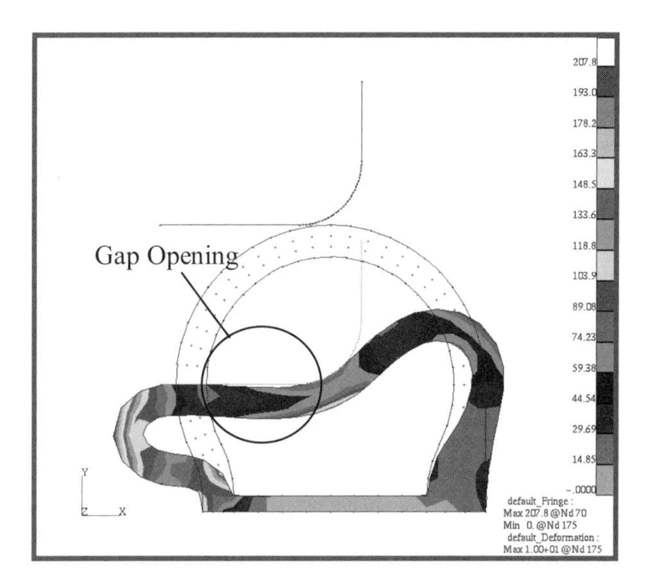

Figure 10.14. Stress distribution for door seal problem.

imposed between interfaces. Analysis is carried out with 100 load steps by a displacement driven procedure.

If friction between the interfaces is ignored, then all rubber materials bulge out on the right side without providing an effective seal. If friction is considered in this problem, the frictional condition significantly affects the analysis results. A constant value of 0.25 is used as the dry frictional coefficient for all contact surfaces. Figure 10.14 shows a contour plot of the second invariant of deviatoric Cauchy stress (von Mises stress) with deformed geometry at the final configuration when friction exists. As the rigid body moves down, the contact region initially increases. After a certain amount of deformation, the middle of the contact region begins to separate. Since there is almost no slip between contact surfaces in this problem, all contact regions are in the stick condition. A high stress concentration exists at the highly distorted region.

The geometry of the structure is parameterized using 26 shape design parameters. The design parameters are shown in Fig. 10.13. The last design parameter is the radius of the rounded corner of the rigid body (door). Even though this design parameter does not change the shape of the structure, it can be treated using the material derivative, as discussed in Section 10.3.3. The remaining 25 design parameters are the parametric coordinates of the boundary shape. The boundaries of the seal are represented by cubic spline curves. The control points or slopes of the spline curves are chosen as design parameters. First, the design velocity field at the boundary is obtained by perturbing the boundary curve corresponding to the design variable, and the domain design velocity field is computed using an isoparametric mapping method [41].

Nine performance measures are chosen: the area of the structure, seven von Mises stresses from the high stress concentrated region, and the square sum of normal gap distances at the discrete points in the opening region, as shown in Fig. 10.14. Note that stress is proportional to the service life and normal gap distances are related to the performance of the seal.

Design sensitivity analysis is carried out at each converged load step to compute the material derivatives of the displacement. Since there are 26 design parameters, 26 linear systems of equations are solved at each load step. The sensitivity coefficients of the performance measures are computed at the final converged load step using the material derivatives of the displacement. The total sensitivity computation cost is about 55% of a response analysis. Since 26 design parameters are considered in this problem, each design parameter takes less than 3% of analysis computation time. Such numerical performance is very efficient compared with that required for the finite difference method.

The accuracy of the sensitivity is compared with forward finite difference results for the perturbation size $\tau = 10^{-6}$. Table 10.16 shows the accuracy of the sensitivity results. The second column, $\Delta\psi$, indicates finite difference results, while the third column represents the percentage of change in performance from the presented method. As can be seen, extremely accurate results are obtained.

Deepdrawing problem

Design optimization of the deepdrawing process includes parameterization of the design, nonlinear meshfree analysis, shape design sensitivity analysis, and optimization algorithm. MSC/PATRAN [42], which uses a parametric representation, is used as the geometric modeling tool. An efficient method for computing the design velocity in parametric space was proposed by Choi and Chang [41].

Figure 10.15 illustrates the simulation setting and design parameterization of the deepdrawing process. Only half of the model is solved due to the symmetric conditions given by the plane strain problem. The blank is modeled with 303 meshfree particles. The von Mises yield criterion is used with an isotropic hardening model. A constant frictional coefficient is used, taken from the modified Coulomb law. The draw die is fixed during the punch motion stage, while the blank holder exerts force to prevent any vertical movement of the blank. After simulating the maximum downstroke of punch (30 mm), the punch, die, and blank holder are removed to calculate springback.

The first two design parameters control the horizontal and vertical position of the punch. Horizontal movement is very important since it controls the gap between punch and draw die. The third and fourth parameters are round radii of the punch and draw die corners. A sharp corner may increase the plastic strain while reducing the amount of springback. The fifth parameter changes the thickness of the blank, which involves changes in the workpiece's shape. The sixth parameter controls the gap between the blank holder and die and allows the frictional force on the blank to be changed.

Table 10.16. Accuracy of sensitivity results.

ψ	$\Delta\psi$	ψ'	$\Delta\psi/\psi' \times 100\%$
u_1			
Area	-.163895E-5	-.163895E-5	100.00
σ_{75}	-.501565E-6	-.501563E-6	100.00
σ_{86}	-.255777E-5	-.255775E-5	100.00
σ_{44}	-.247860E-6	-.247893E-6	99.99
σ_{114}	.525571E-6	.525554E-6	100.00
σ_{31}	-.149431E-6	-.149300E-6	100.09
Σg_n^2	-.114879E-8	-.114878E-8	100.00
u_2			
Area	.163894E-5	.163895E-5	100.00
σ_{75}	.514388E-6	.514395E-6	100.00
σ_{86}	-.268130E-5	-.268129E-5	100.00
σ_{44}	.292610E-4	.292609E-4	100.00
σ_{114}	.126237E-4	.126237E-4	100.00
σ_{31}	.947482E-6	.947679E-6	99.98
Σg_n^2	.223116E-7	.223116E-7	100.00
u_3			
Area	-.405671E-5	-.405671E-5	100.00
σ_{75}	-.858554E-7	-.858371E-7	100.02
σ_{86}	-.361270E-5	-.361266E-5	100.00
σ_{44}	-.481447E-6	-.481452E-6	100.00
σ_{114}	.143501E-5	.143499E-5	100.00
σ_{31}	.869462E-7	.868284E-7	100.14
Σg_n^2	-.311421E-9	-.311370E-9	100.02
u_4			
Area	-.351300E-5	-.351300E-5	100.00
σ_{75}	-.105863E-4	-.105864E-4	100.00
σ_{86}	-.130646E-4	-.130647E-4	100.00
σ_{44}	.122614E-5	.122615E-5	100.00
σ_{114}	-.329776E-5	-.329777E-5	100.00
σ_{31}	-.243310E-6	-.243378E-6	99.97
Σg_n^2	-.210381E-8	-.210383E-8	100.00
u_5			
Area	.447486E-5	.447486E-5	100.00
σ_{75}	.629273E-5	.629276E-5	100.00
σ_{86}	.835460E-5	.835467E-5	100.00
σ_{44}	.122614E-5	.122615E-5	100.00
σ_{114}	.143353E-5	.143352E-5	100.00
σ_{31}	.116624E-6	.116676E-6	99.96
Σg_n^2	-.430791E-9	-.430725E-9	100.02

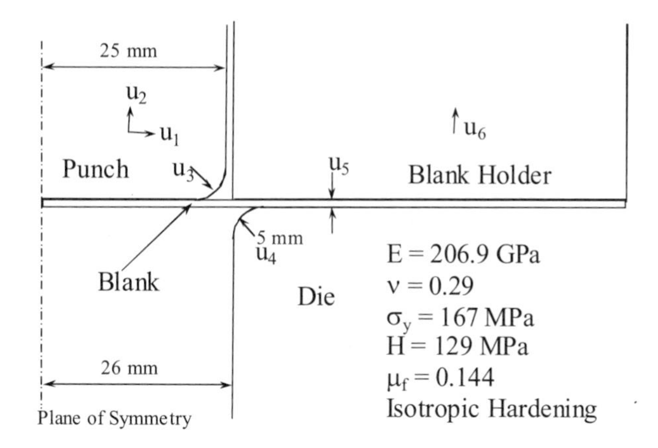

Figure 10.15. Design parameterization of deepdrawing process.

Nonlinear meshfree analysis [49] is carried out to simulate spingback in the deepdrawing process. Rigid materials are assumed for the punch, draw die, and blank holder. Thus, numerical integration is involved only for the workpiece material. A displacement driven method is used so that the position of the punch is given at each time step. A converged configuration is found using the implicit Newton-Raphson method. For stress computation, an elastic predictor followed by a plastic return-mapping is used for the principal Kirchhoff stress. After finding a converged configuration, the factorized tangent stiffness matrix is stored to be used later for design sensitivity analysis purposes.

A slave-master concept is used for the contact problem in order to impose a penalty regularization. Rigid surfaces (punch, draw die, and blank holder) are modeled using piecewise linear master segments. Using linear discretization, a very simple expression of $b_\Gamma(z,\bar{z})$ can be obtained, since the normal and tangential vectors on the contact surface remain constant. However, the possibility exists of a convergence problem at kinked corners of the adjacent linear master segments. A line search algorithm is used when the convergence problem occurs. The contact search is carried out for particles on the domain boundary. If penetration into the rigid surface is detected, then a penalty is imposed using $b_\Gamma(z,\bar{z})$. Stick/slip conditions are determined by measuring the amount of motion relative to two adjacent configurations.

Figure 10.16 shows the results of nonlinear analysis at maximum deformation and after springback. A significant amount of material sliding is observed between the workpiece and the draw die despite a considerable amount of friction. Springback occurs when the punch, draw die, and blank holder are removed. Although the amount of elastic springback is relatively small at individual points on the blank, the total displacement at the edge increases due to the rotational effect.

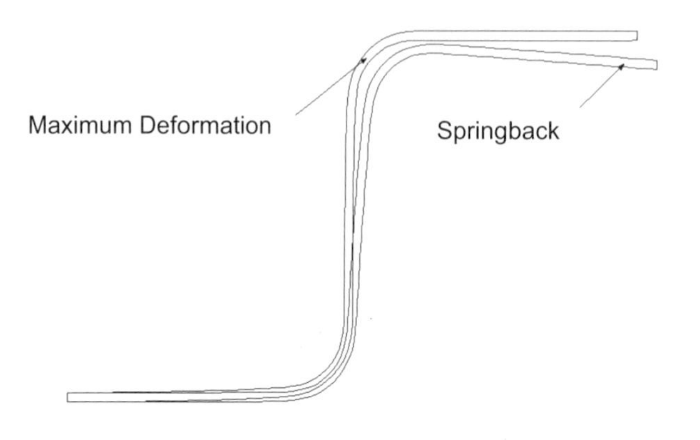

Figure 10.16. Deepdrawing analysis results with springback.

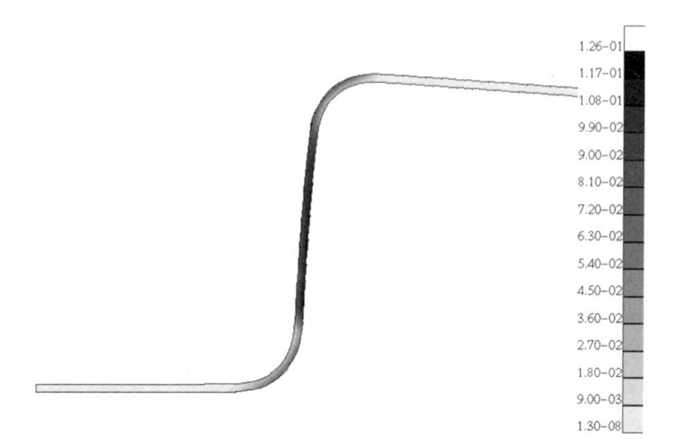

Figure 10.17. Effective plastic strain distribution.

Figure 10.17 provides a contour plot of effective plastic strain at the final configuration. High plastic strain distribution is observed in the vertical section. A design constraint is imposed for the maximum allowable amount of plastic strain to prevent material failure due to excessive plastic deformation. For this study, the maximum allowable amount of effective plastic strain is assumed to be 0.2.

Since there are six design parameters, design sensitivity equation (10.155) is solved six times, at each converged load step. Thus, an efficient method for solving the linear system of equations is very important in terms of computational cost. The performance measures are chosen for the effective plastic strain e^p and the shape difference ψ between the maximum deformation and after springback, defined as

$$\psi(\boldsymbol{x}) = \int_{\Gamma_C} \|\pi(\boldsymbol{x}) - \boldsymbol{x}\|^2 \, d\Gamma,$$

where $\pi(x)$ is the orthogonally projected position of the particle x on the desired final workpiece shape. Since the effective plastic strain is a path-dependent variable and its sensitivity is updated at each configuration, no additional computation is required to compute the sensitivity of e^p. The shape difference ψ is a function of the material points at the final configuration. Thus, the sensitivity of ψ can be calculated using $\dot{z}$ and the chain rule of differentiation as,

$$\frac{d}{d\tau}[\psi(x)]\bigg|_{\tau=0} = \frac{\partial \psi}{\partial x}^T (V + \dot{z}).$$

The accuracy of sensitivity results can be compared with the finite difference result by slightly perturbing the design and re-solving the same structural problem. The finite difference method computes the sensitivity of the performance measure ψ by

$$\Delta\psi \approx \frac{\psi(x + \Delta\tau V) - \psi(x)}{\Delta\tau} \tag{10.166}$$

for small $\Delta\tau$, which strongly depends on the accuracy of the structural analysis and on a knowledge of the machine operational error.

The continuum-based design sensitivity method presented in this paper yields very accurate and efficient results. Table 10.17 compares the accuracy of the sensitivity ψ' of various performance measures with $\Delta\psi$, with excellent agreement. A very small perturbation ($\Delta\tau = 10^{-6}$) is used for the finite difference results.

In Table 10.17, the design sensitivity of performance ψ does not agree as well as other performance measures. The reason for this is that the magnitude of the performance change is large when compared with the other performance measures. For example, the sensitivity of ψ is 10^3 times larger than the other performance measures for u_6. Thus, the finite difference method in (10.166) contains a large approximation error. The size of the error will decrease if perturbation size is reduced, although this may result in other performance measure inaccuracies due to numerical error. In short, it is very difficult to choose an appropriate perturbation size for the finite difference method.

The meshfree analysis required 8082 sec to solve the deepdrawing problem in Fig. 10.17 using a HP Exemplar workstation, whereas design sensitivity analysis required 1843 seconds for six design parameters, which corresponds to a 3.8% analysis cost per design parameter. Such efficiency is expected since sensitivity analysis uses a decomposed tangent stiffness matrix, and no iteration is required for sensitivity computations.

10.4 Dynamic Problems

A Newmark time integration scheme is used that includes both an implicit and explicit method. The sensitivity expression for the explicit time integration method is simpler and less expensive than for the implicit method. The cost of the sensitivity computation, however, may not be less than the response analysis time. Thus, the merit of a sensitivity computation is reduced when compared with the finite difference method for the explicit time integration method. On the other hand, since the design sensitivity equation using an implicit time integration method does not require iteration, the sensitivity computation is significantly less expensive than with the finite difference method. In this section, shape design sensitivity analysis method for structural transient dynamics is presented. It is well known that for the transient dynamic problem with an initial condition the adjoint

Table 10.17. Accuracy of sensitivity results.

ψ	$\Delta\psi$	ψ'	$\Delta\psi/\psi'\times100$
u_1			
e^P_{41}	1.48092E-08	1.48111E-08	99.99
e^P_{45}	1.39025E-09	1.38995E-09	100.02
e^P_{55}	2.92573E-08	2.92558E-08	100.01
e^P_{142}	6.42704E-09	6.42645E-09	100.01
e^P_{147}	8.75082E-09	8.75167E-09	99.99
e^P_{152}	-4.88503E-08	-4.88486E-08	100.00
e^P_{157}	-2.08880E-08	-2.08875E-08	100.00
G	-4.31897E-05	-4.37835E-05	98.64
u_2			
e^P_{41}	-7.51065E-10	-8.28304E-10	90.68
e^P_{45}	-8.63393E-09	-8.77018E-09	98.45
e^P_{55}	-3.46566E-08	-3.32805E-08	104.13
e^P_{142}	-1.02204E-08	-1.02214E-08	99.99
e^P_{147}	2.30561E-09	2.27053E-09	101.55
e^P_{152}	5.58534E-08	5.58827E-08	99.95
e^P_{157}	2.49171E-08	2.41362E-08	103.24
G	2.51654E-05	2.57379E-05	97.78
u_3			
e^P_{41}	-1.81265E-09	-1.81292E-09	99.99
e^P_{45}	-8.32899E-10	-8.33645E-10	99.91
e^P_{55}	-1.60858E-08	-1.60891E-08	99.98
e^P_{142}	-4.17814E-09	-4.17970E-09	99.96
e^P_{147}	-8.43008E-10	-8.43061E-10	99.99
e^P_{152}	2.65440E-08	2.65487E-08	99.98
e^P_{157}	1.14224E-08	1.14229E-08	99.99
G	1.50596E-05	1.55745E-05	96.69
u_4			
e^P_{41}	-1.64206E-08	-1.64212E-08	100.00
e^P_{45}	-1.78461E-08	-1.78462E-08	100.00
e^P_{55}	-2.06306E-08	-2.06324E-08	99.99
e^P_{142}	5.10648E-09	5.10589E-09	100.01
e^P_{147}	-9.75899E-09	-9.75881E-09	100.00
e^P_{152}	-1.46721E-08	-1.46709E-08	100.01
e^P_{157}	-1.65305E-08	-1.65318E-08	99.99
G	5.16216E-05	5.47740E-05	94.24
u_5			
e^P_{41}	6.32293E-07	6.33335E-07	99.84
e^P_{45}	-3.42150E-07	-3.41924E-07	100.07
e^P_{55}	1.01051E-06	1.01077E-06	99.97
e^P_{142}	4.27720E-07	4.28280E-07	99.87
e^P_{147}	-6.64596E-07	-6.64791E-07	99.97
e^P_{152}	-2.43935E-06	-2.43989E-06	99.98
e^P_{157}	-1.54750E-06	-1.54754E-06	100.00
G	-8.01735E-04	-8.51401E-04	94.17
u_6			
e^P_{41}	5.21628E-07	5.22109E-07	99.91
e^P_{45}	-2.04683E-07	-2.05656E-07	99.53
e^P_{55}	1.20942E-06	1.20331E-06	100.51
e^P_{142}	4.41449E-07	4.42033E-07	99.87
e^P_{147}	-5.38989E-07	-5.39373E-07	99.93
e^P_{152}	-2.39527E-06	-2.39169E-06	100.15
e^P_{157}	-1.42736E-06	-1.42780E-06	99.97
G	-1.19663E-03	-1.25937E-03	95.02

variable method becomes a terminal-value problem, where a terminal condition is given for an adjoint equation. Thus, the adjoint equation cannot be solved simultaneously with the response analysis. This fact significantly complicates calculations associated with transient dynamic design sensitivity analysis using the adjoint variable method. Thus, the direct differentiation method is used instead of the adjoint variable method.

For a linear problem it has been shown that displacement z is Fréchet differentiable with respect to the design [50]. Without mathematical support, it is assumed that displacement z is formally differentiable for the nonlinear structural dynamic problem. If it is assumed that the response analysis is converged up to time t_{n+1} and that the sensitivity equation is solved up to time t_n, then the design sensitivity equation at time t_{n+1} can be obtained by taking the material derivative of (8.199), as

$$\left[d_{\Omega}({}^{n+1}z_{,tt},\bar{z})\right]' + \left[a_{\Omega}({}^{n+1}z,\bar{z})\right]' = \left[\ell_{\Omega}(\bar{z})\right]', \qquad \forall \bar{z} \in Z. \tag{10.167}$$

The material derivative formulas for the structural energy and load linear form are discussed in detail in previous sections. The material derivative of the kinetic energy form is

$$\begin{aligned}\left[d({}^{n+1}z_{,tt},\bar{z})\right]' &= \int_{\Omega} \rho \bar{z}^{T}\,{}^{n+1}\dot{z}_{,tt}\, d\Omega + \int_{\Omega} \rho \bar{z}^{T}\,{}^{n+1}z_{,tt} \mathrm{div}\boldsymbol{V}\, d\Omega \\ &\equiv d({}^{n+1}\dot{z}_{,tt},\bar{z}) + d'_{V}({}^{n+1}z_{,tt},\bar{z}),\end{aligned} \tag{10.168}$$

where $d'_{V}({}^{n+1}z_{,tt},\bar{z})$ is called the kinetic energy fictitious form. Since no spatial derivative is involved in the kinetic energy form, the material derivative formula in (10.168) is very simple compared with the structural energy and load forms.

10.4.1 Implicit Method

Acceleration Form

Using (10.168), design sensitivity equation (10.167) can be expressed as

$$\begin{aligned} d_{\Omega}({}^{n+1}\dot{z}_{,tt},\bar{z}) &+ a^{*}_{\Omega}({}^{n+1}z;\,{}^{n+1}\dot{z},\bar{z}) \\ &= \ell'_{V}(\bar{z}) - a'_{V}({}^{n+1}z,\bar{z}) - d'_{V}({}^{n+1}z_{,tt},\bar{z}), \quad \forall \bar{z} \in Z.\end{aligned} \tag{10.169}$$

This design sensitivity equation is another initial boundary-value problem, which requires initial conditions. The material derivatives vanish for the initial conditions of the response analysis in (8.202) and (8.203). Thus, the initial conditions for the sensitivity equation become

$$\dot{z}(\boldsymbol{x},0) = 0, \qquad \boldsymbol{x} \in \Omega \tag{10.170}$$

$$\dot{z}_{,t}(\boldsymbol{x},0) = 0, \qquad \boldsymbol{x} \in \Omega, \tag{10.171}$$

and the time integration of material derivatives of the kinematic variables follow the same procedure as the response analysis,

$${}^{n+1}\dot{z}_{,t} = {}^{n+1}\dot{z}_{,t} + \gamma \Delta t\, {}^{n+1}\dot{z}_{,tt} \tag{10.172}$$

$${}^{n+1}\dot{z} = {}^{pr}\dot{z} + \beta \Delta t^{2}\, {}^{n+1}\dot{z}_{,tt}, \tag{10.173}$$

where the material derivatives of the predictors can be obtained from the results at the previous time by

$$ {}^{pr}\dot{z}_{,t} = {}^{n}\dot{z}_{,t} + (1-\gamma)\Delta t\, {}^{n}\dot{z}_{,tt} \tag{10.174} $$

$$ {}^{pr}\dot{z} = {}^{n}\dot{z} + \Delta t\, {}^{n}\dot{z}_{,t} + (\tfrac{1}{2}-\beta)\Delta t^{2}\, {}^{n}\dot{z}_{,tt}, \tag{10.175} $$

with (10.170) and (10.171) taken as the initial conditions. By using (10.173), (10.169) is expressed in terms of the acceleration sensitivity as

$$ \begin{aligned} & d_{\Omega}({}^{n+1}\dot{z}_{,tt},\bar{z}) + \beta\Delta t^{2} a^{*}_{\Omega}({}^{n+1}z;\, {}^{n+1}\dot{z}_{,tt},\bar{z}) \\ & = \ell'_{V}(\bar{z}) - a'_{V}({}^{n+1}z,\bar{z}) - d'_{V}({}^{n+1}z_{,tt},\bar{z}) - a^{*}_{\Omega}({}^{n+1}z;\, {}^{pr}\dot{z},\bar{z}), \quad \forall \bar{z} \in Z. \end{aligned} \tag{10.176} $$

The left side of (10.176) has the same form as the response analysis but with a different right side, which is the fictitious load. The sensitivity equation is solved a number of times for each design parameter at a given time. Thus, decomposition of the matrix is important for efficiency. Note that the sensitivity equation in (10.176) does not require any convergence iteration, whereas the response analysis is solved iteratively to obtain the converged configuration in (8.208). Thus, a linear sensitivity equation is effectively solved at each converged time step.

Comparing the sensitivity equation with the quasi-static problem for elastic material yields an interesting observation. It is well known that the sensitivity equation is solved only once at the final converged time for the quasi-static problem, as detailed in Section 10.1. However, the sensitivity equation in (10.176) is solved at each converged time step for the dynamic problem regardless of the constitutive model, since it becomes a path-dependent problem. As a result, the sensitivity computation for the dynamic problem with elastic material is less effective than the quasi-static problem.

As was shown in Section 10.2.1, when the constitutive model is based on rate-form elastoplasticity, the sensitivity equation solves for the material derivative of the incremental displacement. Consequently, the sensitivity equation for the dynamic problem can be modified to an incremental form. Material derivatives of the kinetic and structural energy form are expressed in terms of $\Delta\dot{z}_{,tt}$ and $\Delta\dot{z}$, respectively, as

$$ \begin{aligned} \tfrac{d}{d\tau}\left[d_{\Omega}({}^{n+1}z_{,tt},\bar{z})\right] &= \int_{\Omega} \rho\bar{z}^{T}\Delta\dot{z}_{,tt}\, d\Omega + \int_{\Omega} \bar{z}^{T}\, {}^{n}\dot{z}_{,tt}\, d\Omega + \int_{\Omega} \rho\bar{z}^{T}\, {}^{n+1}z_{,tt} \operatorname{div}\boldsymbol{V}\, d\Omega \\ &\equiv d_{\Omega}(\Delta\dot{z}_{,tt},\bar{z}) + d'_{V}({}^{n+1}z_{,tt},\bar{z}). \end{aligned} \tag{10.177} $$

In (10.177), $d'_{V}({}^{n+1}z_{,tt},\bar{z})$ is different from those terms in the total form, although the same variational notations are used to denote the same functionality. The incremental version of the design sensitivity equation becomes

$$ \begin{aligned} & d_{\Omega}(\Delta\dot{z}_{,tt},\bar{z}) + a^{*}_{\Omega}({}^{n+1}z;\Delta\dot{z},\bar{z}) \\ & = \ell'_{V}(\bar{z}) - a'_{V}({}^{n+1}z,\bar{z}) - d'_{V}({}^{n+1}z_{,tt},\bar{z}), \quad \forall \bar{z} \in Z, \end{aligned} \tag{10.178} $$

with the same initial conditions as those given in (10.170) and (10.171). The kinematic relations are expressed incrementally as

$$ \Delta\dot{z} = \beta\Delta t^{2}\Delta\dot{z}_{,tt} \tag{10.179} $$

$$ {}^{n+1}\dot{z} = {}^{n}\dot{z} + \Delta\dot{z}. \tag{10.180} $$

Thus, the acceleration form of the design sensitivity equation becomes

$$\begin{aligned} & d_\Omega(\Delta\dot{z}_{,tt},\bar{z}) + \beta\Delta t^2 \overset{*}{a}_\Omega({}^{n+1}z_{,tt};\Delta\dot{z},\bar{z}) \\ & \quad = \ell'_V(\bar{z}) - a'_V({}^{n+1}z,\bar{z}) - d'_V({}^{n+1}z_{,tt},\bar{z}), \quad \forall\bar{z}\in Z. \end{aligned} \tag{10.181}$$

This incremental sensitivity equation is first solved for $\Delta\dot{z}_{,tt}$, and $\Delta\dot{z}$ and ${}^{n+1}\dot{z}$ are then obtained using (10.179) and (10.180), respectively. Even if (10.181) solves for the material derivative of the incremental acceleration, the computational cost of the sensitivity equation is almost the same as the total form sensitivity equation.

Displacement Form

The displacement form of the design sensitivity equation starts from (10.169), and ${}^{n+1}\dot{z}_{,tt}$ is expressed in terms of ${}^{n+1}\dot{z}$ as

$${}^{n+1}\dot{z}_{,tt} = {}^{pr}\dot{z}_{,tt} + \frac{1}{\beta\Delta t^2}\,{}^{n+1}\dot{z}, \tag{10.182}$$

where ${}^{pr}\dot{z}_{,tt}$ is the material derivative of the acceleration predictor, defined as

$$\begin{aligned} {}^{pr}\dot{z}_{,tt} &= -\frac{1}{\beta\Delta t^2}\left[{}^{n}\dot{z} + \Delta t\,{}^{n}\dot{z}_{,t} + (\tfrac{1}{2}-\beta)\Delta t^2\,{}^{n}\dot{z}_{,tt}\right] \\ &= -\frac{1}{\beta\Delta t^2}\,{}^{pr}\dot{z}. \end{aligned} \tag{10.183}$$

By substituting the kinematic relation of (10.182) into (10.169), the design sensitivity equation for the displacement form is obtained as

$$\begin{aligned} & \frac{1}{\beta\Delta t^2} d_\Omega({}^{n+1}\dot{z},\bar{z}) + \overset{*}{a}_\Omega({}^{n+1}z;{}^{n+1}\dot{z},\bar{z}) \\ & \quad = \ell(\bar{z}) - a'_V({}^{n+1}z,\bar{z}) - d'_V({}^{n+1}z_{,tt},\bar{z}) - d_\Omega({}^{pr}\dot{z}_{,tt},\bar{z}), \quad \forall\bar{z}\in Z. \end{aligned} \tag{10.184}$$

The difference between (10.176) and (10.184) is apparent. Expression of the fictitious load for the displacement form is simpler than that for the acceleration form. In (10.176), a different $\overset{*}{a}_\Omega({}^{n+1}z;{}^{pr}\dot{z},\bar{z})$ needs to be implemented for each constitutive model, whereas in (10.184), $d_\Omega({}^{pr}\dot{z}_{,tt},\bar{z})$ is independent of the constitutive model and has a simple expression. However, it is possible to show that (10.176) and (10.184) are equivalent by using the relations in (10.182) and (10.183).

10.4.2 Explicit Method

Design sensitivity analysis of the explicit time integration method is equally straightforward. The material derivative of the total displacement is approximated based on the same procedure used for the response analysis at configuration time t_{n+1}

$${}^{n+1}\dot{z} = {}^{n}\dot{z} + \Delta t\,{}^{n}\dot{z}_{,t} + \tfrac{1}{2}\Delta t^2\,{}^{n}\dot{z}_{,tt}. \tag{10.185}$$

The sensitivity equation for the explicit integration method is always a total form regardless of the constitutive model, as

$$d_\Omega({}^{n+1}\dot{z}_{,tt},\bar{z}) = \ell(\bar{z}) - \overset{*}{a}_\Omega({}^{n+1}z;{}^{n+1}\dot{z},\bar{z}) - a'_V({}^{n+1}z,\bar{z}) - d'_V({}^{n+1}z_{,tt},\bar{z}), \quad \forall\bar{z}\in Z. \tag{10.186}$$

After computing the material derivative of the acceleration, the material derivative of the velocity is updated using

$$ {}^{n+1}\dot{z}_{,t} = {}^{n}\dot{z}_{,t} + (1-\gamma)\Delta t\, {}^{n}\dot{z}_{,tt} + \gamma\Delta t\, {}^{n+1}\dot{z}_{,tt}. \quad (10.187) $$

From the expression of the fictitious load form in (10.186), it is necessary to compute $a^{*}_{\Omega}({}^{n+1}z; {}^{n+1}\dot{z}, \bar{z})$, which the response analysis does not require. This computation of the linearized structural variational form may increase design sensitivity computation costs more than that of the response analysis.

10.4.3 Numerical Examples

Bumper Contact Problem

A vehicle bumper is installed to protect the body from impact and to absorb the impact energy through plastic deformation. Vehicle design regulations require that the bumper be safe up to a 5-mph impact. In this section, the design optimization of a vehicle bumper structure is performed to reduce plastic deformation. The design optimization problem based on a quasi-static assumption was presented in Kim et al. [36]. However, an actual impact happens in a very short time period even with a 5-mph impact. The same problem is also solved with a transient dynamic assumption in this section.

The cross section of the metal bumper is approximated using 144 meshfree particles and 71 integration zones, as shown in Fig. 10.18. The elastoplastic material constants are Young's modulus E = 206.9 GPa, Poisson's ratio ν = 0.29, plastic hardening modulus H = 1.1 GPa, and initial yield stress σ_y = 0.5 GPa. Linear isotropic hardening is considered where the plastic consistency condition can be solved explicitly without iteration. The Newmark implicit time integration method is used with parameters β = 0.26 and γ = 0.5, and a density ρ = 7800 kg/m^3. Response analysis is carried out up to 0.01 sec. Since the problem includes impact (i.e., contact) with a rigid surface, the following linearized system of the equation is solved instead of (8.211):

$$ \begin{aligned} d(\Delta z^{k+1}_{,tt}, \bar{z}) + \beta\Delta t^2 \left[a^{*}(z^k; \Delta z^{k+1}_{,tt}, \bar{z}) + b^{*}(z^k; \Delta z^{k+1}_{,tt}, \bar{z}) \right] \\ = \ell(\bar{z}) - a(z^k, \bar{z}) - b(z^k, \bar{z}) - d(z^k_{,tt}, \bar{z}), \qquad \forall \bar{z} \in Z, \end{aligned} $$

where $b(z,\bar{z})$ and $b^{*}(z;\Delta z,\bar{z})$ are the frictional contact variational form and its linearization.

Figure 10.19 shows the contour plots of the effective plastic strain and von Mises stress at the final time. The vertical coordinates of the lower contact point remain almost steady for the quasi-static problem because of frictional effects, whereas in the transient dynamic problem, the contact point moves up, since no friction is applied during the oscillating period between a contact and noncontact situation. The majority of contact forces are retained by two endpoints. The concentration of plastic strain appears at the upper region of the bumper, and the stress concentration magnitude is similar at upper and lower regions. Figure 10.20 shows the time history of displacement, velocity, and acceleration of the upper contact point. A very high response frequency is observed for acceleration.

The boundary shape of the bumper is represented by a cubic spline curve and each particle point on the boundary has a unique parametric representation. The locations of the control points of each boundary curve are chosen as shape design parameters. The design velocity vector corresponding to the particle point can be computed using a parametric representation. Since the bumper is usually manufactured by a sheet metal stamping process, it is inappropriate to change the thickness of each section. To maintain a constant thickness of 0.5 cm, design parameters are linked in the thickness direction corresponding to the inner/outer control points. Sixteen shape design parameters are chosen as shown in Fig. 10.18. After choosing the design parameters, the design velocity field is obtained by perturbing the boundary curve in the direction of each design parameter.

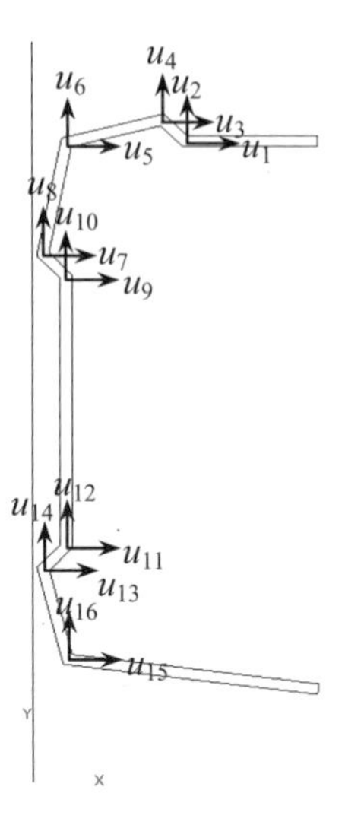

Figure 10.18. Bumper cross-sectional geometry and shape design parameterization.

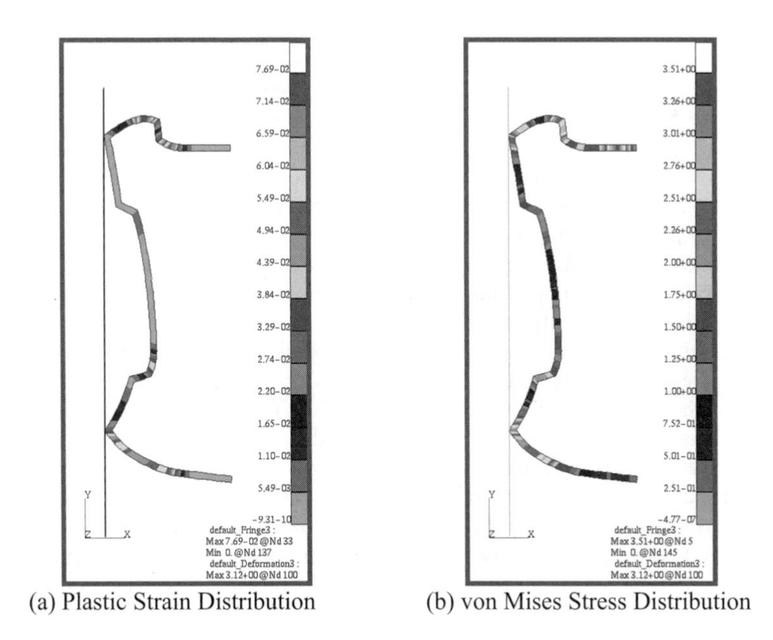

(a) Plastic Strain Distribution (b) von Mises Stress Distribution

Figure 10.19. Meshfree analysis results.

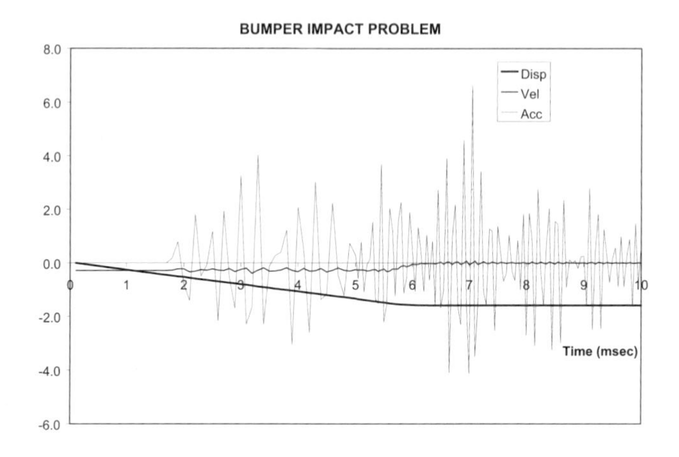

Figure 10.20. Time history of displacement, velocity, and acceleration.

Using the design velocity information, design sensitivity analysis is carried out. Since the problem includes impact (i.e., contact) with a rigid surface, the following linear system of sensitivity equations is solved instead of (10.176):

$$\begin{aligned} d(\dot{z}_{,tt},\bar{z}) + \beta\Delta t^2\left[a^*(z;\dot{z}_{,tt},\bar{z}) + b^*(z;\dot{z}_{,tt},\bar{z})\right] \\ = \ell'_V(\bar{z}) - a'_V(z,\bar{z}) - b'_V(z,\bar{z}) - d'_V(z_{,tt},\bar{z}) \\ - a^*(z;\dot{z}^{pr},\bar{z}) - b^*(z;\dot{z}^{pr},\bar{z}), \qquad \forall\bar{z}\in Z, \end{aligned} \tag{10.188}$$

where $b'_V(z,\bar{z})$ is the contact fictitious load derived in Section 10.3. Equation (10.188) is solved using the factorized tangent stiffness matrix from the response analysis with the fictitious load. No iteration is required to solve the sensitivity equation, but (10.188) is repeatedly solved for each design parameter. This procedure is quite efficient compared with iterative response analysis. The sensitivity coefficients of the performance measures are computed after solving the design sensitivity equation at the final converged load step. Possible performance measures include the displacement, stress tensor, internal plastic variables, reaction force, contact force, and the normal gap distance.

To illustrate the efficiency of the method, the computation time required by response analysis and sensitivity analysis are compared. Response analysis requires 1599 sec, whereas sensitivity analysis requires 853 sec for 16 design parameters, which is less than 3.3% of the response analysis time per design parameter. This ratio is quite efficient compared with the finite difference method. Such efficiency is possible because the sensitivity equation is solved without iteration and the factorized tangent stiffness matrix from response analysis is used. Table 10.18 shows the sensitivity coefficients and provides a comparison of sensitivity and finite difference results. Very accurate sensitivity results are obtained even for highly nonlinear behavior. The second column, $\Delta\psi$, presents first-order sensitivity results from the forward finite difference method using a perturbation of $\Delta\tau = 10^{-6}$, and the third column represents the sensitivity computation results from the presented method. z, e^p, and F_C in the first column are performance measures such as the displacement, effective plastic strain, and contact force, respectively. For example, $e^p{}_{15}$ denotes the effective plastic strain at integration zone 15, and F_{Cx100} denotes the x-directional contact force at slave node 100.

Table 10.18. Comparison of sensitivity accuracy.

ψ	$\Delta\psi$	ψ'	$(\Delta\psi/\psi')\times 100$
u_2			
$e^P{}_{15}$	-.754098E-7	-.754105E-7	100.00
$e^P{}_{65}$	.313715E-7	.313668E-7	100.02
$e^P{}_{29}$	.441192E-7	.441162E-7	100.01
z_{x39}	.790973E-5	.791092E-5	99.98
F_{Cx100}	-.657499E-6	-.657074E-6	100.06
u_4			
$e^P{}_{15}$	.268699E-6	.268712E-6	100.00
$e^P{}_{65}$	-.843101E-9	-.863924E-9	97.59
$e^P{}_{29}$	.123988E-6	.123993E-6	100.00
z_{x39}	-.847749E-5	-.847586E-5	100.02
F_{Cx100}	.410724E-7	.407515E-7	100.79
u_6			
$e^P{}_{15}$	-.317362E-6	-.317349E-6	100.00
$e^P{}_{65}$	-.640031E-7	-.640159E-7	99.98
$e^P{}_{29}$	-.163051E-6	-.163051E-6	100.00
z_{x39}	-.190521E-5	-.190392E-5	100.07
F_{Cx100}	.473040E-6	.472876E-6	100.03
u_8			
$e^P{}_{15}$	.888094E-8	.890589E-8	99.72
$e^P{}_{65}$	.355128E-7	.354794E-7	100.09
$e^P{}_{29}$	-.981276E-8	-.981572E-8	99.97
z_{x39}	-.239706E-5	-.239333E-5	100.16
F_{Cx100}	-.184457E-6	-.183954E-6	100.27
u_{10}			
$e^P{}_{15}$	-.642594E-8	-.643542E-8	99.85
$e^P{}_{65}$	-.151580E-7	-.151527E-7	100.03
$e^P{}_{29}$	.172663E-7	.172698E-7	99.98
z_{x39}	-.154011E-5	-.154125E-5	99.93
F_{Cx100}	-.134372E-6	-.134701E-6	99.76

11 Nonlinear Configuration Design Sensitivity Analysis

The development of configuration design sensitivity analysis is extended from Chapter 7 to include such problems as nonlinear elasticity, nonlinear buckling, and elastoplasticity. The fundamental principle of configuration design is the same with linear problems. Configuration design variables are characterized by shape and orientation changes of structural components. For built-up structures, the effect of configuration design variables on such structural performance measures as displacement, stress, and critical load may be more significant than those of sizing and shape design variables. In this chapter, a configuration design sensitivity formulation is developed for nonlinear structures. The material derivative used to develop the shape design sensitivity is extended to account for the effects of orientation design change.

One of the important conclusions from Chapter 7 is that C^0-continuous design velocity fields must be used for practical design sensitivity applications. The piecewise linear design velocity field (i.e., C^0-continuous) will support configuration design changes of a broad class of built-up structures with beams and plates. To allow use of the C^0-continuous design velocity field, mathematical models of beam and plate bending must be second-order differential equations, so that only first-order derivatives appear on the integrand of the energy equation and, thus, on the integrand of the configuration design sensitivity equation. Since the Timoshenko beam [51] and Mindlin/Reissner plate theories [52] use the displacement and rotation degrees-of-freedom to describe the structural response, mathematical models of beam and plate bending are reduced to second-order differential equations, which nicely fits with the configuration design velocity field requirement. When these theories are used in numerical applications, they involve well-known isoparametric finite element formulations in which displacements and rotations are used as degrees-of-freedom. Consequently, the configuration design sensitivity method will be applicable to nonlinear structures based on Timoshenko beam and Mindlin/Reissner plate theories.

When an updated Lagrangian formulation is used for the large deformation problem, the reference configuration is the current deformed geometry of the structure. However, the reference for the design is always the undeformed, initial geometry. A discrepancy thus exists between analysis and design reference configurations. In the shape design sensitivity formulation in Chapter 10, such a discrepancy was resolved using a transformation between two configurations. However, the corresponding transformation method for the configuration design has not been developed. This is partly because such structural components as beam and plate are used for the configuration design, whereas the continuum solid is used for the shape design. Thus, the conventional design velocity updating method must be employed to calculate the design velocity field at the current configuration. Hopefully, a unified configuration design representation method can be developed in the near future that does not require such an updating process.

11.1 Nonlinear Elastic Problems

In configuration design sensitivity analysis, the orientation angle of the structural component is the design variable. Since a continuum solid component does not have any response variables related to rotation, configuration design is related to truss, beam, and plate components. As with Chapter 7, a line and surface design component will be considered in this section. Two assumptions are used to develop a configuration design sensitivity formulation: that the design component rotates without shape change, and that only a small design perturbation is considered. The first assumption limits the orientation design velocity so that an initially flat component will remain flat during the design change. The second assumption allows rotational movement to be taken separately from shape change.

11.1.1 Configuration Design Sensitivity for Static Response

The material derivative formulas developed in Section 7.1 can be used for configuration design sensitivity analysis for nonlinear structures by assuming that design perturbation occurs in the undeformed configuration. As with Chapter 8, in order to distinguish the current configuration from the undeformed configuration $\boldsymbol{X}$ stands for the material point in the undeformed configuration, while $\boldsymbol{x}$ represents the same material point in the deformed configuration. The relation between two configurations can be made using the displacement $\boldsymbol{z}$, as

$$\boldsymbol{x} = \boldsymbol{X} + \boldsymbol{z}. \tag{11.1}$$

For nonlinear problems, the material derivative formulas in Section 7.1 must be defined in the undeformed configuration, that is, at the material point $\boldsymbol{X}$. If more than two configurations appear in the derivation, a left superscript will be used to distinguish them. For example, ${}^{n}\boldsymbol{z}$ denotes the displacement at time t_n, while ${}^{n-1}\boldsymbol{z}$ denotes time t_{n-1}. In the same notation, the structural domain at the undeformed configuration is indicated by ${}^{0}\Omega$, while the deformed domain is shown by ${}^{n}\Omega$.

In the configuration design, let the original structural domain ${}^{0}\Omega$ be transformed into the perturbed domain ${}^{0}\Omega_{\tau}$. In this case, ${}^{0}\Omega$ is not only the original domain in the design, but also the undeformed domain in the nonlinear problem. As with shape design, design perturbation is denoted using a scalar parameter τ, such that $\tau = 0$ corresponds to the original design. In general, design perturbation contains both shape and orientation changes. From the assumption of a small design perturbation and independence of shape and orientation changes, the decomposition of the design velocity in (7.1) is still valid for nonlinear problems. At the undeformed domain, (7.1) can be rewritten as

$$\boldsymbol{V}(\boldsymbol{X}) = \boldsymbol{V}_{\Omega}(\boldsymbol{X}) + \boldsymbol{V}_{\Theta}(\boldsymbol{X}). \tag{11.2}$$

In the line design component, $\boldsymbol{V}_{\Omega}(\boldsymbol{X})$ moves in the axial direction, while $\boldsymbol{V}_{\Theta}(\boldsymbol{X})$ moves in the perpendicular direction to $\boldsymbol{V}_{\Omega}(\boldsymbol{X})$. In the surface design component, $\boldsymbol{V}_{\Omega}(\boldsymbol{X})$ is in the direction on the plane, and $\boldsymbol{V}_{\Theta}(\boldsymbol{X})$ is normal to the plane. The shape design sensitivity formulation in Chapter 10 will be used when the shape design velocity field is given as $\boldsymbol{V}_{\Omega}(\boldsymbol{X})$. The decomposition in (11.2) is valid only when components are straight or flat.

Direct Differentiation Method

In the direct differentiation method, the displacement sensitivity is first calculated from the design sensitivity equation, and performance sensitivity is then calculated using the chain rule of differentiation. The design velocity in (11.2) corresponds to one design variable. Thus, the design sensitivity equation should be solved for each design variable.

It is clear that the contributions of shape and orientation designs are separated from decomposability in (11.2). It will be shown in the following derivations that the explicitly dependent terms in the design sensitivity equation will also be separated between two designs. Using (8.5), the variational equation of a general nonlinear elastic problem in the perturbed domain can be written as

$$a_{{}^0\Omega_\tau}(z_\tau, \bar{z}_\tau) = \ell_{{}^0\Omega_\tau}(\bar{z}_\tau), \quad \forall \bar{z}_\tau \in Z_\tau, \tag{11.3}$$

where Z_τ is the space of kinematically admissible displacements, which satisfies homogeneous essential boundary conditions in the perturbed domain. Note that the undeformed domain ${}^0\Omega_\tau$ is used in (11.3), which indicates the total Lagrangian formulation of the structural problem.

In many structural problems considered in this text, nonlinearity comes from the energy form $a_{{}^0\Omega_\tau}(\bullet,\bullet)$. Thus, the material derivative of the energy form is accompanied by the linearization procedure. Even if an explicit expression of the energy form is not yet provided, the material derivative of the energy form in (11.3) can be symbolically written as

$$[a_{{}^0\Omega}(z, \bar{z})]' \equiv a^*_{{}^0\Omega}(z; \dot{z}, \bar{z}) + a'_{V_\Omega}(z, \bar{z}) + a'_{V_\theta}(z, \bar{z}), \tag{11.4}$$

where $a^*_{{}^0\Omega}(z; \dot{z}, \bar{z})$ is the linearized energy form developed in (8.20) and implicitly depends on the design through $\dot{z}$, $a'_{V_\Omega}(z, \bar{z})$ is the explicitly dependent term due to the shape design, and $a'_{V_\theta}(z, \bar{z})$ is the explicitly dependent term due to the orientation design. For most elastic materials, $a^*_{{}^0\Omega}(z; \dot{z}, \bar{z})$ is bilinear and symmetric with respect to its arguments. As with the linear problem in (7.14), $\dot{z} = \dot{z}_{V_\Omega} + \dot{z}_{V_\theta}$ includes the effects of both shape and orientation designs. The expressions of those forms on the right side of (11.4) depend on the structural components under consideration, and will be determined in Section 11.1.3.

For simplicity, only the conservative load is considered and, thus, the load form $\ell_{{}^0\Omega_\tau}(\bullet)$ is linear in its argument. The material derivative of the load form can be written as

$$[\ell_{{}^0\Omega}(\bar{z})]' \equiv \ell'_{V_\Omega}(\bar{z}) + \ell'_{V_\theta}(\bar{z}). \tag{11.5}$$

In contrast to the energy form, the material derivative of the load form does not include implicitly dependent terms. This is not true if the load form is not conservative, such as the follow-up or pressure load. For a given design velocity, two forms on the right side of (11.5) can be explicitly calculated. The expressions of these forms depend on the structural components considered and will be determined in Section 11.1.3.

The design sensitivity equation for the configuration design can now be obtained by substituting (11.4) and (11.5) into the material derivative of (11.3) as

$$a^*_{{}^0\Omega}(z; \dot{z}, \bar{z}) = \ell'_{V_\Omega}(\bar{z}) + \ell'_{V_\theta}(\bar{z}) - a'_{V_\Omega}(z, \bar{z}) - a'_{V_\theta}(z, \bar{z}), \quad \forall \bar{z} \in Z, \tag{11.6}$$

where the implicitly dependent term is moved to the left side, while the explicitly dependent terms are placed on the right side. Equation (11.6) is the configuration design sensitivity equation of the nonlinear structural problem in (8.5). When the structural problem is solved for the displacement z and when the design velocity field in (11.2) is given, (11.6) needs to be solved for $\dot{z}$, the material derivative of the displacement. Note that the linearized (11.6) is similar to structural incremental (8.21). Thus, it is very efficient to solve (11.6) using information from the nonlinear structural analysis.

From a computational viewpoint, the sensitivity equation (11.6) solves for the total displacement sensitivity at time t_n. The right side of (11.6), which is called the fictitious load, only depends on the design velocity at t_0 and the total displacement at time t_n. No information between t_0 and t_n is required for sensitivity calculation. Thus, even if the nonlinear structural problem is solved incrementally and iteratively as described in (8.21), the design sensitivity equation (11.6) needs to be solved once at time t_n without any iteration. This procedure provides significant efficiency in the calculation of sensitivity. When an elastic material is used, the structural problem is path-independent and so is the design sensitivity analysis. However, in the case of the elastoplastic material, the structural problem as well as the design sensitivity analysis is path-dependent.

Consider a structural performance measure, defined in the undeformed domain as

$$\psi = \iint_{{}^0\Omega_\tau} g(z, \nabla_0 z)\, d^0\Omega, \tag{11.7}$$

where function g is continuously differentiable with respect to its arguments. The functional ψ includes the first-order gradient of the state variable z. The functional defined on the deformed domain ${}^n\Omega$ can easily be converted into the form in (11.7) by using the relation of $d^n\Omega = Jd^0\Omega$, where J is the determinant of the deformation gradient defined in (8.2). Taking the first variation of ψ and using the relations in (6.37) and (7.23), the variation of (11.7) can be obtained as

$$\psi' = \psi'_{V_\Omega} + \psi'_{V_\theta} = \iint_{{}^0\Omega} [g_{,z}(z'_{V_\Omega} + z'_{V_\theta}) + g_{,\nabla_0 z} : (\nabla_0 z'_{V_\Omega} + \nabla_0 z'_{V_\theta}) + div(g\boldsymbol{V}_\Omega)]\, d^0\Omega. \tag{11.8}$$

Using (6.8) and (7.11) and using the fact that $\dot{z} = \dot{z}_{V_\Omega} + \dot{z}_{V_\theta}$, (11.8) becomes

$$\begin{aligned}
\psi' = &\iint_{{}^0\Omega} [g_{,z}\dot{z} + g_{,\nabla_0 z} : \nabla_0 \dot{z}]\, d^0\Omega \\
&- \iint_{{}^0\Omega} [g_{,z}(\boldsymbol{V}_\theta z) + g_{,\nabla_0 z} : \nabla_0(\boldsymbol{V}_\theta z)]\, d^0\Omega \\
&- \iint_{{}^0\Omega} [g_{,\nabla_0 z} : (\nabla_0 z \nabla_0 \boldsymbol{V}_\Omega) - g\, div \boldsymbol{V}_\Omega]\, d^0\Omega,
\end{aligned} \tag{11.9}$$

where $\boldsymbol{V}_\theta$ is the design velocity matrix for orientation change, that is, it is the same as (7.12) for the line component and (7.19) for the surface component. Using the given design velocity in (11.2), the solution z to the structural problem in (8.5), and the solution $\dot{z}$ to the design sensitivity equation (11.6), the sensitivity of the performance measure can be calculated.

Note that the first integral in (11.9) is the contribution from the implicitly dependent term, the second integral is the contribution from the orientation design velocity, and the last term is the contribution from the shape design velocity. In (11.9), the displacement sensitivity $\dot{z}$ contains the first-order gradient. Since $\dot{z}$ is the solution to the variational (11.6), it has the same regularity as the displacement z. Thus, $\nabla_0 \dot{z}$ is well defined. In the second integral, the first-order gradient of $\boldsymbol{V}_\theta$ appears. From its expression in (7.12) or (7.19), $\boldsymbol{V}_\theta$ includes the first-order gradient of the design velocity. Thus, the second integral of (11.9) has the second-order derivatives of the orientation design velocity. However, from the assumption that the flat surface remains flat and the straight line remains straight, the second-order derivatives of the orientation design velocity vanish, i.e., $\nabla_0 \boldsymbol{V}_\theta = 0$. Thus, the second integral of (11.9) is also well defined if the C^0-continuous design velocity field is defined. Finally, the last integral of (11.9) contains the first-order gradient of the shape design velocity. Again, the C^0-continuous design velocity field provides enough smoothness in defining the integrands of (11.9). As a result, the C^0-continuous design velocity field with a constraint on flatness or straightness is necessary to calculate the sensitivity of the performance measure in (11.9).

Adjoint Variable Method

As with Chapters 9 and 10, the adjoint problem is defined using the implicitly dependent terms in (11.9). The displacement sensitivity $\dot{z}$ in the first integral of (11.9) is replaced with the virtual displacement $\bar{\lambda}$ and equated with the linearized energy form to define the adjoint problem as

$$a^*_{^0\Omega}(z;\lambda,\bar{\lambda}) = \iint_{^0\Omega}(g_{,z}\bar{\lambda} + g_{,\nabla_0 z} : \nabla_0\bar{\lambda})d^0\Omega, \quad \forall\bar{\lambda} \in Z. \tag{11.10}$$

Note that the above adjoint equation satisfies for any $\bar{\lambda}$ that belongs to Z. Since $\dot{z}$ is the solution to the variational equation (11.6), it belongs to Z. Thus, it is possible to replace $\bar{\lambda}$ with $\dot{z}$, to yield

$$a^*_{^0\Omega}(z;\lambda,\dot{z}) = \iint_{^0\Omega}(g_{,z}\dot{z} + g_{,\nabla_0 z} : \nabla_0\dot{z})d^0\Omega, \tag{11.11}$$

where the right side is the same as the first integral in (11.9). The object is to calculate the right side without evaluating $\dot{z}$. Note that λ belongs to Z, since it is the solution to the variational (11.10). Thus, it is possible to replace $\bar{z}$ in (11.6) with λ to obtain the following relation:

$$a^*_{^0\Omega}(z;\dot{z},\lambda) = \ell'_{V_\Omega}(\lambda) + \ell'_{V_\theta}(\lambda) - a'_{V_\Omega}(z,\lambda) - a'_{V_\theta}(z,\lambda). \tag{11.12}$$

From the symmetric property of the linearized energy form, the left sides of (11.11) and (11.12) are the same. Thus, by equating the right sides of two equations we can express the first integral of (11.9) in terms of the adjoint response, as

$$\iint_{^0\Omega}(g_{,z}\dot{z} + g_{,\nabla_0 z} : \nabla_0\dot{z})d^0\Omega = \ell'_{V_\Omega}(\lambda) + \ell'_{V_\theta}(\lambda) - a'_{V_\Omega}(z,\lambda) - a'_{V_\theta}(z,\lambda). \tag{11.13}$$

Thus, the implicitly dependent terms are expressed using the adjoint response and design velocity. Substituting (11.13) into (11.9) yields

$$\begin{aligned}\psi' &= \ell'_{V_\Omega}(\lambda) + \ell'_{V_\theta}(\lambda) - a'_{V_\Omega}(z,\lambda) - a'_{V_\theta}(z,\lambda) \\ &\quad - \iint_{^0\Omega}[g_{,z}(V_\theta z) + g_{,\nabla_0 z} : \nabla_0(V_\theta z)]d^0\Omega \\ &\quad - \iint_{^0\Omega}[g_{,\nabla_0 z} : (\nabla_0 z \nabla_0 V_\Omega) - g\,divV_\Omega]d^0\Omega.\end{aligned} \tag{11.14}$$

In order to evaluate the sensitivity of performance measure ψ, the structural response z from (8.5), the adjoint response λ from (11.10), and the design velocity field are required. Since it is only necessary to solve one adjoint problem, the adjoint variable method is efficient when the number of design variables is greater than the number of performance measures.

It should be remembered that the prime (′) notation is used for the material derivative of the functional ψ instead of a dot (·). A difference between the prime and the dot only appears when a local function is defined in the domain. In the case of a functional that is defined as an integral over the domain, there is no difference between the prime and the dot. The prime notation is used on the left side of (11.4) for this same reason.

In the derivations of direct differentiation and adjoint variable methods, symbolic notations are used for $a'_{V_\Omega}(z,\bar{z})$, $a'_{V_\theta}(z,\bar{z})$, $\ell'_{V_\Omega}(\bar{z})$, and $\ell'_{V_\theta}(\bar{z})$. In Section 11.1.3, their expressions are developed for structural components.

11.1.2 Configuration Design Sensitivity for Critical Load

In this section, a configuration design sensitivity formulation for the critical load factor is developed. Even if the one-point and two-point approaches in Section 8.1.2 may provide an accurate approximation of the actual critical load factor, their sensitivity may not converge to the sensitivity of the actual load factor, as discussed in Section 8.1.2. It is always dangerous to differentiate the approximated function. Thus, only the sensitivity of the actual critical load factor is developed in this section.

In linear problems, the eigenvalue sensitivity of the structural problem does not require the sensitivity of the eigenfunction. Thus, there is no need to solve the sensitivity equation for the direct differentiation method or the adjoint problem for the adjoint variable method. This same principle can be applied to the sensitivity of the nonlinear critical load. The sensitivity expression of the critical load includes the state response and the eigenfunction.

From the definition of the critical load factor given in (8.58), the stability equation for the nonlinear structural problem at its final configuration time t_n for the perturbed domain ${}^0\Omega_\tau$ is given as

$$a_{{}^0\Omega_\tau}(z_\tau, \bar{z}_\tau) = \beta_\tau \ell_{{}^0\Omega_\tau}(\bar{z}_\tau), \quad \forall \bar{z}_\tau \in Z_\tau, \tag{11.15}$$

where β_τ is the actual critical load factor at the perturbed design, which can only be evaluated after the critical load is known. The critical load at the perturbed design can be written, using (8.59), as

$$\boldsymbol{p}_{cr_\tau} = \beta_\tau \boldsymbol{p}_0. \tag{11.16}$$

Note that the total applied load vector $\boldsymbol{p}_0$ does not depend on design perturbation. As the design changes, the system has a new critical load and consequently, a new critical load factor. Thus, it is clear that $\boldsymbol{p}_{cr}$ and β depend on the design.

Assuming the differentiability of the critical load factor with respect to the configuration design variable, the first variation of (11.15) becomes

$$\begin{aligned} &a^*_{{}^0\Omega}(z; \dot{z}, \bar{z}) + a'_{V_\Omega}(z, \bar{z}) + a'_{V_\theta}(z, \bar{z}) \\ &\quad = \beta' \ell_{{}^0\Omega}(\bar{z}) + \beta \ell'_{V_\Omega}(\bar{z}) + \beta \ell'_{V_\theta}(\bar{z}), \quad \forall \bar{z} \in Z, \end{aligned} \tag{11.17}$$

where the contribution of $\dot{\bar{z}}$ is ignored for the reason explained in (6.63). In contrast to the static problem, the variation of the load factor appears on the right side. In fact, β' includes the contribution from shape and orientation design changes. Let $\boldsymbol{y}$ be the eigenfunction associated with the simple, smallest eigenvalue of the linear eigenvalue problem. Since (11.17) satisfies for any $\bar{z} \in Z$, and since the eigenfunction $\boldsymbol{y} \in Z$, substituting $\bar{z}$ for $\boldsymbol{y}$ yields

$$a^*_{{}^0\Omega}(z; \dot{z}, \boldsymbol{y}) + a'_{V_\Omega}(z, \boldsymbol{y}) + a'_{V_\theta}(z, \boldsymbol{y}) = \beta' \ell_{{}^0\Omega}(\boldsymbol{y}) + \beta \ell'_{V_\Omega}(\boldsymbol{y}) + \beta \ell'_{V_\theta}(\boldsymbol{y}). \tag{11.18}$$

Equation (11.18) still requires the evaluation of $\dot{z}$, which is the solution to the design sensitivity equation. At the critical limit point, the slope of the load-displacement curve becomes zero, as shown in Figure 8.5. Thus, the linearized energy form in (11.18) vanishes at this point, represented as

$$a^*_{{}^0\Omega}(z; \dot{z}, \boldsymbol{y}) = 0. \tag{11.19}$$

After solving, the sensitivity of the critical load factor becomes

$$\beta' = \frac{1}{\ell_{^0\Omega}}\left\{a'_{V_\Omega}(z,y) + a'_{V_\theta}(z,y) - \beta[\ell'_{V_\Omega}(y) + \ell'_{V_\theta}(y)]\right\}. \tag{11.20}$$

In addition, after differentiating (11.16) and substituting (11.20), the sensitivity of the critical load can be obtained as

$$p'_{cr} = \beta' p_0 = \frac{p_0}{\ell_{^0\Omega}}\left\{a'_{V_\Omega}(z,y) + a'_{V_\theta}(z,y) - \beta[\ell'_{V_\Omega}(y) + \ell'_{V_\theta}(y)]\right\}. \tag{11.21}$$

As mentioned earlier, the sensitivity of the critical load does not require the solution to the design sensitivity equation or the solution to the adjoint problem. The structural response at the critical load and the eigenfunction of the linear eigenvalue problem can be used to calculate p'_{cr} from (11.21). Thus, sensitivity calculation of the critical load is more efficient than that of the static response.

11.1.3 Analytical Examples

Truss Component

A two-dimensional truss structure is considered as a line design component. From (8.32), the structural energy and load forms of the truss component at the perturbed design can be written as

$$a_{^0\Omega_\tau}(z_\tau, \bar{z}_\tau) = \int_0^{l_\tau} S_{11_\tau} \bar{E}_{11_\tau}\, dX_1 \tag{11.22}$$

$$\ell_{^0\Omega_\tau}(\bar{z}_\tau) = \int_0^{l_\tau} \bar{z}_{1_\tau} f_{1_\tau}\, dX_1, \tag{11.23}$$

where S_{11} is the second Piola-Kirchhoff stress in one dimension and $\bar{E}_{11} = \bar{z}_{1,1}(1 + z_{1,1})$ is the variation of the Green-Lagrange strain.

Although the truss component changes along its axial direction, it is assumed that the truss component is part of the built-up structure. Thus, the X_2-directional displacement and orientation design velocity exist, i.e., $z = [z_1, z_2]^T$ and $V = [V_1, V_2]^T$. Let the truss component be perturbed in the direction of design velocity $V = [V_1, V_2]^T$, in which the shape design velocity is $V_\Omega = [V_1]$ and the orientation design velocity is $V_\Theta = [V_2]$. In order to remain straight during design perturbation, the orientation design velocity V_2 must be a linear function of X_1. Since only the axial component (X_1-direction) is considered, only the first row of the matrix V_θ in (7.12) can be taken to define

$$V_\theta = [0, \ -V_{2,1}]. \tag{11.24}$$

In order to differentiate the structural energy form, let us consider the material derivative of the Green-Lagrange strain [see (8.34)], as

$$\dot{E}_{11} = (1 + z_{1,1})(\dot{z}_{1,1} - z_{1,1}V_{1,1} + V_{2,1}z_{2,1}), \tag{11.25}$$

where $z_{1,1}$ is the gradient of z_1 with respect to the X_1-coordinate, i.e., $z_{1,1} = \partial z_1/\partial X_1$. Using (11.25), the material derivative of the second Piola-Kirchhoff stress defined in (8.34) can be obtained as

$$\dot{S}_{11} = EA(1 + z_{1,1})(\dot{z}_{1,1} - z_{1,1}V_{1,1} + V_{2,1}z_{2,1}). \tag{11.26}$$

In addition, the material derivative of the Green-Lagrange strain variation can be obtained as

$$\dot{\bar{E}}_{11} = (1+z_{1,1})(-\bar{z}_{1,1}V_{1,1} + V_{2,1}\bar{z}_{2,1}) + \bar{z}_{1,1}(\dot{z}_{1,1} - z_{1,1}V_{1,1} + V_{2,1}z_{2,1}). \tag{11.27}$$

Substituting (11.26) and (11.27) into the variation of the structural energy form yields

$$\begin{aligned}[a_{^0\Omega}(z,\bar{z})]' = & \int_0^l [EA\bar{z}_{1,1}\dot{z}_{1,1}(1+z_{1,1})^2 + S_{11}\bar{z}_{1,1}\dot{z}_{1,1}]dX_1 \\ & - \int_0^l [EA(1+z_{1,1})^2\bar{z}_{1,1}z_{1,1} + S_{11}\bar{z}_{1,1}z_{1,1}]V_{1,1}\,dX_1 \\ & + \int_0^l [EA(1+z_{1,1})^2\bar{z}_{1,1}z_{2,1} + S_{11}(\bar{z}_{2,1} + \bar{z}_{1,1}\bar{z}_{2,1} + \bar{z}_{2,1}z_{1,1})]V_{2,1}\,dX_1. \end{aligned} \tag{11.28}$$

The first integral includes the material derivative of the displacement and is the same as the linearized energy form in (8.37) by replacing $\dot{z}$ with Δz, which can be denoted as $a^*_{^0\Omega}(z;\dot{z},\bar{z})$. The second integral is the contribution from the shape design change, while the third integral is from the orientation design change. Thus, the following structural fictitious loads can be defined:

$$a'_{V_\Omega}(z,\bar{z}) = -\int_0^l [EA(1+z_{1,1})^2\bar{z}_{1,1}z_{1,1} + S_{11}\bar{z}_{1,1}z_{1,1}]V_{1,1}\,dX_1 \tag{11.29}$$

$$a'_{V_\theta}(z,\bar{z}) = \int_0^l [EA(1+z_{1,1})^2\bar{z}_{1,1}z_{2,1} + S_{11}(\bar{z}_{2,1} + \bar{z}_{1,1}z_{2,1} + \bar{z}_{2,1}z_{1,1})]V_{2,1}\,dX_1. \tag{11.30}$$

The load form in (11.23) is linear in its argument due to the conservative load. After separating shape and orientation parts, the variation of the load form in (11.23) can be obtained as

$$\ell'_{V_\Omega}(\bar{z}) = \int_0^l (\bar{z}_1 f_{1,1}V_1 + \bar{z}_1 f_1 V_{1,1})dX_1 \tag{11.31}$$

$$\ell'_{V_\theta}(\bar{z}) = -\int_0^l \bar{z}_2 f_1 V_{2,1}\,dX_1. \tag{11.32}$$

It is noted that the expressions of $a'_{V_\Omega}(z,\bar{z})$, $a'_{V_\theta}(z,\bar{z})$, $\ell'_{V_\Omega}(\bar{z})$, and $\ell'_{V_\theta}(\bar{z})$ include the first-order derivatives of displacement and design velocity. Thus, a C^0-continuous design velocity field is enough to define those forms.

Beam Component

As discussed earlier, only the C^0 (Timoshenko) beam component is discussed in this section because technical beam theory requires the additional continuity of design velocity. In Timoshenko beam theory, the structural energy and load forms at the perturbed design can be written as

$$a_{^0\Omega_\tau}(z_\tau,\bar{z}_\tau) = \int_0^{l_\tau} [EI\bar{\theta}_{,1_\tau}\theta_{,1_\tau} + k\mu A(\bar{z}_{2,1_\tau} - \bar{\theta}_\tau)(z_{2,1_\tau} - \theta_\tau)]dX_1 \tag{11.33}$$

$$\ell_{^0\Omega_\tau}(\bar{z}_\tau) = \int_0^{l_\tau} \bar{z}_{2_\tau} f_{2_\tau}\,dX_1, \tag{11.34}$$

where, for two dimensional beam structure, the state response is $z = [z_1, z_2, \theta]^T$ and θ is the rotation in the X_3-coordinate. For convenience, let parameters E, I, k, μ, and A be independent of shape and orientation design change.

As with the truss component, design velocity $\boldsymbol{V} = [V_1, V_2]^T$ is given, in which the shape design velocity is $\boldsymbol{V}_\Omega = [V_1]$ and the orientation design velocity is $\boldsymbol{V}_\Theta = [V_2]$. Although the X_1-directional displacement z_1 does not appear in the state response $\boldsymbol{z}$, it is assumed that the beam component is part of the built-up structure. Thus, X_1-directional displacement z_1 exists. Using this observation, the 6×6 matrix $\boldsymbol{V}_\theta$ in (7.12) can be reduced to a 2×3 matrix (in fact, the second and sixth rows are taken), defined as

$$\boldsymbol{V}_\theta = \begin{bmatrix} V_{2,1} & 0 & 0 \\ 0 & 0 & 0 \end{bmatrix} \tag{11.35}$$

and, thus, the orientation design velocity term can be written as

$$\boldsymbol{V}_\theta \boldsymbol{z} = \begin{Bmatrix} V_{2,1} z_1 \\ 0 \end{Bmatrix}. \tag{11.36}$$

Before differentiating the structural energy form, consider the material derivatives of displacement and rotation gradients, as

$$\left(z_{2,1}\right)^{\boldsymbol{\cdot}} = \dot{z}_{2,1} - z_{2,1}V_{1,1} - z_{1,1}V_{2,1} \tag{11.37}$$

$$\left(\theta_{,1}\right)^{\boldsymbol{\cdot}} = \dot{\theta}_{,1} - \theta_{,1}V_{1,1}. \tag{11.38}$$

It is interesting to note that the orientation effect does not appear in the material derivative of the rotation. This is true because the orientation change of the component in the X_1-X_2 plane should not affect the rotation in the X_3-direction.

Using the shape and orientation design sensitivity formulas and (11.37) and (11.38), the variation of the structural energy form in (11.33) is taken, to obtain

$$\begin{aligned} [a_{{}^0\Omega}(\boldsymbol{z},\overline{\boldsymbol{z}})]' &= \int_0^l [EI\overline{\theta}_{,1}\dot{\theta}_{,1} + k\mu A(\overline{z}_{2,1} - \overline{\theta})(\dot{z}_{2,1} - \dot{\theta})]\,dX_1 \\ &\quad - \int_0^l [EI\overline{\theta}_{,1}\theta_{,1} + k\mu A(\overline{z}_{2,1}z_{2,1} - \overline{\theta}\theta)]V_{1,1}\,dX_1 \\ &\quad - \int_0^l k\mu A[\overline{z}_{1,1}(z_{2,1} - \theta) + (\overline{z}_{2,1} - \overline{\theta})z_{1,1}]V_{2,1}\,dX_1, \end{aligned} \tag{11.39}$$

where the first integral is the same as the linearized energy form in (8.39) by replacing $\dot{\boldsymbol{z}}$ with $\Delta\boldsymbol{z}$, which can be denoted as $a^*_{{}^0\Omega}(\boldsymbol{z};\dot{\boldsymbol{z}},\overline{\boldsymbol{z}})$. The second integral is the contribution from the shape design change, while the third integral is the contribution from the orientation design change. Thus, the following structural fictitious loads can be defined:

$$a'_{V_\Omega}(\boldsymbol{z},\overline{\boldsymbol{z}}) = -\int_0^l [EI\overline{\theta}_{,1}\theta_{,1} + k\mu A(\overline{z}_{2,1}z_{2,1} - \overline{\theta}\theta)]V_{1,1}\,dX_1 \tag{11.40}$$

$$a'_{V_\theta}(\boldsymbol{z},\overline{\boldsymbol{z}}) = -\int_0^l k\mu A[\overline{z}_{1,1}(z_{2,1} - \theta) + (\overline{z}_{2,1} - \overline{\theta})z_{1,1}]V_{2,1}\,dX_1. \tag{11.41}$$

The load form in (11.34) is linear in its argument due to the conservative load. After separating shape and orientation parts, the variation of the load form in (11.34) can be obtained as

$$\ell'_{V_\Omega}(\overline{\boldsymbol{z}}) = \int_0^l (\overline{z}_2 f_{2,1} V_1 + \overline{z}_2 f_2 V_{1,1})\,dX_1 \tag{11.42}$$

$$\ell'_{V_\theta}(\bar{z}) = \int_0^l \bar{z}_1 f_2 V_{2,1}\, dX_1. \tag{11.43}$$

It is noted that the expressions of $a'_{V_\Omega}(z,\bar{z})$, $a'_{V_\theta}(z,\bar{z})$, $\ell'_{V_\Omega}(\bar{z})$, and $\ell'_{V_\theta}(\bar{z})$ include the first-order derivatives of displacement and design velocity. Thus, the C^0-continuous design velocity field provides enough smoothness for defining those forms.

Plane Stress Component

In the plane stress component, the shape design velocity is given in the X_1-X_2 plane, while the orientation design velocity is given in X_3-direction. Thus, the surface component in Section 7.1.2 can be used for the orientation design. However, since only the z_1 and z_2 components of displacement are used for the structural problem, the first two rows of matrix $\boldsymbol{V}_\theta$ in (7.19) are used. Due to the restriction that the initial surface must remain flat during the configuration design change, V_3 must be a linear function of the X_1- and X_2-coordinates. Thus, the second-order derivatives of V_3 with respect to the X_1- and X_2-coordinates vanish. In addition, although the z_3 component of displacement does not appear in the structural problem, the plane stress component is assumed to be part of the built-up structure and, thus, the z_3 component of displacement appears in the explicitly dependent term on the configuration design.

The plane stress component is perturbed in the direction of design velocity. The structural energy form for the plane stress component at the perturbed design can be written as

$$a_{{}^0\Omega_\tau}(z_\tau, \bar{z}_\tau) = \iint_{{}^0\Omega_\tau} h \boldsymbol{S}_\tau : \bar{\boldsymbol{E}}_\tau \, d^0\Omega, \tag{11.44}$$

where h is the thickness of the component, $\boldsymbol{S}$ is 2×2 second Piola-Kirchhoff stress, and $\bar{\boldsymbol{E}}$ is the variation of 2×2 Green-Lagrange strain whose definition is given in (8.8). The nonlinearity of the structural energy form is a result of the constitutive relation between stress and strain.

In order to differentiate the second Piola-Kirchhoff stress, let us first differentiate the deformation gradient, as

$$\begin{aligned} \dot{\boldsymbol{F}} &= (\boldsymbol{I} + \nabla_0 z)^{\boldsymbol{\cdot}} \\ &= (\nabla_0 \dot{z}) - \nabla_0 z \nabla_0 \boldsymbol{V}_\Omega - \boldsymbol{V}_\theta \nabla_0 z. \end{aligned} \tag{11.45}$$

The first two terms on the right side are in exactly the same form as the shape design problem in (10.21). The last term on the right side is the contribution from the orientation change. Its expression for the plane stress component is

$$\boldsymbol{V}_\theta \nabla_0 z = \begin{bmatrix} -V_{3,1} z_{3,1} & -V_{3,1} z_{3,2} \\ -V_{3,2} z_{3,1} & -V_{3,2} z_{3,2} \end{bmatrix}. \tag{11.46}$$

Using (11.46) and its definition in (8.4), the material derivative of the Green-Lagrange strain can be obtained as

$$\begin{aligned} \dot{\boldsymbol{E}} &= sym[(\nabla_0 \dot{z})^T \boldsymbol{F}] - sym[(\nabla_0 z \nabla_0 \boldsymbol{V}_\Omega)^T \boldsymbol{F}] - sym[(\boldsymbol{V}_\theta \nabla_0 z)^T \boldsymbol{F}] \\ &\equiv \Delta \boldsymbol{E}(\dot{z}) + \boldsymbol{E}_{V_\Omega}(z) + \boldsymbol{E}_{V_\theta}(z), \end{aligned} \tag{11.47}$$

where the three terms on the right is defined by the corresponding three terms in the middle of the equation. As expected, the contributions from shape and orientation designs are explicitly separated. It is interesting to note that even if the Green-Lagrange strain is nonlinear, its material derivative in (11.47) is a linear function of design velocity $\boldsymbol{V}_\Omega$ and

V_θ. By differentiating the second Piola-Kirchhoff stress and using the expression in (11.47), the following material derivative can be obtained:

$$\dot{S} = C : \Delta E(\dot{z}) + C : E_{V_\Omega}(z) + C : E_{V_\theta}(z), \tag{11.48}$$

where C is the fourth-order constitutive tensor defined in (8.16) for the isotropic material.

Next, the material derivative of the variation of Green-Lagrange strain can be obtained by using the formula in (11.45) as

$$\begin{aligned}\dot{\bar{E}} &= sym(\nabla_0\bar{z}^T\nabla_0\dot{z}) - sym[F^T(\nabla_0\bar{z}\nabla_0 V_\Omega) + \nabla_0\bar{z}^T(\nabla_0 z\nabla_0 V_\Omega)] \\ &\quad - sym[F^T(V_\theta\nabla_0\bar{z}) + \nabla_0\bar{z}^T(V_\theta\nabla_0 z)] \\ &\equiv \Delta\bar{E}(\dot{z},\bar{z}) + \Delta\bar{E}_{V_\Omega}(z,\bar{z}) + \Delta\bar{E}_{V_\theta}(z,\bar{z}).\end{aligned} \tag{11.49}$$

As with (11.47), the contributions from shape and orientation designs are separated and are linear functions of the design velocity.

Using (11.47) through (11.49), the variation of the structural energy form can be obtained as

$$\begin{aligned}[a_{{}^0\Omega}(z,\bar{z})]' &= \iint_{{}^0\Omega}[\bar{E} : C : \Delta E(\dot{z}) + S : \Delta\bar{E}(\dot{z},\bar{z})]\,d^0\Omega \\ &\quad + \iint_{{}^0\Omega}[\bar{E} : C : E_{V_\Omega}(z) + S : \Delta\bar{E}_{V_\Omega}(z,\bar{z}) + S : \bar{E}divV_\Omega]\,d^0\Omega \\ &\quad + \iint_{{}^0\Omega}[\bar{E} : C : E_{V_\theta}(z) + S : \Delta\bar{E}_{V_\theta}(z,\bar{z})]\,d^0\Omega.\end{aligned} \tag{11.50}$$

The first integral in (11.50) is the same as the linearized energy form in (8.40) when $\dot{z}$ is replaced with Δz, which constitutes the implicitly dependent terms in the design. The second integral is the same as the structural fictitious load in (10.27) when V is considered as V_Ω and is thus defined as

$$a'_{V_\Omega}(z,\bar{z}) = \iint_{{}^0\Omega}[\bar{E} : C : E_{V_\Omega}(z) + S : \Delta\bar{E}_{V_\Omega}(z,\bar{z}) + S : \bar{E}divV_\Omega]\,d^0\Omega. \tag{11.51}$$

The last integral in (11.50) is the contribution from orientation changes and is thus denoted by

$$a'_{V_\theta}(z,\bar{z}) = \iint_{{}^0\Omega}[\bar{E} : C : E_{V_\theta}(z) + S : \Delta\bar{E}_{V_\theta}(z,\bar{z})]\,d^0\Omega. \tag{11.52}$$

From the assumption of a conservative load, the load form does not have any implicitly dependent term. The load form for the plane stress component at the perturbed design can be written as

$$\ell_{{}^0\Omega_\tau}(\bar{z}_\tau) = \iint_{{}^0\Omega_\tau}\bar{z}_\tau^T f_\tau\,d^0\Omega, \tag{11.53}$$

where $f = [f_1, f_2]^T$ is two-dimensional body force. For simplicity, surface traction force is not considered. Using the material derivative of shape and orientation designs, the load form can be differentiated to yield

$$\begin{aligned}[\ell_{{}^0\Omega}(\bar{z})]' &= \iint_{{}^0\Omega}[\bar{z}^T(\nabla_0 f V_\Omega) + \bar{z}^T f divV_\Omega]\,d^0\Omega \\ &\quad - \iint_{{}^0\Omega}[(V_\theta\bar{z})^T f]\,d^0\Omega,\end{aligned} \tag{11.54}$$

where it is assumed that $f'_{V_\Omega} = f'_{V_\theta} = 0$ and

$$V_\theta \bar{z} = \begin{Bmatrix} -V_{3,1}\bar{z}_3 \\ -V_{3,2}\bar{z}_3 \end{Bmatrix}. \tag{11.55}$$

Using (11.54), the shape and orientation design contribution to the load form can be written as

$$\ell'_{V_\Omega}(\bar{z}) = \iint_{^0\Omega} [\bar{z}^T(\nabla_0 f V_\Omega) + \bar{z}^T f div V_\Omega] d^0\Omega \tag{11.56}$$

$$\ell'_{V_\theta}(\bar{z}) = -\iint_{^0\Omega} [(V_\theta \bar{z})^T f] d^0\Omega. \tag{11.57}$$

Plate Component

Between two plate formulations, the Mindlin/Reissner formulation (C^0) is used in this section to derive the expressions of structural fictitious loads. As discussed with the beam component, the thin plate (C^1) formulation may not be appropriate for shape and orientation design problems due to the requirement of higher-order derivatives of design velocity fields. The surface design component in Section 7.1.2 can be used to define orientation design velocity. Thus, $V_\Omega = [V_1, V_2]^T$ and $V_\Theta = [V_3]$. Conventionally, the plate component has three degrees-of-freedom: vertical displacement and two rotations on the plane of the component, i.e., $[z_3, \theta_1, \theta_2]^T$. However, it is assumed that the plate component is part of the built-up structure. Thus, the other degrees-of-freedom (z_1, z_2, and θ_3) appear in the fictitious load of the orientation design. Within the matrix V_θ in (7.19), the third, fourth, and fifth rows are used in the plate component, as

$$V_\theta = \begin{bmatrix} V_{3,1} & V_{3,2} & 0 & 0 & 0 & -V_{3,1} - V_{3,2} \\ 0 & 0 & 0 & 0 & 0 & -V_{3,1} \\ 0 & 0 & 0 & 0 & 0 & -V_{3,2} \end{bmatrix}, \tag{11.58}$$

with $z = [z_1, z_2, z_3, \theta_1, \theta_2, \theta_3]^T$.

Using (8.44), the structural energy form of the plate component at the perturbed design can be written as

$$a_{^0\Omega_\tau}(z_\tau, \bar{z}_\tau) = \iint_{^0\Omega_\tau} \kappa(\bar{z}_\tau)^T C^b \kappa(z_\tau) d^0\Omega + \iint_{^0\Omega_\tau} \gamma(\bar{z}_\tau)^T C^s \gamma(z_\tau) d^0\Omega. \tag{11.59}$$

The first integral represents bending energy, while the second represents transverse shear energy. In the shape and orientation design change, it is assumed that material parameters C^b and C^s are constant. In order to differentiate the structural energy form, let us consider the material derivative of the curvature vector in the first integral. The curvature vector has two rotations as arguments, and first-order derivatives are involved. Using shape and orientation derivative formulas, the material derivatives of the curvature vector and its variation can be obtained as

$$(\kappa(z))^{\cdot} \equiv \kappa(\dot{z}) + \kappa_{V_\Omega}(z) + \kappa_{V_\theta}(z) \tag{11.60}$$

$$(\kappa(\bar{z}))^{\cdot} \equiv \kappa_{V_\Omega}(\bar{z}) + \kappa_{V_\theta}(\bar{z}), \tag{11.61}$$

where the implicitly dependent term $\dot{z}$ only appears in (11.60), while $\dot{\bar{z}}$ is ignored in (11.61) for the same reason explained in (6.63). The explicitly dependent term on the shape design is defined as

$$\boldsymbol{\kappa}_{V_\Omega}(z) = -\begin{Bmatrix} \theta_{1,\alpha}V_{\alpha,1} \\ \theta_{2,\alpha}V_{\alpha,2} \\ \theta_{1,\alpha}V_{\alpha,2} + \theta_{2,\alpha}V_{\alpha,1} \end{Bmatrix}, \tag{11.62}$$

while the explicitly dependent term on the orientation design is defined as

$$\boldsymbol{\kappa}_{V_\theta}(z) = \begin{Bmatrix} \theta_{3,1}V_{3,1} \\ \theta_{3,2}V_{3,2} \\ \theta_{3,1}V_{3,2} + \theta_{3,2}V_{3,1} \end{Bmatrix}, \tag{11.63}$$

where the summation rule is used for the repeated indices α = 1, 2. It is clear from its definition that $\boldsymbol{\kappa}_{V_\theta}(z)$ vanishes when there is no X_3-directional rotation in the built-up structure.

The transverse shear vector can also be differentiated using the same procedure outlined for the curvature vector to obtain

$$(\gamma(z))^{\cdot} \equiv \gamma(\dot{z}) + \gamma_{V_\Omega}(z) + \gamma_{V_\theta}(z) \tag{11.64}$$

$$(\gamma(\bar{z}))^{\cdot} \equiv \gamma_{V_\Omega}(\bar{z}) + \gamma_{V_\theta}(\bar{z}), \tag{11.65}$$

where the explicitly dependent terms on the shape and orientation design are defined as

$$\gamma_{V_\Omega}(z) = -\begin{Bmatrix} z_{3,\alpha}V_{\alpha,2} \\ z_{3,\alpha}V_{\alpha,1} \end{Bmatrix} \tag{11.66}$$

and

$$\gamma_{V_\theta}(z) = -\begin{Bmatrix} V_{3,\alpha}(z_{\alpha,2} - \theta_{3,2}) \\ V_{3,\alpha}(z_{\alpha,1} - \theta_{3,1}) \end{Bmatrix}. \tag{11.67}$$

Note that the material derivatives of the curvature and shear vectors include the first-order derivative of the design velocities. Thus, a piecewise linear design velocity (C^0) field is enough to define the shape and orientation design changes.

Using the material derivative formulas from (11.60) through (11.67), the structural energy form can be differentiated with respect to shape and orientation designs to obtain

$$\begin{aligned}
[a_{^0\Omega}(z,\bar{z})]' = &\iint_{^0\Omega}[\boldsymbol{\kappa}(\bar{z})^T \boldsymbol{C}^b \boldsymbol{\kappa}(\dot{z}) + \gamma(\bar{z})^T \boldsymbol{C}^s \gamma(\dot{z})]d^0\Omega \\
&+ \iint_{^0\Omega}[\boldsymbol{\kappa}_{V_\Omega}(\bar{z})^T \boldsymbol{C}^b \boldsymbol{\kappa}(z) + \boldsymbol{\kappa}(\bar{z})^T \boldsymbol{C}^b \boldsymbol{\kappa}_{V_\Omega}(z)]d^0\Omega \\
&+ \iint_{^0\Omega}[\gamma_{V_\Omega}(\bar{z})^T \boldsymbol{C}^s \gamma(z) + \gamma(\bar{z})^T \boldsymbol{C}^s \gamma_{V_\Omega}(z)]d^0\Omega \\
&+ \iint_{^0\Omega}[\boldsymbol{\kappa}(\bar{z})^T \boldsymbol{C}^b \boldsymbol{\kappa}(z) + \gamma(\bar{z})^T \boldsymbol{C}^s \gamma(z)] divV_\Omega \, d^0\Omega \\
&+ \iint_{^0\Omega}[\boldsymbol{\kappa}_{V_\theta}(\bar{z})^T \boldsymbol{C}^b \boldsymbol{\kappa}(z) + \boldsymbol{\kappa}(\bar{z})^T \boldsymbol{C}^b \boldsymbol{\kappa}_{V_\theta}(z)]d^0\Omega \\
&+ \iint_{^0\Omega}[\gamma_{V_\theta}(\bar{z})^T \boldsymbol{C}^s \gamma(z) + \gamma(\bar{z})^T \boldsymbol{C}^s \gamma_{V_\theta}(z)]d^0\Omega,
\end{aligned} \tag{11.68}$$

where, by replacing $\dot{z}$ with Δz ,the first integral is the same as the linearized energy form in (8.47), which constitutes the implicitly dependent terms in the design. The fictitious loads for shape and configuration designs can be written as

$$
\begin{aligned}
a'_{V_\Omega}(z,\bar{z}) = & \iint_{{}^0\Omega}[\boldsymbol{\kappa}_{V_\Omega}(\bar{z})^T \boldsymbol{C}^b \boldsymbol{\kappa}(z) + \boldsymbol{\kappa}(\bar{z})^T \boldsymbol{C}^b \boldsymbol{\kappa}_{V_\Omega}(z)]d^0\Omega \\
& + \iint_{{}^0\Omega}[\gamma_{V_\Omega}(\bar{z})^T \boldsymbol{C}^s \gamma(z) + \gamma(\bar{z})^T \boldsymbol{C}^s \gamma_{V_\Omega}(z)]d^0\Omega \\
& + \iint_{{}^0\Omega}[\boldsymbol{\kappa}(\bar{z})^T \boldsymbol{C}^b \boldsymbol{\kappa}(z) + \gamma(\bar{z})^T \boldsymbol{C}^s \gamma(z)] div \boldsymbol{V}_\Omega \, d^0\Omega
\end{aligned}
\tag{11.69}
$$

and

$$
\begin{aligned}
a'_{V_\theta}(z,\bar{z}) = & \iint_{{}^0\Omega}[\boldsymbol{\kappa}_{V_\theta}(\bar{z})^T \boldsymbol{C}^b \boldsymbol{\kappa}(z) + \boldsymbol{\kappa}(\bar{z})^T \boldsymbol{C}^b \boldsymbol{\kappa}_{V_\theta}(z)]d^0\Omega \\
& + \iint_{{}^0\Omega}[\gamma_{V_\theta}(\bar{z})^T \boldsymbol{C}^s \gamma(z) + \gamma(\bar{z})^T \boldsymbol{C}^s \gamma_{V_\theta}(z)]d^0\Omega.
\end{aligned}
\tag{11.70}
$$

The load form for the plate component at the perturbed design can be written as

$$
\ell_{{}^0\Omega_\tau}(\bar{z}_\tau) = \iint_{{}^0\Omega_\tau} \bar{z}_{3_\tau} f_{3_\tau} \, d^0\Omega,
\tag{11.71}
$$

where f_3 is body force applying in the vertical direction. For simplicity, surface traction force is not considered. Using the material derivative of shape and orientation designs, the load form can be differentiated to yield

$$
\begin{aligned}
[\ell_{{}^0\Omega}(\bar{z})]' = & \iint_{{}^0\Omega}[\bar{z}_3(\nabla_0 f_3^T \boldsymbol{V}_\Omega) + \bar{z}_3 f_3 div \boldsymbol{V}_\Omega]d^0\Omega \\
& - \iint_{{}^0\Omega} V_{3,\alpha}(\bar{z}_\alpha - \bar{\theta}_3) f_3 \, d^0\Omega,
\end{aligned}
\tag{11.72}
$$

where it is assumed that $f_3' = 0$. Using (11.72), the contributions from shape and orientation designs to the load form can be written as

$$
\ell'_{V_\Omega}(\bar{z}) = \iint_{{}^0\Omega}[\bar{z}_3(\nabla_0 f_3^T \boldsymbol{V}_\Omega) + \bar{z}_3 f_3 div \boldsymbol{V}_\Omega]d^0\Omega
\tag{11.73}
$$

$$
\ell'_{V_\theta}(\bar{z}) = -\iint_{{}^0\Omega} V_{3,\alpha}(\bar{z}_\alpha - \bar{\theta}_3) f_3 \, d^0\Omega.
\tag{11.74}
$$

11.1.4 Numerical Examples

Four numerical examples are provided to show the accuracy of the configuration design sensitivity results. Although the formulations in Sections 11.1.1 and 11.1.2 can be applied to a general elastic material model, for simplicity a linear elastic material is used in all examples.

Two-Member Beam Model

A two-member beam structure subject to a conservative force at node 2 is considered, as shown in Fig. 11.1. A finite element model of the structure is created using the ABAQUS beam element B21H [40]. For analysis data, a constant cross-sectional area A = 0.8 in^2 and a Young's modulus of $E = 1.0 \times 10^7$psi are used. In this example, the applied force is 2 lb. The selected performance measure is the displacement in the X_2-direction at node 2.

Member 1 in Fig. 11.1 is treated as a design component to describe shape and configuration design changes. For shape design, the length of this member is changed, as shown in Fig. 11.2. The position of node 1 is perturbed by δL_1. In this case, the design velocity field becomes

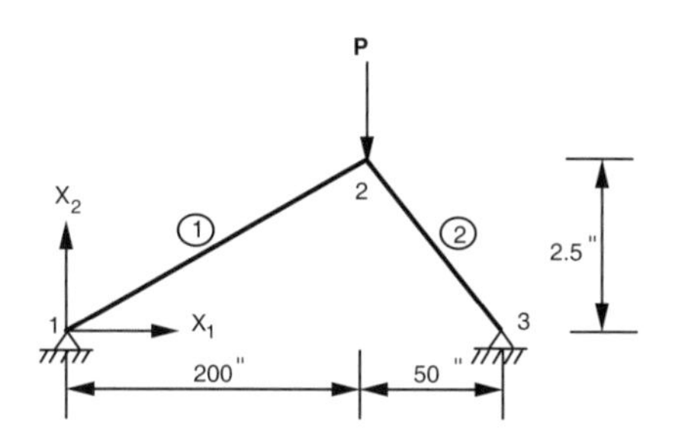

Figure 11.1. Two-member beam model.

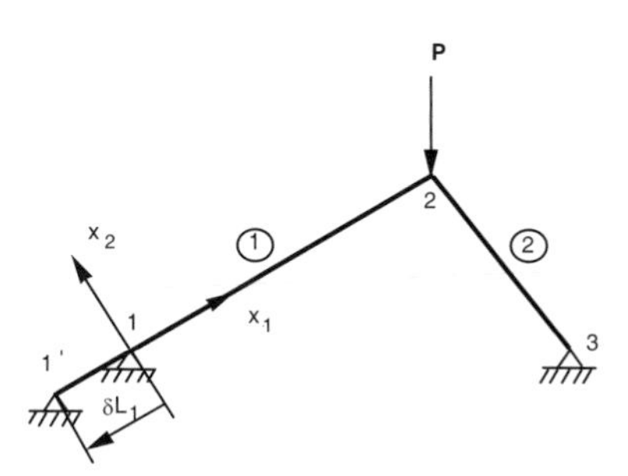

Figure 11.2. Shape design perturbation.

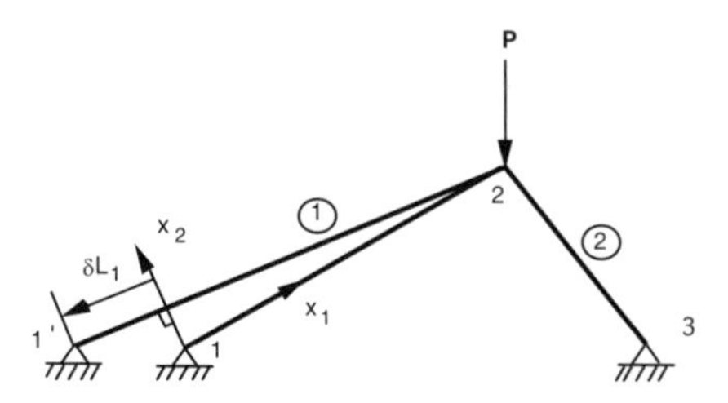

Figure 11.3. Configuration design perturbation.

$$V_1 = \frac{X_1}{L_1}\delta L_1 - \delta L_1, \quad X_1 \in [0,\ 200.0156], \tag{11.75}$$

where L_1 is the original length of member 1.

In order to describe shape and orientation changes, it is necessary to change both the length and the orientation of member 1, as shown in Fig. 11.3. The position of node 1 is perturbed by δL_1. In this case, the design velocity field becomes

$$V_1 = \frac{X_1}{L_1}\delta L_1 - \delta L_1 \tag{11.76}$$

$$V_2 = \frac{-2.5X_1}{L_1^2}\delta L_1 - \frac{2.5}{L_1}\delta L_1. \tag{11.77}$$

Table 11.1 provides design sensitivity results of the displacement performance measure, where analysis results with backward and forward design perturbations are listed in the third and fourth columns, respectively. The fifth column is the central finite difference $\Delta\psi = [\psi(b + \Delta b) - \psi(b - \Delta b)]/2$, and the sixth column is the predicted change by the proposed sensitivity calculation method. The agreement between $\Delta\psi$ and ψ' in the last column of Table 11.1 is 96.4% for 1 in of design perturbation, which is good.

Table 11.2 provides configuration design sensitivity results for the critical load. Design sensitivity results for an applied load of 12.13 lb are compared with results using the finite difference method for 1.0, 0.1, and 0.01% design perturbations. The agreement between $\Delta^{cr}p$ and ${}^{cr}p'$ in the last column of Table 11.2 is 101.6% for a 1.0% design perturbation, 101.2% for a 0.1% design perturbation, and 101.2% for a 0.01% perturbation. Favorable agreement with finite difference results indicates that the proposed method yields accurate design sensitivity.

Seven-Member Truss Model

A seven-member truss model subject to equal conservative forces at nodes 1, 3, and 5 is considered, as shown in Fig. 11.4. A finite element model of the structure is created using the ABAQUS truss element C1D2 [40]. For analysis data, a constant cross-sectional area $A = 0.8$ in^2 and Young's modulus $E = 1.0 \times 10^7$ psi are used. In this example, ABAQUS analysis results showed that the critical load level is 525 lb.

The perturbed design shown in Fig. 11.5 is defined so that nodal point 1 is moved upward in the X_2-direction by 1 in. In the perturbed design, both the shape (i.e., length) and the orientation of truss design components may be changed. Based on linear design velocity fields, design sensitivities of the critical load are computed.

Table 11.1. Configuration design sensitivity results of displacement of two-member beam model.

Δb (in)	$\psi(b)$	$\psi(b-\Delta b)$	$\psi(b+\Delta b)$	$\Delta\psi$	ψ'	$\psi'/\Delta\psi$
2	1.564E2	1.545E2	1.583E2	1.92E2	1.85E2	96.4%

Table 11.2. Configuration design sensitivity results of critical load of two-member beam model.

Length(in)	Pert	${}^{cr}p(L_1-\Delta L_1)$	${}^{cr}p(L_1+\Delta L_1)$	$\Delta^{cr}p$	${}^{cr}p'$	${}^{cr}p'/\Delta^{cr}p$
200.0156	–	12.132	12.132			
202.0158	1.0%	12.384	11.888	–0.248	–0.252	101.6%
200.2156	0.1%	12.157	12.107	–0.0249	–0.0252	101.2%
200.0356	0.01%	12.135	12.1295	–0.00249	–0.00252	101.2%

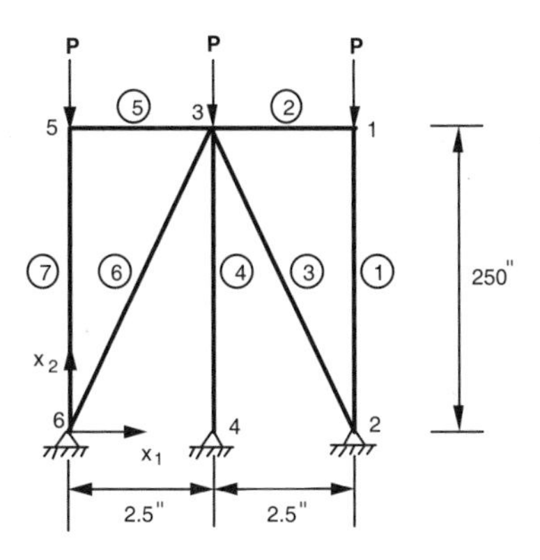

Figure 11.4. Seven-member truss model.

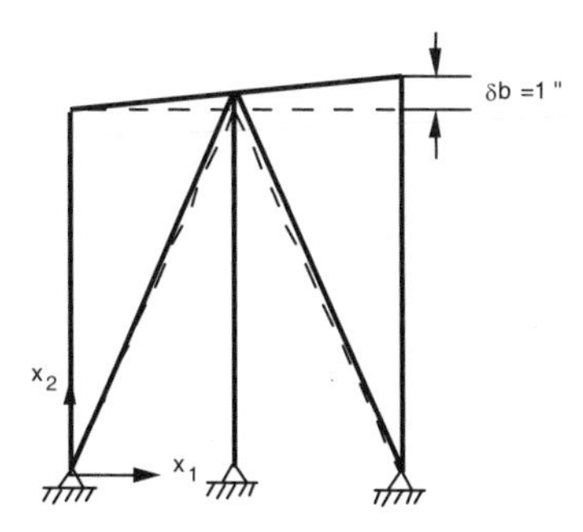

Figure 11.5. Perturbed design of the seven-member truss model.

Table 11.3. Design sensitivity results of critical load of seven-member truss model.

Δb	${}^{cr}p$	${}^{cr}p(b\text{-}\Delta b)$	${}^{cr}p(b+\Delta b)$	$\Delta^{cr}p$	${}^{cr}p'$	${}^{cr}p'/\Delta^{cr}p$
1	525	527.1	522.9	-2.1	-2.0998	99.9

Table 11.3 provides design sensitivity results for the critical load. In Table 11.3, backward and forward design results are listed in the third and fourth columns, respectively. The fifth column is the central finite difference $\Delta p_{cr} = [p_{cr}(b + \Delta b) - p_{cr}(b - \Delta b)]/2$, and the sixth column is the predicted change using the design sensitivity analysis method. The agreement between Δp_{cr} and p'_{cr} in the last column of the table is 99.9% for a 1 in design perturbation.

Table 11.4 shows two contributions to configuration design sensitivity results of the critical load. The third column presents the amount contributed from the shape effect, and the fourth column presents the amount contributed from the orientation effect. The second column is the summation of these two effects. Table 11.4 shows that shape effects are dominant for elements 1 and 4, while orientation effects are dominant for elements 2 and 5. For elements 3 and 6, both shape and orientation effects are important.

Table 11.4. Contributions from shape and orientation changes to design sensitivity of critical load of seven-member truss model.

Element	${}^{cr}p'$	${}^{cr}p'$ (Shape)	${}^{cr}p'$ (Orientation)
1	1.2118E–7	1.2118E–7	0.0E+0
2	1.3333E–3	0.0E+0	1.333E–3
3	–1.9998E-3	–9.9996E–3	–9.9987E–3
4	–1.1551E-16	–1.155E–16	0.0E+0
5	–1.3333E-3	0.0E+0	–1.3333E–3
6	–1.9998E-3	–9.9996E–3	–9.9987E–3
7	0.0E+0	0.0E+0	0.0E+0

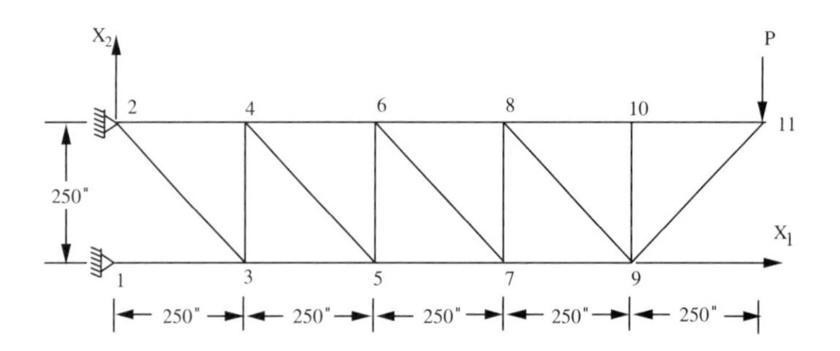

Figure 11.6. Crane model.

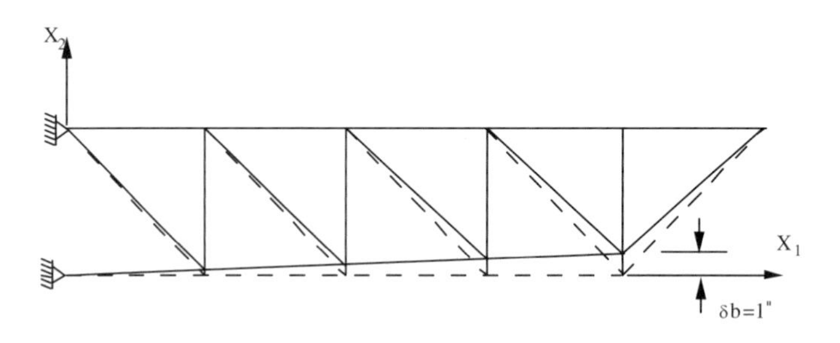

Figure 11.7. Configuration design perturbation of crane model.

Crane Model

A crane model that is subjected to a conservative force at node 11 is considered, as shown in Figure 11.6. In this example, the applied force is 1000 lb. A finite element model of the structure is created using the ABAQUS truss element C1D2 [40]. For analysis data, a constant cross-sectional area $A = 10 \text{ in}^2$ and Young's modulus $E = 1.0 \times 10^7$ psi are used. There are 18 truss elements and 11 nodal points. Each finite element is treated as a single design component. The perturbed design shown in Figure 11.7 is obtained by moving nodal points 9, 7, 5, and 3 upward in the X_2-direction by 1, 0.75, 0.5,

Table 11.5. Configuration design sensitivity results of displacement of crane model.

Δb (in)	$\psi(b)$	$\psi(b-\Delta b)$	$\psi(b+\Delta b)$	$\Delta\psi$	ψ'	$\psi'/\Delta\psi$
1	−0.6982	−0.6978	−0.6986	−0.0004	−0.00043	107.5%

and 0.25 in, respectively. Note that both shape and orientation are changed to obtain the perturbed design. A linear design velocity field is used for each truss design component; that is,

$$V_{\Omega}(X) = \frac{X}{L_i}\delta L_i, \tag{11.78}$$

where x is the local coordinate of the ith design component and L_i and δL_i are the length and the variation of the length of the ith design component, respectively. Based on these linear design velocity fields in (11.78), design sensitivity of the displacement performance measure in the X_2-direction at node 11 is computed.

Table 11.5 provides the design sensitivity results of the displacement performance measure, where backward and forward design results are listed in the third and fourth columns, respectively. The fifth column is the central finite difference $\Delta\psi = [\psi(b + \Delta b) - \psi(b - \Delta b)]/2$, and the sixth column is the predicted change by the proposed sensitivity calculation method. The agreement between $\Delta\psi$ and ψ' in the last column of Table 11.5 is 107.5% for 1 in of design perturbation at node 9. The design sensitivity results are good.

Roof Model

A roof structure subject to a conservative force at the center of the edge is considered for the displacement performance measure, as shown in Figure 11.8. A finite element model of the structure is created using the ABAQUS shell element S4R [40]. For analysis data, a constant thickness t = 1 in and Young's modulus $E = 1.0 \times 10^7$ psi are used. In this example, the applied force is 100 lb at node 7.

The perturbed design shown in Fig. 11.9 is defined so that nodal points 1 and 3 are moved away in the X_1-direction by 1 in. In the perturbed design, both the shape and orientation of the shell design components may be changed. In this case, the design velocity field on the left side of the roof (1-2-5-4) becomes

$$V_1 = \frac{\delta b}{400}\cos\theta X_1' - \delta b\cos\theta \tag{11.79}$$

$$V_2 = 0 \tag{11.80}$$

and

$$V_3 = -\frac{\delta b}{400}\sin\theta X_1' + \delta b\sin\theta, \tag{11.81}$$

and the design velocity field on the right-hand side of the roof (2-3-6-5) becomes

$$V_1 = \frac{\delta b}{400}\cos\theta X_1'' \tag{11.82}$$

$$V_2 = 0 \tag{11.83}$$

and

$$V_3 = \frac{\delta b}{400} \sin\theta X_1''. \tag{11.84}$$

Based on these linear design velocity fields, Table 11.6 shows the design sensitivity results of the displacement performance measure in the X_3-direction at node 7. The agreement between $\Delta\psi$ and ψ' in the last column of Table 11.6 is 96.6% for 1 in of design perturbation, which is good.

Next, ABAQUS analysis results showed that the critical load level is 2296 lb. Table 11.7 shows design sensitivity results of the critical load. In Table 11.7, analysis results with backward and forward design perturbations are listed in the third and fourth columns, respectively. The fifth column is the central finite difference $\Delta^{cr}p = [^{cr}p(b+\Delta b) - {}^{cr}p(b-\Delta b)]/2$, and the sixth column is the predicted design sensitivity. The agreement between $\Delta^{cr}p$ and p'_{cr} in the last column of the table is 92.5% for 1 in of design perturbation, which is good.

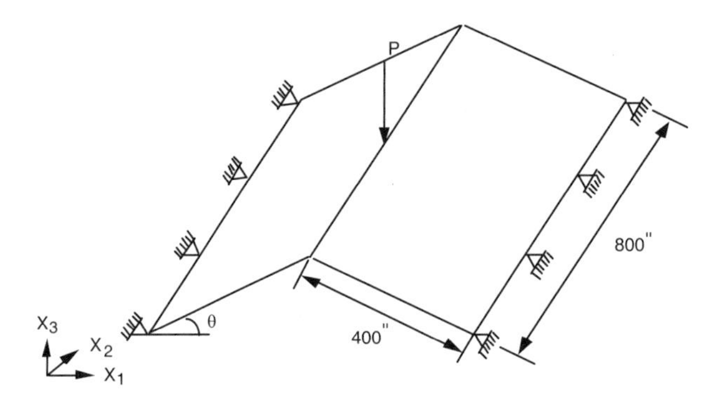

Figure 11.8. Roof model.

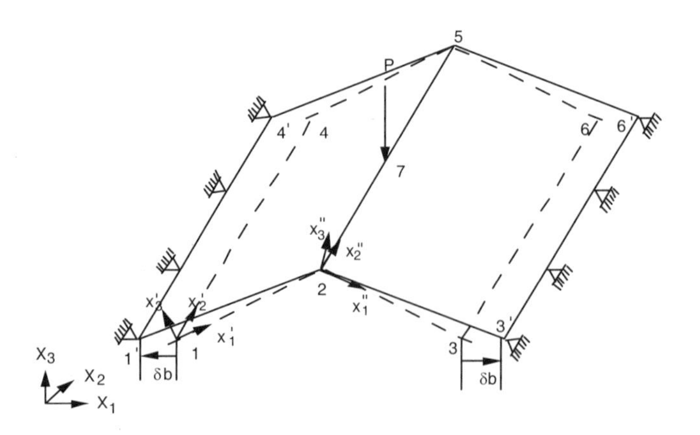

Figure 11.9. Configuration design perturbation of roof model.

Table 11.6. Configuration design sensitivity results of displacement of roof model.

Δb (in)	$\psi(b)$	$\psi(b-\Delta b)$	$\psi(b+\Delta b)$	$\Delta\psi$	ψ'	$\psi'/\Delta\psi$
145	3.323E2	3.255E2	3.393E2	6.95E2	6.72E2	96.6%

Table 11.7. Configuration design sensitivity results of critical load of roof model.

Δb (in)	${}^{cr}p(b)$	${}^{cr}p(b-\Delta b)$	${}^{cr}p(b+\Delta b)$	$\Delta^{cr}p$	${}^{cr}p'$	${}^{cr}p'/\Delta^{cr}p$
1	2296	2320	2272	–24	–22.2	92.5%

11.2 Elastoplastic Problems

The configuration design sensitivity analysis for elastoplastic materials is a current field of research. It is difficult to develop a general analysis procedure with elastoplastic material because different structural components use different formulations. Instead of developing a general configuration sensitivity formulation for elastoplastic materials, an elastoplastic truss component is considered. Readers who are interested in more general formulations for beam and shell structures are referred to the literature [39] and [53].

11.2.1 Configuration Design Sensitivity for Elastoplastic Truss

The elastoplastic analysis described in Section 8.2.1 is for three-dimensional continuum structures. The second-order tensor notation has been used to describe stress and strain. In the truss component, however, since the axial components of stress and strain are only nonzero quantities, a scalar notation can be used. Thus, most tensors in Section 8.2.1 can be considered scalars for the truss component. In addition, from the assumption of a small deformation, only material nonlinearity is considered. Thus, it is unnecessary to distinguish between deformed and undeformed configurations. All configurations are referred to the undeformed initial configuration.

As discussed in previous chapters, the adjoint variable method is very inefficient in such path-dependent problems as elastoplasticity, since the material derivative of the path-dependent variable must be calculated throughout the entire domain. Thus, only the direct differentiation method is developed in this section. By solving the design sensitivity equation, the material derivative of state response ($\dot{z}$) is calculated. Then, the material derivatives of path-dependent variables (stress, effective plastic strain) are updated using the value of $\dot{z}$.

For elastoplastic material, it is convenient to distinguish between time steps because analysis is carried out in discrete time steps. Let left superscript n denote the time step t_n. It is assumed that the analysis has been carried out until time step t_n, and that a new equilibrium at time step t_{n+1} is required. The variational equation of the truss component at time step t_{n+1} can be written as

$$a_\Omega({}^{n+1}z,\bar{z}) \equiv \int_0^l \varepsilon(\bar{z}_1)\,{}^{n+1}\sigma\, A dx = \int_0^l \bar{z}_1\,{}^{n+1}f\, dx \equiv \ell_\Omega(\bar{z}), \quad \forall \bar{z} \in Z, \tag{11.85}$$

where A is the cross-sectional area of the truss, and ${}^{n+1}f$ is the distributed body force at the current time step, in the axial direction. Equation (11.85) is the same as the linear truss component if the stress-strain relation is linear elastic, i.e., ${}^{n+1}\sigma = E\varepsilon(z_1)$. However, nonlinearity appears due to the elastoplastic constitutive relation.

For simplicity, let us consider a two-dimensional truss component, such that the design velocity field is given as $\boldsymbol{V} = [V_1, V_2]^T$, in which $\boldsymbol{V}_\Omega = [V_1]$ contributes to the shape design change, while $\boldsymbol{V}_\Theta = [V_2]$ contributes to the orientation design change. In order to remain straight during design perturbation, the orientation design velocity V_2 must be a linear function of x_1. Since only the axial degree-of-freedom (the x_1-direction) is considered, the matrix $\boldsymbol{V}_\theta$ is given in (11.24) with $\boldsymbol{z} = [z_1, z_2]^T$.

The strain term in (11.85) is the same as the truss component in the linear problem. Thus, its material derivative can be easily obtained as

$$\begin{aligned}(\varepsilon(\bar{z}_1))^{\cdot} &= -\bar{z}_{,1}V_{1,1} + \bar{z}_{2,1}V_{2,1} \\ &\equiv \varepsilon_{V_\Omega}(\bar{z}_1) + \varepsilon_{V_\theta}(\bar{z}_1).\end{aligned} \tag{11.86}$$

For the same reason explained in (6.63), $\dot{\bar{z}}_1$ is ignored in (11.86). Although z_2 does not appear in the variational equation (11.85), it is assumed to be part of the built-up structure so that transverse displacement z_2 appears in (11.86).

The material derivative of stress is not straightforward because of the complicated constitutive relations in the elastoplastic material. From (8.75) and (8.76), the stress at time t_{n+1} can be calculated by

$$\begin{aligned}{}^{n+1}\sigma &= {}^{n}\sigma + \Delta\sigma \\ &= {}^{n}\sigma + E(\Delta\varepsilon - \hat{\gamma}),\end{aligned} \tag{11.87}$$

where $\Delta\varepsilon = \varepsilon(\Delta z_1)$ is the increment of strain and $\hat{\gamma}$ is the solution to the nonlinear yield condition defined in (8.93). The material derivative of the stress involves that of incremental strain and $\hat{\gamma}$. Similar to (11.86), the material derivative of the incremental strain can be obtained as

$$\begin{aligned}(\Delta\varepsilon)^{\cdot} &= \Delta\dot{z}_{1,1} - \Delta z_1 V_{1,1} + \Delta z_{2,1}V_{2,1} \\ &\equiv \varepsilon(\Delta\dot{z}_1) + \varepsilon_{V_\Omega}(\Delta z_1) + \varepsilon_{V_\theta}(\Delta z_1),\end{aligned} \tag{11.88}$$

where the first term on the right side will be solved as an unknown in the design sensitivity equation, while the other two explicitly dependent terms can be calculated from the given design velocity fields.

The material derivative of the plastic consistency parameter $\hat{\gamma}$ can be obtained by differentiating (8.93) with respect to the design as

$$\dot{\hat{\gamma}} = \frac{E(\Delta\varepsilon)^{\cdot} + {}^{n}\dot{\eta} - (H_{\alpha,e_p}\gamma + \sqrt{\tfrac{2}{3}}\kappa_{,e_p})\,{}^{n}\dot{e}_p}{E + H_\alpha + \sqrt{\tfrac{2}{3}}H_{\alpha,e_p}\hat{\gamma} + \tfrac{2}{3}\kappa_{,e_p}}, \tag{11.89}$$

where $H_\alpha(e_p)$ is the plastic modulus for kinematic hardening, and $\kappa(e_p)$ is the radius of the elastic domain determined by isotropic hardening. In the response analysis described in Section 8.2.1, the yield condition in (8.93) is iteratively solved to calculate $\hat{\gamma}$. However, in the sensitivity calculation, the material derivative $\dot{\hat{\gamma}}$ can be calculated without any iteration. Sensitivity calculations do not require any return-mapping iteration, since information is used at the return-mapped point.

In the case of the linear, combined isotropic/kinematic hardening material described in (8.102) and (8.103), (11.89) can be further simplified to

$$\dot{\hat{\gamma}} = \frac{E(\Delta\varepsilon)^{\cdot} + {}^{n}\dot{\eta} - \sqrt{\tfrac{2}{3}}(1-\beta)H\,{}^{n}\dot{e}_{p}}{E + \tfrac{2}{3}H}, \tag{11.90}$$

where H is the linear hardening modulus, and $\beta \in [0,1]$ is a parameter used to consider the Baushinger effect. In (11.90), ${}^{n}\dot{\eta} = {}^{n}\dot{\sigma} - {}^{n}\dot{\alpha}$ and ${}^{n}\dot{e}_{p}$ are the material derivatives at the previous time step t_n. In order to calculate the material derivative of stress, the material derivatives of these path-dependent variables must be calculated and saved at the previous time step.

By combining (11.88) and (11.90), the material derivative of the stress can be obtained as

$$\begin{aligned} {}^{n+1}\dot{\sigma} &= {}^{n}\dot{\sigma} + E[(\Delta\varepsilon)^{\cdot} - \dot{\hat{\gamma}}] \\ &= C^{alg}[\varepsilon(\Delta\dot{z}_1) + \varepsilon_{V_\Omega}(\Delta z_1) + \varepsilon_{V_\theta}(\Delta z_1)] + \sigma^{fic}, \end{aligned} \tag{11.91}$$

where C^{alg} is the algorithmic modulus for the elastoplastic truss component, defined by

$$C^{alg} = E - \frac{E^2}{E + \dfrac{2}{3}H}, \tag{11.92}$$

and σ^{fic} is the contribution from the path-dependent terms, defined as

$$\sigma^{fic} = {}^{n}\dot{\sigma} - \frac{E}{E + \dfrac{2}{3}H}[{}^{n}\dot{\eta} + \sqrt{\frac{2}{3}}(1-\beta)H\,{}^{n}\dot{e}_{p}]. \tag{11.93}$$

Using (11.86) and (11.91), the structural energy form on the left side of (11.85) can be differentiated with respect to the design, to obtain

$$\begin{aligned} [a_\Omega(z,\bar{z})]' &= \int_0^l \varepsilon(\bar{z}_1)C^{alg}\varepsilon(\Delta\dot{z}_1)\,dx \\ &+ \int_0^l [\varepsilon(\bar{z}_1)C^{alg}\varepsilon_{V_\Omega}(\Delta z_1) + \varepsilon_{V_\Omega}(\bar{z}_1)\,{}^{n+1}\sigma + \varepsilon(\bar{z}_1)\,{}^{n+1}\sigma V_{1,1}]\,dx \\ &+ \int_0^l [\varepsilon(\bar{z}_1)C^{alg}\varepsilon_{V_\theta}(\Delta z_1) + \varepsilon_{V_\theta}(\bar{z}_1)\,{}^{n+1}\sigma]\,dx \\ &+ \int_0^l [\varepsilon(\bar{z}_1)\sigma^{fic}]\,dx. \end{aligned} \tag{11.94}$$

As expected, the first integral constitutes the implicitly dependent term, which will be solved in the design sensitivity equation; the second integral is the contribution from the shape design; the third integral consists of the contribution from the orientation design; and the last integral is made up of path-dependent terms.

The first integral in (11.94) is the same as the linearized energy form for the truss component and can be denoted by $a^{*}_{\Omega}(z;\Delta\dot{z},\bar{z})$. For notational convenience, the following fictitious loads are defined:

$$a'_{V_\Omega}(z,\bar{z}) = \int_0^l [\varepsilon(\bar{z}_1)C^{alg}\varepsilon_{V_\Omega}(\Delta z_1) + \varepsilon_{V_\Omega}(\bar{z}_1)\,{}^{n+1}\sigma + \varepsilon(\bar{z}_1)\,{}^{n+1}\sigma V_{1,1}]\,dx \tag{11.95}$$

$$a'_{V_\theta}(z,\bar{z}) = \int_0^l [\varepsilon(\bar{z}_1)C^{alg}\varepsilon_{V_\theta}(\Delta z_1) + \varepsilon_{V_\theta}(\bar{z}_1)\,{}^{n+1}\sigma]\,dx \tag{11.96}$$

and

$$a'_P(z,\bar{z}) = \int_0^{l}[\varepsilon(\bar{z}_1)\sigma^{fic}]dx. \tag{11.97}$$

The material derivative of the load form on the right side of (11.85) is the same as the truss component in the linear problem in Section 7.2.3. Since the load form is independent of material nonlinearity, its material derivative is the same as (7.60) and (7.61).

Now, we can differentiate structural (11.85) with respect to shape and configuration design. After differentiating (11.85) and using (11.94), (7.60), and (7.61), the design sensitivity equation for the direct differentiation method can be obtained as

$$a^*_{\Omega}(z;\Delta\dot{z},\bar{z}) = \ell'_{V_\Omega}(\bar{z}) + \ell'_{V_\Omega}(\bar{z}) - a'_{V_\Omega}(z,\bar{z}) - a'_{V_\theta}(z,\bar{z}) - a'_P(z,\bar{z}), \quad \forall\bar{z}\in Z, \tag{11.98}$$

where the incremental material derivative $\Delta\dot{z}_1$ is required. Note that the incremental displacement Δz_1 is calculated iteratively until the residual force vanishes. However, sensitivity equation (11.98) does not require iteration to solve the material derivative $\Delta\dot{z}_1$. After the response analysis converges at time step t_{n+1}, the linear problem in (11.98) is solved for $\Delta\dot{z}_1$. After solving the incremental material derivative, the total material derivative of displacement can be updated using the following formula:

$$^{n+1}\dot{z}_1 = {}^{n}\dot{z}_1 + \Delta\dot{z}_1. \tag{11.99}$$

As mentioned before, for sensitivity computation the path-dependent variables must be updated at the next time step. After calculating $\Delta\dot{z}_1$, it is necessary to calculate $\dot{\hat{\gamma}}$ from (11.90). The material derivative of stress at time t_{n+1} can be updated using the formula in (11.91). The material derivatives of back stress and effective plastic strain can be updated using the following formulas:

$$^{n+1}\dot{\alpha} = {}^{n}\dot{\alpha} + \frac{2}{3}\beta H\dot{\hat{\gamma}} \tag{11.100}$$

and

$$^{n+1}\dot{e}_p = {}^{n}\dot{e}_p + \sqrt{\frac{2}{3}}\dot{\hat{\gamma}}. \tag{11.101}$$

Due to its path-dependent properties, the design sensitivity formulation of elastoplasticity is not as efficient as compared with its use for nonlinear elastic problems. The material derivatives of path-dependent variables must be calculated at each time step and are used for sensitivity calculation at the next time step. However, the sensitivity formulation developed in this section does not require any iteration. Thus, the cost of such a sensitivity calculation is still considerably less than that of a response analysis.

11.2.2 Numerical Examples

Elastoplastic Three-Bar Truss

An elastoplastic three-bar truss structure is used to verify the accuracy of the sensitivity calculation and distinguish the contributions from shape and orientation designs. A three-bar truss structure subjected to a concentrated load at node 4 is considered, as shown in Figure 11.10. For analysis data, constant cross-sectional area $A_1 = A_2 = A_3 = 0.1$ in^2, plastic modulus $H_1 = H_2 = H_3 = 1.0 \times 10^4$, and Young's modulus $E_1 = E_2 = E_3 = 1.0 \times 10^4$ psi are used. The yield stress of each member is 50 psi.

A perturbed design is shown in Figure 11.10 in which nodal point 4 is moved downward in the x_2-direction by 1 in. In the perturbed design, both the shape and the orientation of truss design components are changed. Finite element analysis results show that element 2 is in the plastic stress range at the load level 8.55, and elements 1 and 3 are in the plastic stress range at the load level 12.54. Table 11.8 shows configuration design sensitivity results of displacement z_2 at node 4 using the proposed method. Note that the sign of the design sensitivity changes from positive to negative between load levels of 5 and 10 lb. In addition, the sign of design sensitivity changes from negative to positive between load levels of 15 and 20 lb. In this example, note that tip displacement decreases as the orientations of members 1 and 3 change to become vertical, whereas tip displacement increases as the length of each member increases. Table 11.9 illustrates that the contribution from an orientation change dominates with load levels between a 10 and 15 lb, while the contribution from shape change dominates with load levels between 20 and 30 lb.

Table 11.8. Configuration design sensitivity results of displacement of three-member truss model.

Load*	$z_2(b)$**	$z_2\,(b+\Delta b)$	Δz_2	z_2'	$z_2'/\Delta z_2$
5	2.929E–4	2.940E–4	1.107E–6	1.108E–6	100.09%
10	6.832E–4	6.811E–4	–2.189E–6	–2.241E–6	102.38%
15	2.592E–3	2.581E–3	–1.079E–5	–1.022E–5	94.72%
20	5.802E–3	5.803E–3	1.558E–6	1.339E–6	85.94%
25	9.024E–3	9.038E–3	1.379E–5	1.349E–5	97.82%
30	1.224E–2	1.227E–2	2.609E–5	2.557E–5	98.01%

* Unit in lb.
** Unit in in.

Table 11.9. Contributions from shape and orientation changes to design sensitivity results of displacement of three-member truss model.

Load*	Shape	Orientation
5	1.727E–6	–6.180E–7
10	9.771E–7	–3.218E–6
15	2.033E–5	–3.116E–5
20	3.901E–5	–3.764E–5
25	5.702E–5	–4.356E–5
30	7.599E–5	–5.045E–5

* Unit in lb.

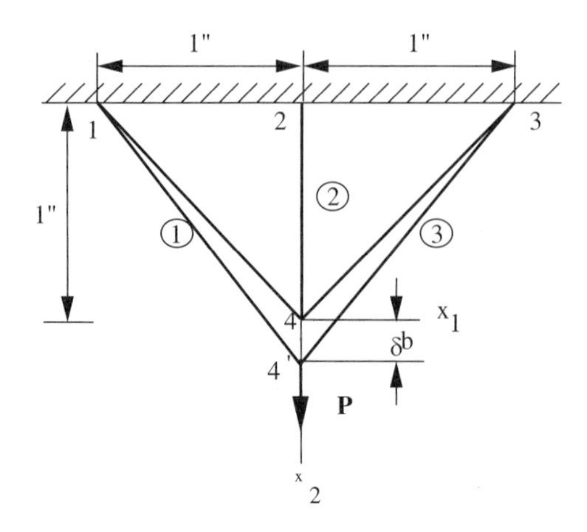

Figure 11.10. Configuration design perturbation of three-member truss model.

PART IV
Numerical Implementation and Applications

12
Design Parameterization

Design parameterization is an essential step in the structural design process. The purpose of design parameterization is to define geometric parameters, as well as to collect a subset of these parameters as design variables that can vary during the design process. In this chapter, types of sizing and shape design variables and their use are introduced. During the design process, the design engineer will vary the design variables in order to improve structural performances. The selection of geometric parameters as design variables must take both design and manufacturing considerations into account. Geometric and finite element modeling cannot be complete until design variables are defined, since the element section properties in the analysis model must be consistent with the design model.

12.1 Sizing Design Parameterization

In this section, several possible design parameterizations, such as constant and linear designs as described in Fig. 12.1, are introduced for the line and surface design components. However, these are not at all the only possible design parameters and other more complicated design parameterizations can be used. However, the method presented in this chapter together with the numerical implementation method in Chapter 13 can be extended to other complicated design parameterizations. One important thing to consider when more complicated design parameterizations are used, is that the finite element model must be sophisticated enough to support the design parameterization method used. Geometric parameters can be defined at the end grid points of a line, or at the corner points of a surface. A bilinear thickness distribution can be used to characterize a surface design component, as shown in Fig. 12.1(b). Note that each dimension that defines the cross-sectional shape, such as width and height in Fig. 12.1(a), can be treated as a design variable, and be allowed to vary to the same degree as the corresponding variable at the other end (constant parameterization), or to a different degree (linear parameterization).

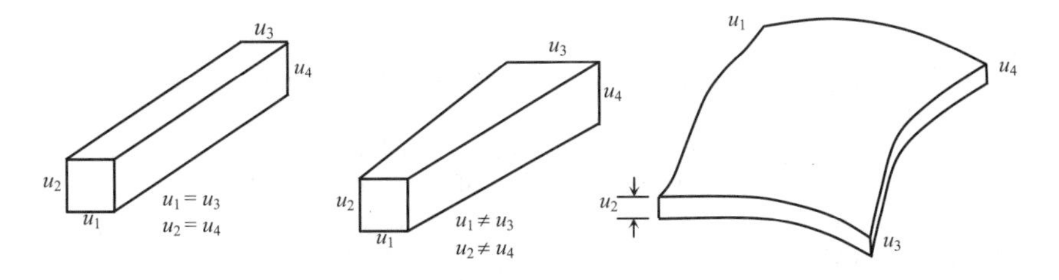

Constant Parameterization Linear Parameterization Bilinear Parameterization

(a) Line Design Component (b) Surface Design Component

Figure 12.1. Line and surface design parameterization.

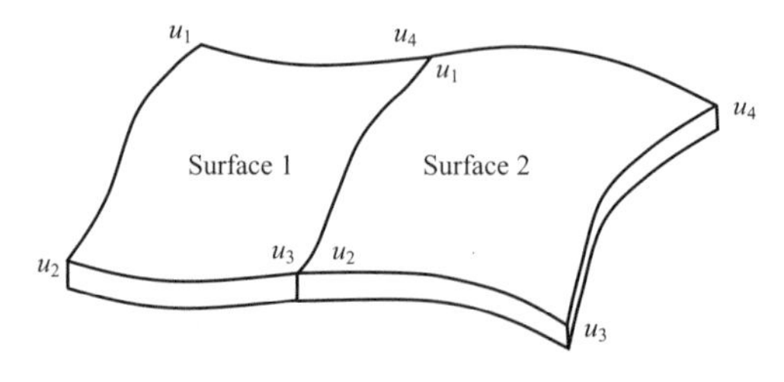

Figure 12.2. Design variable linking.

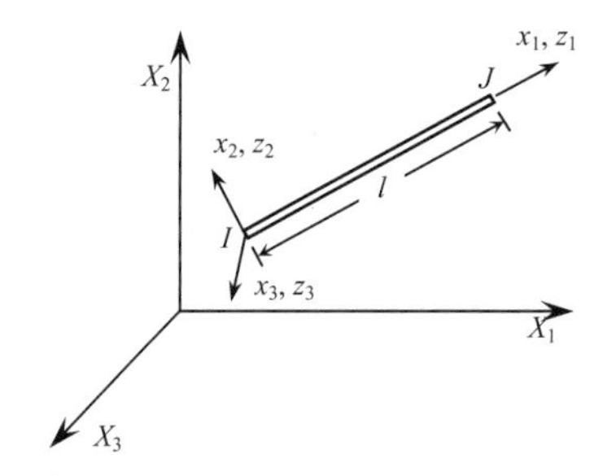

Figure 12.3. Truss design component.

In order to maintain design continuity for a symmetric design, or to reduce the number of design variables, design variables can change either independently of, or proportional to, certain variables across design components through design variable linking, as shown in Fig. 12.2.

12.1.1 Line Design Components

Truss Design Component

Consider a three-dimensional truss design component that can handle tensile and compressive load and that may be composed of several truss finite elements. There are three cross-section types for this design component: symmetric, unsymmetric, and general. A linearly tapered cross-sectional shape can be considered within the design component. Figure 12.3 illustrates the geometry of a truss design component.

Material properties such as mass density and Young's modulus, and geometric parameters that define the cross-sectional shape can be taken as design variables along the axial axis x_1. All material property design variables are assumed to be constant along the design component's axial axis.

The geometric design variables that can linearly vary along the axis of the design component are the dimensions of each cross section, that is, r for a solid and hollow circular cross section, and b and h for other cross sections, as shown in Tables 12.1 and 12.2.

Table 12.1. Symmetric cross sections.

Cross-sectional Shape	Design Variable* 1	2	3	4	5	6
Solid circular	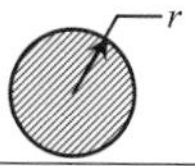 r	E	ρ	–	–	–
Solid rectangular	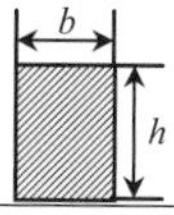 h	b	E	ρ	–	–
Hollow circular	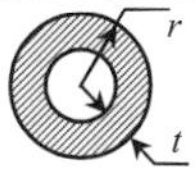 r	t	E	ρ	–	–
Hollow rectangular	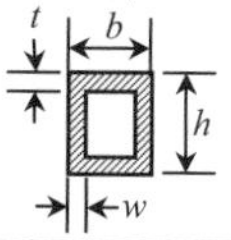 h	b	t	w	E	ρ
I-section	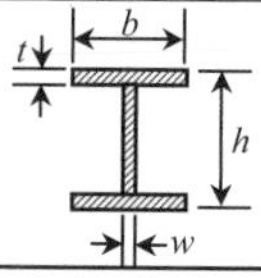 h	b	t	w	E	ρ

* E is Young's modulus and ρ is the mass density.

Table 12.2. Unsymmetric cross sections.

Cross-sectional Shape	Design Variable 1	2	3	4	5	6	7
Channel	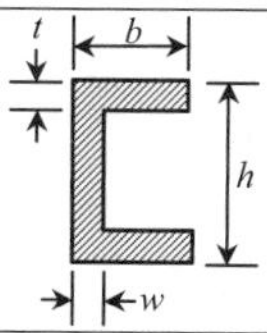 h	b	t	w	E	ρ	–
Angle	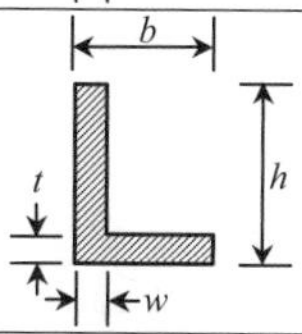 h	b	t	w	E	ρ	–
T-section	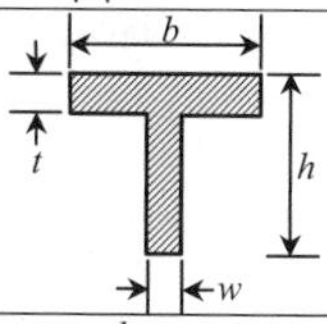 h	b	t	w	E	ρ	–
Unsymmetric I-section	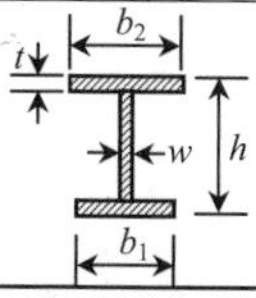 h	b_1	b_2	t	w	E	ρ

Table 12.3. General cross sections.

General Cross-sectional Shape	Design Variable					
	1	2	3	…	$2n$+1	$2n$+2
s_1, t_1; s_2, t_2; s_3, t_3; s_n, t_n	s_1	t_1	s_1	…	E	ρ

Proportionality can be used between two geometric design variables within a design component through design variable linking. Such proportional design variables are used to meet local buckling requirements for a structural design specification, to maintain geometric proportionality, or to reduce the number of design variables. The pairs of geometric design variables that may be considered for proportionality purposes are: (t,b) and (w,h) for a hollow rectangle and an I-section, (b,h) for a solid rectangle, and (t,r) for a hollow circle.

The symmetric cross section has two perpendicular axes of symmetry, as shown in Table 12.1. These two axes are the principal axes of the cross section, and are usually taken as two component coordinates. The component's axial axis is assumed to be the same as the centroidal axis. Solid and hollow circles, solid and hollow rectangles, and the I–section are all placed in this category.

The unsymmetric cross section has either only one axis of symmetry, or no symmetry with some shape regularity, as shown in Table 12.2. The channel, angle, T-section, and unsymmetric I-section all fit into this category. The component coordinates x_2 and x_3 may be different from the principal axes x_{p2} and x_{p3}, respectively. The centroidal axis is assumed to coincide with the axial axis x_1 of the truss design component.

The general cross section has no regular shape and may be composed of several thin-walled segments, as shown in Table 12.3.

Two-Dimensional Beam Design Component

Consider a two-dimensional beam component that can handle tensile, compressive, and bending load. In the case of bending load, only axial stress is treated using the technical beam theory. The design component may be composed of several two-dimensional beam finite elements, and can be classified into one of three types according to its cross-sectional shape: symmetric, unsymmetric, and general. A linearly tapered cross-sectional shape can be considered within the design component. Figure 12.4 describes the geometry of a two-dimensional beam design component. For such a component, the component coordinate x_2 is a symmetric axis of the cross section, and the X_1-X_2 plane or the x_1-x_2 plane is the plane of loading.

The symmetric cross section has two perpendicular axes, x_2 and x_3, which coincide with the principal axes of the cross section. Solid and hollow circles, solid and hollow rectangles, and the I-section are all placed in this category.

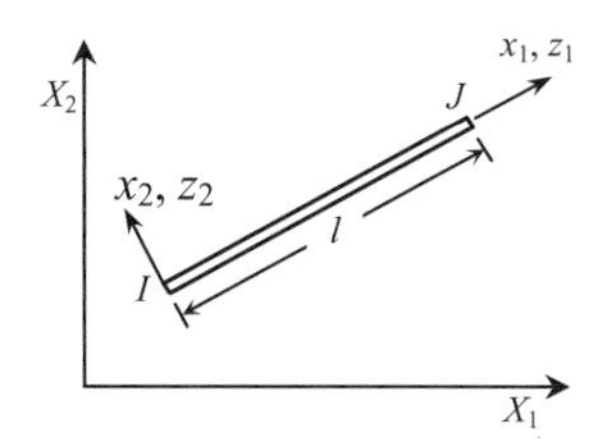

Figure 12.4. Two-dimensional beam design component.

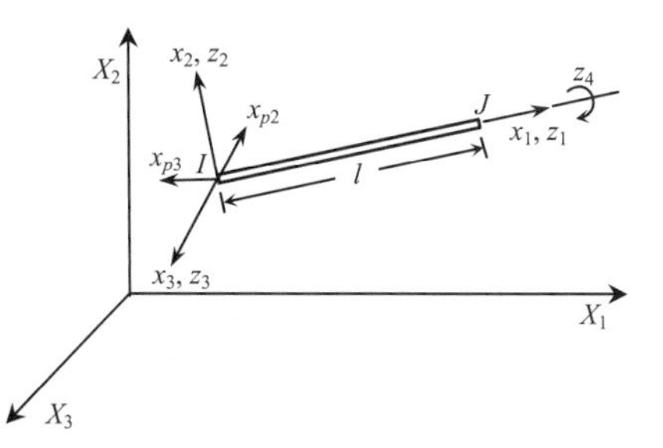

Figure 12.5. Three-dimensional beam design component.

The unsymmetric cross section has only one symmetric axis, which lies on the plane (X_1-X_2 or x_1-x_2) of loading, as shown in Table 12.2. The channel, T-section, and unsymmetric I-section are all placed in this category. The perpendicular axes x_2 and x_3 coincide with the principal axes. The centroidal axis is the same as axial axis x_1 of the truss design component.

The general cross section for the two-dimensional beam component has one axis of symmetry in the plane of loading, and is composed of several thin-walled segments, as shown in Table 12.3. The perpendicular axes x_2 and x_3 coincide with the principal axes. The centroidal axis is the same as axial axis x_1 of the design component.

Three-Dimensional Beam Design Component

Consider a three-dimensional beam component that can handle tensile, compressive, bending, and torsional load. For the bending load, only the axial stress is treated, using the technical beam theory. For a three-dimensional beam design component, torsional stress is also taken into account. The beam design component may be composed of several three-dimensional finite elements. This design component has three types, according to its cross-sectional shape: symmetric, unsymmetric, and general. A linearly tapered cross-sectional shape can be considered within the design component. Figure 12.5 describes the geometry of the three-dimensional beam design component. For general cross sections, the principal axes x_{p2} and x_{p3} may be different from the component coordinates x_2 and x_3, as shown in Fig. 12.3.

The same design variables used for three-dimensional truss components can be used for three-dimensional beam components. Thus, all design variables in Tables 12.1 through 12.3 can be used.

12.1.2 Surface Design Components

Membrane and Shell Design Component

A three-dimensional membrane component can handle both an in-plane tensile and a compressive load. A three-dimensional shell component can handle an in-plane tensile, compressive, and bending load. The design component may be composed of several membrane/shell finite elements. Figure 12.6 illustrates the geometry of a membrane/shell design component.

Surface component design variables include thickness, mass density, and Young's modulus. Surface thickness is parameterized using a bilinear shape function. Four geometric design variables are defined for each surface design component: thickness h_I, h_J, h_K, and h_L at grid points I, J, K, and L, respectively, as shown in Fig. 12.6 and Table 12.4.

The four-node quadrilateral surface component can be reduced to a triangular surface component by defining duplicate node numbers for the third and fourth (K and L) node locations. If node L is not defined, then it defaults to node K. The design component thickness is assumed to vary bilinearly inside the design component.

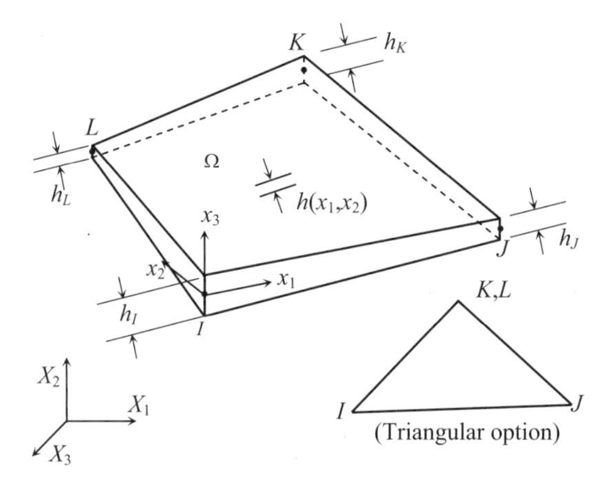

Figure 12.6. Three-dimensional membrane design component.

Table 12.4. Design variables of surface design component.

Design Variables					
1	2	3	4	5	6
E	ρ	h_I	h_J	h_K	h_L

12.2 Shape Design Parameterization

Shape design parameterization, which describes the boundary shape of a structure as a function of the design variables, is an essential step in the shape design process. Inappropriate parameterization can lead to unacceptable shapes [54] and [55]. To parameterize the structural boundaries and to achieve optimum shape design, boundary shape can be described in three ways: (1) by using boundary nodal coordinates, (2) by using polynomials, and (3) by using spline blending functions. However, it is important to point out that there are many different methods of parameterization and that the methods presented in this section are only part of those methods, including very complicated parameterization methods developed in commercial CAD tools. One important aspect of shape design parameterization is the connection of the design parameter to the computation of the design velocity field, as explained in Chapter 6.

In the first method, boundary nodal coordinates of the finite element model are used as shape design variables. Although the method is simple and easy to use, it has the following drawbacks: (1) the number of design variables tends to become very large, which may lead to high computational costs and optimization problems that are difficult to solve; (2) the first derivative of the design boundary is not continuous across boundary nodes, which may lead to an unacceptable or impractical design; and (3) computational accuracy is not ensured, since it is difficult to maintain an adequate finite element mesh during the optimization process. One can use coordinates of selected master nodes as shape design variables and employ an isoparametric mapping to generate a finite element mesh.

Several methods have been developed to parameterize the design boundary with polynomials. Bhavikatti and Ramakrishnan [56] used a fifth-degree polynomial, with the coefficients taken as design variables, to parameterize the boundary shape of a rotating disk. Prasad and Emerson [57] used a similar approach to optimize an engine connecting rod. In a more general approach, such as that used by Kristensen and Madsen [58] and Pedersen and Laursen [59], the boundary is parameterized using a linear combination of shape functions, with the coefficients as design variables. The total number of shape design variables can be reduced using polynomials for shape representation. However, using high-order polynomials to represent the boundary shape may result in oscillating boundaries.

Splines eliminate the problem of oscillating boundaries since they are composed of low-order polynomial segments that are combined to maximize the smoothness of the boundaries. Yang and Choi [60], Luchi et al. [61], and Weck and Steinke [62] used a cubic spline to define the boundary geometry. The spline representation was shown to yield better sensitivity accuracy than a piecewise linear representation of the boundary [60]. Braibant et al. [63] used Bezier and B-spline blending functions to describe the design boundary. Blending functions provide great flexibility for geometrical description. Using B-splines, Braibant and Fleury [64] optimized a beam in bending, a fillet, and a hole in a plate. Finally, Yao and Choi [65] through [67] used a Bezier surface to optimize an engine bearing cap and an arch dam.

The shape design parameterization method presented in this section deals with geometric features. A geometric feature is a subset of the geometric boundaries of a structural component. For example, a fillet or a circular hole is a geometric feature that contains characteristics associated with it and is likely to be chosen as a design variable. A geometric feature whose design variables are defined is known as a parameterized geometric feature, and is treated as a single entity in the shape design process. For example, a circular hole with its radius and center location defined as design variables is a parameterized geometric feature. Such a parameterized circular hole can be moved around in the structure with a varied size due to design changes. However, the shape of the circular hole itself remains constant.

Two steps are involved in the design parameterization process: geometric modeling and defining the design variables. A geometric model is first generated in the modeling process, with all its dimensions defined. Geometric features that can be varied in the design process need to be identified by both design and manufacturing engineers at the beginning of the design process. The design engineer then parameterizes the geometric model, using the geometric feature that is consistent with both engineering requirements and manufacturing limitations. The design parameterization developed in this section is based on the assumption that the geometric model has already been created and its dimensions defined.

In general, structural shape design problems can be classified into four types, in terms of the characteristics of the design boundary. In the first type, the shape of an arbitrary open or closed boundary is determined, such as a fillet [37] or a dam surface [65]. In the second type, the dimensions of predefined shapes are determined, such as the radius of a circular hole, the major and minor axes of an elliptic hole, the dimensions of a slot, the length of a rectangular membrane, or the radius of a rounded corner. In the third type, the design boundary location is determined, such as the center location of a circular hole, an elliptic hole, an arbitrary shaped hole, or a slot. In the final type, a rotation angle of the design boundary, either arbitrary or predefined, is treated as the design variable. In this section, shape design parameterization of the first three types of design problems is considered.

Details of the three-step design parameterization procedure are described in Section 12.2.2. Since shape design parameterization is built on the geometric modeler, the geometry representation method, utilized for model generation, design parameterization, and the geometry update, is introduced in Section 12.2.1. The advantages and limitations of geometric modeling are described from the viewpoint of shape design parameterization. Basic shape design parameterization is defined for such geometric entities as curves and suthreefaces. Descriptions of the basic geometric entities appropriate for shape design are given in Section 12.2.3 for two-dimensional structural components, and in Section 12.2.4 for three-dimensional components. Finally, design variable linking procedures, necessary for creating parameterized geometric features, are discussed in Section 12.2.5.

12.2.1 Representation of Geometry in Parametric Space

In general, geometric entities can be represented using parametric cubic (PC) lines, patches (surfaces), and hyperpatches (solids). A parametric cubic line is represented by three functions

$$\left.\begin{aligned} x &= x(u) \\ y &= y(u) \\ z &= z(u) \end{aligned}\right\}, \tag{12.1}$$

where u is the parametric direction of the line with domain [0,1]. Each of these functions is, at most, a cubic polynomial of the form

$$z(u) = s_3u^3 + s_2u^2 + s_1u + s_0. \tag{12.2a}$$

The first and second derivatives of (12.2)a can be written as

$$z_{,u}(u) = 3s_3u^2 + 2s_2u + s_1 \tag{12.2b}$$

$$z_{,uu}(u) = 6s_3u + 2s_2. \tag{12.2)c}$$

It is noted from (12.2)b and (12.2)c that the most the slope of the cubic line can change its sign is twice, and that the curve can only have one inflection point. Consequently, parametric cubic (PC) entities such as PC lines and PC patches minimize the possibility of yielding oscillating boundaries during the design process [55]. However, geometric entities with predefined or sophisticated shapes, such as a circular hole, cannot be represented by a single cubic curve. To minimize modeling errors, such a boundary can be broken into small pieces. These pieces are then "glued" together in the design process as one geometric feature by appropriately linking design variables. For shape design, planar parametric cubic lines and spatial parametric bi-cubic patches are utilized to represent the design boundaries of two-dimensional and three-dimensional structural components, respectively.

There are a number of methods for creating geometric entities, for example, defining four control points to create a Bezier curve, or constructing four edge curves to create a surface. Although geometric entities have different characteristics depending on the way they are created, they are nevertheless always represented by polynomials with the same order regardless of the way they are created. Consequently, a single geometric entity can be created using a variety of methods. For example, the planar curve shown in Fig. 12.7(a) is created by defining four distinct points in the plane, and is thus called a four-point curve.

The mathematical expression of the four-point curve is given as

$$\left.\begin{aligned} x(u) &= 3u \\ y(u) &= 9u^3 - 13.5u^2 + 6.5u \end{aligned}\right\}. \tag{12.3}$$

The same curve shown in Fig. 12.7(b) can also be created by giving the position and slope at the endpoints, referred to as geometric coefficients:

$$\left.\begin{aligned} \boldsymbol{p}_0 &= [0,0] \\ \boldsymbol{p}_1 &= [3,2] \\ \boldsymbol{p}_0^u &= [3,6.5] \\ \boldsymbol{p}_1^u &= [3,6.5] \end{aligned}\right\}. \tag{12.4}$$

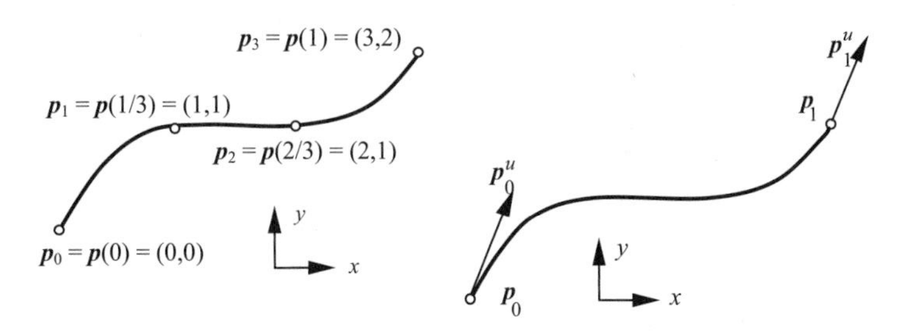

(a) Curve Created by Four Points (b) Curve Created by Geometric Coefficients

Figure 12.7. Planar parametric cubic curves.

Therefore, one curve can be represented using different methods. However, these representations of curves and patches are mathematically equivalent and one can be transformed into another by using certain linear transformations. For example, the geometric coefficients of a curve can be transformed into a four-point format, so the shape of the curve can be controlled according to the position of the four points. In fact, geometric coefficients are selected as unified geometric data, independent of the methods used to create geometric entities. Because design parameterization has the versatility of representing the same geometric entity in different ways, it can be systematically developed to provide the design engineer with sufficient resources for solving a large variety of shape design problems.

12.2.2 Shape Design Parameterization Method

The design parameterization method presented in this section is a three-step process, as illustrated in Fig. 12.8. The first step is to create a geometric feature by grouping a number of interconnected geometric entities together, and by defining the type of geometric feature. The design engineer identifies the geometric entities that form the geometric features.

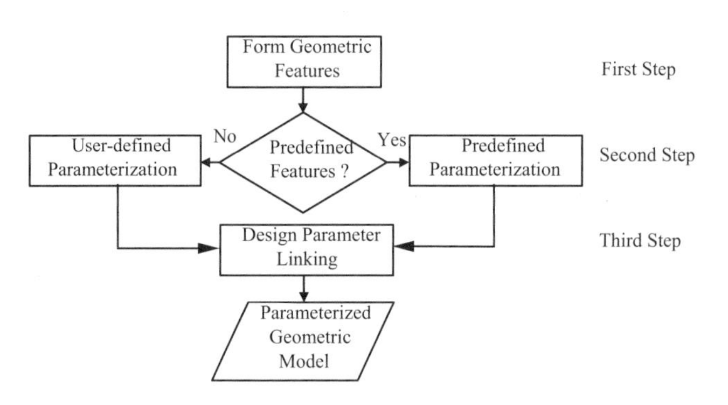

(a) Overall Design Parameterization Process

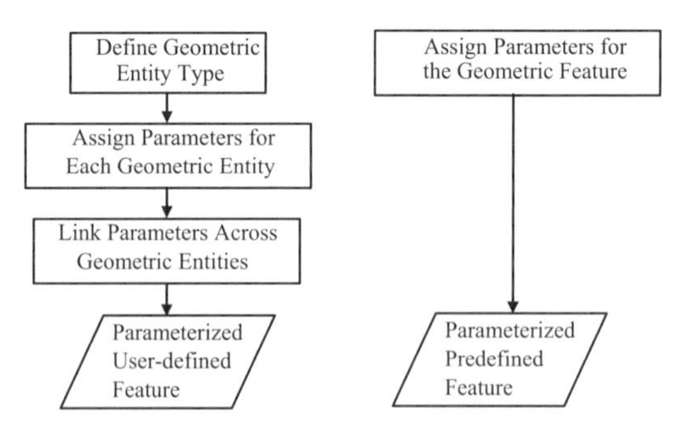

(b) Parameterization for a User-defined Feature (c) Parameterization for a Predefined Feature

Figure 12.8. Shape design parameterization of features.

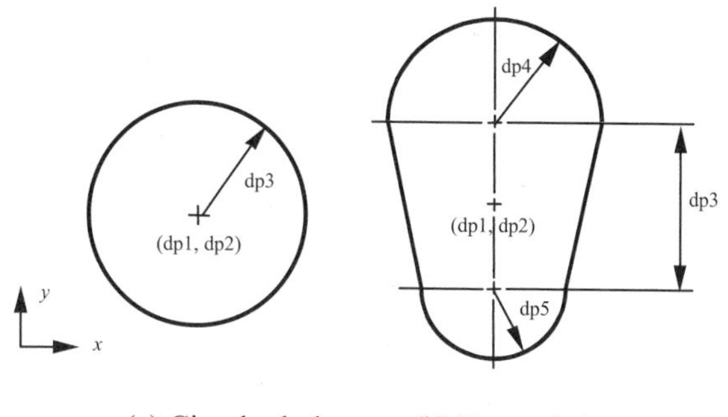

(a) Circular hole (b) Tapered slot

Figure 12.9. Predefined geometric features.

The second step is to define the design variables within each geometric feature. Geometric features that are frequently used in the construction of structural components can be put in the library of predefined geometric features. The design engineer can then parameterize these predefined features simply by selecting associated predefined shape design variables. To demonstrate the use of the library, two predefined geometric features have been defined, a circular hole and a tapered slot, as shown in Fig. 12.9. For the circular hole, which is formed by connecting a number of circular arcs end to end with the same center point and radius, both the radius and center point of the circle can be defined as shape design variables. For the tapered slot, which is formed by connecting a number of straight lines and circular arcs, length dp3, radii dp4 and dp5, and center point dp1 and dp2 can all be defined as shape design variables. The design parameterization flow for the predefined geometric feature is illustrated in Fig. 12.8(c).

A geometric feature that is not included in the library can be seen as a user-defined feature. To generate the latter, the design engineer can define the design variable by using geometric entities, and can then link these variables across the entities. For example, suppose a cantilever beam is to be parameterized such that its length H can be varied, as shown in Fig. 12.10(a). To parameterize the beam, the following procedure can be used:

(1) Identify lines #101, #102, and #103 to form the geometric feature, that is, the edges of the beam, as shown in Fig. 12.10(b).

(2) Define line #102 as a straight line and define the x-coordinate at end 1 of line 102, that is, grid #3, as the free design variable dp1, and the x-coordinate at end 2 of line #102, that is, grid #2, to be proportional to dp1 with a proportionality of 1.0.

(3) Define lines #101 and #103 as straight lines, and define the x-coordinate at end 2 of lines #101 and #103, that is, grids #2 and #3, as the free design variables dp2 and dp3, respectively.

(4) Link dp2 and dp3 to dp1, with a proportionality of 1.0.

After design variable linking, only one design variable dp1, the beam length represented by the x-coordinate of grid #3 in line #102, is allowed to vary independently. The parameterization flow for the user-defined feature is shown in Fig. 12.8(b).

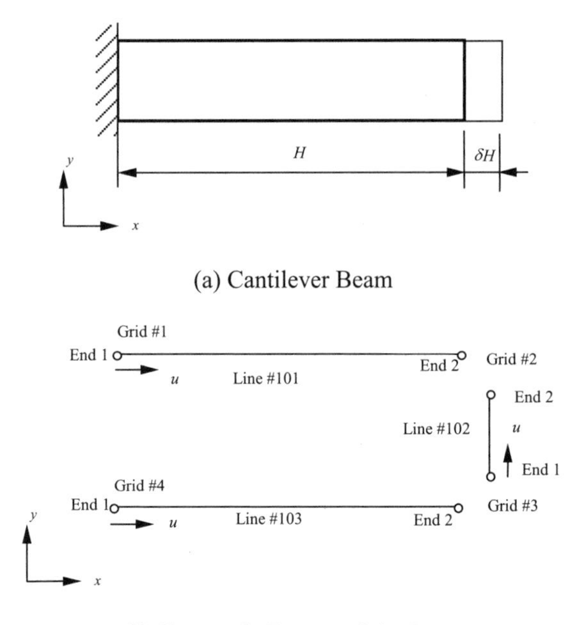

Figure 12.10. Parameterization of a cantilever beam.

The shape design parameterization procedure for a user-defined geometric feature, as illustrated above, is summarized as follows:

1. Identify the types of geometric entities that are to be used to construct the user-defined geometric feature.

2. Parameterize the geometric entities by defining free and proportional design variables in each geometric entity.

3. Generate the parameterized geometric feature by linking free design variables across geometric entities.

Each predefined geometric feature that can be included in the feature library, such as the circular hole and tapered slot, is preconstructed by using the above procedure.

If necessary, the third step is designed to link design variables across geometric features. For example, a cantilever beam with a circular hole, as shown in Fig. 12.11, is to be parameterized so that the position of the hole is proportional to the beam length. The x-coordinate of the circular hole H_h can be parameterized by using the predefined parameterization process, as described in Fig. 12.9(a). The length of beam H_b can be parameterized as a user defined feature, as illustrated in Fig. 12.10. With the two parameterized features, the x-coordinate of the hole can be linked to the beam length.

As described above, fundamental shape design parameterization is defined within geometric entities, and the parameterized features are created by using geometric entities. A hierarchy of the design parameterization method to build a parameterized geometric model is shown in Fig. 12.12.

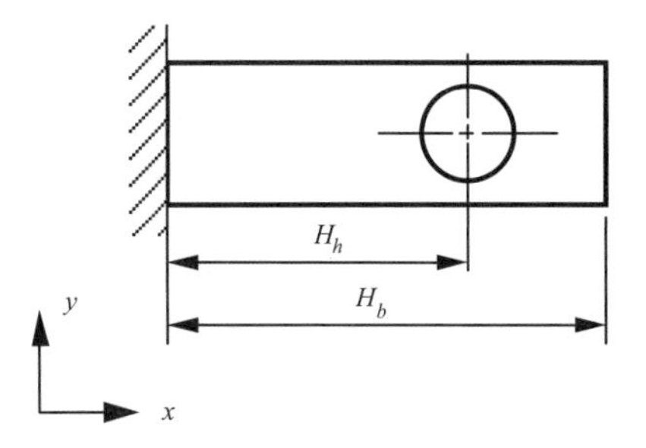

Figure 12.11. Design variable linking across parameterized geometric features.

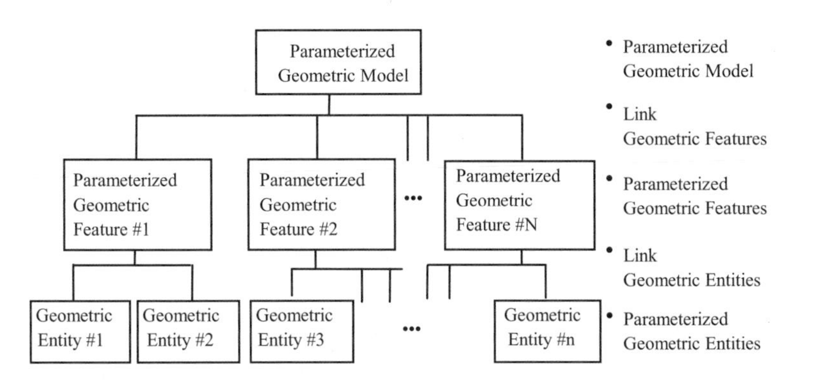

Figure 12.12. Hierarchy of shape design parameterization.

12.2.3 Curve Design Parameterization

In this section, shape design variables are defined for a parametric cubic curve, whose geometric representation and design freedoms will also be discussed. Furthermore, any smoothness requirements on the shape of the design boundary will be considered.

For a two-dimensional shape design, the boundaries are planar curves. In general, there are eight degrees of freedom for a planar cubic curve, as expressed in (12.1) and (12.2), with $z(u)$ serving as the constant. Planar curves with eight degrees of freedom are designated as basic curves, while predefined curves that are constrained, such as a circular arc, are designated as specialized curves. A specialized curve has fewer degrees of freedom since some of the basic degrees of freedom are linked (constrained) in order to define the required characteristics of the curve. For handling two-dimensional shape design problems, six basic curves are discussed: algebraic, geometric, four-point, Bezier, spline, and B-spline; and three specialized curves are introduced: straight line, conic, and arc.

Basic Curves

From a computational point of view, algebraic and geometric curves are the most interesting among the six basic types of curves. As explained in Section 12.2.2, all parametric cubic entities can be transformed into various other formats by using certain

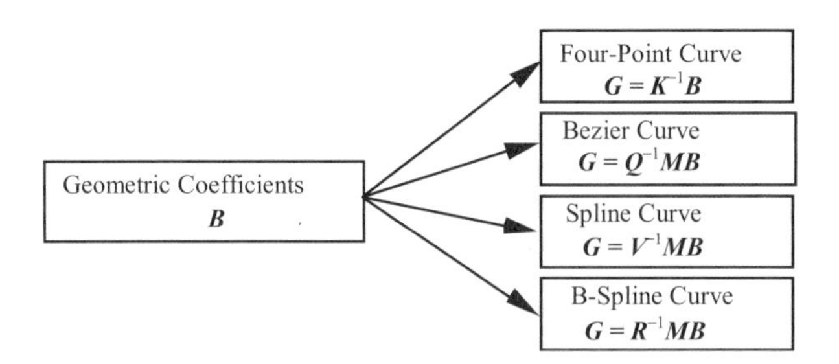

(a) Transformations from **B** to **G**

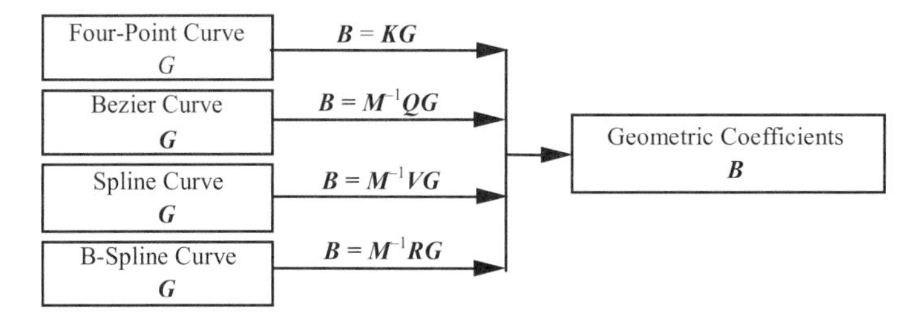

(b) Transformations from $\boldsymbol{G}$ to $\boldsymbol{B}$

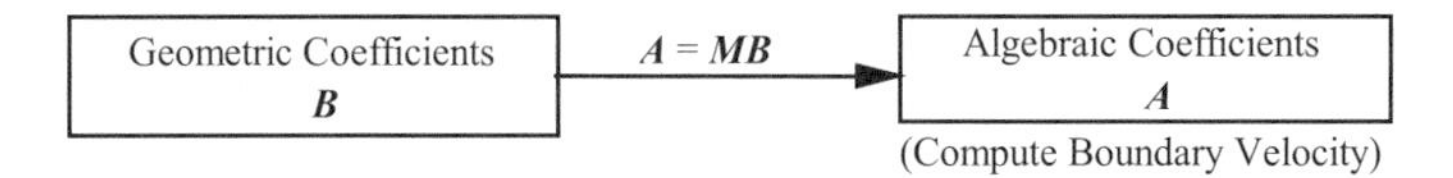

(c) Transformations from $\boldsymbol{B}$ to $\boldsymbol{A}$

Figure 12.13. Curve format transformations for two-dimensional shape design.

linear transformations. For shape design, three major transformations are necessary: (1) from the geometric coefficient matrix $\boldsymbol{B}$ to the design variable matrix $\boldsymbol{G}$ to compute design variable values, (2) from matrix $\boldsymbol{G}$ to matrix $\boldsymbol{B}$ to update geometric entities for a perturbed design shape (discussed in Chapter 6), and (3) from matrix $\boldsymbol{G}$ to the algebraic coefficient matrix $\boldsymbol{A}$ to compute the boundary velocity field (discussed in Chapter 13). For the basic curves, the transformation from matrix $\boldsymbol{G}$ to matrix $\boldsymbol{B}$ for each curve format can be described by their corresponding 4×4 constant matrices. The curve format transformations are summarized in Fig. 12.13.

Algebraic Curve

The algebraic curve is the most basic type of curve. However, it cannot be used for practical shape design purposes because it does not provide any geometric visualization of the relationship between design boundary change and design variable change, which are the algebraic coefficients of the polynomial. The usefulness of the algebraic curve format in terms of shape design is in the computation of the boundary velocity. The mathematical representation of the algebraic curve is given as

$$\begin{aligned} \boldsymbol{p}(u) &= \boldsymbol{a}_3 u^3 + \boldsymbol{a}_2 u^2 + \boldsymbol{a}_1 u + \boldsymbol{a}_0 \\ &= [u^3 \quad u^2 \quad u \quad 1] \begin{bmatrix} \boldsymbol{a}_3 \\ \boldsymbol{a}_2 \\ \boldsymbol{a}_1 \\ \boldsymbol{a}_0 \end{bmatrix}_{4\times 2} \\ &= \boldsymbol{UA} \qquad\qquad u \in [0,1], \end{aligned} \tag{12.5}$$

where $\boldsymbol{p}(u) = [p_x, p_y]$ and $\boldsymbol{a}_i = [a_{ix}, a_{iy}]$, $i = 0, \ldots, 3$, are the algebraic coefficients of the curve.

Geometric Curve

The geometric curve is represented by the position vectors and tangent vectors at its two endpoints, as shown in Fig. 12.14. The cubic form of this curve is often called the Hermite curve. To parameterize the geometric curve, eight geometric coefficients in matrix $\boldsymbol{B}$ can be defined as shape design variables, as

$$\boldsymbol{G} = \boldsymbol{B} = \begin{bmatrix} \boldsymbol{p}_0 \\ \boldsymbol{p}_1 \\ \boldsymbol{p}_0^u \\ \boldsymbol{p}_1^u \end{bmatrix}_{4\times 2}, \tag{12.6}$$

where $\boldsymbol{p}_0 = [p_x, p_y]_{u=0}$, $\boldsymbol{p}_1 = [p_x, p_y]_{u=1}$, $\boldsymbol{p}_0^u = [dp_x/du, dp_y/du]_{u=0}$, and $\boldsymbol{p}_1^u = [dp_x/du, dp_y/du]_{u=1}$. The transformation between the geometric and algenraic coefficient matrices can be obtained using $\boldsymbol{A} = \boldsymbol{MB}$ where

$$\boldsymbol{M} = \begin{bmatrix} 2 & -2 & 1 & 1 \\ -3 & 3 & -2 & -1 \\ 0 & 0 & 1 & 0 \\ 1 & 0 & 0 & 0 \end{bmatrix}.$$

One important aspect of the geometric curve for shape design purposes is that C^0- and C^1-continuities can be maintained at the joining point of two adjacent curves by linking shape design variables, as shown in Fig. 12.15.

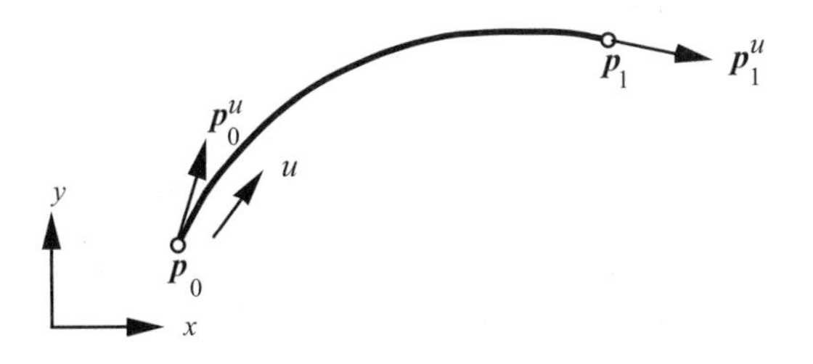

Figure 12.14. Geometric curve.

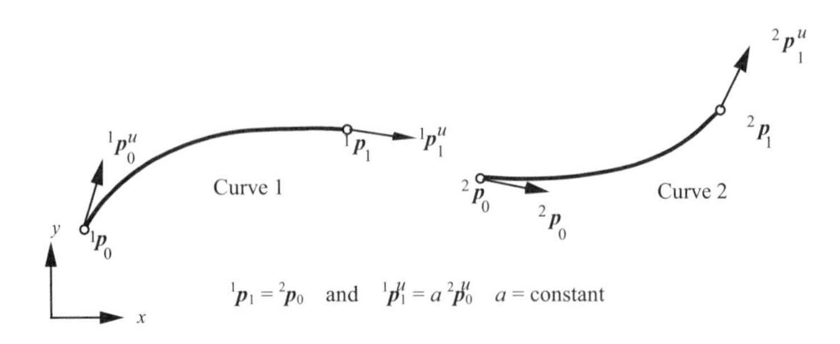

Figure 12.15. C^0- and C^1-continuities across geometric curves.

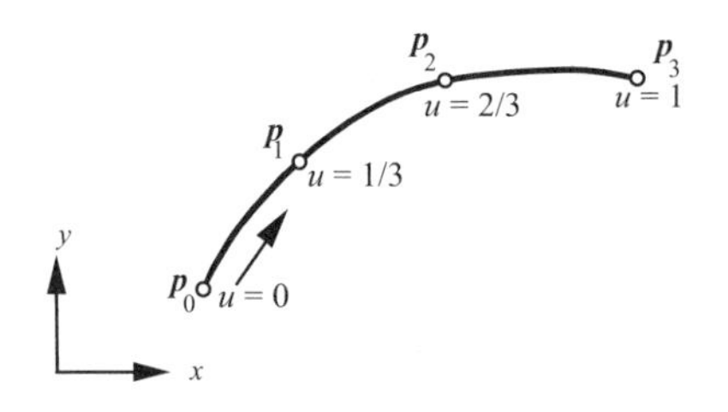

Figure 12.16. Four-point curve.

Four-Point Curve

The four-point curve is geometrically represented by the positions of points at $u = 0$, 1/3, 2/3, and 1 in the parametric direction, as shown in Fig. 12.16. The curve actually passes through these four points. The positions of the four points given in the associated $\boldsymbol{G}$ matrix can be defined as shape design variables, as

$$\boldsymbol{G} = \begin{bmatrix} \boldsymbol{p}_0 \\ \boldsymbol{p}_1 \\ \boldsymbol{p}_2 \\ \boldsymbol{p}_3 \end{bmatrix}_{4\times 2} . \tag{12.7}$$

Since the four points are on the curve, the design engineer will have a clear understanding of how to change or control the shape of the curve by moving the four points.

Bezier Curve

The Bezier curve is geometrically represented by the position of four control points, which determine the shape of the curve by using Bernstein basis functions [68], as

$$\boldsymbol{p}(u) = \sum_{i=0}^{3} \boldsymbol{p}_i B_{i,3}(u), \qquad u \in [0,1], \tag{12.8}$$

where $\boldsymbol{p}_i$, $i = 0, \ldots, 3$, are the position vectors of the four control points, and $B_{i,3}(u)$ is the third-order Bernstein polynomial. The four control points are the vertices of the characteristic polygon of the Bezier curve, as shown in Fig. 12.17.

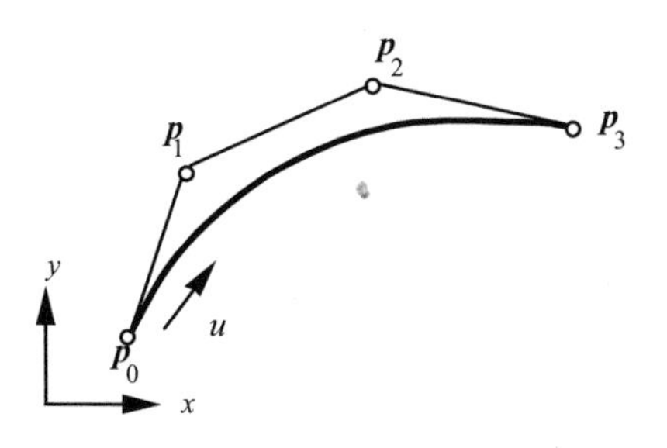

Figure 12.17. Bezier curve.

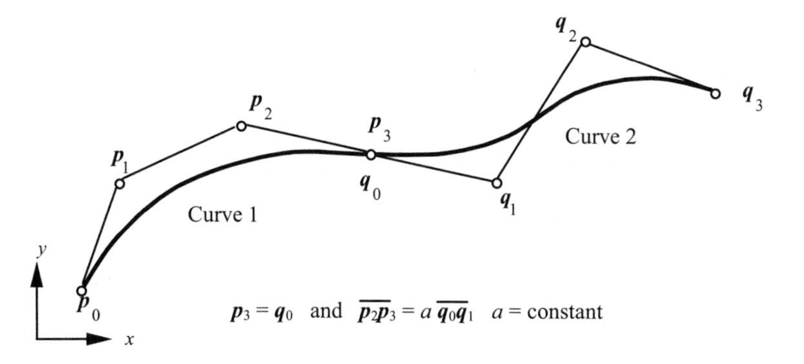

Figure 12.18. C^0- and C^1-continuities across Bezier curves.

Three important characteristics of the Bezier curve are stated as follows: (1) the Bezier curve passes through the two end control points, (2) the tangents at the endpoints $\boldsymbol{p}_0$ and $\boldsymbol{p}_3$ are the vectors $\overrightarrow{\boldsymbol{p}_0\boldsymbol{p}_1}$ and $\overrightarrow{\boldsymbol{p}_2\boldsymbol{p}_3}$, respectively, and (3) the shape of the Bezier curve is controlled by moving the control points. To parameterize a Bezier curve, the position of the four control points given in the associated $\boldsymbol{G}$ matrix can all be defined as shape design variables, as

$$\boldsymbol{G} = \begin{bmatrix} \boldsymbol{p}_0 \\ \boldsymbol{p}_1 \\ \boldsymbol{p}_2 \\ \boldsymbol{p}_3 \end{bmatrix}_{4\times 2} . \tag{12.9}$$

As with the geometric curve, C^0- and C^1-continuities at the joining point of two adjacent Bezier curves can be obtained by linking the positions of joining points to the nearest control points, so that the tangent vectors at the joining point are determined as multiples of each other, as shown in Fig. 12.18.

Spline Curve

The spline curve is geometrically represented by the position and curvature of the two endpoints, as shown in Fig. 12.19. Since the spline curve behaves exactly like a beam

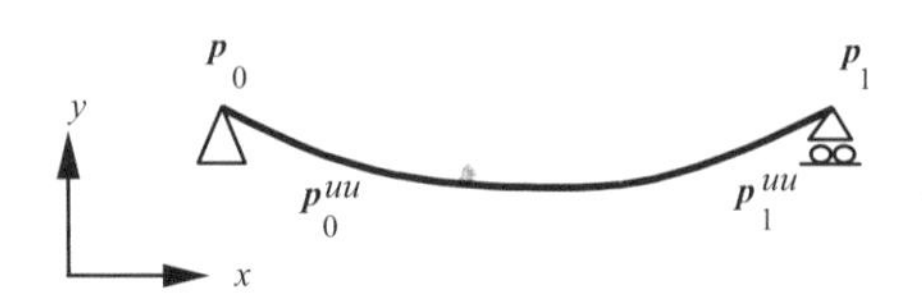

Figure 12.19. Spline curve.

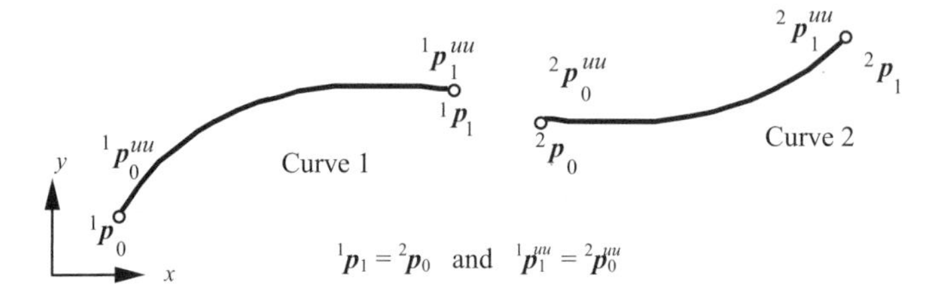

Figure 12.20. C^0-, C^1-, and C^2-continuities across spline curves.

bending that form a smooth curve, it is often called the elastic curve, or the minimum-energy curve. To parameterize a spline curve, the position and curvature at the two endpoints of the curve given in the associated **G** matrix can be defined as shape design variables, as

$$\boldsymbol{G} = \begin{bmatrix} \boldsymbol{p}_0 \\ \boldsymbol{p}_1 \\ \boldsymbol{p}_0^{uu} \\ \boldsymbol{p}_1^{uu} \end{bmatrix}_{4\times 2}. \tag{12.10}$$

One important aspect of the spline curve for shape design is that C^0-, C^1-, and C^2-continuities across curves can be retained by linking the position and curvature at the joining point, as shown in Fig. 12.20.

B-Spline Curve

The (cubic) B-spline curve is created by using at least five control points, producing two B-spline curves. The mathematical expression for the B-spline curve is

$$\boldsymbol{p}(u) = \sum_{i=0}^{n} \boldsymbol{p}_i N_{i,k}(u), \qquad u \in [0,1], \tag{12.11}$$

where $N_{i,k}(u)$ are the B-spline blending functions [68]. With such a curve, the degree of the polynomial is controlled by parameter k, which has a value of four in the case of a cubic polynomial. Moreover, k is independent of the number of control points, which is $n + 1$. The number of B-spline curves is determined by the degree of the polynomial and the number of control points, as

$$m = n - k + 2. \tag{12.12}$$

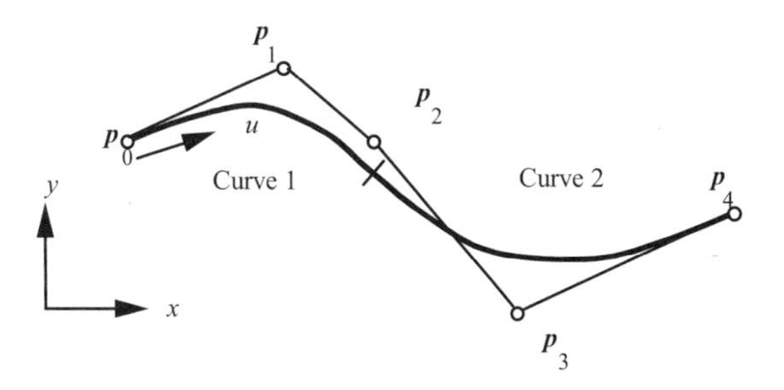

Figure 12.21. B-spline curves.

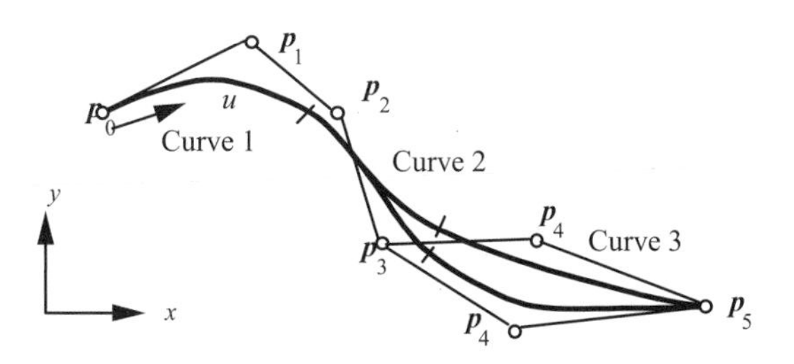

Figure 12.22. Local control behavior of the B-spline curve.

The first and last control points are the start- and endpoints of the first and last B-spline curves. Similar to the Bezier curve, the B-spline curve is represented by a characteristic polygon, whose vertices are the control points, as shown in Fig. 12.21. The tangents at endpoints $\boldsymbol{p}_0$ and $\boldsymbol{p}_4$ are the vectors $\overrightarrow{\boldsymbol{p}_0\boldsymbol{p}_1}$ and $\overrightarrow{\boldsymbol{p}_3\boldsymbol{p}_4}$, respectively.

To parameterize B-spline curves, at least two curves must be considered together. The positions of all control points can be defined as shape design variables, as

$$\boldsymbol{p}_i(u) = \boldsymbol{U}\boldsymbol{R}\begin{bmatrix} \boldsymbol{p}_{i-1} \\ \boldsymbol{p}_i \\ \boldsymbol{p}_{i+1} \\ \boldsymbol{p}_{i+2} \end{bmatrix}, \quad \text{for } i \in [1, n-2]. \tag{12.13}$$

One important characteristic of the (cubic) B-spline curve is that C^2-continuity across curves is naturally retained, without any artificial design variable linking [68]. Another important characteristic is the B-spline's "local control" behavior. As is clear from (12.13), a control point only has an effect on neighboring curves. For example, a change in the position of control point $\boldsymbol{p}_4$ does not have any influence on curve 1, as shown in Fig. 12.22. The influence of control point $\boldsymbol{p}_4$ is restricted to curves 2 and 3.

Certain predefined geometric shapes frequently used to describe structural boundaries, such as a straight edge, a circular arc, and an elliptic hole, are difficult to parameterize using the basic curves discussed in this section. Certain design variables must be linked to force the curve to behave as desired. Three specialized curves, the

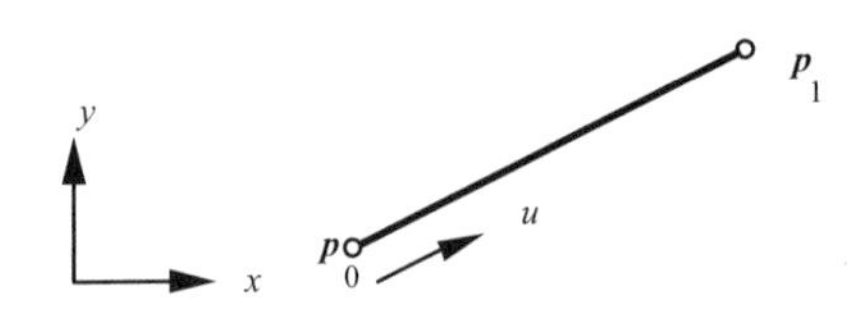

Figure 12.23. Straight line.

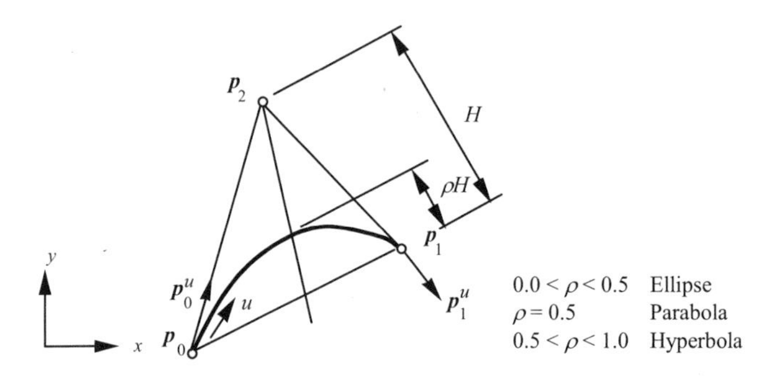

Figure 12.24. Conic curve.

straight line, conic, and arc, can be employed in order to parameterize predefined geometric boundaries. From the given $\boldsymbol{B}$ matrices, the computation of shape design variables for these curves is not as systematic as for the basic types of curves. As a result, the methods for computing such variables for specialized curves must be considered on an individual basis. Notice that the transformation from matrix $\boldsymbol{B}$ to matrix $\boldsymbol{A}$ is the same for the specialized curve as for the basic type of curve, as shown in Fig. 12.13(c). In fact, specialized curves are nothing more than a specialized form of geometric curves.

Straight Line

The straight line is useful in the design of straight boundaries, for example, the length of a beam, as shown in Fig. 12.10. From a shape design point of view, possible shape design variables for a straight line include the locations of the endpoints, as shown in Fig. 12.23. A planar straight line has four degrees of freedom, as expressed in matrix $\boldsymbol{G}$, as

$$\boldsymbol{G} = \begin{bmatrix} \boldsymbol{p}_0 \\ \boldsymbol{p}_1 \end{bmatrix}_{2\times 2}. \tag{12.14}$$

Conics

The conic curve is utilized to design the boundaries for hyperbolas, parabolas, and ellipses. The conic curve can be created by specifying three distinct points and one constant ρ. A conic curve has seven degrees of freedom. The relative altitude ρ determines the characteristics of the conic curve, as shown in Fig. 12.24. A parametric cubic line is utilized to approximate the conic curve shape.

From a shape design point of view, the locations of the three vertices and the relative altitude ρ are possible shape design variables for a conic curve. The design variable matrix $\boldsymbol{G}$ is

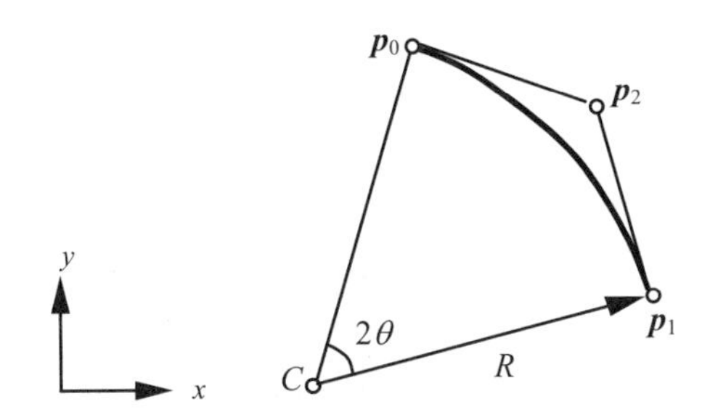

Figure 12.25. Circular arc.

$$G = \begin{bmatrix} p_0 \\ p_1 \\ p_2 \end{bmatrix}_{3\times 2} \cup \{\rho\}. \tag{12.15}$$

Circular Arc

A circular arc can be used to design circular boundaries, such as a circular hole. A circular arc can be created in a plane by specifying the angle, radius, and position of the center of the arc. The parametric cubic arc deviates slightly from a true circular arc. However, the ratio of maximum deviation in the radius of the arc is less than 0.000036% when $\theta \leq 30°$.

From a shape design point of view, the design variables for a circular arc are the location of the center and the radius, as shown in Fig. 12.25. The design variable matrix $\boldsymbol{G}$ of a circular arc is

$$\boldsymbol{G} = \boldsymbol{C}_{1\times 2} \cup \{R\}. \tag{12.16}$$

12.2.4 Surface Design Parameterization

For three-dimensional shape design, design boundaries are surfaces in space. In general, there are 48 degrees of freedom for a parametric bi-cubic surface. For a parametric bi-cubic surface, the x-, y-, and z-components can be expressed using the three functions, as

$$\left.\begin{aligned} x &= x(u,w) \\ y &= y(u,w) \\ z &= z(u,w) \end{aligned}\right\}, \tag{12.17}$$

where u and w are the parametric directions of the geometric entity, and $(u, w) \in [0,1] \times [0,1]$. Each of these functions is, at most, a bi-cubic function of the form

$$\begin{aligned} z(u,w) &= a_{33}u^3w^3 + a_{32}u^3w^2 + a_{31}u^3w + a_{30}u^3 \\ &+ a_{23}u^2w^3 + a_{22}u^2w^2 + a_{21}u^2w + a_{20}u^2 \\ &+ a_{13}u^1w^3 + a_{12}u^1w^2 + a_{11}u^1w^1 + a_{10}u \\ &+ a_{03}w^3 + a_{02}w^2 + a_{01}w + a_{00} \\ &= \sum_{i,j=0}^{3} a_{ij}u^iw^j. \end{aligned} \tag{12.18}$$

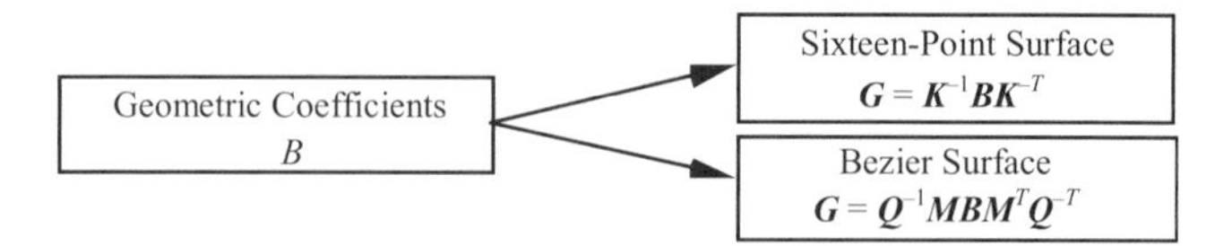

(a) Transformations from $\boldsymbol{B}$ to $\boldsymbol{G}$

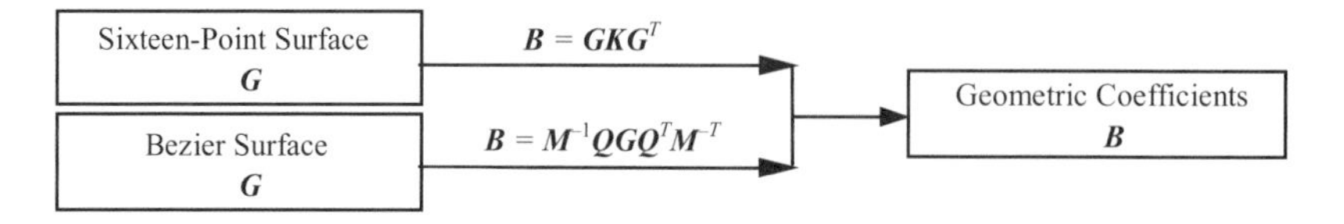

(b) Transformations from $\boldsymbol{G}$ to $\boldsymbol{B}$

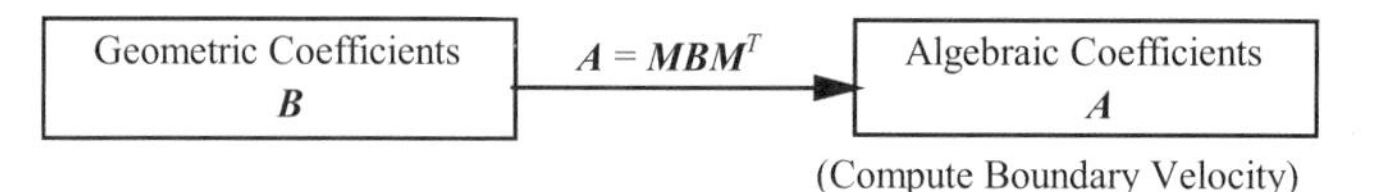

(c) Transformations from $\boldsymbol{B}$ to $\boldsymbol{A}$

Figure 12.26. Surface format transformations for three-dimensional shape design.

Surfaces with 48 degrees-of-freedom are defined as basic surfaces, while specialized surfaces are constrained to represent predefined conditions. As with the specialized curve, the specialized surface has fewer degrees of freedom. There are four basic surfaces: algebraic, geometric, 16-point, and Bezier, and there are also four specialized surfaces: plane, cylindrical, ruled, and surface of revolution, which have been developed to handle three-dimensional shape design problems. The B-spline surface is not recommended for three-dimensional shape design since, in contrast to the B-spline curve, the control points at the edges of the polyhedron are not on the B-spline surface. Therefore, the physical surface boundary does not closely resemble the characteristic polyhedron and, consequently, it is difficult to use in geometric modeling and design.

Basic Surfaces

Similar to the basic types of curves, transformations between matrices $\boldsymbol{G}$, $\boldsymbol{A}$, and $\boldsymbol{B}$ of the basic surface can be used to obtain shape design variables, to update geometric entities, and to compute the boundary velocity field. The various surface format transformations are categorized in Fig. 12.26.

Algebraic Surface

The most basic surface is algebraic. As with the algebraic curve, the algebraic surface is not useful for creating geometric surfaces for practical shape design parameterization purposes, but it is an important tool for computing the velocity field. The mathematical expression of the algebraic surface is

$$
\begin{aligned}
\boldsymbol{p}(u,w) &= \sum_{i,j=0}^{3} \boldsymbol{a}_{ij} u^i w^j \\
&= [u^3 \quad u^2 \quad u \quad 1] \begin{bmatrix} \boldsymbol{a}_{33} & \boldsymbol{a}_{32} & \boldsymbol{a}_{31} & \boldsymbol{a}_{30} \\ \boldsymbol{a}_{23} & \boldsymbol{a}_{22} & \boldsymbol{a}_{21} & \boldsymbol{a}_{20} \\ \boldsymbol{a}_{13} & \boldsymbol{a}_{12} & \boldsymbol{a}_{11} & \boldsymbol{a}_{10} \\ \boldsymbol{a}_{03} & \boldsymbol{a}_{02} & \boldsymbol{a}_{01} & \boldsymbol{a}_{00} \end{bmatrix} \begin{bmatrix} w^3 \\ w^2 \\ w \\ 1 \end{bmatrix} \\
&= \boldsymbol{UAW}^T, \qquad (u,w) \in [0,1] \times [0.1],
\end{aligned}
\tag{12.19}
$$

where $\boldsymbol{p}(u,w) = [p_x, p_y, p_z]$, and $\boldsymbol{a}_{ij} = [a_{ijx}, a_{ijy}, a_{ijz}]$ are the algebraic coefficients of the surface.

Geometric Surface

The geometric surface is represented by the position vectors, tangent vectors, and twist vectors at the four corner points of the surface, as shown in Fig. 12.27. To parameterize the surface, all 48 coefficients in the $\boldsymbol{B}$ matrix can be defined as shape design variables, as

$$
\boldsymbol{G} = \boldsymbol{B} = \begin{bmatrix} \boldsymbol{p}_{00} & \boldsymbol{p}_{01} & \boldsymbol{p}_{00}^{w} & \boldsymbol{p}_{01}^{w} \\ \boldsymbol{p}_{10} & \boldsymbol{p}_{11} & \boldsymbol{p}_{10}^{w} & \boldsymbol{p}_{11}^{w} \\ \boldsymbol{p}_{00}^{u} & \boldsymbol{p}_{01}^{u} & \boldsymbol{p}_{00}^{uw} & \boldsymbol{p}_{01}^{uw} \\ \boldsymbol{p}_{10}^{u} & \boldsymbol{p}_{11}^{u} & \boldsymbol{p}_{10}^{uw} & \boldsymbol{p}_{11}^{uw} \end{bmatrix}_{4\times4\times3}.
\tag{12.20}
$$

In the **B** matrix, the first two rows are the geometric coefficients of auxiliary curves 1 and 3; the first two columns are the geometric coefficients of auxiliary curves 4 and 2, respectively, and $\boldsymbol{p}^{uw} = [\partial^2 p_x/\partial u\partial w, \partial^2 p_y/\partial u\partial w, \partial^2 p_z/\partial u\partial w]^T$ is the twist vector at each corner.

To ensure C^0- and C^1-continuity across two adjacent surfaces, the intersurface curve that is shared by the two surfaces must be the same, and the twist vectors and tangent vectors at the ends of the joining curve shared by the two adjacent surfaces must be co-linear. In notational form, the requirements are as follows: $\boldsymbol{p}_{10} = \boldsymbol{q}_{00}$, $\boldsymbol{p}_{10}^{w} = \boldsymbol{q}_{00}^{w}$, $\boldsymbol{p}_{11} = \boldsymbol{q}_{01}$, $\boldsymbol{p}_{11}^{w} = \boldsymbol{q}_{01}^{w}$, $\boldsymbol{q}_{00}^{u} = a\boldsymbol{p}_{10}^{u}$, $\boldsymbol{q}_{01}^{u} = a\boldsymbol{p}_{11}^{u}$, $\boldsymbol{q}_{00}^{uw} = a\boldsymbol{p}_{10}^{uw}$, and $\boldsymbol{q}_{01}^{uw} = a\boldsymbol{p}_{11}^{uw}$, as shown in Fig. 12.28.

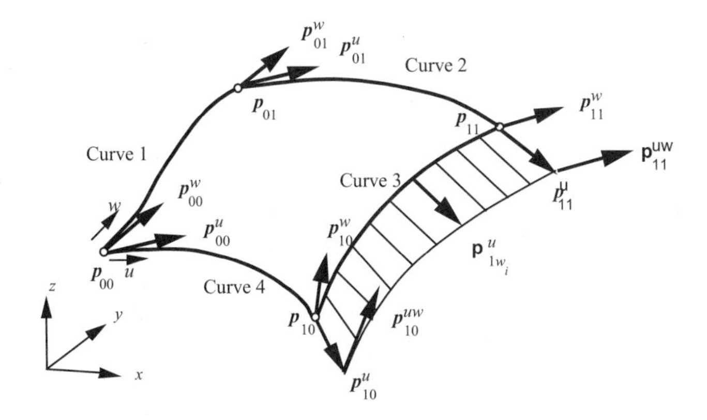

Figure 12.27. Geometric surface.

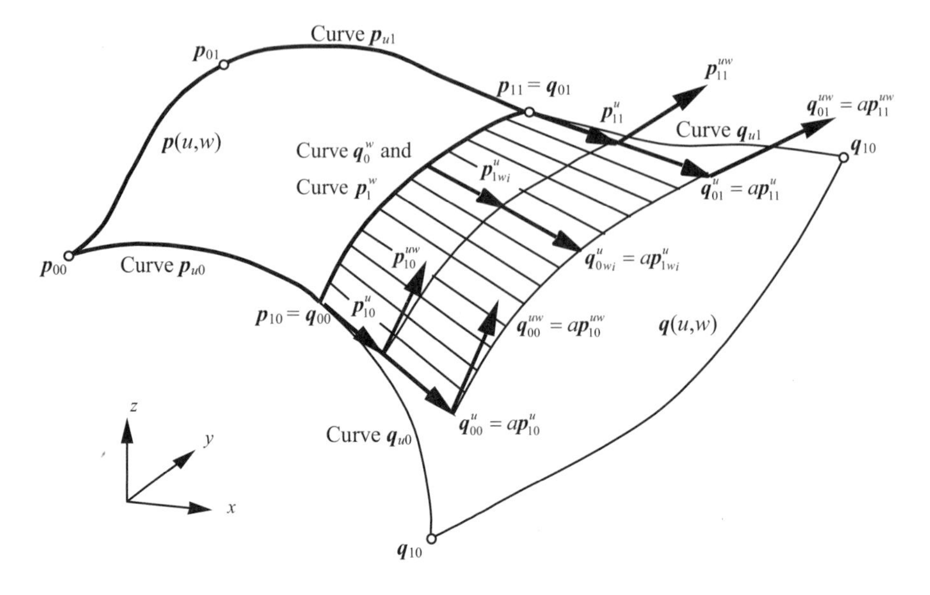

Figure 12.28. C^0- and C^1-continuities across geometric surfaces.

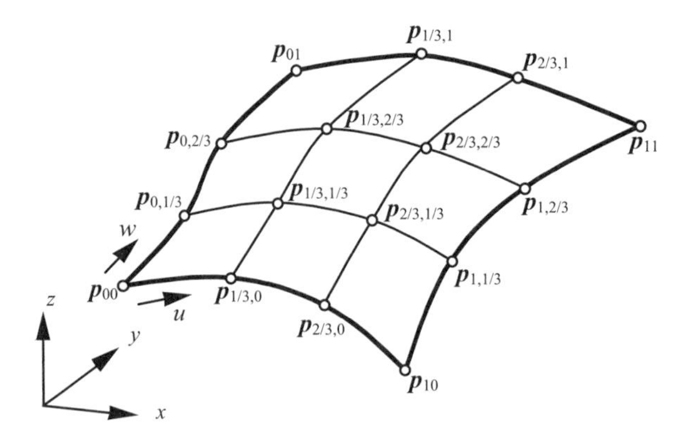

Figure 12.29. Sixteen-point surface.

Sixteen-Point Surface

The 16-point surface is geometrically represented by the positions of 16 points at u, w = 0, 1/3, 2/3 and 1 in the parametric direction of the surface, as shown in Fig. 12.29. As with the four-point curve, the 16-point surface actually passes through these 16 points. For shape design parameterization, the positions of the sixteen points given in the associated $\boldsymbol{G}$ matrix can be defined as the shape design variables, as

$$
G = \begin{bmatrix} p(0,0) & p(0,\frac{1}{3}) & p(0,\frac{2}{3}) & p(0,1) \\ p(\frac{1}{3},0) & p(\frac{1}{3},\frac{1}{3}) & p(\frac{1}{3},\frac{2}{3}) & p(\frac{1}{3},1) \\ p(\frac{2}{3},0) & p(\frac{2}{3},\frac{1}{3}) & p(\frac{2}{3},\frac{2}{3}) & p(\frac{2}{3},1) \\ p(1,0) & p(1,\frac{1}{3}) & p(1,\frac{2}{3}) & p(1,1) \end{bmatrix}_{4\times4\times3} . \tag{12.21}
$$

Bezier Surface

As with the Bezier curve discussed in Section 12.2.3, the Bezier surface is geometrically represented by the position of sixteen control points that determine surface shape, using Bernstein basis functions to arrive at the following:

$$
p(u,w) = \sum_{i=0}^{3}\sum_{j=0}^{3} p_{ij} B_{i,3}(u) B_{j,3}(w), \qquad (u,w) \in [0,1]\times[0,1], \tag{12.22}
$$

where p_{ij} are vertices of the characteristic polyhedron that form a 4×4 array of control points, as shown in Fig. 12.30, and $B_{i,3}(u)$ and $B_{j,3}(w)$ are the Bernstein basis functions. To parameterize a Bezier surface, degrees of freedom of the 16 control points in the associated $\boldsymbol{G}$ matrix can be defined as shape design variables, as

$$
G = \begin{bmatrix} p_{11} & p_{12} & p_{13} & p_{14} \\ p_{21} & p_{22} & p_{23} & p_{24} \\ p_{31} & p_{32} & p_{33} & p_{34} \\ p_{41} & p_{42} & p_{43} & p_{44} \end{bmatrix}_{4\times4\times3} . \tag{12.23}
$$

Similar to the role that specialized curves play for two-dimensional shape design as described in Section 12.2.3, a number of specialized surfaces, such as planes, cones, and cylinders, are also necessary to support three-dimensional shape design to determine the constrained boundaries. Four specialized surfaces can be used to support three-dimensional shape design: plane surface, cylindrical surface, ruled surface, and surface of revolution.

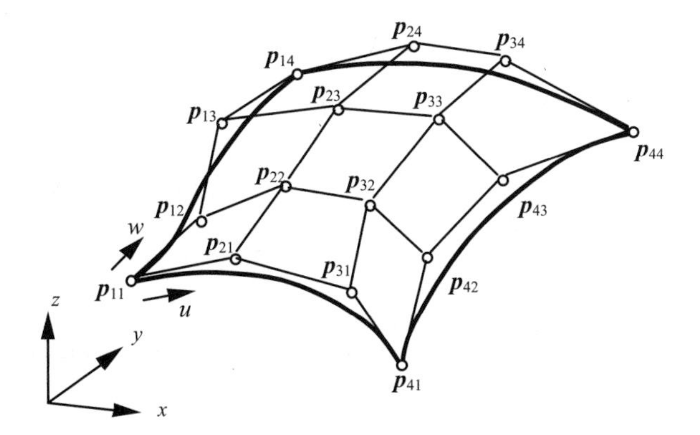

Figure 12.30. Bezier surface.

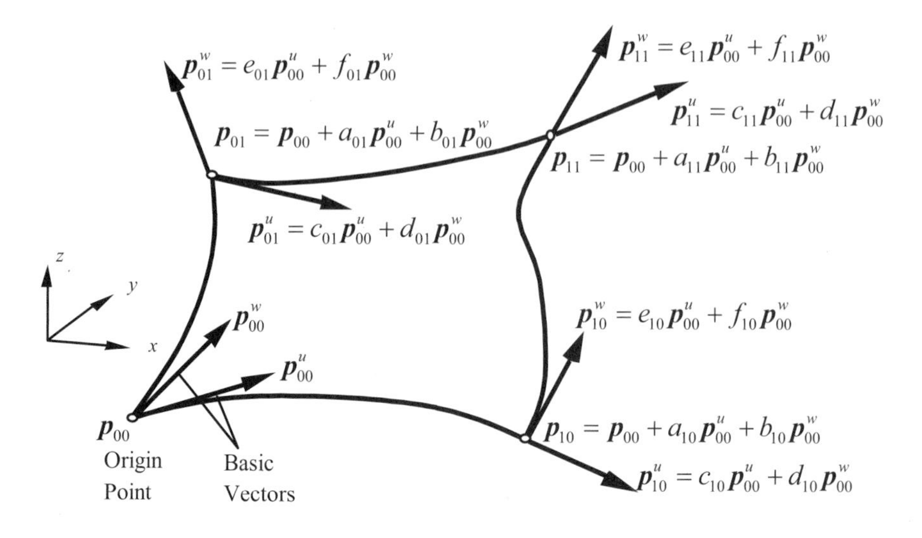

Figure 12.31. General plane surface in space.

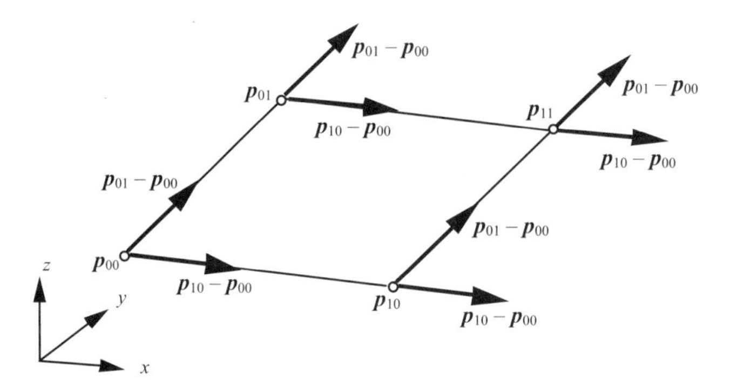

Figure 12.32. Plane surface with parallel straight edges.

Planar Surface

The plane surface is important in the design of an outer surface. One necessary condition for a plane surface is that all twist vectors are zero. However, it is sufficient if the position and tangent vectors at the four corners remain on the same plane. A general plane surface is shown in Fig. 12.31. From a shape design point of view, possible shape design variables for a general plane surface would include the location of the origin points, the location of the base vectors, and the 12 constants, given in matrix $\boldsymbol{G}$ as

$$\boldsymbol{G} = \begin{bmatrix} \boldsymbol{p}_{00} \\ \boldsymbol{p}_{00}^{u} \\ \boldsymbol{p}_{00}^{w} \end{bmatrix}_{3\times 3} \cup \begin{bmatrix} a_{10} & b_{10} & c_{10} & d_{10} \\ a_{01} & b_{01} & c_{01} & d_{01} \\ a_{11} & b_{11} & c_{11} & d_{11} \end{bmatrix}_{3\times 4}. \tag{12.24}$$

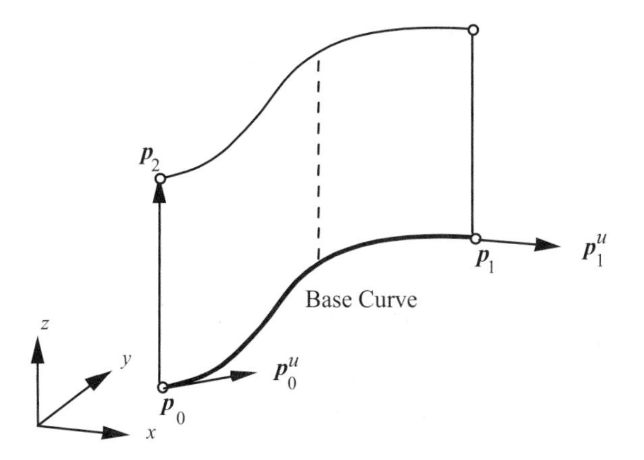

Figure 12.33. Cylindrical surface.

A specialized plane surface containing parallel straight edges is shown in Fig. 12.32. Such a plane surface can be determined by three distinct points. Consequently, the locations of these three points can be defined as the shape design variables for the surface, namely,

$$G = \begin{bmatrix} \boldsymbol{p}_0 \\ \boldsymbol{p}_{10} \\ \boldsymbol{p}_{01} \end{bmatrix}_{3\times 3} . \tag{12.25}$$

Cylindrical Surface

A cylindrical surface is utilized to design such boundaries as cylindrical holes. The cylindrical surface is a specialized form of the ruled surface. Possible shape design variables for the cylindrical surface would include the base curve shape and the position of point $\boldsymbol{p}_2$, as shown in Fig. 12.33. Matrix $\boldsymbol{G}$ can be expressed as

$$G = \begin{bmatrix} \boldsymbol{p}_0 \\ \boldsymbol{p}_1 \\ \boldsymbol{p}_0^u \\ \boldsymbol{p}_1^u \end{bmatrix}_{4\times 3} \cup \boldsymbol{p}_2 . \tag{12.26}$$

Ruled Surface

A ruled surface is a generalization of the cylindrical surface. The ruled surface is created by defining two distinct base curves $\boldsymbol{p}(u)$ and $\boldsymbol{q}(u)$ and then joining each point on $\boldsymbol{p}(u)$ with a straight line to a corresponding point on $\boldsymbol{q}(u)$, for all values of u. The ruled surface can be utilized in the design of cylindrical boundaries, such as tapered holes or cones. Possible shape design variables for the ruled surface would include the geometric shape of the two base curves, as shown in Fig. 12.34. Matrix $\boldsymbol{G}$ is defined as

$$G = \begin{bmatrix} p_0 \\ p_1 \\ p_0^u \\ p_1^u \end{bmatrix}_{4\times 3} \cup \begin{bmatrix} q_0 \\ q_1 \\ q_0^u \\ q_1^u \end{bmatrix}_{4\times 3} . \quad (12.27)$$

Surface of Revolution

The surface of revolution is generated by revolving a planar curve around a line in its plane. The planar curve is also called a profile curve. At various positions around the axis, this curve creates meridians. From a shape design point of view, shape design variables for the surface of revolution would include the variables that define the shape of the profile curve, as shown in Fig. 12.35. Matrix G is defined as

$$G = \begin{bmatrix} p_{00} \\ p_{10} \\ p_{00}^u \\ p_{10}^u \end{bmatrix}_{4\times 2} . \quad (12.28)$$

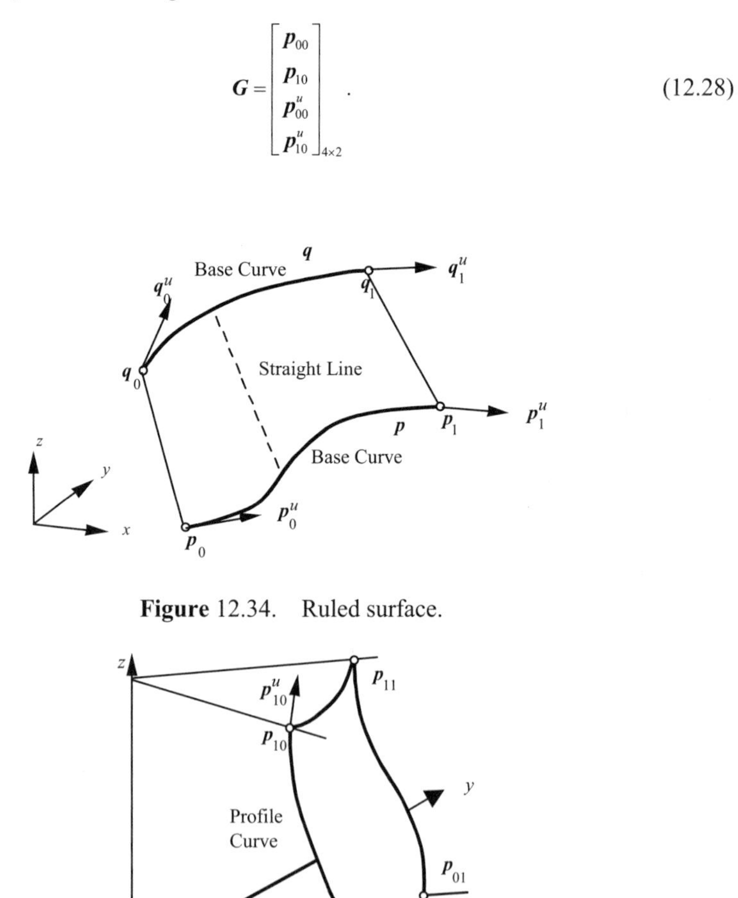

Figure 12.34. Ruled surface.

Figure 12.35. Surface of revolution.

12.2.5 Design Variable Linking Across Geometric Entities

There are three methods for categorizing the shape design problem: (1) by identifying the arbitrary boundary shape, (2) by determining the dimensions of the predefined shape, and (3) by finding the locations of the boundaries. A design variable linking process is discussed in this section in order to support and ensure continuity between boundaries, as well as to retain predefined boundary shapes. For a geometric feature that is formed by a set of B-spline curves, C^2-continuity across curves is naturally retained.

As described in the previous sections, for the first type of shape design boundary any basic curve or surface can be utilized for parameterization. If constrained boundary shapes are required for two-dimensional structures, such as a straight edge or circular arc, then that specific type of curve must be used. For three-dimensional structures, the constrained boundaries can be predefined, such as a plane, cylinder, ball, or surface of revolution. If the geometric feature is composed of more than one curve or surface, continuity across curves or surfaces must be retained by linking shape design variables at the joint point or curve. As mentioned before, C^0-, C^1-, and C^2-continuities can be retained for two-dimensional structures by using the appropriate curve type and design variable linking. For three-dimensional problems, C^2-continuity is difficult to maintain. The following three examples illustrate how the design variable linking process can be used to support the three types of shape design problems.

Type One Problems—Identifying the Arbitrary Boundary Shape

A two-dimensional beam with arbitrarily shaped holes, shown in Fig. 12.36(a), illustrates a design variable linking process that supports the first type of shape design problem.

Figure 12.36(b) shows that hole A is formed by three curves. To parameterize this hole, the curves are defined as geometric curves. The endpoints and tangent vectors at the endpoints of these connected curves are linked to retain C^0- and C^1-continuities, respectively.

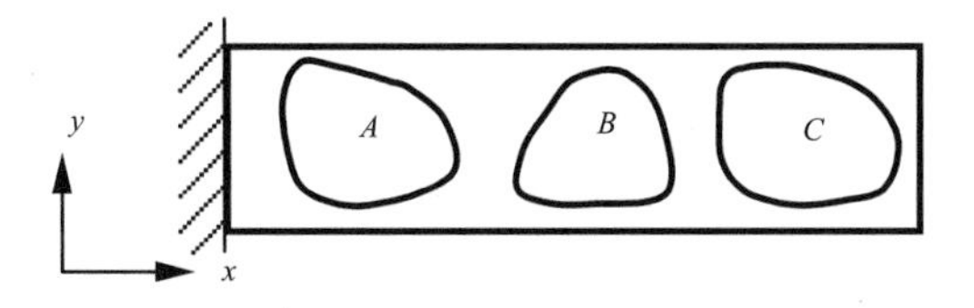

(a) Beam with Arbitrarily Shaped Holes

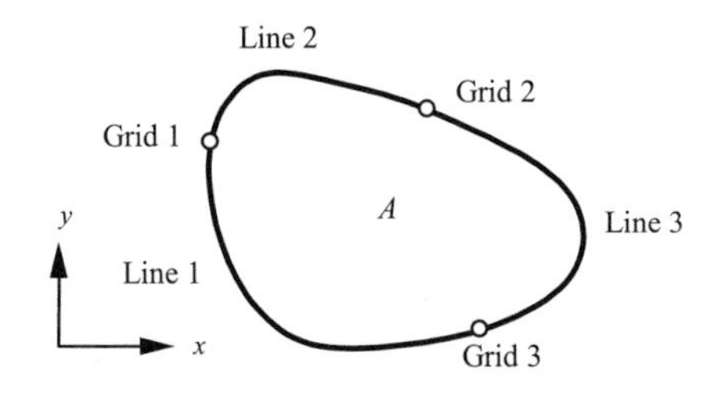

(b) Hole A Formed by Three Curves

Figure 12.36. Parameterization of arbitrarily shaped holes.

The shape of the hole can be changed due to endpoint movement, length, or the direction of the tangent vector, as shown in Fig. 12.37. Hole shape change due to movement δdp_1 of grid 3 in the negative y-direction is shown in Fig. 12.37(a), such that the tangent vectors of lines 1 and 3 at grid 3 are kept the same. Moreover, hole shape change due to scaling the tangent vector of line 3 at grid 3 by δdp_2, shown in Fig. 12.37(b), is such that the location of grid 3 and the direction of the tangent vector are kept the same. Finally, hole shape change due to change δdp_3 in the tangent vector direction of lines 1 and 3 is shown in Fig. 12.37(c).

In addition to the geometric curves used to parameterize hole A, other basic curves described in Section 12.2.3, such as the Bezier curve, can be used to parameterize other holes.

To avoid meaningless designs, for example, a hole that penetrates the boundary edges of a beam, the geometric boundaries can be displayed after the design change. Furthermore, numerical limits can be defined for design variables in order to restrict the degree of design perturbation.

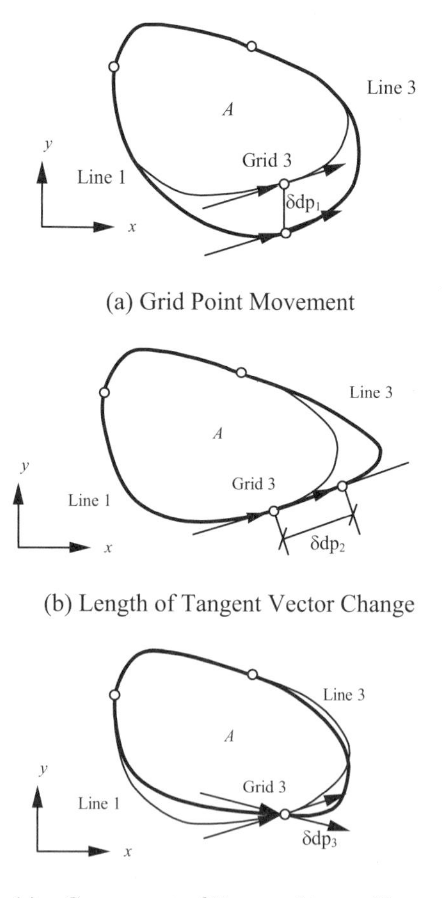

Figure 12.37. Shape variation of a hole A.

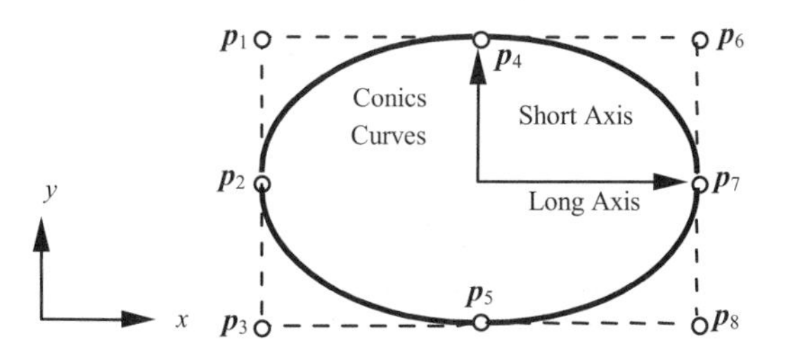

Figure 12.38. Parameterization of an elliptic hole.

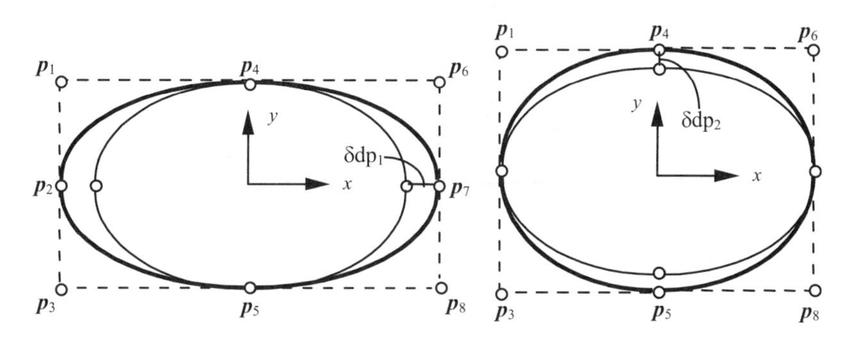

Figure 12.39. Shape variation of an elliptic hole.

Type Two Problems—Determining the Dimensions of a Predefined Shape

For the second type of shape design problem, specialized curves and surfaces are used to parameterize geometric entities. In addition, design variable linking may be carried out in the design process to group curves and surfaces that define the geometric feature.

The design variable linking method for parameterizing the second type of shape design boundary can be described using an elliptic hole formed by four conic curves, as shown in Fig. 12.38. Each conic curve has three points to control its shape; however, relative altitude ρ is kept constant.

To vary the major axis, the x-coordinate of point $\boldsymbol{p}_7$ is defined as the independent design variable dp1. Points $\boldsymbol{p}_6$ and $\boldsymbol{p}_8$ are linked to dp1, with a proportionality of 1.0. In addition, the x-coordinates of points 1, 2, and 3 are linked to dp1 with a proportionality of –1.0. The elliptic hole varies in shape when the major axis is perturbed by δdp1, as shown in Fig. 12.39(a). Similarly, for a design change in the minor axis, the y-coordinate of point $\mathbf{p}_4$ is defined as the independent design variable dp2. Points $\boldsymbol{p}_1$ and $\boldsymbol{p}_6$ are linked to dp2, with a proportionality of 1.0. Also, the y-coordinates of points $\boldsymbol{p}_3$, $\boldsymbol{p}_5$, and $\boldsymbol{p}_8$ are linked to dp2, with a proportionality of –1.0. Shape change due to the perturbation δdp2 of the minor axis is shown in Fig. 12.39(b).

For three-dimensional problems, this type of shape design can be the radius of a ball, a shell, or a cylindrical hole. For example, the tapered semicylindrical surface shown in

Fig. 12.40 can be parameterized by linking the design variables defined in the two ruled surfaces that form the tapered semicylindrical surface. The first ruled surface $\boldsymbol{p}_1\boldsymbol{p}_2\boldsymbol{p}_4\boldsymbol{p}_5$ is created using circular arcs $\boldsymbol{p}_1\boldsymbol{p}_2$ and $\boldsymbol{p}_4\boldsymbol{p}_5$. The other ruled surface $\boldsymbol{p}_2\boldsymbol{p}_3\boldsymbol{p}_5\boldsymbol{p}_6$ is created using circular arcs $\boldsymbol{p}_2\boldsymbol{p}_3$ and $\boldsymbol{p}_5\boldsymbol{p}_6$.

To vary the radius R_1, the radius of arc $\boldsymbol{p}_1\boldsymbol{p}_2$ is defined as the independent design variable dp1, and the radius of arc $\boldsymbol{p}_2\boldsymbol{p}_3$ is linked to dp1 with a proportionality of 1.0. Similarly, the design variable dp2 can be defined for radius R_2. The shape of the tapered semicylindrical surface is defined by the shape design variables dp1 and dp2, as shown in Fig. 12.41(a) and (b), respectively.

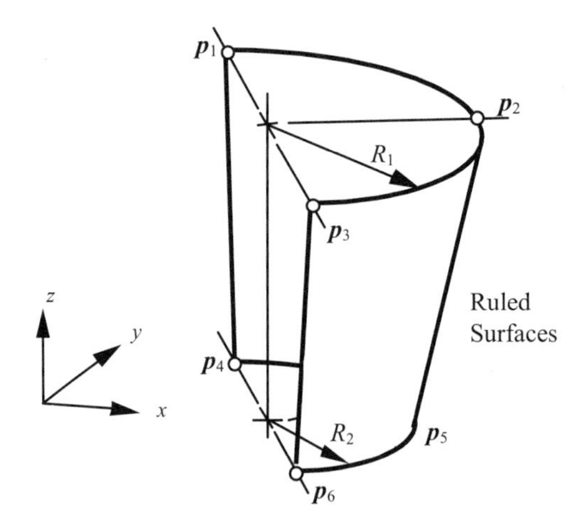

Figure 12.40. A semitapered cylinder formed by two ruled surfaces.

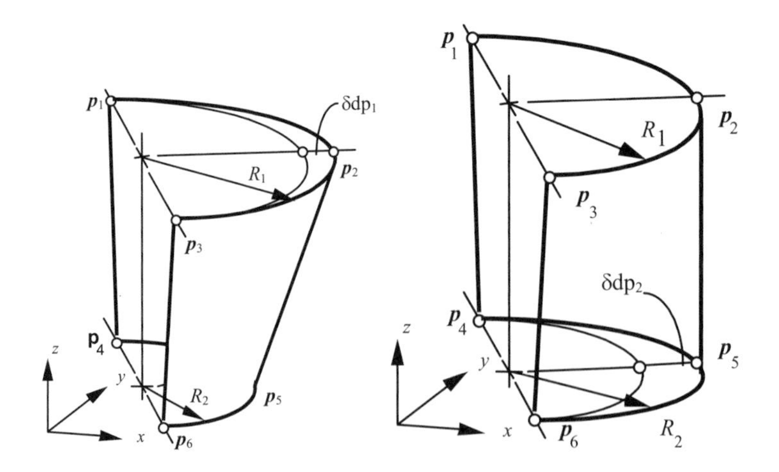

(a) Shape Variation Due to δdp1 (b) Shape Variation Due to δdp2

Figure 12.41. Shape variation of a tapered semicylindrical surface.

Type Three Problems—Finding the Locations of the Boundaries

To support the third type of shape design problem, such geometric entities as curves or surfaces that form the geometric feature are linked together so that only *x*-, *y*-, and *z*-movements of the parameterized geometric feature are allowed. For example, the *x*-coordinate of points $\boldsymbol{p}_1$ to $\boldsymbol{p}_8$ of the elliptic hole shown in Fig. 12.40 can be linked, with a proportionality of 1.0, so that the four conic curves can be moved together to form an elliptic hole.

13 Numerical Implementation of Sensitivity Analysis

The design sensitivity formulations that were presented in Chapters 5 through 7 and Chapters 9 through 11 are based on the continuum method where the design variation is applied to the continuum structure. If this continuum design sensitivity equation is analytically solved, then the exact design sensitivity can be obtained. As was shown in Fig. 1.13 of Chapter 1, this approach is called the continuum-continuum method. However, the analytical design sensitivity is available only for very simple structural problems. Most engineering applications cannot be solved analytically, but require approximated solutions, and thus, approximated design sensitivities. The continuum design sensitivity equation must be solved by approximation in the same way that structural problems are solved. Since the design variation is taken at the continuum domain and is then followed by discretization, this is called the continuum-discrete method, which is the focus of this chapter.

During past five decades, the finite element method has dramatically been developed, and many engineering applications have been solved using this method. A domain approximation using simple-shaped finite elements and sound mathematical formulations using the variational principle are major advantages of the finite element method. In this chapter, a finite element approximation of the continuum design sensitivity equation is developed. Two different types of designs are developed in separate sections: sizing and shape designs. Although the derivations of design sensitivity equations for these two types of designs are quite different, the numerical implementations of them are relatively similar. Numerical implementation method for the sizing design is presented in Section 13.1 and the shape design is presented in Section 13.2.

Design velocity field computation is an important step in computing the shape design sensitivity and updating the finite element mesh during the shape design optimization process. Applying inappropriate design velocity field for shape design sensitivity analysis and optimization may yield inaccurate sensitivity results or distorted finite element mesh, and thus fail in achieving an optimal solution. In Section 6.2.7, theoretical regularity and practical requirements of the design velocity field were discussed. The crucial step of using the design velocity field to update finite element mesh in the design optimization process is emphasized. In Section 13.3, available design velocity field computation methods in the literature are summarized and their applicability for shape design sensitivity analysis and optimization are discussed. It is identified that a combination of isoparametric mapping and boundary displacement methods is ideal for the design velocity field computation.

13.1 Sizing Design Sensitivity Computation

The key idea of the design sensitivity approximation is to use the same approximation method as the structural problem. If the structural problem is approximated using the finite element method, the same approximation method must be used for the purpose of design sensitivity computation. In this section, the design sensitivity with respect to sizing design parameter is approximated using the finite element method. Although the finite element approximation of the static response is developed in this section, other types of structural problems can be approximated similarly.

13.1.1 Finite Element Approximation

The general expression of structural problems in Chapter 3 and design sensitivity equations in Chapter 5 can be written, respectively, as

$$a_u(z,\bar{z}) = \ell_u(\bar{z}), \quad \forall \bar{z} \in Z \tag{13.1}$$

$$a_u(z',\bar{z}) = \ell'_{\delta u}(\bar{z}) - a'_{\delta u}(z,\bar{z}), \quad \forall \bar{z} \in Z, \tag{13.2}$$

where Z is the space of kinematically admissible displacement that satisfy the essential boundary conditions. The forms that appear in (13.1) and (13.2) include integration over the domain and spatial derivatives of the state variable. Thus, the numerical approximation of above equations involves the domain discretization and interpolation of the state variable.

The finite element approximation of (13.1) was developed in Section 3.3 for various structural components, which requires the discretization of structural domain and the approximation of the state variable. The displacement z within element I is approximated by

$$z(\xi) = \boldsymbol{N}(\xi)^T \boldsymbol{d}_I, \tag{13.3}$$

where ξ is the reference coordinate vector of a finite element, $\boldsymbol{N}(\xi)$ is the shape function matrix, and $\boldsymbol{d}_I$ is the nodal displacement vector of element I. The dimension of vectors in (13.3) is different for different finite elements, which will be discussed in detail in Section 13.1.2. In the Galerkin approximation, the displacement variation $\bar{z}$ is also approximated using the same shape function as in (13.3). In the isoparametric element, the geometry of the element is approximated in the same method as the displacement. Thus, the material point $\boldsymbol{x}(\xi)$ within the element can be approximated by

$$\boldsymbol{x}(\xi) = \boldsymbol{N}(\xi)^T \boldsymbol{X}_I, \tag{13.4}$$

where $\boldsymbol{X}_I$ is the vector of nodal coordinates of element I. Equation (13.4) can be used to calculate the Jacobian relation between the physical and the reference coordinates. That is,

$$\boldsymbol{J}_I \equiv \frac{\partial \boldsymbol{x}(\xi)}{\partial \xi} = \frac{\partial \boldsymbol{N}(\xi)^T \boldsymbol{X}_I}{\partial \xi}. \tag{13.5}$$

This Jacobian relation is very useful to calculate the spatial derivative of state variable and to transform the physical domain into the reference domain.

In addition to the interpolation of the state variable, the forms in (13.1) and (13.2) include the spatial derivative of state variable. However, the interpolation in (13.4) depends on the reference coordinate ξ, instead of the spatial coordinate $\boldsymbol{x}$. Thus, the spatial derivative is calculated using the chain rule of differentiation, as

$$\frac{\partial z}{\partial \boldsymbol{x}} = \frac{\partial z}{\partial \xi}\frac{\partial \xi}{\partial \boldsymbol{x}} = \left[\frac{\partial \boldsymbol{N}(\xi)^T \boldsymbol{d}_I}{\partial \xi}\right] \boldsymbol{J}_I^{-1}, \tag{13.6}$$

where J_I^{-1} is the inverse of the Jacobian matrix in (13.5). Depending on structural components, second-order derivatives of the state variable are sometimes required, which can also be obtained by applying the additional chain rule of differentiation to (13.6).

The domain integration of the forms in (13.1) and (13.2) can be carried out analytically or numerically. The analytical domain integration, however, is limited for simple one-dimensional elements. Most cases, a numerical integration is preferred in the finite element formulation. For example, the numerical integration of a function $g(\xi,\eta)$ over a two-dimensional finite element $\Omega_I \subset \Omega$ can be carried out using a Gauss integration rule as

$$\begin{aligned} \iint_{\Omega_I} g(\xi,\eta)\,d\Omega^e &= \int_{-1}^{1}\int_{-1}^{1} g(\xi,\eta)|\boldsymbol{J}_I|\,d\xi d\eta \\ &\approx \sum_{i=1}^{NG}\sum_{j=1}^{NG} \omega_i\omega_j g(\xi_i,\eta_j)|\boldsymbol{J}_I|, \end{aligned} \tag{13.7}$$

where NG is the number of integration points in each reference coordinate direction; ξ_i and η_i are the Gauss integration points; and ω_i is the integration weights. A summary of integration points and weights is presented in Table 3.1.

In this chapter, a vector in the global level is denoted with { • }, while a matrix in the global level is denoted with [•]. With these notations in hand, the discrete variational equation can be obtained by applying (13.6) and (13.7) to the structural variational equation (13.1) and by assembling every finite element, to obtain

$$\{\bar{z}_g\}^T[\boldsymbol{K}_g]\{z_g\} = \{\bar{z}_g\}^T\{\boldsymbol{F}_g\}, \quad \forall\{\bar{z}_g\} \in Z_h, \tag{13.8}$$

where $[\boldsymbol{K}_g]$ is the generalized global stiffness matrix, $\{\boldsymbol{F}_g\}$ is the generalized global force vector, and Z_h is the discrete space of kinematically admissible displacements. The expressions of $[\boldsymbol{K}_g]$ and $\{\boldsymbol{F}_g\}$ for various structural components have been presented in Section 3.3.

Direct Differentiation Method

As mentioned before, the domain discretization and state variable interpolation of design sensitivity equation must follow the same methods with the structural problem. By following the same procedure as in (13.8), the approximation of design sensitivity (13.2) in the global level can be obtained as

$$\{\bar{z}_g\}^T[\boldsymbol{K}_g]\{z'_g\} = \{\bar{z}_g\}^T\{\mathcal{F}_g\}\delta u, \quad \forall\{\bar{z}_g\} \in Z_h, \tag{13.9}$$

where $\{z'_g\}$ is the global sensitivity vector of the displacement, and $\{\mathcal{F}_g\}$ is the global fictitious force vector, which represents the right side of (13.2). If (13.8) and (13.9) are compared, then the only difference is the right sides, $\{\boldsymbol{F}_g\}$ and $\{\mathcal{F}_g\}$. Thus, design sensitivity analysis requires constructing the fictitious force $\{\mathcal{F}_g\}$ and solving the variational equation, which is similar to the structural problem in (13.8).

Note that (13.9) includes δu in the last part. From its definition in (5.5), the displacement variation is defined by $z' = (\partial z/\partial u)\,\delta u$. Thus, (13.9) can be rewritten as

$$\{\bar{z}_g\}^T[\boldsymbol{K}_g]\left\{\frac{\partial z_g}{\partial u}\right\}\delta u = \{\bar{z}_g\}^T\{\mathcal{F}_g\}\delta u, \quad \forall\{\bar{z}_g\} \in Z_h, \tag{13.10}$$

and the coefficients of δu in (13.10), which is the design sensitivity, can be solved in the numerical implementation, by assuming $\delta u = 1$.

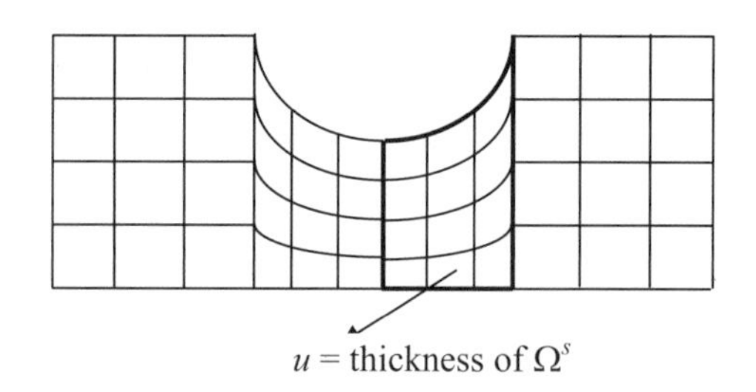

Figure 13.1. Finite element approximation of domain Ω and parameterized subdomain Ω^s.

The construction of fictitious force $\{\mathcal{F}_g\}$ requires the domain integration. However, in most cases, the integration does not need to be carried out over the entire structural domain. From its definition, the term $\ell'_{\delta u}(\bar{z})-a'_{\delta u}(z,\bar{z})$ vanishes for those finite elements that do not belong to the design component. To explain it in detail, let a structural domain Ω is divided by 48 finite elements, as shown in Fig. 13.1. If the sizing design parameter u involves the thickness change of 12 elements, then the term $\ell'_{\delta u}(\bar{z})-a'_{\delta u}(z,\bar{z})$ is nonzero only for those 12 elements on which the design u is defined. This term vanishes for other 36 elements. Thus, the numerical integration needs to be carried out only for a small subdomain rather than the entire structural domain.

Since the continuum structural equation is differentiated first, and then approximated using the finite element method, (13.9) represents the continuum-discrete method, as was categorized in Fig. 1.13. Unlike the discrete design sensitivity formulation, the continuum-discrete formulation in (13.9) does not require the derivative of the stiffness matrix. However, the construction of $\{\mathcal{F}_g\}$ involves in the finite element approximation and numerical integration. The expression of $\{\mathcal{F}_g\}$ is different for different structural components and different design parameters, which is derived in Section 13.1.2.

As discussed in Chapter 4, the generalized stiffness matrix $[\boldsymbol{K}_g]$ is not positive definite because the boundary condition is not imposed yet. Thus, it is inconvenient to solve (13.8) and, thus, (13.9) directly. Instead of solving the discrete variational equation, the following reduced system of matrix equation is solved after applying the essential boundary conditions:

$$[\boldsymbol{K}]\{z'\} = \{\mathcal{F}\}\delta u. \tag{13.11}$$

If simple boundary conditions exist, then the reduced system of matrix equation can be constructed by simply removing the rows and columns of those degrees-of-freedom whose displacements are constrained. When the multipoint constraints exist, however, applying the boundary conditions is not straightforward. The reduced stiffness matrix $[\boldsymbol{K}]$ is now positive definite, thus a unique solution $\{z'\}$ can be expected from (13.11).

During the solution procedure of the structural problem, the stiffness matrix $[\boldsymbol{K}]$ is factorized. The cost of factorizing the $N \times N$ stiffness matrix is proportional to N^3. After the factorized matrix is obtained, the cost of solution procedure is proportional to N^2. Thus, for large N, the major part of the computational cost is involved in factorizing the stiffness matrix. If the structural problem is already solved, it means the solution $\{z_g\}$ as well as the factorized stiffness matrix are available. In the commercially available finite element codes, this factorized stiffness matrix is used when reanalysis procedure is invoked. When many load cases exist, this factorized stiffness matrix is repeatedly used. The design sensitivity equation (13.11) can be viewed as a reanalysis with different force vector (fictitious force). Thus, solving (13.11) is much more efficient than solving the original structural problem from which the efficiency of design sensitivity analysis comes.

After calculating $\{z'\}$, the global displacement sensitivity vector $\{z'_g\}$ can be constructed by considering the imposition of boundary conditions. For example, if simple, homogeneous boundary conditions exist in the space Z, then $\{z'_g\}$ is the same as $\{z'\}$ with zeros for those degrees-of-freedom whose displacements are constrained. Finally, the displacement sensitivity within an element can be approximated using the same interpolation as in (13.3). Let $\boldsymbol{d}'_I$ be the displacement sensitivity of the nodes that belong to the element. $\boldsymbol{d}'_I$ can be extracted from the global displacement sensitivity vector $\{z'_g\}$. The displacement sensitivity at the reference coordinate ξ can be interpolated by

$$z'(\xi) = \boldsymbol{N}(\xi)^T \boldsymbol{d}'_I. \tag{13.12}$$

Note that the shape function matrix $\boldsymbol{N}(\xi)$ is independent of the design. The displacement sensitivity of (13.12) corresponds to one design parameter. Thus, the design sensitivity equation must be solved for each design parameter, as different design parameters yield different $\{\mathcal{F}_g\}$ in (13.9).

Consider a general performance measure, which is a function of design parameter, displacement and its gradient, as

$$\psi = \iint_{\Omega^s} g(z, \nabla z, u)\, d\Omega^s, \tag{13.13}$$

where Ω^s is the subdomain of Ω. The function g is continuously differentiable with respect to its arguments. By using the chain rule of differentiation, the sensitivity expression of ψ can be derived as

$$\psi' = \iint_{\Omega^s} (g_{,z} z' + g_{,\nabla z} : \nabla z' + g_{,u} \delta u)\, d\Omega^s. \tag{13.14}$$

From the definition of function g, the expressions of $g_{,z}$, $g_{,\nabla z}$, and $g_{,u}$ are assumed to be known. Thus, using the calculated displacement sensitivity z' from (13.12), the sensitivity ψ' can be calculated through domain integration. In many cases, the integration domain in (13.13) involves a small portion of the whole domain, or even one finite element. Thus, the numerical integration of (13.14) is not at all an expensive procedure.

The procedure to obtain the design sensitivity ψ' using (13.14) is called the direct differentiation method. This method is conceptually straightforward, but it is usually more expensive than the adjoint variable method because the number of design parameters is greater than the number of constraints in most design optimization problems, as discussed in Section 4.1.4.

Adjoint Variable Method

If the adjoint variable method is used to calculate the design sensitivity coefficient, then the adjoint problem in (5.11) must be approximated. For the general performance measure provided in (13.13) and its sensitivity in (13.14), the adjoint equation that is independent of the design parameter can be written as

$$a_u(\lambda,\bar{\lambda}) = \iint_{\Omega^s} (g_{,z}\bar{\lambda} + g_{,\nabla z} : \nabla\bar{\lambda})d\Omega^s, \quad \forall\bar{\lambda} \in Z. \tag{13.15}$$

The form $a_u(\bullet,\bullet)$ on the left side of (13.15) is exactly the same as the left side of (13.1) and (13.2). Thus, in the discrete form, the same generalized stiffness matrix with (13.8) can be used. From the definition of the function g, the right side of (13.15), the adjoint load, can be calculated using domain integration. The discrete version of (13.15) will then be obtained as

$$\{\bar{\lambda}_g\}^T[\boldsymbol{K}_g]\{\lambda_g\} = \{\bar{\lambda}_g\}^T\{\boldsymbol{G}_g\}, \quad \{\bar{\lambda}_g\} \in Z_h, \tag{13.16}$$

where $\{\boldsymbol{G}_g\}$ is the global adjoint force vector. Although (13.16) is in the global form, the calculation of vector $\{\boldsymbol{G}_g\}$ only involves in integration over subdomain Ω^s, which is smaller than the structural domain Ω. Often, Ω^s represents one element. For example, when a displacement at a node is the performance measure, the adjoint force is nothing but a unit force at the node. In that case, $\{\boldsymbol{G}_g\} = [0, \ldots, 0, 1, 0, \ldots, 0]^T$ where "1" is at the location of the node. Thus, it is not expensive to calculate the adjoint force vector $\{\boldsymbol{G}_g\}$. In addition, as will be shown in the following section, the adjoint load in (13.15) is the same for the sizing and shape design parameters. Thus, the same (13.16) can be used for all types of design parameters, which makes the numerical implementation convenient.

Since $\bar{\lambda}$ belongs to the same space Z of kinematically admissible displacements, the discrete adjoint equation (13.16) can be reduced to the similar form as (13.11) after imposing the essential boundary condition, to obtain

$$[\boldsymbol{K}]\{\lambda\} = \{\boldsymbol{G}\}. \tag{13.17}$$

Again, the factorized stiffness matrix $[\boldsymbol{K}]$ from structural analysis can be used to solve (13.17) efficiently.

After solving for $\{\lambda\}$, the generalized global adjoint vector $\{\lambda_g\}$ can be constructed by considering the imposition of the boundary condition. For example, if simple, homogeneous boundary conditions exist in the space Z, then $\{\lambda_g\}$ is the same as $\{\lambda\}$ with zeros for those degrees-of-freedom whose displacements are constrained. Finally, the adjoint variable within an element can be approximated using the same interpolation as (13.3).

The sensitivity expression of the performance measure ψ in the adjoint variable method can be written with the displacement z and the adjoint variable λ as in (5.15). By using the definition of the global fictitious force $\{\mathcal{F}_g\}$ in (13.9), the approximated ψ' can be written as

$$\begin{aligned}\psi' &= \ell'_{\delta u}(\lambda) - a'_{\delta u}(z,\lambda) + \iint_{\Omega^s} g_{,u}\delta u\, d\Omega^s \\ &\approx \{\lambda_g\}^T\{\mathcal{F}_g\}\delta u + \iint_{\Omega^s} g_{,u}\delta u\, d\Omega^s.\end{aligned} \tag{13.18}$$

Note that the expression of $\{\mathcal{F}_g\}$ is different for each design parameter, but $\{\lambda_g\}$ is not. Thus, numerical integration in (13.18) must be carried out for each design parameter. However, the adjoint problem in (13.16) or (13.17), which takes the most computational effort, needs to be solved once per each performance measure.

As mentioned before, the expression $\{\lambda_g\}^T\{\mathcal{F}_g\}\delta u$ does not have to be calculated in the global level. After interpolating z and λ within an element, this term can be calculated using the numerical integration in the elements that are affected by the design parameter.

In this section, the finite element approximation of the design sensitivity computation is presented in the general expression. Since the finite element analysis of the structural problem is already carried out, we assume that such information as $[\boldsymbol{K}_g]$ and $\{z_g\}$, or equivalently $[\boldsymbol{K}]$ and $\{z\}$, is available. Thus, for the design sensitivity computation purposes, vector $\{\mathcal{F}_g\}$ needs to be calculated for each design parameter. In addition, different performance measures have different vector $\{\boldsymbol{G}_g\}$. In the following sections, these two terms are derived for different structural components and different performance measures.

13.1.2 Structural Components

In this section, the numerical approximations of the design sensitivity equation, fictitious load vector, and sensitivity coefficients are presented for different structural components. The numerical process is explained using the simplest sizing design parameterizations presented in Chapter 12. However, extensions of this procedure to more complicate sizing design parameterizations of Chapter 12 can be easily done once the approximation process for simple design parameterization process is understood.

Beam Design Component

The design sensitivity formulation of the beam component in Section 5.1.3 involves two kinds of sizing design parameters: the material property and the cross-sectional geometry, such as cross-sectional area or moment of inertia. Only the cross-sectional area is considered in this section, such that the design parameter vector is chosen by $u = [A]$ and its variation becomes $\delta u = [\delta A]$. Theoretically, there is no difference whether the design parameter is the cross-sectional area or the cross-sectional dimension.

The structural fictitious load for the beam problem is provided by (5.19) and (5.20). Without loss of generality, let the beam's moment of inertia can be represented by $I(x) = \alpha A(x)^2$ where α is a positive constant. This is possible if the cross-sectional dimensions of the beam are similar in all directions. The reason of introducing the relation of $I(x) = \alpha A(x)^2$ is that we want to express the moment of inertia as a function of the cross section, which is the design parameter. If the cross-sectional dimensions are not similar in all directions, then the cross-sectional dimensions described in Section 12.1 should be chosen as design parameters. In addition, the load linear form consists of the externally applied load $F(x)$ and the self-weight $\gamma A(x)$ per unit length, where γ is the weight density of the beam. Using this notation, the explicitly dependent terms in (5.19) and (5.20) is rewritten here, as

$$a'_{\delta u}(z,\bar{z}) = \int_0^l 2E\alpha A\delta A z_{,11}\bar{z}_{,11}\,dx \tag{13.19}$$

$$\ell'_{\delta u}(\bar{z}) = \int_0^l \gamma\delta A\bar{z}\,dx. \tag{13.20}$$

In (13.19) and (13.20), scalar function z and $\bar{z}$ are used because the transverse displacement is the only unknown function in the technical beam problem.

Let the beam component is approximated by a set of beam finite elements, whose formulation is presented in Section 3.3.2. In the technical beam theory, the transverse displacement and rotation are nodal degrees-of-freedom, which are related by $\theta_1 = z_{1,1}$ and $\theta_2 = z_{2,1}$. Thus, the nodal displacement vector is defined as $\boldsymbol{d}_I = [z_1,\ \theta_1,\ z_2,\ \theta_2]_I^T$.

Among N_E number of beam finite elements, let the cross-sectional area of the N_u ($\ll N_E$) number of elements is changed due to the design change. The fictitious load in (13.9) can be obtained for the cross-sectional area design parameter, as

$$\{\mathcal{F}_g\}\delta A = \mathcal{A}\left[\sum_{I=1}^{N_u}\int_0^{l_I}\{\gamma \boldsymbol{N} - 2E\alpha A\boldsymbol{B}_I^T\boldsymbol{B}_I\boldsymbol{d}_I\}\delta A\,dx\right]. \tag{13.21}$$

where $\mathcal{A}$ [•] represents the global assembly process of finite element formulation described in Section 3.4.1; l_I is the length of element I; $\boldsymbol{N}$ is the shape function of interpolation given in (3.132); and $\boldsymbol{B}_I$ is the discrete displacement-strain matrix provided in (3.133). From the solution to the structural problem, the nodal displacement $\boldsymbol{d}_I$ can be obtained. In the case of the beam element, the integrals in (13.21) can be obtained analytically.

First, consider a displacement performance measure at a point $\hat{x}$, which can be achieved using the Dirac delta measure, as

$$\psi_1 = \int_0^l z\delta(x-\hat{x})\,dx. \tag{13.22}$$

Throughout this text, we used the integral to define a general performance measure, even if the performance measure is defined at a point. It may look like such complication is unnecessary. However, the integral representation of a performance measure is required in the continuum formulation because the variation of performance measure is used as an adjoint load, which is defined as an integral. Nonetheless, the physical meaning of ψ_1 is the displacement at point $\hat{x}$, and the adjoint load of ψ_1 is the unit force at point $\hat{x}$, as explained in (5.37). If point $\hat{x}$ is a nodal point, then the adjoint force vector in the finite element approximation can be written as

$$\{\boldsymbol{G}_g\} = [0 \quad \cdots \quad 0 \quad 1 \quad 0 \quad \cdots \quad 0]^T, \tag{13.23}$$

where the value of "1" is in the location of the node corresponding to $\hat{x}$. After calculating the adjoint vector $\{\lambda_g\}$ as a solution to (13.5), the sensitivity of ψ_1 can be calculated from (13.18), as

$$\psi' = \{\lambda_g\}^T\{\mathcal{F}_g\}\delta A. \tag{13.24}$$

Note that, unlike (13.18), there is no explicitly dependent term in (13.24).

Example 13.1. Displacement Sensitivity of Clamped-Clamped Beam. As an example that can be compared with the analytical solution, consider a uniform clamped-clamped beam with $A = A_0 = 0.005$ m^2, $E = 2 \times 10^5$ MPa, $\alpha = 1/6$, $F = 49.61$ kN/m, and $\gamma = 77{,}126$ N/m^3. The analytical displacement under the given distributed load is $z(x) = 2.5 \times 10^{-3}[x^2(1-x)^2]$, as presented in Section 5.1.3. In order to solve the finite element equation (13.8), the commercial finite element code MSC/NASTRAN is used to numerically calculate the displacement vector $\{z_g\}$. Although the CBAR element of NASTRAN supports the transverse shear deformation, transverse shear stiffness is removed to satisfy the technical beam theory. If the design sensitivity of displacement at the center of the beam is desired, then $\hat{x} = ½$. Thus, the adjoint force in (13.23) is applied to the node at the center of the beam. After calculating the global displacement $\{z_g\}$ and adjoint vector $\{\lambda_g\}$, the sensitivity of (13.24) can be calculated. Instead of calculating the sensitivity in the global level, (13.24) is modified to the form before domain integration, as

$$\psi_1' = \int_0^l[\gamma \boldsymbol{N}^T\boldsymbol{\eta}_I - 2E\alpha A\boldsymbol{\eta}_I^T\boldsymbol{B}_I^T\boldsymbol{B}_I\boldsymbol{d}_I]\delta A\,dx, \tag{13.25}$$

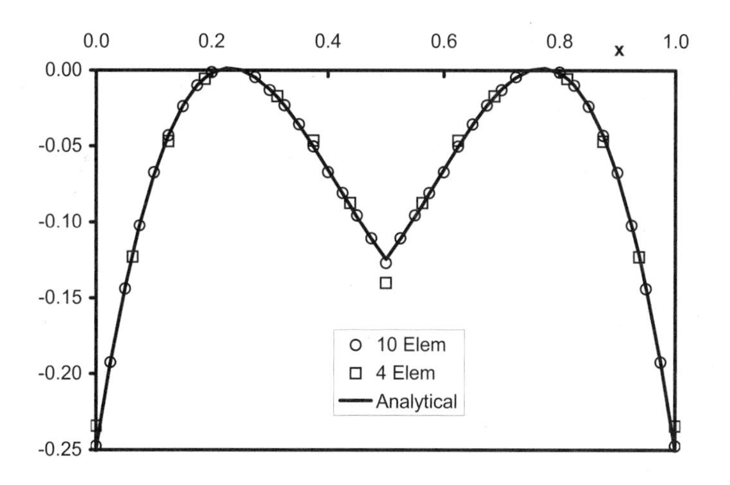

Figure 13.2. Displacement sensitivity for clamped-clamped beam.

where $\boldsymbol{\eta}_I$ is the adjoint response that is extracted from the global $\{\lambda_g\}$. In order to evaluate the numerical results, the coefficient of δA in the integral of (13.25) is plotted in Fig. 13.2 with analytical results from Fig. 5.2. When the entire structure is approximated by four finite elements (CBAR element), the result at the middle of the structure is slightly deviated from the analytical result. However, the result with 10 finite elements becomes very accurate.

Next, consider a compliance functional, which is a measure of structural flexibility, defined as

$$\psi_2 = \int_0^l [F + \gamma A] z \, dx. \tag{13.26}$$

Unlike the displacement functional, the variation of ψ_2 contains both the implicitly and explicitly dependent terms, as

$$\psi_2' = \int_0^l [F + \gamma A] z' \, dx + \int_0^l \gamma z \delta A \, dx. \tag{13.27}$$

In the adjoint variable method, the first integral of (13.27) is used to define the adjoint load. For the compliance performance measure, the discrete adjoint force vector can be obtained by

$$\{\boldsymbol{G}_g\} = \boldsymbol{\mathcal{A}} \left[\sum_{I=1}^{N_E} \int_0^{l_I} [F + \gamma A] \boldsymbol{N} \, dx \right]. \tag{13.28}$$

Note that the adjoint force vector $\{\boldsymbol{G}_g\}$ is identical to the initial load vector $\{\boldsymbol{F}_g\}$ in this special functional. Thus, the structural response $\{z_g\}$ and the adjoint response $\{\lambda_g\}$ are the same. The sensitivity is then

$$\psi_2' = \int_0^l [2\gamma \boldsymbol{N}^T \boldsymbol{d}_I - 2E\alpha A (\boldsymbol{B}_I \boldsymbol{d}_I)^2] \delta A \, dx. \tag{13.29}$$

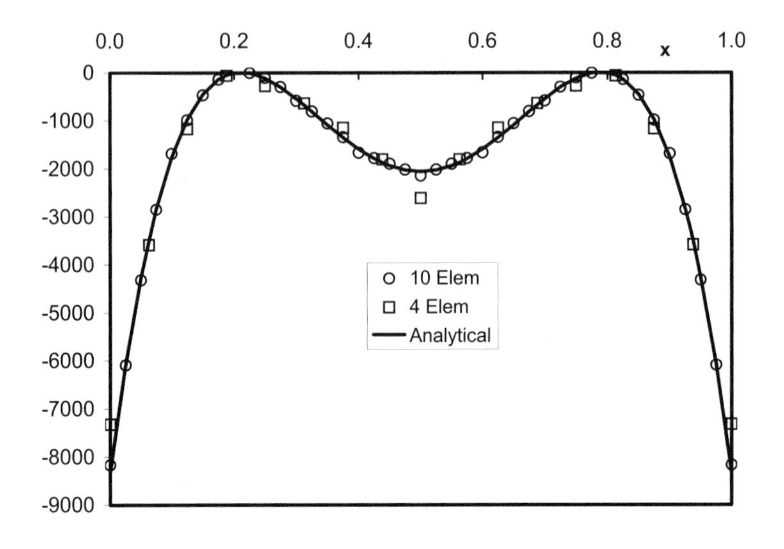

Figure 13.3. Compliance sensitivity for clamped-clamped beam.

Example 13.2. Compliance Sensitivity of Clamped-Clamped Beam. To illustrate how this result could be employed, consider the clamped-clamped beam of the previous example. The numerically calculated values of the coefficient of δA in the integral of (13.29) are compared with the analytical results in Fig. 13.3. Similar to Example 13.1, the results with 10 finite elements are close enough to the analytical results.

Plate Design Component

Even if the plate component is the extension of the one-dimensional beam to the two-dimensional space, its mathematical expression is quite complicated. Previous four-parameter beam element is conforming, while 12-parameter rectangular plate element is nonconforming [69]. However, this element is often used in the engineering applications with Hermite shape functions. The finite element interpolation of plate element I consists of the approximation of the transverse displacement $z(\xi)$, as

$$z(\xi) = \mathbf{N}(\xi)^T \mathbf{d}_I, \tag{13.30}$$

where $\mathbf{d}_I = [z_1, \theta_{x1}, \theta_{y1}, \ldots, z_4, \theta_{x4}, \theta_{y4}]^T$ is the nodal displacement, and $N(\xi)$ is the Hermite shape function of approximation given in (3.155). The definitions of parameters in (13.30) are illustrated in Fig. 3.19.

The plate formulation in Section 3.1.3 includes the curvature vector $\boldsymbol{\kappa}$, as defined in (3.39). Before approximating the curvature vector, we need to derive the Jacobian relation first. Even if the displacement is interpolated using the nodal displacement and rotation, the geometry is approximated using the nodal coordinate alone. The relation between the physical and reference coordinates becomes

$$\begin{bmatrix} x \\ y \end{bmatrix} = \begin{bmatrix} a\xi \\ b\eta \end{bmatrix}. \tag{13.31}$$

Thus, the Jacobian relation in (13.5) can be obtained as

$$\boldsymbol{J}_I = \begin{bmatrix} a & 0 \\ 0 & b \end{bmatrix}, \tag{13.32}$$

from which the inverse matrix $\boldsymbol{J}_I^{-1}$ can be obtained.

Using the Jacobian relation of (13.32) and using the interpolation relation in (13.30), the curvature vector in (3.39) can be approximated as

$$\boldsymbol{\kappa}(z) = \begin{bmatrix} z_{,11} \\ z_{,22} \\ 2z_{,12} \end{bmatrix} = \boldsymbol{B}_I \boldsymbol{d}_I, \tag{13.33}$$

where the displacement-strain matrix $\boldsymbol{B}_I$ is defined by

$$\boldsymbol{B}_I = \begin{bmatrix} \dfrac{1}{a^2}\boldsymbol{N}_{,\xi\xi}^T \\ \dfrac{1}{b^2}\boldsymbol{N}_{,\eta\eta}^T \\ \dfrac{2}{ab}\boldsymbol{N}_{,\xi\eta}^T \end{bmatrix}_{3\times 12}. \tag{13.34}$$

The explicit expression of the matrix $\boldsymbol{B}_I$ can be found Section 6.6 of Shames and Dym [70].

The thickness of plate is chosen as a design parameter for the plate problem, such that $u = [h]$. In addition, the load linear form consists of the externally applied load $F(x)$ and the self-weight $\gamma h(\boldsymbol{x})$ per unit area, where γ is the weight density of the plate. Using this notation, the fictitious load in (5.51) and (5.52) can be rewritten here, as

$$a'_{\delta u}(z,\bar{z}) = \iint_\Omega \frac{Eh^2\,\delta h}{4(1-\nu^2)} \boldsymbol{\kappa}(\bar{z})^T \boldsymbol{C}\boldsymbol{\kappa}(z)\,d\Omega \tag{13.35}$$

$$\ell'_{\delta u}(\bar{z}) = \iint_\Omega \gamma \bar{z}\,\delta h\,d\Omega. \tag{13.36}$$

Among N_E number of plate elements, let the thickness of N_u number of elements change due to the design changes. Using the approximation of the curvature in (13.33), the discrete fictitious load can be obtained as

$$\{\mathcal{F}_g\}\delta h = \boldsymbol{\mathcal{A}}\left[\sum_{I=1}^{N_u} \iint_{\Omega_I} \left(\gamma \boldsymbol{N} - \frac{Eh^2}{4(1-\nu^2)} \boldsymbol{B}_I^T \boldsymbol{C}\boldsymbol{B}_I \boldsymbol{d}_I\right) \delta h\,d\Omega\right]. \tag{13.37}$$

An analytical or numerical integration using Gauss quadrature can be employed to calculate the integral in (13.37).

After calculating the fictitious load from (13.37), the discrete design sensitivity (13.9) can be solved for discrete displacement sensitivity $\{z'_g\}$ in the direct differentiation method, from which the sensitivity of performance measure is calculated as in (13.14). In the adjoint variable method, the adjoint load that depends on the performance measure is calculated first. After solving the adjoint (13.16), the sensitivity of performance measure is calculated through the domain integration in (13.18), which also requires the fictitious load or a similar procedure.

For example, the displacement functional and its variation at point $\hat{x}$ of the plate can be defined using the Dirac delta measure, as

$$\psi = \iint_{\Omega} z_3 \delta(\boldsymbol{x} - \hat{\boldsymbol{x}}) d\Omega \tag{13.38}$$

$$\psi' = \iint_{\Omega} z_3' \delta(\boldsymbol{x} - \hat{\boldsymbol{x}}) d\Omega. \tag{13.39}$$

In the adjoint variable method, the adjoint load is the unit force at point $\hat{x}$, which is similar to (13.23). Using this adjoint load, the discrete adjoint matrix equation of (13.17) is solved for $\{\boldsymbol{\lambda}\}$. After solving the adjoint response $\{\boldsymbol{\lambda}_g\}$ and state response $\{z_g\}$, the displacement sensitivity in (13.39) is evaluated by

$$\psi' = \sum_{I=1}^{N_u} \iint_{\Omega_I} \left[\gamma \boldsymbol{N}^T \boldsymbol{\eta}_I - \frac{Eh^2}{4(1-\nu^2)} \boldsymbol{d}_I^T \boldsymbol{B}_I^T \boldsymbol{C} \boldsymbol{B}_I \boldsymbol{\eta}_I \right] \delta h \, d\Omega. \tag{13.40}$$

The design sensitivity formula in the continuum formulation is exact if the analytical results of displacement and adjoint response are used. However, the sensitivity result from the continuum-discrete formulation in (13.40) is not exact because of the approximation errors in state and adjoint responses and the numerical domain integration involved in (13.40). However, the theoretical study of the finite element method guarantees that the approximated result will approach to the analytical result as the size of finite element reduces. In the following example, the result of continuum-discrete sensitivity analysis also is shown to approach to the result of continuum sensitivity analysis as the size of finite element reduces.

Example 13.3. Displacement Sensitivity of a Clamped Plate. Figure 13.4 shows a clamped plate with a concentrated force at the center point. For material properties, Young's modulus of E = 3 × 10^6 and Poisson's ratio of ν = 0.3 are used. The plate thickness is h = 0.8, and the dimensions of the plate are a = b = 80. A unit force of P = 1.0 is applied at the center of the plate. To see the convergent behavior of the sensitivity results, four different models are considered for finite element discretization, i.e., 5 × 5, 9 × 9, 13 × 13, and 17 × 17 nodes. Since the geometry and boundary conditions are symmetrical, only a quarter of the plate is modeled.

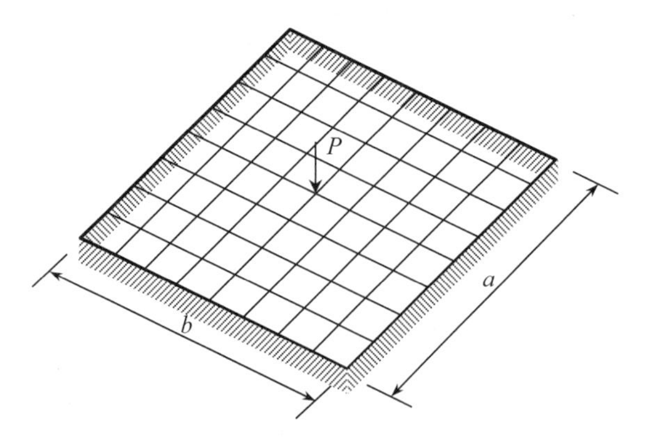

Figure 13.4. Finite element model of clamped plate.

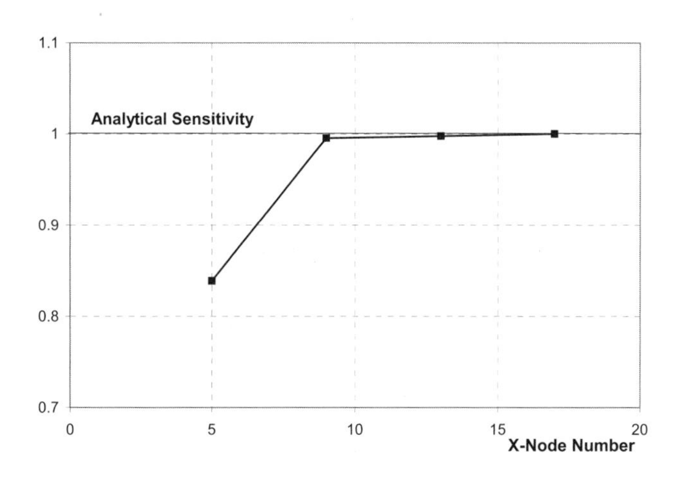

Figure 13.5. Design sensitivity convergence of center point deflection.

The analytical solution for center point deflection can be found in Timoshenko [71]. In the case of the clamped, rectangular plate, the maximum deflection appears at the center of the plate whose magnitude is

$$z_{\max} = \alpha \frac{Pa^2}{D}, \tag{13.41}$$

where α is the function of a/b, that is, the ratio of the plate dimension, and D is the flexural rigidity, defined as

$$D = \frac{Eh^3}{12(1-\nu^2)}. \tag{13.42}$$

Since this analytical solution is the explicit function of plate thickness, the analytical design sensitivity for center point deflection can easily be obtained. From (13.41) and (13.42), it is clear that center point deflection is a function of plate thickness through the flexural rigidity. Thus, the sizing design sensitivity of (13.41) can be obtained as

$$\dot{z}_{\max} = -\alpha \frac{3Pa^2}{Dh}. \tag{13.43}$$

The design sensitivity result in (13.40) is then compared with the analytical design sensitivity in (13.43) for different number of nodes. Only the second term of the integral in (13.40) is used because the body force is ignored in this example. Figure 13.5 plots the convergent behavior of normalized design sensitivity results. The continuum-discrete sensitivity method yields almost exactly the same result with the analytical sensitivity after 13 × 13 number of nodes. Center point deflection also shows a similar convergent behavior.

Plane Solid Design Component

A plane solid component often appears in the sizing design problem coupled with the plate component to represent the membrane effect. The finite element formulation of the three-dimensional solid component in Section 3.3.4 can be directly used by reducing the dimension to two. The explicitly dependent terms for the energy bilinear and load linear forms are derived in (5.75) and (5.76), respectively, which are rewritten here

$$a'_{\delta u}(z,\lambda)=\iint_{\Omega}\sigma(z)^T\varepsilon(\lambda)\delta h\,d\Omega \tag{13.44}$$

$$\ell'_{\delta u}(\lambda)=\iint_{\Omega}f^T\lambda\delta h\,d\Omega. \tag{13.45}$$

The triangular and quadrilateral elements are often used in the numerical approximation of the plane solid component. In this section, the four-node quadrilateral element is used to approximate the design sensitivity equation. In element I, the strain and stress vectors in (13.44) can be approximated by

$$\varepsilon(\lambda)=B_I\eta_I \tag{13.46}$$

$$\sigma(z)=CB_Id_I, \tag{13.47}$$

where C is the stiffness matrix of plane stress problem, $\eta_I=[\lambda_1^1,\ \lambda_2^1,\ \ldots,\ \lambda_1^4,\ \lambda_2^4]_I^T$ is the nodal adjoint response of element I, $d_I=[z_1^1,\ z_2^1,\ \ldots,\ z_1^4,\ z_2^4]_I^T$ is the nodal displacement vector, and B_I is the strain-displacement matrix of element I. For the four-node quadrilateral element, B_I is defined by

$$B_I=\begin{bmatrix} N_{1,1} & 0 & N_{2,1} & 0 & N_{3,1} & 0 & N_{4,1} & 0 \\ 0 & N_{1,2} & 0 & N_{2,2} & 0 & N_{3,2} & 0 & N_{4,2} \\ N_{1,2} & N_{1,1} & N_{2,2} & N_{2,1} & N_{3,2} & N_{3,1} & N_{4,2} & N_{4,2} \end{bmatrix}_I, \tag{13.48}$$

where N_J is the shape function corresponding to node J. Using the finite element approximation described in (13.46) and (13.48), the structural fictitious loads in (13.44) and (13.45) can be discretized by

$$a'_{\delta u}(z,\lambda)\approx\sum_{I=1}^{N_u}\iint_{\Omega_I}[d_I^TB_I^TCB_I\eta_I]\delta h\,d\Omega \tag{13.49}$$

$$\ell'_{\delta u}(\lambda)\approx\sum_{I=1}^{N_u}\iint_{\Omega_I}[\eta_I^TNf_I]\delta h\,d\Omega, \tag{13.50}$$

where f_I is the applied force vector of element I. In the adjoint variable method, the performance sensitivity is calculated using (13.49) and (13.50). In the direct differentiation method, the sensitivity matrix equation is solved first using the fictitious force vector. The global fictitious force vector can be obtained by combining the coefficients of η_I in (13.49) and (13.50), after the assembly process, as

$$\{\mathcal{F}_g\}\delta h=\mathcal{A}\left[\sum_{I=1}^{N_u}\iint_{\Omega_I}\left\{Nf_I-B_I^TCB_I\eta_I\right\}\delta h\,d\Omega\right]. \tag{13.51}$$

As with other structural components, the adjoint force vector for the displacement performance measure is the unit force at the node where the sensitivity is evaluated. After solving the adjoint problem, the sensitivity coefficient is evaluated through the numerical integration in (13.49) and (13.50).

The stress performance measure for the plane solid component in Section 5.2.3 is formulated using a characteristic function m_p that is defined in the subdomain Ω^s. This is necessary because mathematically, the stress cannot be defined pointwise. However, in engineering applications the stress value at a point is often of interesting. In such a situation, the characteristic function m_p is assumed to be the Dirac delta measure. Consider a von Mises stress function at a point of element I. The performance measure can be defined as

$$\psi = \sigma_{vM}(\hat{x}) = \iint_{\Omega_I} \sqrt{\sigma_{11}^2 + \sigma_{22}^2 - \sigma_{11}\sigma_{22} + 3\sigma_{12}^2}\,\delta(\boldsymbol{x} - \hat{\boldsymbol{x}})\,d\Omega. \tag{13.52}$$

Since there is no explicitly dependent term exists, the first-order variation of ψ with respect to the sizing design parameter can be obtained as

$$\begin{aligned} \psi' &= \iint_{\Omega_I} \frac{1}{2\sigma_{vM}} \boldsymbol{q}^T \boldsymbol{\sigma}'\,\delta(\boldsymbol{x} - \hat{\boldsymbol{x}})\,d\Omega \\ &= \iint_{\Omega_I} \frac{1}{2\sigma_{vM}} \boldsymbol{q}^T \boldsymbol{C}\boldsymbol{B}_I \boldsymbol{d}_I'\,\delta(\boldsymbol{x} - \hat{\boldsymbol{x}})\,d\Omega, \end{aligned} \tag{13.53}$$

where $\boldsymbol{q} = [2\sigma_{11} - \sigma_{22},\ 2\sigma_{22} - \sigma_{11},\ 6\sigma_{12}]^T$ is the coefficient vector, and the point $\hat{\boldsymbol{x}}$ belongs to element I. Note that matrices $\boldsymbol{C}$ and $\boldsymbol{B}_I$ are independent of the design. In this special performance measure, the sensitivity of von Mises stress cannot be evaluated if the stress value is zero.

In the adjoint variable method, the adjoint load is the coefficient of $\boldsymbol{d}_I'$ in (13.53). Thus, after the global assembly process, the discrete adjoint force can be obtained as

$$\{\boldsymbol{G}\} = \mathcal{A}\left[\iint_{\Omega_I} \frac{1}{2\sigma_{vM}} \boldsymbol{B}_I^T \boldsymbol{C}\boldsymbol{q}\,\delta(\boldsymbol{x} - \hat{\boldsymbol{x}})\,d\Omega \right]. \tag{13.54}$$

After obtaining the adjoint force vector, the adjoint equation of (13.17) is solved for the adjoint response $\{\boldsymbol{\lambda}\}$. Using the displacement vector $\{\boldsymbol{z}\}$ and the adjoint response $\{\boldsymbol{\lambda}\}$, the sensitivity of the stress performance measure is calculated through the numerical integration as

$$\psi' = \sum_{I=1}^{N_u} \iint_{\Omega_I} [\boldsymbol{\eta}_I^T \boldsymbol{N} \boldsymbol{f}_I - \boldsymbol{d}_I^T \boldsymbol{B}_I^T \boldsymbol{C}\boldsymbol{B}_I \boldsymbol{\eta}_I]\,\delta h\,d\Omega. \tag{13.55}$$

The numerical approximations of the design sensitivity equation, fictitious load vector, and sensitivity coefficient are discussed in this section. Those matrices that are generated for state response finite element analysis are still used for the purposes of sizing design sensitivity analysis. Thus, the implementation effort of design sensitivity analysis is minimal.

13.2 Shape Design Sensitivity Computation

Apart from the sizing sensitivity computation, the shape sensitivity computation requires the design velocity field corresponding to the design parameter. The calculation of design velocity field in conjunction with the design parameterization is developed in Section 13.3 that is consistent with the theoretical and practical requirements presented in Section 6.2.7. In this section, it is assumed that the design velocity field $\boldsymbol{V}(\boldsymbol{x})$ corresponding to the current design parameter is already available at each node of finite elements. For the node $\boldsymbol{x}$ that does not change its location during the design perturbation, $\boldsymbol{V}(\boldsymbol{x}) = \boldsymbol{0}$.

As with the sizing design problem, the development of the finite element method in

Section 3.3 can be used for the approximation of the state equation. In the numerical calculation, the design velocity field also needs to be discretized. Since the design velocity field requires the same regularity as the displacement, as discussed in Section 6.2.7, it is meaningful to approximate the design velocity field using the same approximation method as the displacement.

13.2.1 Finite Element Approximation

The general expression of the structural problem is the same as (13.1). The design sensitivity equation for the shape design problem has been developed in Section 6.2.1. The continuum-based design sensitivity (6.67) is rewritten here, as

$$a_{\Omega}(\dot{z},\bar{z}) = \ell'_V(\bar{z}) - a'_V(z,\bar{z}), \quad \forall \bar{z} \in Z. \tag{13.56}$$

Since the left side of (13.56) is the same as the energy bilinear form, its discretization also has the same expression as the left side of (13.9). The right side of (13.56) is the explicitly dependent terms on the design, or specifically, on the design velocity field $\boldsymbol{V}(\boldsymbol{x})$. This is the reason that the subscribed $\boldsymbol{V}$ is used in the definitions of $\ell'_V(\bullet)$ and $a'_V(\bullet,\bullet)$. The right side can be calculated if the result z of the state equation and the design velocity $\boldsymbol{V}(\boldsymbol{x})$ are available.

In addition to the interpolation of the displacement z in (13.3), the shape design sensitivity analysis also requires the interpolation of the design velocity field $\boldsymbol{V}(\boldsymbol{x})$. As discussed in Section 6.2.7, this design velocity field must have the same regularity with the displacement function. Thus, it is natural to use the same interpolation function with the displacement to approximate the design velocity field. In element I, the design velocity is interpolated by

$$\boldsymbol{V}(\xi) = \boldsymbol{N}(\xi)^T \boldsymbol{V}_I, \tag{13.57}$$

where $\boldsymbol{V}_I$ is the nodal design velocity vector of element I. The shape parameterization method in Section 12.2 and the design velocity calculation method in Section 13.3 provide the design velocity field at each node. If a structural point remains fixed during shape design change, then the design velocity field corresponding to that point is zero.

After using the same approximation method and the same domain integration method with the state equation, the shape design sensitivity (13.56) can be approximated as

$$\{\bar{z}_g\}^T [\boldsymbol{K}_g]\{\dot{z}_g\} = \{\bar{z}_g\}^T \{\mathcal{F}_g\}, \quad \forall \{\bar{z}_g\} \in Z_h, \tag{13.58}$$

where Z_h is the discrete space of kinematically admissible displacements; $\{\dot{z}_g\}$ is the global vector of displacement sensitivity; $\{\mathcal{F}_g\}$ is the global fictitious force, which is the approximation of the right side of (13.56). The calculation of $\{\mathcal{F}_g\}$ depends on the structural component, which will be described in Section 13.2.2.

As with the sizing design sensitivity analysis, the construction of fictitious force $\{\mathcal{F}_g\}$ can be carried out using the domain integration over a small subdomain in which the geometry changes due to the shape design parameter changes. To explain it in detail, let a structural domain Ω be divided by 37 finite elements, as shown in Fig. 13.6. If the shape design parameter u has an effect on the geometry of nine elements, then the design velocity $\boldsymbol{V}(\boldsymbol{x})$ is nonzero and the term $\ell'_V(\bar{z}) - a'_V(z,\bar{z})$ has values only for those nine elements. The explicitly dependent terms vanish for other 28 elements. Thus, the numerical integration needs to be carried out only for a small subdomain rather than the entire structural domain.

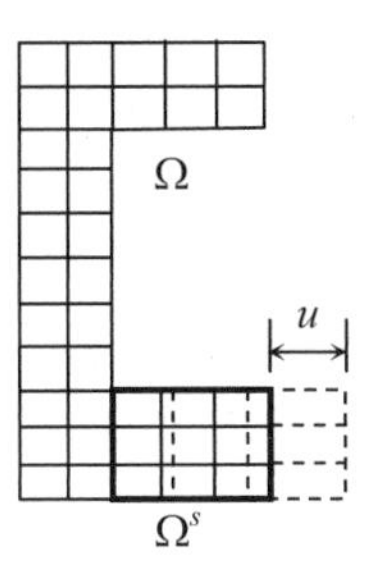

Figure 13.6. Finite Element approximation of domain Ω and shape parameterization in subdomain Ω^s.

Since the generalized stiffness matrix $[\boldsymbol{K}_g]$ in (13.58) is not positive definite, the discrete variational (13.58) is converted to the reduced matrix equation after imposing the boundary condition, to obtain

$$[\boldsymbol{K}]\{\dot{z}\} = \{\mathcal{F}\}, \tag{13.59}$$

where $\{\dot{z}\}$ is the reduced vector of displacement sensitivity, and $\{\mathcal{F}\}$ is the reduced fictitious force vector. Now, the reduced stiffness matrix $[\boldsymbol{K}]$ is positive definite, and the unique solution $\{\dot{z}\}$ can be obtained.

After calculating $\{\dot{z}\}$, the generalized vector of displacement sensitivity $\{\dot{z}_g\}$ can be obtained by considering the imposition of the essential boundary condition. This procedure is essentially the same as the sizing design problem because the shape function of approximation is independent of the sizing and shape design parameters.

The approximation of the displacement sensitivity within a finite element can be carried out using the following relation:

$$\dot{z}(\xi) = \boldsymbol{N}(\xi)^T \dot{\boldsymbol{d}}_I, \tag{13.60}$$

where $\dot{\boldsymbol{d}}_I$ is the nodal displacement sensitivity of element *I*, extracted from $\{\dot{z}_g\}$. Again, the shape function $\boldsymbol{N}(\xi)$ remains constant during shape variation.

The general form of the performance measure and its variation in the shape design problem have been defined in (6.69) and (6.71), respectively, which are rewritten here, as

$$\psi = \int_{\Omega_\tau} g(z_\tau, \nabla z_\tau)\, d\Omega_\tau \tag{13.61}$$

$$\psi' = \int_{\Omega_\tau} (g_{,z}\dot{z} + g_{,\nabla z} : \nabla \dot{z})\, d\Omega - \int_{\Omega} (g_{,\nabla z} : \nabla z \nabla \boldsymbol{V} - g\, div\boldsymbol{V})\, d\Omega. \tag{13.62}$$

The first integral on the right side of (13.62) implicitly depends on the design through $\dot{z}$, while the second integral explicitly depends on the design.

In the direct differentiation method, the approximated z in (13.3), the approximated $\dot{z}$ in (13.60), and the design velocity $\boldsymbol{V}(\xi)$ are used to calculate the integrals in (13.62). Since the analytical integration is possible only for very simple geometry, the numerical integration is often used to calculate (13.62).

In the adjoint variable method, the adjoint load is first defined by replacing the displacement sensitivity $\dot{z}$ with the virtual displacement $\bar{\lambda}$ from the first integral on the right side of (13.62). As mentioned in Section 13.1.1, the adjoint problem is independent of design parameter. In fact, the adjoint equation of the performance measure in (13.61) is the exactly same as (13.15). Thus, the discrete matrix equation of the adjoint problem is also the same as the sizing design problem, given in (13.16) for the discrete variational equation and in (13.17) for the reduced matrix equation.

13.2.2 Structural Components

Beam Design Component

In Section 3.1, two beam theories have been studied: the technical and the Timoshenko beam theories. The former assumes that the bending stress is the dominant contribution to the structure, whereas the latter also takes into account the transverse shear effect. In shape sensitivity analysis in Chapter 6, the technical beam theory requires C^1-continuous design velocity field, which is impractical in complicated engineering applications. Thus, in this section, the Timoshenko beam theory is used to numerically implement the shape design sensitivity equation. Since the variational formulation of this theory only requires the first-order derivative of the state response, the C^0-continuous design velocity is enough, which is more convenient to apply for geometric shape.

In order to derive the discrete matrix equation of design sensitivity analysis, the explicitly dependent terms on the shape design for the beam problem developed in (6.186) and (6.188) are rewritten here as

$$a'_V(z,\bar{z}) = -\int_0^l [EI\bar{\theta}_{,1}\theta_{,1} + k\mu A(\bar{z}_{,1}z_{,1} - \bar{\theta}\theta)]V_{,1}\,dx \tag{13.63}$$

$$\ell'_V(\bar{z}) = \int_0^l (\bar{z}^T f_{,1}V + \bar{z}^T fV_{,1})\,dx, \tag{13.64}$$

where $z = [z,\ \theta]^T$ is the state response, E is Young's modulus, I is the second moment of inertia, k is the shear correction factor, μ is the shear modulus, and $\boldsymbol{f} = [f,\ m]^T$ is the distributed force and moment. As discussed before, (13.63) and (13.64) only contain the first-order derivative of the design velocity. The finite element approximation of (13.63) and (13.64) includes the domain discretization and the interpolation of the state response z and the design velocity V.

The beam component is approximated using a one-dimensional linear beam element as shown in Fig. 13.7. If the element-fixed local coordinate system is used, then the shape design velocity of the beam element has only the x_1-component. Thus, the nodal design velocity of element I is defined by $\boldsymbol{V}_I = [V_1,\ V_2]_I^T$.

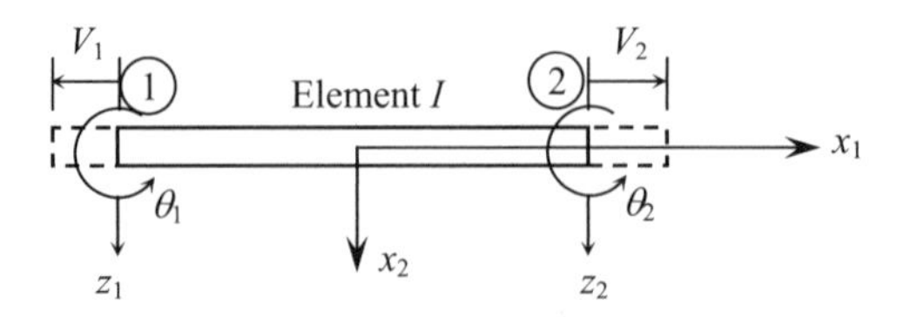

Figure 13.7. Design perturbation of the linear beam element.

In order to approximate the explicitly dependent terms in (13.63) and (13.64), the approximations of θ, $\theta_{,1}$, $z_{,1}$, V, $V_{,1}$, and $\bar{z}$ are required, which can be obtained for the linear beam element in Fig. 13.7, as

$$\theta = [0,\ N_1,\ 0,\ N_2]\boldsymbol{d}_I \equiv \boldsymbol{N}_\theta^T \boldsymbol{d}_I \tag{13.65}$$

$$\theta_{,1} = [0,\ N_{1,1},\ 0,\ N_{2,1}]_I \begin{Bmatrix} z_1 \\ \theta_1 \\ z_2 \\ \theta_2 \end{Bmatrix}_I \equiv \boldsymbol{B}_I^\theta \boldsymbol{d}_I \tag{13.66}$$

$$z_{,1} = [N_{1,1},\ 0,\ N_{2,1},\ 0]_I \boldsymbol{d}_I \equiv \boldsymbol{B}_I^z \boldsymbol{d}_I \tag{13.67}$$

$$V = [N_1,\ N_2]\boldsymbol{V}_I \equiv \boldsymbol{N}_V \boldsymbol{V}_I \tag{13.68}$$

$$V_{,1} = [N_{1,1},\ N_{2,1}]_I \begin{Bmatrix} V_1 \\ V_2 \end{Bmatrix}_I \equiv \boldsymbol{M}_I^T \boldsymbol{V}_I \tag{13.69}$$

$$\bar{z} = \begin{Bmatrix} z \\ \theta \end{Bmatrix} = \begin{bmatrix} N_1 & 0 & N_2 & 0 \\ 0 & N_1 & 0 & N_2 \end{bmatrix} \bar{\boldsymbol{d}}_I \equiv \boldsymbol{N}^T \bar{\boldsymbol{d}}_I, \tag{13.70}$$

where $\boldsymbol{d}_I = [z_1,\ \theta_1,\ z_2,\ \theta_2]_I^T$ is the nodal displacement of element I, and N_1 and N_2 are the linear shape functions of the beam element given in (3.140).

After substituting (13.65) through (13.70) into (13.63) and (13.64), the explicitly dependent terms are approximated as

$$a'_V(z,\bar{z}) = -\sum_{I=1}^{N_u} \int_0^{l_I} \bar{\boldsymbol{d}}_I^T [EI\boldsymbol{B}_I^{\theta T} \boldsymbol{B}_I^\theta + k\mu A(\boldsymbol{B}_I^{zT} \boldsymbol{B}_I^z - \boldsymbol{N}_\theta \boldsymbol{N}_\theta^T)]\boldsymbol{d}_I (\boldsymbol{M}_I^T \boldsymbol{V}_I)\, dx \tag{13.71}$$

$$\ell'_V(\bar{z}) = \sum_{I=1}^{N_u} \int_0^{l_I} [\bar{\boldsymbol{d}}_I^T \boldsymbol{N} f_{,1} (\boldsymbol{N}_V \boldsymbol{V}_I) + \bar{\boldsymbol{d}}_I^T \boldsymbol{N} f (\boldsymbol{M}_I^T \boldsymbol{V}_I)]\, dx. \tag{13.72}$$

As discussed in Section 3.3.2, the Timoshenko beam formulation has a locking problem when a regular two-point Gauss integration is employed. In order to resolve the numerical locking problem, the transverse shear term of the energy bilinear form is integrated using reduced order integration. The explicitly dependent term in (13.71) also follows the same reduced integration method for the second term within the integral.

Plate Design Component

The shape design of plate component is limited to change the two-dimensional integration domain in the component-fixed local coordinate system. The rigid body rotation and curvature change of the component are related to the configuration design. Thus, the finite element approximation of the design sensitivity equation is carried out in the two-dimensional local coordinate system. As has been discussed in Chapter 7, the regularity of design velocity field for the plate component must be the same as that of the displacement function. In the case of Kirchhoff plate theory, the displacement field requires C^1-continuity, and thus, the design velocity field. However, the construction of

C^1-continuous design velocity field is not a trivial procedure and, sometimes, is impractical for complicated built-up structures. Thus, in the numerical implementation of plate component the Mindlin/Reissner plate theory is used in which C^0-continuity is required. In the following approximation, the four-node quadrilateral plate element is taken into account, as shown in Fig. 13.8. A two-dimensional design velocity vector is defined at each node of the element.

The structural fictitious load form for the Mindlin/Reissner plate has been presented in (6.189) of Section 6.2.5 whose continuum form is rewritten here as

$$\begin{aligned} a'_V(z,\bar{z}) = & \iint_\Omega [\boldsymbol{\kappa}^V(\bar{z})\boldsymbol{C}^b\boldsymbol{\kappa}(z) + \boldsymbol{\kappa}(\bar{z})\boldsymbol{C}^b\boldsymbol{\kappa}^V(z) + \boldsymbol{\kappa}(\bar{z})\boldsymbol{C}^b\boldsymbol{\kappa}(z)div\boldsymbol{V}]d\Omega \\ & + \iint_\Omega [\boldsymbol{\gamma}^V(\bar{z})\boldsymbol{C}^s\boldsymbol{\gamma}(z) + \boldsymbol{\gamma}(\bar{z})\boldsymbol{C}^s\boldsymbol{\gamma}^V(z) + \boldsymbol{\gamma}(\bar{z})\boldsymbol{C}^s\boldsymbol{\gamma}(z)div\boldsymbol{V}]d\Omega. \end{aligned} \quad (13.73)$$

In the first integral, $\boldsymbol{\kappa}(z)$ is the curvature vector in (3.48), $\boldsymbol{\kappa}^V(z)$ is the explicitly dependent term of the curvature vector in (6.193), and $\boldsymbol{C}^b$ is the bending stiffness matrix in (3.40). The form $a'_V(z,\bar{z})$ is defined on the body-fixed local coordinate system.

Let the design velocity field be given by $\boldsymbol{V}_I = [V_1^1,\ V_2^1,\ V_1^2,\ V_2^2,\ V_1^3,\ V_2^3,\ V_1^4,\ V_2^4]_I^T$ at element I in the local coordinate system. Using the vector notation, $\boldsymbol{\kappa}^V(z)$ in (13.73) can be approximated by

$$\begin{aligned} \boldsymbol{\kappa}^V(z) &= -\frac{1}{2}\left(\frac{\partial\theta_i}{\partial x_k}\frac{\partial V_k}{\partial x_j} + \frac{\partial\theta_j}{\partial x_k}\frac{\partial V_k}{\partial x_i}\right) \\ &= -\boldsymbol{S}\boldsymbol{\Lambda}_I\boldsymbol{G}_I\boldsymbol{d}_I, \end{aligned} \quad (13.74)$$

where $\boldsymbol{d}_I = [z_3^1,\ \theta_1^1,\ \theta_2^1,\ z_3^2,\ \theta_1^2,\ \theta_2^2,\ z_3^3,\ \theta_1^3,\ \theta_2^3,\ z_3^4,\ \theta_1^4,\ \theta_2^4]_I^T$ is the nodal parameter at element I,

$$\boldsymbol{S} = \begin{bmatrix} 1 & 0 & 0 & 0 \\ 0 & 0 & 0 & 1 \\ 0 & 1 & 1 & 0 \end{bmatrix} \quad (13.75)$$

is the mapping matrix from the tensor to the vector notation,

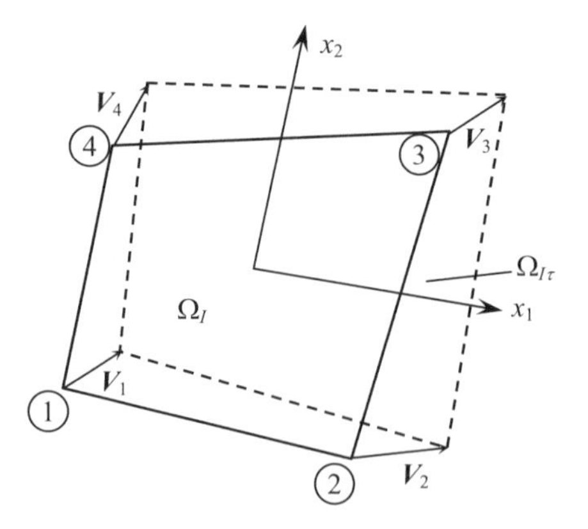

Figure 13.8. Design perturbation of the linear quadrilateral plate element.

$$\boldsymbol{\Lambda}_I = \begin{bmatrix} \boldsymbol{\Xi}_{2\times2}^T & \boldsymbol{0}_{2\times2} \\ \boldsymbol{0}_{2\times2} & \boldsymbol{\Xi}_{2\times2}^T \end{bmatrix}_I \tag{13.76}$$

is the matrix composed by the Jacobian matrix of the design velocity vector,

$$\boldsymbol{\Xi}_{2\times2} = \begin{bmatrix} V_{1,1} & V_{1,2} \\ V_{2,1} & V_{2,2} \end{bmatrix} = \sum_{k=1}^{4} \begin{bmatrix} N_{k,1}V_1^k & N_{k,2}V_1^k \\ N_{k,1}V_2^k & N_{k,2}V_2^k \end{bmatrix}, \tag{13.77}$$

and

$$\boldsymbol{G}_I = \begin{bmatrix} 0 & N_{1,1} & 0 & 0 & N_{2,1} & 0 & 0 & N_{3,1} & 0 & 0 & N_{4,1} & 0 \\ 0 & N_{1,2} & 0 & 0 & N_{2,2} & 0 & 0 & N_{3,2} & 0 & 0 & N_{4,2} & 0 \\ 0 & 0 & N_{1,1} & 0 & 0 & N_{2,1} & 0 & 0 & N_{3,1} & 0 & 0 & N_{4,1} \\ 0 & 0 & N_{1,2} & 0 & 0 & N_{2,2} & 0 & 0 & N_{3,2} & 0 & 0 & N_{4,2} \end{bmatrix} \tag{13.78}$$

is the second kind strain-displacement vector. In addition to the Jacobian matrix, (13.73) requires the divergence of the design velocity, which can be defined as

$$div\boldsymbol{V} = \sum_{i=1}^{2} V_{i,i} = \sum_{k=1}^{4} (N_{k,1}V_1^k + N_{k,2}V_2^k). \tag{13.79}$$

The second integral in (13.73) represents the effect of the transverse shear deformation. In (13.73), $\gamma(z)$ is the shear strain vector in (3.49), $\gamma^V(z)$ is the explicitly dependent term of the curvature vector in (6.194), and $\boldsymbol{C}^s$ is the shear stiffness matrix in (3.52). Using the vector notation, $\gamma^V(z)$ in (13.73) can be approximated by

$$\gamma^V(z) = -\begin{Bmatrix} \dfrac{\partial z_3}{\partial x_k}\dfrac{\partial V_k}{\partial x_1} \\ \dfrac{\partial z_3}{\partial x_k}\dfrac{\partial V_k}{\partial x_2} \end{Bmatrix} = -\boldsymbol{\Xi}_{2\times2}^T \boldsymbol{L}_I \boldsymbol{d}_I, \tag{13.80}$$

where

$$\boldsymbol{L}_I = \begin{bmatrix} N_{1,1} & 0 & 0 & N_{2,1} & 0 & 0 & N_{3,1} & 0 & 0 & N_{4,1} & 0 & 0 \\ 0 & N_{1,2} & 0 & 0 & N_{2,2} & 0 & 0 & N_{3,2} & 0 & 0 & N_{4,2} & 0 \end{bmatrix}_I. \tag{13.81}$$

Using the definitions in (13.74), (13.80) and using the approximations of strains in (3.165) and (3.166), the structural fictitious load can be approximated by

$$\begin{aligned} a'_V(z,\bar{z}) = &-\sum_{I=1}^{N_u} \iint_{\Omega_I} \bar{\boldsymbol{d}}_I^T \left[\boldsymbol{G}_I^T \boldsymbol{\Lambda}_I^T \boldsymbol{S}^T \boldsymbol{C}^b \boldsymbol{\kappa} + \boldsymbol{B}_I^{bT} \boldsymbol{C}^b \boldsymbol{S} \boldsymbol{\Lambda}_J \boldsymbol{G}_J \boldsymbol{d}_J - \boldsymbol{B}_I^{bT} \boldsymbol{C}^b \boldsymbol{\kappa} div\boldsymbol{V} \right] d\Omega \\ &-\sum_{I=1}^{N_u} \iint_{\Omega_I} \bar{\boldsymbol{d}}_I^T \left[\boldsymbol{L}_I^T \boldsymbol{\Xi}_I^T \boldsymbol{C}^s \boldsymbol{\gamma} + \boldsymbol{B}_I^{sT} \boldsymbol{C}^s \boldsymbol{\Xi}_I \boldsymbol{L}_I \boldsymbol{d}_I - \boldsymbol{B}_I^{sT} \boldsymbol{C}^s \boldsymbol{\gamma} div\boldsymbol{V} \right] d\Omega. \end{aligned} \tag{13.82}$$

Although the expression in (13.82) is complicated, the appearance of each term is systematic, and those terms $\boldsymbol{C}^b$, $\boldsymbol{C}^s$, $\boldsymbol{B}^b$, $\boldsymbol{B}^s$, $\boldsymbol{\kappa}$, and $\boldsymbol{\gamma}$ are already available from finite element analysis of the plate problem. Thus, the terms $\boldsymbol{G}$, $\boldsymbol{L}$, $\boldsymbol{\Lambda}$, and $\boldsymbol{L}$ are additionally required for the design sensitivity purposes.

The variation of the load linear from in (6.196) is rewritten here as

$$\ell'_V(\bar{z}) = \iint_\Omega (\bar{z}^T \nabla \boldsymbol{f} \boldsymbol{V} + \bar{z}^T \boldsymbol{f} div\boldsymbol{V}) d\Omega, \tag{13.83}$$

where $\boldsymbol{f} = [f, m_1, m_2]^T$ is the distributed force and moment, $\nabla \boldsymbol{f}$ is the 3×2 matrix, and $\boldsymbol{V} = \boldsymbol{N}_V \boldsymbol{V}_I$ where

$$\boldsymbol{N}_V = \begin{bmatrix} N_1 & 0 & N_1 & 0 & N_1 & 0 & N_1 & 0 \\ 0 & N_2 & 0 & N_2 & 0 & N_2 & 0 & N_2 \end{bmatrix}. \tag{13.84}$$

Thus, the variation of the load linear form in (13.83) is approximated by

$$\ell'_V(\bar{z}) = \sum_{I=1}^{N_u} \iint_{\Omega_I} \bar{\boldsymbol{d}}_I^T (\boldsymbol{N}\nabla \boldsymbol{f}\, \boldsymbol{N}_V \boldsymbol{V}_I + \boldsymbol{N}\boldsymbol{f} div \boldsymbol{V})\, d\Omega. \tag{13.85}$$

Solid Design Component

The solid component is important in shape design sensitivity analysis because it is often used to represent the continuum domain whose shape is the design parameter. Four-node tetrahedron, six-node wedge, and eight-node hexahedron elements can be used in the linear approximation of the three-dimensional solid component. Since the geometry and the displacement are interpolated using the same shape function, these elements are called the isoparametric element. The eight-node hexahedron elements are considered in this section. For other types of elements, refer to the finite element literature [1], [72], and [73]. Instead of the tensor notation, the vector notation is convenient in the numerical implementation, which is used in the following derivation.

The structural fictitious load form for the solid component in (6.201) can be rewritten here

$$a'_V(z,\bar{z}) = \iiint_{\Omega} [\varepsilon^V(\bar{z})^T \sigma(z) + \sigma(\bar{z})^T \varepsilon^V(z) + \varepsilon(\bar{z})^T \sigma(z) div \boldsymbol{V}]\, d\Omega. \tag{13.86}$$

The strain vector is first approximated using the nodal displacement as

$$\varepsilon(z) = \boldsymbol{B}_I \boldsymbol{d}_I, \tag{13.87}$$

where $\boldsymbol{d}_I = [z_1^1,\ z_2^1,\ z_3^1,\ \ldots,\ z_1^8,\ z_2^8,\ z_3^8]_I^T$ is the displacement vector of element I, and

$$\boldsymbol{B}_I = \begin{bmatrix} N_{1,1} & 0 & 0 & N_{2,} & 0 & 0 & & N_{8,1} & 0 & 0 \\ 0 & N_{1,2} & 0 & 0 & N_{2,2} & 0 & & 0 & N_{8,2} & 0 \\ 0 & 0 & N_{1,3} & 0 & 0 & N_{2,3} & \cdots & 0 & 0 & N_{8,3} \\ N_{1,2} & N_{1,1} & 0 & N_{2,2} & N_{2,1} & 0 & & N_{8,2} & N_{8,1} & 0 \\ 0 & N_{1,3} & N_{1,2} & 0 & N_{2,3} & N_{2,2} & & 0 & N_{8,3} & N_{8,2} \\ N_{1,3} & 0 & N_{1,1} & N_{2,3} & 0 & N_{2,1} & & N_{8,3} & 0 & N_{8,1} \end{bmatrix}_I \tag{13.88}$$

is the discrete strain-displacement matrix. Since the shape function $\boldsymbol{N}(\xi)$ depends on the parametric coordinate, the spatial derivative in (13.88) can be obtained using the Jacobian relation in (3.176). The same matrix $\boldsymbol{B}_I$ can be used in the approximation of $\varepsilon(\dot{z})$ and $\varepsilon(\bar{z})$.

In order to discretized the design sensitivity equation, let the design velocity vector be given at eight nodes by $\boldsymbol{V}_I = [V_1^1,\ V_2^1,\ V_3^1,\ldots,V_1^8,\ V_2^8,\ V_3^8]_I^T$ at element I. In the shape design sensitivity formulation in Section 6.2.5, the material derivative of the strain tensor yields the implicitly dependent term $\varepsilon(\dot{z})$, which can be approximated using matrix $\boldsymbol{B}_I$, and explicitly dependent term $\varepsilon^V(z)$. This explicitly dependent strain on the design parameter in (6.200) can be approximated, using the vector notation, as

$$\begin{aligned}\varepsilon^V(z) &= -\frac{1}{2}\left(\frac{\partial z_i}{\partial x_k}\frac{\partial V_k}{\partial x_j}+\frac{\partial z_j}{\partial x_k}\frac{\partial V_k}{\partial x_i}\right)\\ &= -\boldsymbol{S}\boldsymbol{\Lambda}_I\boldsymbol{G}_I\boldsymbol{d}_I,\end{aligned} \tag{13.89}$$

where the constant matrix $\boldsymbol{S}$ maps the second-order tensor to the vector, defined as

$$\boldsymbol{S}=\begin{bmatrix}1&0&0&0&0&0&0&0&0\\0&0&0&1&0&0&0&0&0\\0&0&0&0&0&0&0&0&1\\0&1&0&1&0&0&0&0&0\\0&0&0&0&0&1&0&1&0\\0&0&1&0&0&0&1&0&0\end{bmatrix}, \tag{13.90}$$

and the 9 × 9 matrix $\boldsymbol{\Lambda}_I$ is defined as

$$\boldsymbol{\Lambda}_I=\begin{bmatrix}\boldsymbol{\Xi}^T_{3\times3}&\boldsymbol{0}_{3\times3}&\boldsymbol{0}_{3\times3}\\\boldsymbol{0}_{3\times3}&\boldsymbol{\Xi}^T_{3\times3}&\boldsymbol{0}_{3\times3}\\\boldsymbol{0}_{3\times3}&\boldsymbol{0}_{3\times3}&\boldsymbol{\Xi}^T_{3\times3}\end{bmatrix}_I, \tag{13.91}$$

where $\boldsymbol{\Xi}_{3\times3}$ is the Jacobian matrix of the nodal design velocity. Finally,

$$\boldsymbol{G}_I=\begin{bmatrix}N_{1,1}&0&0&N_{2,1}&0&0&&N_{8,1}&0&0\\N_{1,2}&0&0&N_{2,2}&0&0&&N_{8,2}&0&0\\N_{1,3}&0&0&N_{2,3}&0&0&&N_{8,3}&0&0\\0&N_{1,1}&0&0&N_{2,1}&0&&0&N_{8,1}&0\\0&N_{1,2}&0&0&N_{2,2}&0&\cdots&0&N_{8,2}&0\\0&N_{1,3}&0&0&N_{2,3}&0&&0&N_{8,3}&0\\0&0&N_{1,1}&0&0&N_{2,1}&&0&0&N_{8,1}\\0&0&N_{1,2}&0&0&N_{2,2}&&0&0&N_{8,2}\\0&0&N_{1,3}&0&0&N_{2,3}&&0&0&N_{8,3}\end{bmatrix}_I \tag{13.92}$$

is the second kind strain-displacement matrix. The matrix $\boldsymbol{B}_I$ in (13.88) is used when the symmetric strain relation is required, while the matrix $\boldsymbol{G}_I$ is used when nonsymmetric strain relation is required.

The fictitious load form in (13.86) also includes the divergence of the design velocity. As mentioned before, the design velocity is approximated using the same interpolation function with the displacement, and has the same regularity as the displacement. The divergence of the velocity vector can be approximated by

$$div\boldsymbol{V}=\sum_{i=1}^{3}V_{i,i}=\sum_{k=1}^{8}(N_{k,1}V_1^k+N_{k,2}V_2^k++N_{k,4}V_3^k). \tag{13.93}$$

Using the approximations in (13.88), (13.89) and (13.92), the structural fictitious load can be approximated by

$$a'_V(z,\bar{z})=-\sum_{I=1}^{N_u}\iiint_\Omega \bar{\boldsymbol{d}}_I^T\left[\boldsymbol{G}_I^T\boldsymbol{\Lambda}_I^T\boldsymbol{S}^T\boldsymbol{\sigma}_I+\boldsymbol{B}_I^T\boldsymbol{C}\boldsymbol{S}\boldsymbol{\Lambda}_I\boldsymbol{G}_I\boldsymbol{d}_I+\boldsymbol{B}_I^T\boldsymbol{\sigma}_I div\boldsymbol{V}\right]d\Omega. \tag{13.94}$$

By considering the body force only, the variation of the load linear from in (6.201) is rewritten here as

$$\ell'_V(\bar{z}) = \iint_{\Omega} (\bar{z}^T \nabla \boldsymbol{f} V + \bar{z}^T \boldsymbol{f} divV) d\Omega, \tag{13.95}$$

where $\boldsymbol{f} = [f_1, f_2, f_3]^T$ is the distributed force and moment, and $\nabla \boldsymbol{f}$ is the 3 × 3 Jacobian matrix. The variation of the load linear form is approximated by

$$\ell'_V(\bar{z}) = \sum_{I=1}^{N_u} \iint_{\Omega_I} \bar{\boldsymbol{d}}_I^T (\boldsymbol{N} \nabla \boldsymbol{f} \boldsymbol{N}^T \boldsymbol{V}_I + \boldsymbol{N} \boldsymbol{f} divV) d\Omega. \tag{13.96}$$

The approximations of the structural fictitious loads $a'_V(z, \bar{z})$ and $\ell'_V(\bar{z})$ are developed in this section for beam, plate, and solid components. The fictitious loads are explicitly written as linear functions of the design velocity. Although the expression of the fictitious loads for the shape design problem are more complicated than that of sizing design problem, a systematic matrix representation minimizes the implementation effort.

13.3 Design Velocity Field Computation

The design velocity field represents the direction and relative magnitude of shape design change. In the finite element approximation in Section 13.2, the design velocity in the structural domain is assumed to be available. In addition, the boundary representation method of shape design sensitivity in Section 6.2.3 requires the design velocity on the structural boundary. The velocity computation is composed of boundary velocity field computation and domain velocity field computation. The boundary velocity field is the movement of boundary nodes that result from a boundary shape change of the geometric model. The boundary velocity field is closely related to the shape design parameterization method presented in Chapter 12. To compute the boundary velocity field, the isoparametric mapping method is utilized in which linear dependency of the boundary velocity field on perturbations of shape design parameters is established.

A number of methods have been presented in the literature to compute the domain design velocity field. Four methods for computing domain design velocity fields are summarized, and their application to shape DSA and optimization is discussed. These methods are (1) finite difference method using CAD mesh generator [74] through [76], (2) isoparametric mapping method [77] through [80], (3) boundary displacement method [65], [67], and [81] and fictitious load method [82] through [84], and (4) a hybrid method that combines isoparametric and boundary displacement methods [85].

For the boundary displacement method, an intensive computation is required to solve the auxiliary elasticity problems, to obtain the domain velocity field. On the other hand, the isoparametric mapping method is far more efficient than the boundary displacement method since only a few matrix multiplications are required to compute both the boundary and the domain velocity fields at one time. However, this method is only applicable to the structure or the subdomain of a structure that has a simple geometric shape, so that the accurate isoparametric mapping function can be found. The domain velocity field of the boundary displacement method tends to generate more regular finite element mesh than that of the isoparametric mapping method since, for the former, the finite element matrix equation is used to generate the domain velocity field [65], [67], and [81]. Both methods maintain the linear dependency of the domain velocity field on the design boundary changes. It is found that the hybrid method is ideal for the design velocity field computation.

13.3.1 Boundary Velocity Field Computation

Computation of the boundary velocity field for the discretized finite element model is directly related to the shape design parameterization that is defined for the geometric model. As discussed in Chapter 12, the most primitive design parameters are defined on the geometric entities. In order to parameterize the geometric features, design parameter linking must be carried out across geometric entities. Consequently, the boundary velocity field must be computed for all the geometric entities on which the associated shape design parameters, either independent or dependent, are defined.

The structural boundary of two-dimensional component is a line or curve, whose parametric representation is provided in (5.114). From (6.3), the design velocity field is the variation of this parametric representation due to the shape design parameter. Thus, taking variation of (5.114) yields

$$V_{1\times 2} = [u^3 \quad u^2 \quad u \quad 1] \begin{bmatrix} \delta \boldsymbol{a}_3 \\ \delta \boldsymbol{a}_2 \\ \delta \boldsymbol{a}_1 \\ \delta \boldsymbol{a}_0 \end{bmatrix}_{4\times 2} = \boldsymbol{U}\delta \boldsymbol{A}, \qquad u \in [0,1], \tag{13.97}$$

where the matrix $\delta \boldsymbol{A}$ is the algebraic form of shape design perturbations of the geometric entity that represents the design boundary. During design perturbation, the parametric coordinate u is fixed. Each component of matrix $\delta \boldsymbol{A}$ can serve as a design parameter.

The structural boundary of three-dimensional component is a surface, whose parametric representation is provided in (5.128). By following the same procedure as (13.97), the design velocity field is obtained as

$$\begin{aligned} V_{1\times 3} &= [u^3 \quad u^2 \quad u \quad 1] \begin{bmatrix} \delta \boldsymbol{a}_{33} & \delta \boldsymbol{a}_{32} & \delta \boldsymbol{a}_{31} & \delta \boldsymbol{a}_{30} \\ \delta \boldsymbol{a}_{23} & \delta \boldsymbol{a}_{22} & \delta \boldsymbol{a}_{21} & \delta \boldsymbol{a}_{20} \\ \delta \boldsymbol{a}_{13} & \delta \boldsymbol{a}_{12} & \delta \boldsymbol{a}_{11} & \delta \boldsymbol{a}_{10} \\ \delta \boldsymbol{a}_{03} & \delta \boldsymbol{a}_{02} & \delta \boldsymbol{a}_{01} & \delta \boldsymbol{a}_{00} \end{bmatrix}_{4\times 4\times 3} \begin{bmatrix} w^3 \\ w^2 \\ w \\ 1 \end{bmatrix} \\ &= \boldsymbol{U}\delta \boldsymbol{A}\boldsymbol{W}^T, \qquad (u,w) \in [0,1]\times[0,1]. \end{aligned} \tag{13.98}$$

Vectors $\boldsymbol{U}$ and $\boldsymbol{W}$ are locations of the nodes in the parametric directions of the geometric entity of the design boundary.

The perturbed design parameter matrix $\delta \boldsymbol{G}$ is defined as the variation of the design parameter matrix $\boldsymbol{G}$ of the geometric entity, as described in Chapter 12. Entries in matrix $\delta \boldsymbol{G}$ represent the perturbations of independent and dependent design parameters defined on the geometric entity in the shape design parameterization process. For the basic curves and surfaces, the matrix $\delta \boldsymbol{G}$ can be transformed into geometric format $\delta \boldsymbol{B}$, following the matrix transformations described in Figs. 12.13(b) and 12.26(b). The linear dependence of the boundary velocity field on design parameter perturbations is naturally satisfied since both the isoparametric mapping and matrix transformations are linear operations.

Computational Procedures

The procedure for computing boundary velocity fields for each independent design parameter is as follows:

1. Identify the geometric entities of the design boundary on which the independent and dependent shape design parameters are defined. Information that defines the mapping between the design parameters and the boundary geometric entities is generated in the design parameterization process.

2. For each basic curve or surface entity identified in step (1), form the perturbed design parameter matrix $\delta\boldsymbol{G}$ by perturbing the independent and dependent design parameters defined in its design parameter matrix $\boldsymbol{G}$ by a unit magnitude and a proportionality value, respectively.

3. For each specialized curve or surface identified in step (1), construct the perturbed geometric coefficient matrix $\delta\boldsymbol{B}$ from the perturbed design parameter matrix $\delta\boldsymbol{G}$ of the boundary geometric entity.

4. Transform the perturbed matrix $\delta\boldsymbol{B}$ or $\delta\boldsymbol{G}$ to the perturbed algebraic coefficient matrix $\delta\boldsymbol{A}$. The transformations are described in Figs. 12.13 and 12.26 for line and surface entities, respectively.

5. For each boundary geometric entity obtained in step (1), identify the geometric entity that represents the subdomain of the geometric model to which the boundary geometric entity is connected.

6. For each domain geometric entity identified in step (5), identify the node identifications and their locations in the parametric directions of the boundary geometric entity.

7. Compute the boundary velocity field using (13.97) and (13.98).

A two-dimensional model of two surfaces, where the design boundary is parameterized as two Bezier curves shown in Fig. 13.9, is utilized to explain the boundary velocity field computational procedure. The y-direction movement of the control point $\boldsymbol{p}_1$ of Bezier curve 1 is defined as the independent design parameter b_1. The y-direction movement of the control point $\boldsymbol{q}_1$ of Bezier curve 2 depends on b_1, with a factor of -1.0. Therefore, when the design parameter b_1 is perturbed by a unit magnitude in the positive y-direction, the dependent design parameter b_2 must be perturbed by a magnitude -1.0, as shown in Fig. 13.9.

For the parameterized boundary that is obtained by using two Bezier curves, the perturbed design parameter matrices $\delta\boldsymbol{G}_1$ and $\delta\boldsymbol{G}_2$ for the curves can be written as

$$\delta\boldsymbol{G}_1 = \begin{bmatrix} \delta\boldsymbol{p}_0 \\ \delta\boldsymbol{p}_1 \\ \delta\boldsymbol{p}_2 \\ \delta\boldsymbol{p}_3 \end{bmatrix}_{4\times 2} = \begin{bmatrix} 0 & 0 \\ 0 & 1 \\ 0 & 0 \\ 0 & 0 \end{bmatrix}_{4\times 2} \tag{13.99}$$

$$\delta\boldsymbol{G}_2 = \begin{bmatrix} \delta\boldsymbol{q}_0 \\ \delta\boldsymbol{q}_1 \\ \delta\boldsymbol{q}_2 \\ \delta\boldsymbol{q}_3 \end{bmatrix}_{4\times 2} = \begin{bmatrix} 0 & 0 \\ 0 & -1 \\ 0 & 0 \\ 0 & 0 \end{bmatrix}_{4\times 2}. \tag{13.100}$$

The perturbed design parameter matrices can be transformed to the perturbed algebraic coefficient matrices $\delta\boldsymbol{A}_1$ and $\delta\boldsymbol{A}_2$ by

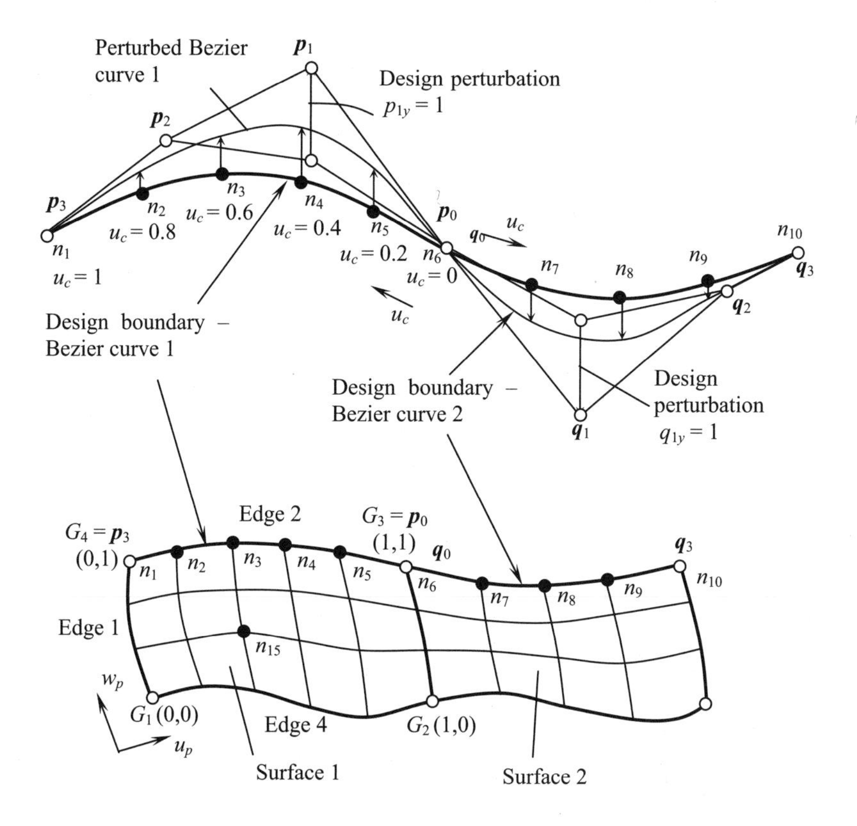

Figure 13.9. Two-dimensional model with design boundary parameterized as bezier curves.

$$\delta A_1 = Q\delta G_1 = \begin{bmatrix} -1 & 3 & -3 & 1 \\ 3 & -6 & 3 & 0 \\ -3 & 3 & 0 & 0 \\ 1 & 0 & 0 & 0 \end{bmatrix}_{4\times4} \begin{bmatrix} 0 & 0 \\ 0 & 1 \\ 0 & 0 \\ 0 & 0 \end{bmatrix}_{4\times2} = \begin{bmatrix} 0 & 3 \\ 0 & -6 \\ 0 & 3 \\ 0 & 0 \end{bmatrix}_{4\times2} \tag{13.101}$$

$$\delta A_2 = Q\delta G_2 = \begin{bmatrix} -1 & 3 & -3 & 1 \\ 3 & -6 & 3 & 0 \\ -3 & 3 & 0 & 0 \\ 1 & 0 & 0 & 0 \end{bmatrix}_{4\times4} \begin{bmatrix} 0 & 0 \\ 0 & -1 \\ 0 & 0 \\ 0 & 0 \end{bmatrix}_{4\times2} = \begin{bmatrix} 0 & -3 \\ 0 & 6 \\ 0 & -3 \\ 0 & 0 \end{bmatrix}_{4\times2}, \tag{13.102}$$

where $\boldsymbol{Q}$ is the transformation matrix from matrices $\boldsymbol{G}$ to $\boldsymbol{A}$.

With a 3 × 5 uniform mesh defined on surface 1, as shown in Fig. 13.9, locations of the boundary nodes n_1 to n_6 in the curve parametric direction are obtained as 1.0, 0.8, 0.6, 0.4, 0.2, and 0.0. Moreover, with a uniform mesh defined on Surface 2, nodes n_6 to n_{10} on Bezier curve 2 have u_c = 0.0, 0.25, 0.5, 0.75, and 1.0, respectively. Then, the boundary velocity field can be computed, for example, at nodes n_3 and n_8, where u_c = 0.6 and 0.5 in Bezier curves 1 and 2, respectively, as

$$\begin{aligned} \boldsymbol{V}_{n_3} &= \boldsymbol{U}\delta\boldsymbol{A}_1 \\ &= [0.6^3 \quad 0.6^2 \quad 0.6 \quad 1]\begin{bmatrix} 0 & 3 \\ 0 & -6 \\ 0 & 3 \\ 0 & 0 \end{bmatrix} \\ &= [0 \quad 0.288] \end{aligned} \tag{13.103}$$

$$\begin{aligned} \boldsymbol{V}_{n_3} &= \boldsymbol{U}\delta\boldsymbol{A}_2 \\ &= [0.5^3 \quad 0.5^2 \quad 0.5 \quad 1]\begin{bmatrix} 0 & -3 \\ 0 & 6 \\ 0 & -3 \\ 0 & 0 \end{bmatrix} \\ &= [0 \quad -0.375]. \end{aligned} \tag{13.104}$$

The above procedures must be performed for each boundary geometric entity with an independent design parameter and for all other boundary geometric entities with the associated dependent design parameters. As a result, perturbation of the design parameters of the parameterized geometric feature, which is created by linking free design parameters defined on the geometric entities that form the geometric feature, will propagate appropriately to the boundary of the geometric feature.

For the boundary velocity field computation, the geometric modeler must provide two types of data: (1) the transformation matrix from the perturbed design parameter matrix $\delta\boldsymbol{G}$ to the corresponding perturbed algebraic coefficient matrix $\delta\boldsymbol{A}$ of the boundary geometric entities and (2) node identifications and their locations in the parametric directions of the boundary geometric entities.

13.3.2 Domain Velocity Field Computation Using Finite Difference Method

The finite difference method generates a design velocity field by subtracting the nodal coordinates of the original mesh from the new mesh that is associated with the perturbed boundary. This method depends on a mesh generator to create a finite element mesh for each design variation. Velocity computation and mesh generation methods used by Godse et al. [74], Yang et al. [75], and Kodiyalam et al. [76] are discussed.

Velocity Generation Using P/CONCEPT Mesh Generator

Godse et al. [74] used the finite difference method with the automatic mesh generator in P/CONCEPT [42] to compute the design velocity field. Two problems were reported in their study: (1) change of mesh topology and (2) loss of linear dependency of the design velocity field. Using a simple cantilever beam example, they showed that it is difficult to retain the finite element topology even for a model with very simple geometry. In addition, linear dependency cannot be maintained since the automatic mesh generator does not produce interior nodes using a mathematical rule that preserves it.

Even though the automatic mesh generator provides the design velocity field efficiently and can be naturally linked to a CAD modeler, it cannot be used since linear dependency of the design velocity field cannot be assured for the design model of the next design iteration. On the other hand, if one imposes linear dependency of the design velocity field to the variation of design parameters by moving nodal points in the direction of the design velocity field obtained from the finite difference method with a proportional step size, the boundary nodes may not lie on the geometric boundary of the design model of the next design iteration. Moreover, the order of the interior nodes may change to yield a different mesh topology.

Boundary Shape Function Approach

Yang et al. [75] proposed the boundary shape function approach. In this approach, shape design parameters are defined at the boundary of the design model through predefined shape functions that define the design boundary. Using the same shape function, the boundary nodal point movements of the analysis model can be obtained for a perturbed design. Interior nodal point locations are determined by the Laplace smoothing technique, which is applied several times to adjust interior nodal point locations and avoid mesh distortion. The domain design velocity field is then obtained using the finite difference method. When the design optimization is performed to arrive at the next design iteration, the nodal point coordinates are updated to reflect the new design parameters, and a Laplace smoothing operation is carried out on all interior nodes to minimize element distortion.

This approach cannot maintain linear dependency of the domain velocity field on the variation of shape design parameters because a mathematical rule does not exist to relate the design variation to the domain design velocity field. Moreover, applying the Laplace smoothing technique for finite element mesh generation is inefficient [86].

Velocity Generation Using TAUGS Mesh Generator

Kodiyalam et al. [76] presented shape design optimization using the TAUGS mesh generator, Laplace smoothing technique, and overall finite difference method to obtain design sensitivity coefficients. Their finite element mesh update method is like the boundary shape function approach but uses TAUGS [76] to locate the boundary nodes. Botkin [87] presented a similar approach. Since the overall finite difference method is used for computing design sensitivity coefficients, design velocity field need not be computed explicitly.

An initial tetrahedral finite element mesh is generated using the OCTREE mesh generator in TAUGS, and then the Laplace smoothing technique is used to reposition nodes to get better mesh. To compute design sensitivity coefficients, the boundary nodes are repositioned to the perturbed boundary using the parametric location of nodes in the geometric boundary. Once all boundary nodes are repositioned, the Laplace smoothing technique is employed to reposition the interior nodes. After a new design is obtained, geometry of the model is updated using TAUGS, and the OCTREE/Laplace mesh generator is employed to create a new mesh. Another design iteration is then carried out. In this operation, the design velocity field is not utilized to update the finite element mesh at the new design. Moreover, the new finite element mesh generated at each design iteration might have different finite element topology.

However, there are three drawbacks in this approach. First, the finite difference method must be used with uncertain step sizes. Second, if movements of nodal points do not depend linearly on the variation of shape design parameters, in addition to nonlinear dependency of structural responses, the overall dependency of performance measures may become highly nonlinear for the finite difference method. For example, for displacement performance measures defined on nodal points 5 and 6 of the two-dimensional beam shown in Fig. 13.10, if the overall finite difference method is used to evaluate sensitivity by perturbing δb_1 in the x_1-direction, the displacements would be evaluated at new nodal points 5' and 6' using FEA. However, if the design change is $k\delta b_1$, then the sensitivity obtained from the overall finite difference will predict displacement at nodal points 5b and 6b instead of 5a and 6a which are the actual nodal points of the new design. Nonlinear dependence of the design velocity field on the variation of shape

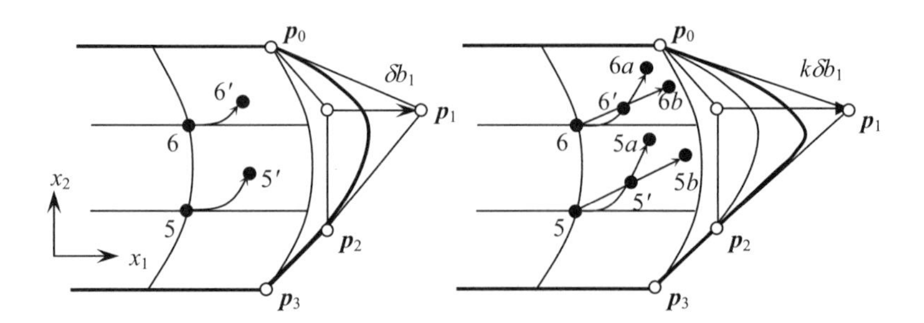

Figure 13.10. Nonlinear dependency of movements of nodal points 5 and 6 on variation of shape design parameter b_1.

design parameters is shown in Fig. 13.10. Third, the new finite element mesh that is generated at each design iteration may have different mesh topology, making it difficult to keep track of performance measure definition in the optimization process.

Discussion of the Finite Difference Method

In this method, if a mesh generator is used to create the mesh for new designs, linear dependency of the design velocity field cannot be maintained. On the other hand, maintaining the linear dependency of the design velocity field, assuming that the nodal point moves in the direction of the design velocity field obtained using a mesh generator with a proportional step size, means that the boundary nodes may not lie on the geometric boundary and the order of the interior nodes may change, yielding a different mesh topology.

Practically, this method is ideal since it connects to the shape design optimization to the CAD modeler, is easy to use and efficient, if the smoothing technique is not used. However, the automatic mesh generator is restricted to tetrahedral and triangular elements, making the method too limited to choosing various finite elements. In addition, the method may cause the finite element and geometric models to be inconsistent at the design boundary if one requires the design velocity field to be linearly dependent on the variation of the design parameters. Moreover, it is difficult to keep track of a location-dependent performance measure, such as displacement at a nodal point or stress at an element, in the design iterations since finite element mesh is not updated using any mathematical rule.

13.3.3 Boundary Displacement Method

The boundary displacement and fictitious load methods compute the design velocity field by solving an auxiliary structure with either specified displacements or specified loads. In both methods, the design velocity field is determined by solving a matrix equation that is obtained by discretizing the governing structural partial differential equation.

Boundary Displacement Method

Perturbation of shape design parameters will result in movement of the design boundary, as discussed in Section 13.3.2. The boundary displacement method proposed by Yao and Choi [65], [67], and [81] generates the domain design velocity field that corresponds to the design perturbation by solving an auxiliary elasticity problem with prescribed displacements on the design boundary, as shown in Fig. 13.11. Movements of nodal points at the design boundary are treated as the prescribed displacement field.

In order to compute the domain design velocity field, the auxiliary problem requires the boundary design velocity field resulting from the shape change and the displacement constraints on the rest boundary. The discretized equilibrium equation of the auxiliary problem can be written as

$$[\boldsymbol{K}]\{\boldsymbol{V}\} = \{\boldsymbol{f}\}, \tag{13.105}$$

where $[\boldsymbol{K}]$ is the reduced stiffness matrix of the auxiliary structure, which is different from that of the original structure; $\{\boldsymbol{V}\}$ is the design velocity vector; and $\{\boldsymbol{f}\}$ is the unknown vector of boundary force that produces the prescribed variation (deformation) of the design boundary. In the partitioned form, (13.105) can be written as

$$\begin{bmatrix} \boldsymbol{K}_{bb} & \boldsymbol{K}_{bd} \\ \boldsymbol{K}_{db} & \boldsymbol{K}_{dd} \end{bmatrix} \begin{Bmatrix} \boldsymbol{V}_b \\ \boldsymbol{V}_d \end{Bmatrix} = \begin{Bmatrix} \boldsymbol{f}_b \\ \boldsymbol{0} \end{Bmatrix}, \tag{13.106}$$

where $\{\boldsymbol{V}_b\}$ is the given movements of nodes on the boundary, $\{\boldsymbol{V}_d\}$ is the nodal velocity vector in the interior domain, and $\{\boldsymbol{f}_b\}$ is the fictitious boundary force that acts on the varying boundary. Equation (13.106) can be rearranged as

$$[\boldsymbol{K}_{dd}]\{\boldsymbol{V}_d\} = -[\boldsymbol{K}_{db}]\{\boldsymbol{V}_b\} \tag{13.107}$$

to define a linear relation between the boundary and domain velocity fields. Thus, the domain design velocity field satisfies the partial differential equation that describes the elasticity of the structure.

The boundary displacement method can be used to compute the domain design velocity field using the prescribed boundary design velocity obtained from various boundary representation methods. As a result, boundary and domain design velocity fields can be computed separately. The same finite element analysis code used for analysis can be used to compute the velocity (displacement) field of the auxiliary model. The solution, that is, the domain design velocity field, naturally satisfies the regularity and linear dependency requirements.

An important characteristic of the elasticity problem, an elliptic partial differential equation, is that the solution trajectory tends to maintain orthogonal property [65] through [67], [88], and [89]. As a result, the updated mesh that is obtained from the velocity field because of design change tends to be orthogonal.

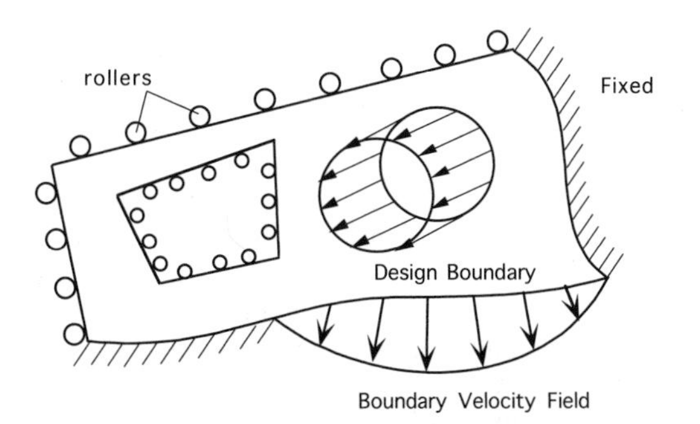

Figure 13.11. Auxiliary model for domain velocity field.

Fictitious Load Method

Rajan and Belegundu [82] through [84] presented a fictitious load method to generate the design velocity field. In this approach, fictitious loads, applied at certain nodes of the design boundary of an auxiliary structure, are defined as shape design parameters. The deformation (design velocity field) produced by these loads is used to update the shape. The design velocity field can be obtained from solving the finite element equation

$$[\boldsymbol{K}]\{\boldsymbol{d}^i\} = \{\boldsymbol{f}^i\}, \tag{13.108}$$

where $[\boldsymbol{K}]$ is the stiffness matrix, $\{\boldsymbol{f}^i\}$ is a unit magnitude fictitious load that is applied at the control point in the direction of design variation, and $\{\boldsymbol{d}^i\}$ is the nodal displacement, which is used as the domain design velocity field for computing design derivatives. Note that, similar to the boundary displacement method, the stiffness matrix in (13.108) is different from that of the original structure.

One restriction in using this method is that the design parameterization must be embedded in the finite element model instead of the geometric model. This means that the finite element model must be optimized, resulting in a design with nonsmooth boundary [54]. Zhang and Belegundu [84] varied the fictitious load method to ensure a smooth design boundary by adding beam and plate elements on the curve and surface design boundaries, respectively. Such a remedy increases complexity of the finite element model of the auxiliary structure and makes it difficult to link shape design optimization to a CAD modeler.

Computational Procedures

The procedure of using the boundary displacement method to compute the domain velocity field is as follows:

1. Identify all the boundary curves or surfaces of the geometric model from the topology information.

2. Identify characteristics of the boundary curves or surfaces for defining displacement constraint equations.

3. Define appropriate constraint equations for movements of the boundary nodes, to keep the velocity field orthogonal by allowing the nodes to move along the boundary.

4. Generate a finite element input data file, using the boundary velocity field and the constraint equations defined in step 3 as the specified boundary displacement field.

5. Perform finite element analysis to compute the domain velocity field.

One of the drawbacks of the boundary displacement method or fictitious load method is that a finite element matrix equation must be solved to generate the velocity field for each associated shape design parameter. The reduced stiffness matrix of the auxiliary finite element model must be formed and decomposed for each design parameter. Furthermore, a process is required to identify straight boundaries for defining appropriate constraints. Nevertheless, to date, this method could be the only one that satisfies the theoretical requirements and is applicable for general shape design problems.

Discussion of Boundary Displacement and Fictitious Load Methods

Both methods solve matrix equations, discretized from the elliptic partial differential equation, to compute the design velocity field. Therefore, interior node movements are determined by the partial differential equation. Since the solution trajectory of the elliptic partial differential equation maintains orthogonal property [65], [88], and [89], finite element mesh updated using the design velocity field tends to be orthogonal and does not distort in the optimization process. For large applications, however, both methods are inefficient since they require solutions of auxiliary finite element models.

Design velocity field generated using these methods depends on design parameters $\boldsymbol{b}$ since the material point location $\boldsymbol{x}$ (function of $\boldsymbol{b}$) determine the $[\boldsymbol{K}]$ matrices of the auxiliary structures, which implies that the design velocity field must be updated at each design iteration. However, using the boundary displacement method, if the boundary design velocity is computed using the isoparametric mapping method, which produces a boundary design velocity independent of design parameters $\boldsymbol{b}$, the overall design velocity field can be used for several design iterations since the boundary velocity field is constant and domain velocity produces regular finite element meshes. However, it is not true for the fictitious load method since the loads, instead of displacements that describe the structural boundary shape, are treated as design parameters, in which boundary velocity field varies due to fictitious loads applied at varied boundary [84] and [90].

For practical applications, the boundary displacement method seems preferable in terms of design parameterization, link to CAD, and generality. Using the boundary displacement method, CAD boundary representation methods, such as Bezier curves/surfaces, are utilized to characterize the design boundary. As a result, CAD parameters can be employed to define the design model, and, therefore, a CAD geometric model is optimized.

On the other hand, the fictitious load method requires that the finite element model be optimized. Beam or plate elements must be introduced to ensure a smooth design boundary, thereby increasing difficulties in tying the method to CAD modelers. In addition, Hou et al. [91] and Rajan [92] pointed out that selection of proper fictitious loads is not easy, and that the fictitious load method is not convenient when applied to problems in which the unknown boundary varies according to geometric constraints, such as circles. Moreover, for design parameters that do not stay on the boundary, such as control points of a Bezier curve, the fictitious load approach is not applicable since the fictitious load must be applied to the finite element nodes or element boundary.

13.3.4 Velocity Field Computation Using Isoparametric Mapping Method

The isoparametric mapping method generates the mesh and computes the design velocity field based on spatial decomposition. That is, the structure to be meshed must be manually broken up into a coarse mesh of hexahedrons. A mesh of smaller elements is then created by specifying subdivisions in the parametric space of the coarse hexahedron [93]. Two isoparametric mapping approaches are discussed in this section: design element [77] through [80] and cubic isoparametric mapping [85].

Design Element Approach

The design element approach was applied for two- and three-dimensional elasticity problems in literature [77] through [80]. In this approach, the geometry of the structure is described by design elements whose key dimensions are associated with shape design parameters. The finite element mesh for analysis is then generated within each design element by an isoparametric mapping technique,

$$\boldsymbol{x}_{3\times1} = \begin{bmatrix} \boldsymbol{N} & \boldsymbol{0} & \boldsymbol{0} \\ \boldsymbol{0} & \boldsymbol{N} & \boldsymbol{0} \\ \boldsymbol{0} & \boldsymbol{0} & \boldsymbol{N} \end{bmatrix}_{3\times3m} \{\boldsymbol{X}\}_{3m\times1}, \tag{13.109}$$

where $\boldsymbol{x} = [x_1, x_2, x_3]^T$ are the coordinates of an interior node; $\{\boldsymbol{X}\}_{3m\times1}$ are coordinates of the corner points of the design elements; $\boldsymbol{N}(u, w, v)_{1\times m}$ is the vector of mapping functions; m is the number of corner points of the design element; and u, w, and v are the parametric locations of the finite element node in the design element. The domain design velocity field can be obtained using the same mapping functions. For example, for the kth shape design parameter, the domain design velocity for a unit design variation $\delta b_k = 1$ and $\delta b_j = 0$ $(j \neq k)$ can be evaluated as

$$\boldsymbol{V}^k(\boldsymbol{x}) = \frac{d}{d\tau}\boldsymbol{x}(\boldsymbol{b} + \tau\delta\boldsymbol{b})\bigg| = \left\{\frac{\partial \boldsymbol{x}}{\partial b_k}\right\}\delta b_k = \begin{bmatrix} \boldsymbol{N} & \boldsymbol{0} & \boldsymbol{0} \\ \boldsymbol{0} & \boldsymbol{N} & \boldsymbol{0} \\ \boldsymbol{0} & \boldsymbol{0} & \boldsymbol{N} \end{bmatrix}\left\{\frac{\partial \boldsymbol{X}}{\partial b_k}\right\}\delta b_k. \tag{13.110}$$

This method yields the C^0-continuous design velocity field with integrable first derivatives. Linear dependency of the design velocity field is also satisfied since $\partial \boldsymbol{X}/\partial b_k$ is a constant vector. However, when the geometric shape of the structure is complicated, it is very difficult to create an appropriate mapping function $\boldsymbol{N}$ that accurately represents the existing decomposed physical domain.

Cubic Isoparametric Mapping Approach

The difficulty of finding an appropriate mapping function $\boldsymbol{N}$ can be alleviated by connecting the mapping method to a geometric modeler that provides a standard mapping function [42]. In addition, one can utilize various options to parameterize the structural geometric model [85].

The geometric representation of a surface and corresponding design velocity field is provided in Section 13.3.2 for the boundary velocity computation. When the geometric boundary is a curve, this parametric surface can serve as a domain of the structure. In a three-dimensional domain, a solid component must be used, whose mathematical representation is given by

$$\begin{aligned} \boldsymbol{x}(u,w,v) &= \sum_{i,j,k=0}^{3} \boldsymbol{a}_{ijk}u^i w^j v^k \\ &= \sum_{i,j,k=1}^{4} \boldsymbol{a}_{(4-i)(4-j)(4-k)}u^{4-i}w^{4-j}v^{4-k} \\ &= \sum_{k=1}^{4} v^{4-k}\boldsymbol{U}\boldsymbol{A}_k\boldsymbol{W}^T, \quad \langle u,w,v\rangle \in [0,1]\times[0,1]\times[0,1], \end{aligned} \tag{13.111}$$

where $\boldsymbol{U} = [u^3, u^2, u, 1]$, $\boldsymbol{W} = [w^3, w^2, w, 1]$, u, w, and v are the parametric locations of the point $\boldsymbol{x}$ in the solid, and matrices $\boldsymbol{A}_k = [\boldsymbol{a}_{ijk}]_{4\times4}$ are the algebraic coefficient matrices of the solid. The design velocity field, therefore, can be obtained from variations of the above equations with unit design variation $\delta b = 1$, that is,

$$\begin{aligned} \boldsymbol{V} &= \delta\boldsymbol{x}(u,w,v) \\ &= \sum_{k=1}^{4} v^{4-k}\boldsymbol{U}\delta\boldsymbol{A}_k\boldsymbol{W}^T, \end{aligned} \tag{13.112}$$

where $\delta\boldsymbol{A}_k = [\delta\boldsymbol{a}_{ijk}]_{4\times4}$ are the perturbed algebraic coefficient matrices of the solid. Thus, the domain design velocity field is generated following an algebraic equation.

Velocity fields that are computed using the isoparametric mapping method satisfy both regularity and linearity requirements. The isoparametric mapping method is far more efficient than the boundary displacement method since the former needs only a few matrix multiplications. However, in the isoparametric method, accuracy of the velocity field depends on the function $\boldsymbol{N}$ that maps nodal points from parametric coordinates of the geometric model to global coordinates. Finding an appropriate mapping function $\boldsymbol{N}$ for accurate velocity field computation is difficult when the geometrical shape of the structure is complicated. However, for a simple geometric model or a small subdomain of the structure, for which an accurate mapping function $\boldsymbol{N}$ can be found, the isoparametric mapping method is attractive. For such a model or subdomain that can be modeled by a single geometric entity, velocity fields can be computed using the isoparametric mapping method.

Computational Procedures

To compute the perturbed algebraic coefficient matrix of the geometric entity that represents the domain or subdomain of the geometric model, the following procedures must be performed for each independent design parameter:

1. Identify the geometric entities of the design boundary in which the independent design parameters and dependent shape design parameters are defined.

2. For each boundary geometric entity identified in step 1, identify the geometric entity that represents the subdomain that is connected to the boundary geometric entity.

3. Form the perturbed design matrix $\delta\boldsymbol{G}$ or perturbed geometric coefficient matrix $\delta\boldsymbol{B}$ for a geometric entity of the design boundary, similar to steps 2 and 3 in Section 13.3.2. The orientations of the parametric directions of the boundary geometric entity and the connected domain geometric entity must be checked to determine if they coincide, so that a consistent design perturbation from boundary to domain geometric entities can be obtained. Accordingly, the perturbed design parameter matrix $\delta\boldsymbol{G}$ must be formed.

4. Translate $\delta\boldsymbol{G}$ into $\delta\boldsymbol{B}$ for the basic curves or surfaces, using the transformation matrices described in Figs. 12.13(b) and 12.26(b).

5. Read the edge or face number of the boundary geometric entity on the domain geometric entity.

6. Form the perturbed geometric coefficient matrix $\delta\boldsymbol{B}$ for the surface or solid, based on the edge or face number obtained in step 5 and the convention of entries defined in the geometric coefficient matrix $\boldsymbol{B}$.

7. Transform $\delta\boldsymbol{B}$ to $\delta\boldsymbol{A}$ for two- and three-dimensional structures by [68]

$$\delta\boldsymbol{A} = \boldsymbol{M}\delta\boldsymbol{B}\boldsymbol{M}^T \tag{13.113}$$

and

$$\delta\boldsymbol{A}_k = M_{ki}\boldsymbol{M}\delta\boldsymbol{B}_i\boldsymbol{M}^T, \tag{13.114}$$

where $\boldsymbol{A}_k = [\boldsymbol{a}_{ijk}]$, M_{ki} is a single specific element of the matrix $\boldsymbol{M}$, and $\boldsymbol{B}_1$, $\boldsymbol{B}_2$, $\boldsymbol{B}_3$, and $\boldsymbol{B}_4$ are the geometric coefficient matrices.

8. Compute velocity field using (13.113).

The parameterized model shown in Fig. 13.9 is again utilized to illustrate the procedure. For Bezier curve 1, the perturbed design parameter matrix $\delta\boldsymbol{G}$ is given as

$$\delta\boldsymbol{G} = \begin{bmatrix} 0 & 0 \\ 0 & 0 \\ 0 & 1 \\ 0 & 0 \end{bmatrix}_{4\times 2} . \tag{13.115}$$

Note that (13.115) is not the same as (13.100) since the parametric directions u_c and u_p of the boundary and domain geometric entities, respectively, are reversed. Here, the perturbed geometric coefficient matrix can be obtained as

$$\begin{aligned} \delta\boldsymbol{B} &= \boldsymbol{M}^{-1}\boldsymbol{Q}\delta\boldsymbol{G} \\ &= \begin{bmatrix} 0 & 0 & 0 & 1 \\ 1 & 1 & 1 & 1 \\ 0 & 0 & 1 & 0 \\ 3 & 2 & 1 & 0 \end{bmatrix} \begin{bmatrix} -1 & 3 & -3 & 1 \\ 3 & -6 & 3 & 0 \\ -3 & 3 & 0 & 0 \\ 1 & 0 & 0 & 0 \end{bmatrix} \begin{bmatrix} 0 & 0 \\ 0 & 0 \\ 0 & 1 \\ 0 & 0 \end{bmatrix} \\ &= \begin{bmatrix} 0 & 0 \\ 0 & 0 \\ 0 & 0 \\ 0 & -3 \end{bmatrix}, \end{aligned} \tag{13.116}$$

where $\boldsymbol{Q}$ and $\boldsymbol{M}$ are transformation matrices.

According to the $\boldsymbol{B}$ matrix convention, the perturbed geometric coefficient matrix $\delta\boldsymbol{B}$ of Surface 1, resulting from perturbation of its second edge, that is, Bezier curve 1, can be formed as

$$\delta\boldsymbol{B}_x = \boldsymbol{0} \tag{13.117}$$

and

$$\delta\boldsymbol{B}_y = \begin{bmatrix} 0 & 0 & 0 & 0 \\ 0 & 0 & 0 & 0 \\ 0 & 0 & 0 & 0 \\ 0 & -3 & 0 & 0 \end{bmatrix} . \tag{13.118}$$

Then, $\delta\boldsymbol{A}$ is transformed from $\delta\boldsymbol{B}$ as

$$\delta\boldsymbol{A}_x = \boldsymbol{M}\delta\boldsymbol{B}_x\boldsymbol{M}^T = \boldsymbol{0} \tag{13.119}$$

and

$$\delta \boldsymbol{A}_y = \boldsymbol{M} \delta \boldsymbol{B}_y \boldsymbol{M}^T$$

$$= \begin{bmatrix} 2 & -2 & 1 & 1 \\ -3 & 3 & -2 & -1 \\ 0 & 0 & 1 & 0 \\ 1 & 0 & 0 & 0 \end{bmatrix} \begin{bmatrix} 0 & 0 & 0 & 0 \\ 0 & 0 & 0 & 0 \\ 0 & 0 & 0 & 0 \\ 0 & -3 & 0 & 0 \end{bmatrix} \begin{bmatrix} 2 & -3 & 0 & 1 \\ -2 & 3 & 0 & 0 \\ 1 & -2 & 1 & 0 \\ 1 & -1 & 0 & 0 \end{bmatrix} \tag{13.120}$$

$$= \begin{bmatrix} 6 & -9 & 0 & 0 \\ -6 & 9 & 0 & 0 \\ 0 & 0 & 0 & 0 \\ 0 & 0 & 0 & 0 \end{bmatrix}.$$

Equation (13.112) can be used to compute nodal point velocity, for example, for node n_{15} in surface 1, as shown in Fig. 13.9, $u_p = (2/5, 1/3)$, and the velocity of the node is

$$V_x = 0 \tag{13.121}$$

and

$$V_y = \begin{bmatrix} 0.4^3 & 0.4^2 & 0.4 & 1 \end{bmatrix} \begin{bmatrix} 6 & -9 & 0 & 0 \\ -6 & 9 & 0 & 0 \\ 0 & 0 & 0 & 0 \\ 0 & 0 & 0 & 0 \end{bmatrix} \begin{bmatrix} (\frac{1}{3})^3 \\ (\frac{1}{3})^2 \\ (\frac{1}{3}) \\ 1 \end{bmatrix} \tag{13.122}$$

$$= 0.0746.$$

Also, for the boundary node n_3 at edge 2 of surface 1, where $(u_p, w_p) = (2/5,1)$, the velocity field can be obtained as

$$V_x = 0 \tag{13.123}$$

and

$$V_y = \begin{bmatrix} 0.4^3 & 0.4^2 & 0.4 & 1 \end{bmatrix} \begin{bmatrix} 6 & -9 & 0 & 0 \\ -6 & 9 & 0 & 0 \\ 0 & 0 & 0 & 0 \\ 0 & 0 & 0 & 0 \end{bmatrix} \begin{bmatrix} 1 \\ 1 \\ 1 \\ 1 \end{bmatrix} \tag{13.124}$$

$$= 0.288,$$

which is the same as the one obtained in (13.103).

Discussion of Isoparametric Mapping Method

Using the isoparametric mapping method, both regularity and linear dependency requirements of the design velocity field, described in Section 6.2.7, are satisfied for two- and three-dimensional elastic solids. An algebraic equation, such as (13.110) or (13.112), specifies a mathematical rule to determine the interior node movements. Note that the design velocity field generated using the mapping method does not depend on shape design parameters since the parametric locations of the finite element nodes are prespecified. Hence, one design velocity field can be used for all design optimization iterations.

Practically, this method can be connected to a geometric modeler and is very efficient since it needs only matrix multiplications. Moreover, finite element mesh generated using the mapping method is not restricted to tetrahedral or triangular elements, but is able to support a broad range of elements. In addition, the isoparametric mapping guarantees that the boundary nodes lie on the geometric boundary. Since an algebraic equation specifies interior node movements, the method does not change mesh topology.

The biggest obstacle in using the mapping method is the need for spatial decomposition of the structural domain. For a structure with complicated geometry, spatial decomposition is tedious and difficult. In addition, using the mapping method, regular finite element mesh is difficult to ensure, as shown in the two-dimensional fillet in the following.

Two-Dimensional Fillet

A two-dimensional fillet [65], shown in Fig. 13.12, is employed here to illustrate that finite element meshes updated using the boundary displacement method are of higher quality than those updated using isoparametric mapping. Figure 13.12(b) shows that, using isoparametric mapping, finite element mesh could be severely distorted after a large design change, for example, element *A*. However, as shown in Fig. 13.12(c), finite element mesh obtained using the boundary displacement method tends to be regular. The fillet example also verifies that the design velocity field is reusable for several design iterations [65].

13.3.5 Combination of Isoparametric Mapping and Boundary Displacement Methods

Chang and Choi [85] presented to systematically calculate the design velocity field using a hybrid method that combines the cubic isoparametric mapping for the boundary design velocity field and boundary displacement method for the domain design velocity field.

A parametric surface can be represented mathematically as a bi-cubic polynomial

(a) Original Mesh

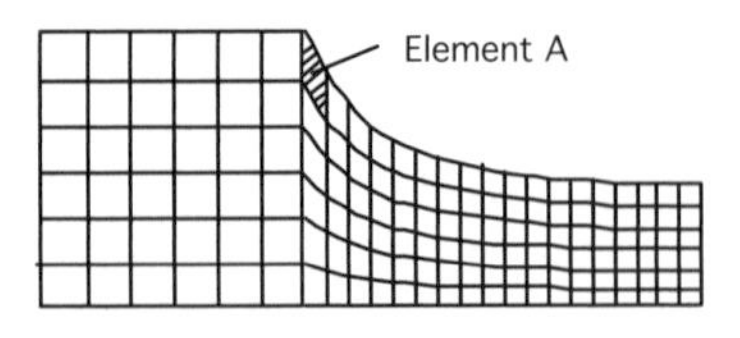

(b) Isoparametric Mapping Method

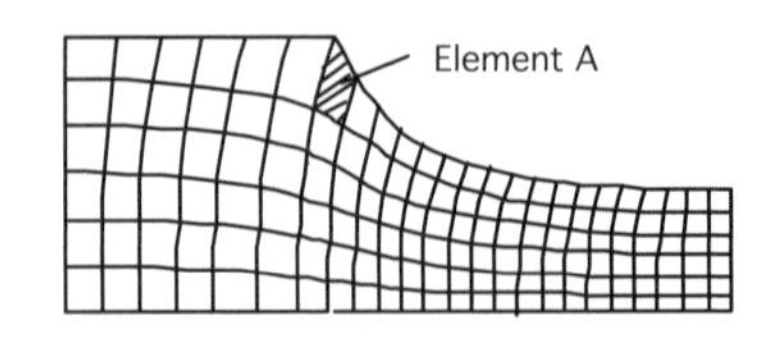

(c) Boundary Displacement Method

Figure 13.12. Two-dimensional fillet example.

$$
\begin{aligned}
\boldsymbol{x}(u,w) &= \sum_{j,k=0}^{3} \boldsymbol{a}_{jk} u^j w^k \\
&= \{u^3, u^2, u, 1\} \begin{bmatrix} \boldsymbol{a}_{33} & \boldsymbol{a}_{32} & \boldsymbol{a}_{31} & \boldsymbol{a}_{30} \\ \boldsymbol{a}_{23} & \boldsymbol{a}_{22} & \boldsymbol{a}_{21} & \boldsymbol{a}_{20} \\ \boldsymbol{a}_{13} & \boldsymbol{a}_{12} & \boldsymbol{a}_{11} & \boldsymbol{a}_{10} \\ \boldsymbol{a}_{03} & \boldsymbol{a}_{02} & \boldsymbol{a}_{01} & \boldsymbol{a}_{00} \end{bmatrix}_{4\times4\times3} \begin{Bmatrix} w^3 \\ w^2 \\ w \\ 1 \end{Bmatrix} \\
&= \boldsymbol{U}\boldsymbol{A}\boldsymbol{W}^T, \quad \langle u,w\rangle \in [0,1]\times[0,1],
\end{aligned}
\tag{13.125}
$$

where $\boldsymbol{U} = [u^3, u^2, u, 1]$, $\boldsymbol{W} = [w^3, w^2, w, 1]$, u and w are the parametric locations of the point $\boldsymbol{x}$ in the surface, and matrix $\boldsymbol{A}$ is the algebraic coefficient matrix of the surface. Therefore, using the isoparametric mapping method, the boundary design velocity field for three-dimensional structural components can be obtained by taking the variation of (13.125) as

$$
\begin{aligned}
\boldsymbol{V}^k_{1\times3} &= \delta\boldsymbol{x}(u,w) \\
&= \sum_{j,k=0}^{3} \delta\boldsymbol{a}_{jk} u^j w^k \\
&= \{u^3, u^2, u, 1\} \begin{bmatrix} \delta\boldsymbol{a}_{33} & \delta\boldsymbol{a}_{32} & \delta\boldsymbol{a}_{31} & \delta\boldsymbol{a}_{30} \\ \delta\boldsymbol{a}_{23} & \delta\boldsymbol{a}_{22} & \delta\boldsymbol{a}_{21} & \delta\boldsymbol{a}_{20} \\ \delta\boldsymbol{a}_{13} & \delta\boldsymbol{a}_{12} & \delta\boldsymbol{a}_{11} & \delta\boldsymbol{a}_{10} \\ \delta\boldsymbol{a}_{03} & \delta\boldsymbol{a}_{02} & \delta\boldsymbol{a}_{01} & \delta\boldsymbol{a}_{00} \end{bmatrix}_{4\times4\times1} \begin{Bmatrix} w^3 \\ w^2 \\ w \\ 1 \end{Bmatrix} \\
&= \boldsymbol{U}\,\delta\boldsymbol{A}\,\boldsymbol{W}^T, \quad \langle u,w\rangle \in [0,1]\times[0,1],
\end{aligned}
\tag{13.126}
$$

where the matrix $\delta\boldsymbol{A}$ is the algebraic form of the shape design variations of the surfaces that represent the design boundary. Note that the boundary design velocity field linearly depends on variations of shape design parameters since both the isoparametric mapping and matrix transformations are linear operations. In order to maintain regular finite element mesh, boundary nodes other than the design boundary are also allowed to move in certain directions. Once the boundary node movements are specified on the auxiliary finite element model, a finite element analysis code is used to solve the domain design velocity field. Note that this method is applicable to all three types of design applications discussed earlier.

Based on the discussion of the velocity field computation methods and examples evaluated, the design velocity computation methods are compared in Table 13.1. The comparison is based on basic theoretical and practical requirements discussed earlier.

As can be seen in Table 13.1, the hybrid method, a combination of the isoparametric mapping and boundary displacement methods, best satisfies the theoretical and practical requirements described in Section 6.2.7. Although this method is not computationally efficient, reusability and good mesh quality of the design velocity make the method attractive for practical applications. Another good method is isoparametric mapping. One drawback in applying the isoparametric mapping method is that the structure must be decomposed into a coarse hexahedron mesh. However, this method is very efficient. The fictitious load method satisfies the theoretical requirements but cannot easily be connected to a CAD geometric modeler for general applications. The finite difference method is simple but easily destroys the linear dependency of the design velocity field and lends to change mesh topology.

Table 13.1. Comparison of velocity field computation methods.

Requirements	Finite Difference	Isoparam. Mapping	Boundary Displ.	Fictitious Load	Hybrid
Regularity[§]	Yes	Yes	Yes	Yes	Yes
Linearity	No	Yes	Yes	Yes	Yes
FE Topology	No	Yes	Yes	Yes	Yes
Boundary Match	No	Yes	Yes	No	Yes
Mapping Rule	No	Algeb[†]	PDE[††]	PDE	Algeb+PDE
Mesh Quality	No	OK	Good	Good	Good
CAD Link	Yes	Yes	No	No	Yes
Efficient	Yes	Yes	No	No	No
Reusability	No	Yes	Yes	No	Yes
General	Yes/No[§§]	Yes[§§§]	Yes	No	Yes
Design Model	Geometric	Geometric	Geometric	FE Model	Geometric

† Algebraic equation.
†† Partial differential equation.
§ Only for two- and three-dimensional elastic solids.
§§ Depending on capabilities of the CAD.
§§§ If connected to the CAD.

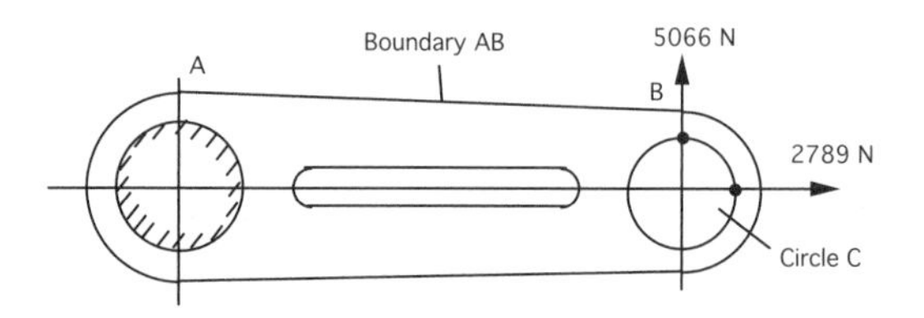

(a) Geometric Model

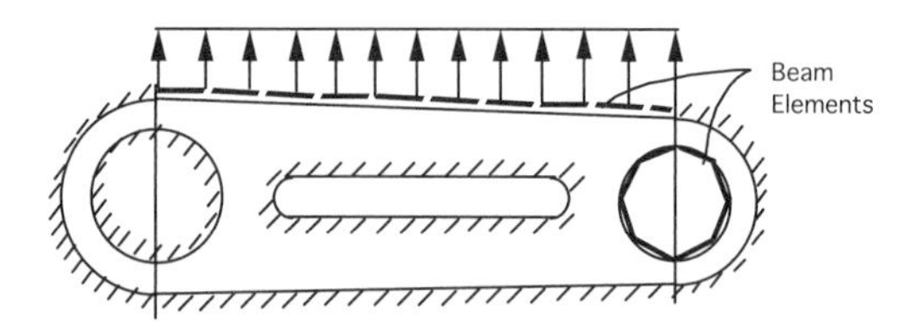

(b) Auxiliary Model for Velocity Field Computation Using Fictitious Load

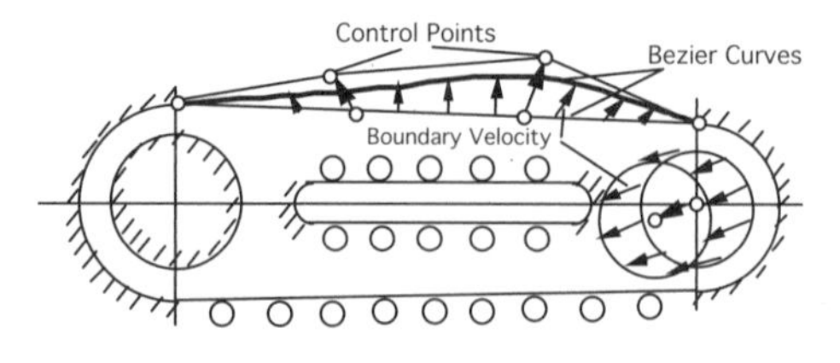

(c) Auxiliary Model for Velocity Field Computation Using Hybrid Method

Figure 13.13. Two-dimensional control arm.

Two-Dimensional Control Arm

The two-dimensional control arm example [84] shown in Fig. 13.13 contains the first (boundary *AB*) and the third types (circle *C*) of design applications. If the fictitious load method is used, beam elements must be added at boundary *AB* and circle *C* to maintain smoothness, and a load needs to be applied at the design boundary, as shown in Fig. 13.13(b). Note that the fictitious load method cannot handle design of movement of circle *C* since it is impractical to generate a set of loads to move the circle.

If a combination of the isoparametric mapping and boundary displacement methods is used, the design boundary can be parameterized using parameters defined in a CAD modeler, for example, control points that characterize the Bezier curve *AB*. In this hybrid method, the isoparametric mapping method is used to obtain the boundary velocity based on a CAD geometric representation, for example, (13.126). Constraints are defined at the nondesign boundary to complete the auxiliary model. Then, the boundary displacement method can be used to compute the domain design velocity field for a given boundary design velocity using (13.107). As illustrated in Fig. 13.13(c), nodal points on a straight boundary are constrained to move along the boundary to minimize finite element mesh distortion at the new design [85]. Similar approach can be applied to compute the design velocity field that governs the movement of nodal points along circle C. This example demonstrates that the hybrid method can be connected naturally to a CAD modeler.

14
Design Applications

In this chapter, various structural design optimization problems are applied using design sensitivity analysis methods of previous chapters. For each design problem, the following procedures are explained: structural modeling, design parameterization, structural analysis, design sensitivity analysis, trade-off analysis, the what-if study, and design optimization. The design sensitivity analysis and optimization tool (DSO) [94] is used in the design procedure. Interactive design optimizations as well as automated optimization procedures are used. In the integrated design environment, it is very important to construct a seamless interface between applications to minimize human effort during design optimization process.

For linear structural problems, commercial finite element analysis (FEA) codes are used for structural analysis purposes, and design sensitivity analysis is carried out using postprocessing data and the reanalysis capability of FEA. For nonlinear structural problems, however, it is very expensive to store stiffness matrices at each load step. In addition, most FEA codes do not provide enough information for a sensitivity analysis. Thus, for nonlinear applications, design sensitivity analysis has to be implemented within a nonlinear analysis code, resulting in extremely accurate and efficient design sensitivity results even for nonlinear structural problems.

14.1 Sizing Design Applications

The sizing design problem is related to a system's parameters. The sizing design variable appears as thickness for a shell structure and cross-sectional area for a truss/beam structure. In this section, interactive/automated design optimization is introduced with respect to sizing design parameters. Design parameters are defined using the parameters of the geometric model. Sizing design update is relatively simple, since the geometry of the structure remains fixed, compared with the shape design parameters that will be discussed in the next section.

14.1.1 Design of Wheel Structure

The road wheel of a tracked vehicle, shown in Fig. 14.1 [95], demonstrates the sizing design procedure of a structural problem. The road wheel supports the vehicle from the ground pressure load, which is generated by the vehicle's weight. By varying the thickness of the road wheel, it is possible to minimize the wheel's volume, while permitting only a certain amount of deformation at the contact point.

Geometric Model Generation

Because the road wheel is symmetric, only half of it is used for design and analysis purposes. The outer diameter of the wheel is 25 in, with two cross-sectional thicknesses: 1.25 at the ring, and 0.58 at the hub, as illustrated in Fig. 14.2. The structural modeling of the road wheel is carried out using the commercial geometric modeler PATRAN [42]. To model the wheel, 216 patches (surfaces) and 432 triangular finite elements are defined in geometric and finite element models.

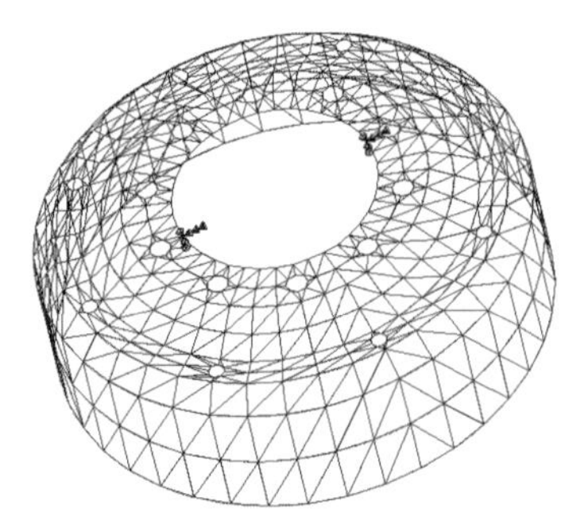

Figure 14.1. Road wheel of tracked vehicle.

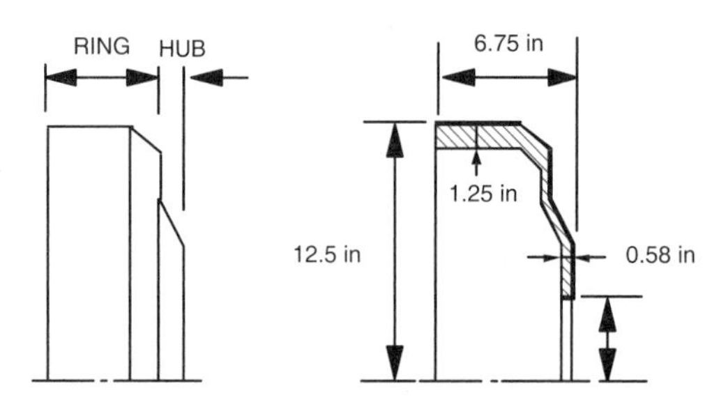

Cross Section of Initial Design

Figure 14.2. Geometric model of road wheel.

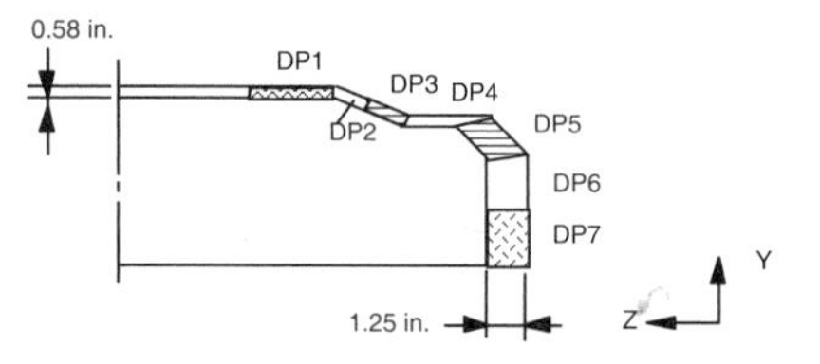

(a) Design Parameters of Road Wheel

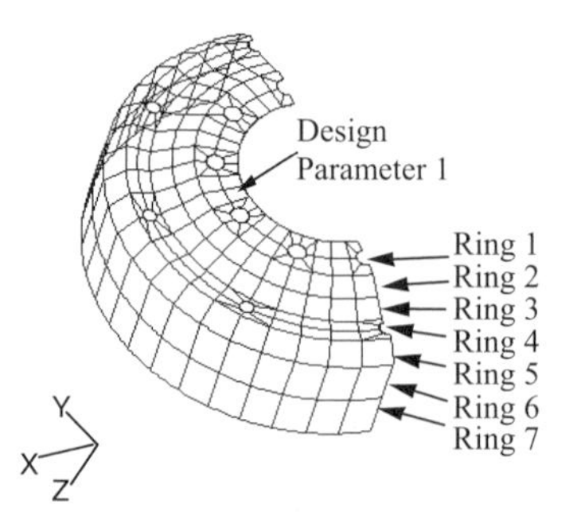

(b) Design Parameters of the Road Wheel

Figure 14.3. Design parameter definition of road wheel.

Design Parameterization

Wheel thicknesses are defined as design parameters, which are linked along the wheel circumferential direction to maintain a symmetric design. Seven design parameters are defined for the road wheel along the radial direction, as shown in Fig. 14.3.

Analysis Model Generation

A commercial FEA code ANSYS [28] is used for structural analysis. Plate element STIF63 of ANSYS is employed for finite element analysis. There are 432 triangular plate elements and 1650 degrees of freedom defined in the model, as shown in Fig. 14.4(a). This wheel is made of aluminum with Young's modulus, $E = 10.5 \times 10^6$ psi, shear modulus, $G = 3.947 \times 10^6$ psi, and Poisson's ratio, $\nu = 0.33$.

Symmetric boundary conditions are applied as shown in Fig. 14.4(a). Displacements and rotations are fixed at those nodes corresponding to the mounting points. The semicircular area where the shaft enters is restricted to x- and y-directions. The applied load from the ground is considered a surface pressure at the base of the wheel, as shown in Table 14.1.

The deformed wheel shape, obtained from the ANSYS analysis result, is displayed for evaluation in Fig. 14.4(b). The analysis results show that the maximum displacement occurs at the contact point, node 266, in the y-direction and with a magnitude of 0.108173 in. The volume of the structure is 361.94 in^3.

Table 14.1. Pressure loading condition.

Element ID	Normal Pressure (psi)
393	−1,602.0
394	−1,602.0
395	−401.0
396	−401.0
413	−1,602.0
414	−1,602.0
415	−401.0
416	−401.0

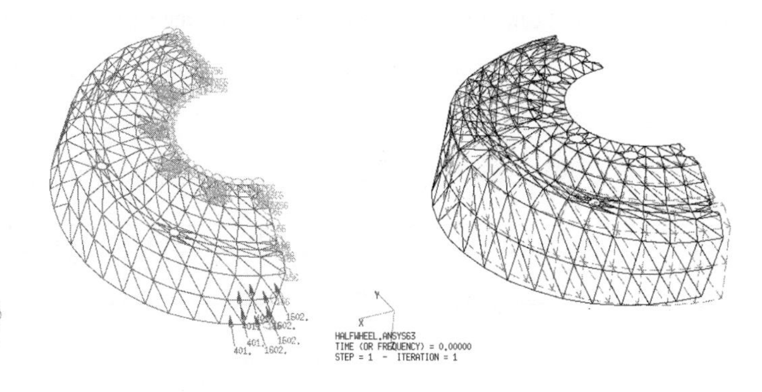

(a) Finite Element Model (b) Deformed Shape

Figure 14.4. Finite element model of road wheel.

Performance Measure Definition

The maximum displacement at node 266 in the y-direction, and the structural volume are defined as the wheel's performance measures. Displacement data can be obtained from the analysis results, and the volume can be computed by taking the surface area of the plate and multiplying it by its thickness.

Design Sensitivity Results and Display

The volume performance measure can be explicitly obtained without recourse to structural analysis. In fact, the sensitivity coefficient of the volume performance measure is the surface area. The adjoint variable method is used to compute the design sensitivity of the displacement performance measure whose adjoint load is the unit force at node 266 in the y-direction. From the analysis results and from the ANSYS reanalysis capability, the sensitivity coefficient of the displacement performance measure can be computed. Table 14.2 lists the design sensitivity coefficients.

It should be noted that these sensitivity coefficient are all negative, which means that increasing the wheel thicknesses reduces the displacement performance measure. However, all of the volume sensitivity coefficients are positive, so that increasing the wheel thicknesses will also increase the wheel volume. Note that the design sensitivity coefficients in Table 14.2 are obtained from the linear approximation of the performance measure with respect to the design. Since volume performance measure has a linear dependence on the design, the sensitivities in Table 14.2 are exact. The displacement performance measure, however, has a nonlinear dependence on the design. Thus, Table 14.2 can accurately predict the displacement performance measure only when the perturbation is small. As the perturbation increases in size, nonlinear terms increasingly dominate over linear terms.

A visual display of sensitivity results is very helpful to understand the effect of design parameters on performance measures. Figure 14.5 shows the design sensitivity coefficients of the displacement performance measure using a color plot and a bar chart. The contour plot indicates that the most significant way to reduce the displacement performance measure is to increase the thickness at the outer edge, which is design parameter dp7. As indicated in the spectrum and y-coordinate in the bar chart, increasing the wheel thickness by a unit magnitude at the outer edge reduces the displacement by 0.0993 in. This influence increasingly diminishes as one moves from the outer to the inner edges of the wheel. However, at the innermost edge, the influence of unit magnitude thickness change increases to about 40% of the maximum value. A similar contour plot and bar chart can also be generated for the volume performance measure.

Table 14.2. Design sensitivity coefficients.

Performance	dp1	dp2	dp3	dp4	dp5	dp6	dp7
Displacement	–.045865	–.01553	–.011015	–.019515	–.019420	–.056306	–.099343
Volume	35.1301	26.2068	29.8408	34.8080	47.7519	93.8567	92.4915

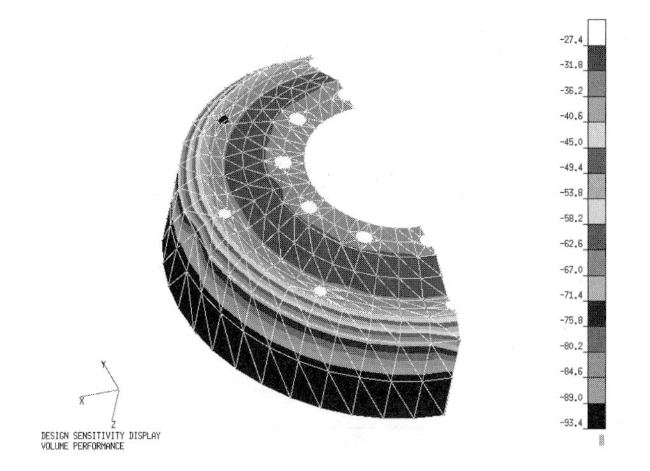

(a) Design Sensitivity Contour Plot ($\times 10^{-3}$)

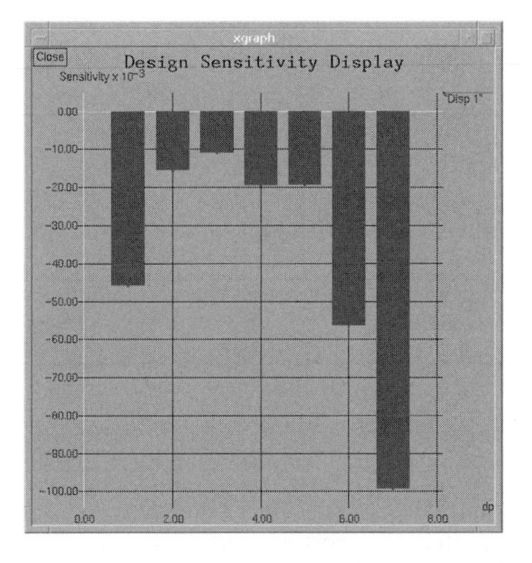

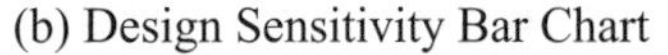
(b) Design Sensitivity Bar Chart

Figure 14.5. Design sensitivity of displacement performance measure.

Trade-off Determination

As the design sensitivity results show, a conflict exists between reducing the structural volume and reducing the maximum deformation. A design change that increases the volume reduces the displacement, and vice versa. A trade-off study is therefore performed to find the best design direction. For this study, the volume performance measure is selected as the cost function, and the displacement performance measure is used to define the constraint function, with an upper bound of 0.1 in. Note that at the current design, the displacement performance measure is 0.108173 in, which is greater than the upper bound. Therefore, the current design is infeasible. The side constraints are defined for all the design parameters with bounds 0.1 and 10.0 in, as listed in Table 14.3.

Table 14.3. Definitions of side constraints for the road wheel (unit: in).

Dpid	Lower Bound	Current Value	Upper Bound
1	0.1	0.58	10.0
2	0.1	0.58	10.0
3	0.1	0.58	10.0
4	0.1	0.58	10.0
5	0.1	1.25	10.0
6	0.1	1.25	10.0
7	0.1	1.25	10.0

Table 14.4. Design direction obtained from trade-off determination.

dpid	Current value	Direction	Perturbation	%
1	0.58	0.3596	0.03596	6.20
2	0.58	0.1218	0.01218	2.10
3	0.58	0.0864	0.00864	1.49
4	0.58	0.1530	0.01530	2.64
5	1.25	0.1523	0.01523	1.22
6	1.25	0.4415	0.04415	3.53
7	1.25	0.7790	0.07790	6.23

Table 14.5. What-if results and verification.

Cost/Constraint	Current Value	Predicted Value	FEA Results	Accuracy
Cost	361.941	376.345	376.387	100.3
Constraint	0.108173	0.095420	0.096425	101.0

Since the design is unfeasible, a constraint correction algorithm [96] is used for the trade-off study, by neglecting the cost function. The direction of design change is determined by forming the QP (quadratic programming) subproblem, which is solved using the Stanford QP solver [97]. Table 14.4 shows the design direction obtained from the QP solver.

What-if Study

Following the design direction suggested by the trade-off determination (Table 14.4), a what-if study is again carried out using a step size of 0.1 in. Table 14.5 shows the approximate cost and constraint values using design sensitivity coefficients and design perturbation. In addition, finite element analysis is carried out at the perturbed design to verify the accuracy of the what-if predictions. Note that the constraint violation is removed while the cost function slightly increases. This procedure can be repeated interactively to improve the cost function, while maintaining the constraint conditions.

Design Optimization

The same cost, constraint, and side constraint definitions used in trade-off determination are utilized in the design optimization, which combines design optimization tool (DOT)

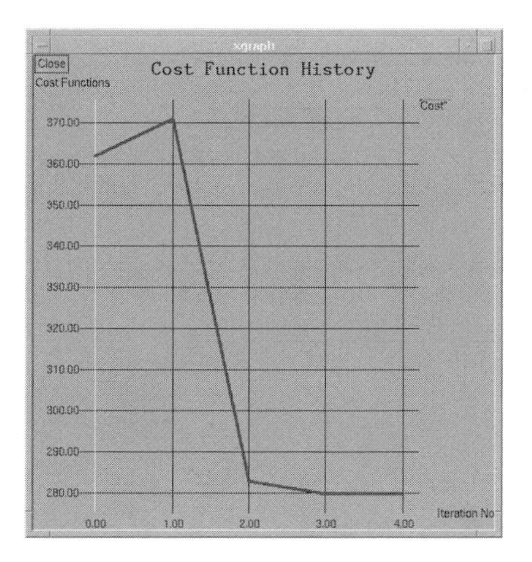

(a) Cost Function History

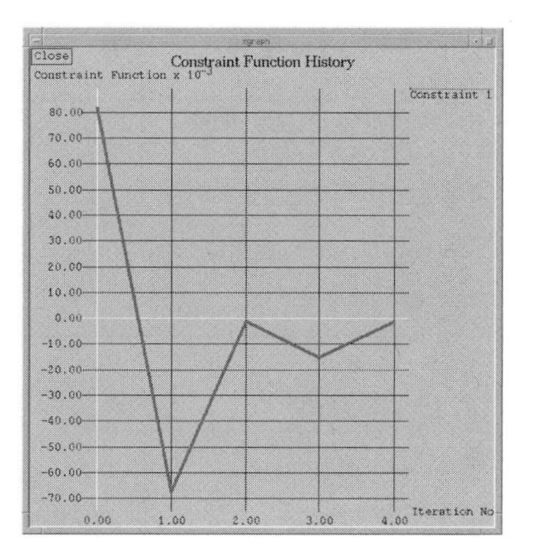

(b) Constraint Function History (scaled)

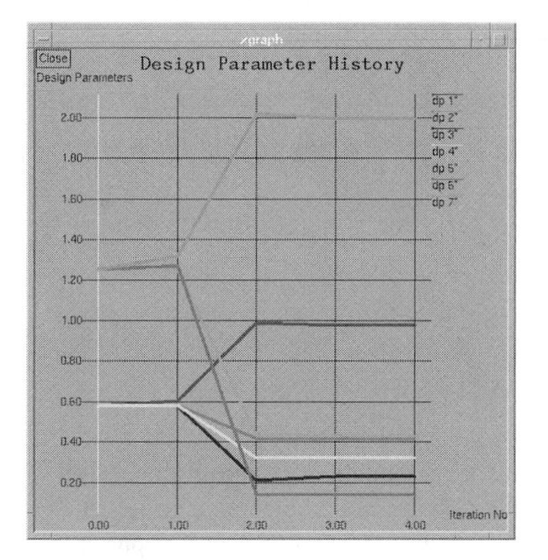

(c) Design Parameter History

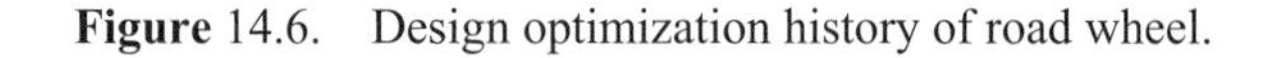
Figure 14.6. Design optimization history of road wheel.

[98], ANSYS, and local sensitivity computation and model update programs. The DOT determines the amount of design change using analysis results and sensitivity results. It is also necessary to updates structural and computational models corresponding to the new design parameters. After four iterations, a local minimum is obtained. The optimization histories for cost, constraint, and design parameters are shown in Figs. 14.6(a), (b), and (c), respectively.

Figure 14.6(a) shows that the cost function starts near 362 in^3 and immediately jumps to 371 in^3 to correct for the constraint violation. The cost is then reduced until it achieves a minimum at 279.53 in^3. The constraint function history plot, Fig. 14.6(b), shows that an 80% violation is found at the initial design, but that the violation is reduced to 65% below the upper bound at the first iteration. The constraint function then stabilizes and stays

Table 14.6. Design parameter values at optimum design.

dpid	Initial design	Optimum design
1	0.58	0.761
2	0.58	0.606
3	0.58	0.533
4	0.58	0.416
5	1.25	0.237
6	1.25	0.165
7	1.25	1.951

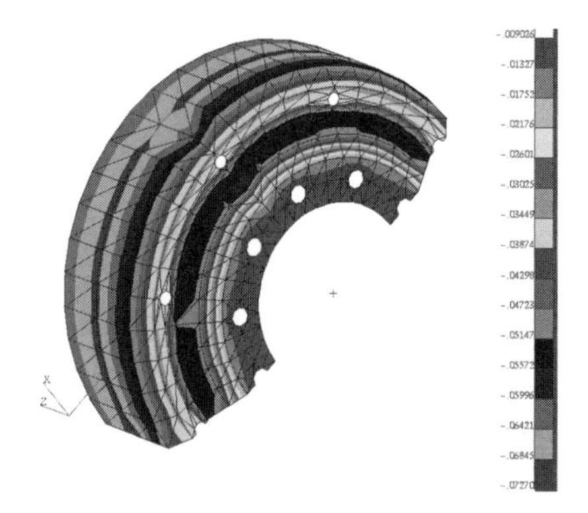

Figure 14.7. Design sensitivity of displacement performance at optimum design.

feasible for the remaining iterations. At the optimum design, the constraint is 0.291% above the upper bound, the maximum displacement becomes 0.100291 in, and the design is acceptable. Interestingly, in Fig. 14.6(c) all design parameters descend in the design process except for dp1 and dp7. Dp1 increases from 0.58 to 0.761 in at the optimum design. However, the most significant design change is dp7, which changes from 1.25 to 1.951 in to resist deformation. As shown in the design sensitivity plot, this is the most effective way to strengthen the wheel, since a decrease in any other design parameter contributes to a reduction in wheel volume. It is important to note that the optimum solution agrees with the design trend at the initial design (as shown in Fig. 14.5). Initial and optimum design parameter values are listed in Table 14.6.

Post-Optimum Study

At the optimum point, the designer can still use the four-step design process to obtain significant information about structural behavior. The design sensitivity plot for the displacement performance measure at optimum design is shown in Fig. 14.7. The plot shows that thickness at the outer rim significantly affects maximum displacement at the contact point. However, sensitivity coefficients are comparatively small in other regions. The plot suggests that a relatively small tolerance needs to be applied to the thickness at the outer rim during the manufacturing process, since a small error made in the outer rim will affect the displacement significantly.

Table 14.7. Design direction obtained from trade-off determination.

dp ID	Current Value	Direction	New Design	%
dp1	0.761	–0.137	0.754	–0.898
dp2	0.606	–0.154	0.598	–1.275
dp3	0.533	0.677	0.567	6.345
dp4	0.416	0.515	0.442	6.189
dp5	0.237	0.0453	0.239	0.956
dp6	0.165	0.0596	0.168	1.818
dp7	1.951	–0.479	1.927	–1.226

Table 14.8. What-if results and verification.

Function	Description	Current Value	Predicted Value	% Change
Cost	Volume	279.526	279.158	–0.132
Constraint	Node 266, *y*-displ	0.100291	0.100291	0.05

Trade-off and what-if studies are performed at the optimum design to obtain a better understanding of the design's structural behavior. Since all constraints are satisfied, the cost reduction algorithm is utilized in trade-off analysis to see if cost can be further reduced. This algorithm finds the design direction that reduces cost by decreasing the values of dps 1 to 6, as shown in Table 4.7.

A what-if study is again carried out following the design direction suggested by the trade-off determination, using a step size of 0.05 in, as indicated in Table 14.7. Table 14.8 shows the approximation of cost and constraint values using design sensitivity coefficients and design perturbation.

14.1.2 Design of Wing Structure

A wing model [99], shown in Fig. 14.8, is used in the sizing design problem when design parameters are chosen from both the cross-sectional area of the truss and from the thickness of a membrane. In addition, both static and eigenvalue analyses are used to evaluate performance measures of the design problem. The wing is made of aluminum and is composed of truss and membrane elements.

The design objective is to minimize the structural volume of the wing (equivalently, the weight of the wing). Displacements, stresses, volume, mass, and natural frequency are defined as performance measures.

Geometric Modeling

The geometric modeler PATRAN is used to create the geometric and finite element models of the wing. The geometric model has 84 line and 126 surface (patch) design entities, which are grouped into five assemblies: FRONT PANEL, BACK PANEL, SHEAR PANEL, LSPAR CAPS, and VSPAR CAPS, as shown in Fig. 14.9. In the initial design, the patches in the first half (close to the fuselage) of the skin panels have a thickness of 0.2 in, and the patches in the second half of the skin panels (close to the wing tip) have a thickness of 0.1 in. The patches in the shear panel are 0.2 in thick. The longitudinal spar caps are designed as solid circular cross sections with an area of 0.02 in^2. The vertical spar caps have the same cross sections, but with an area of 0.2 in^2. The thickness of the surface design entities on the skin panels and the radius of the line design entities on the longitudinal spar caps are defined as the design parameters. However, the thickness of the shear panel and the cross-sectional area of the lines in the vertical spar caps are fixed.

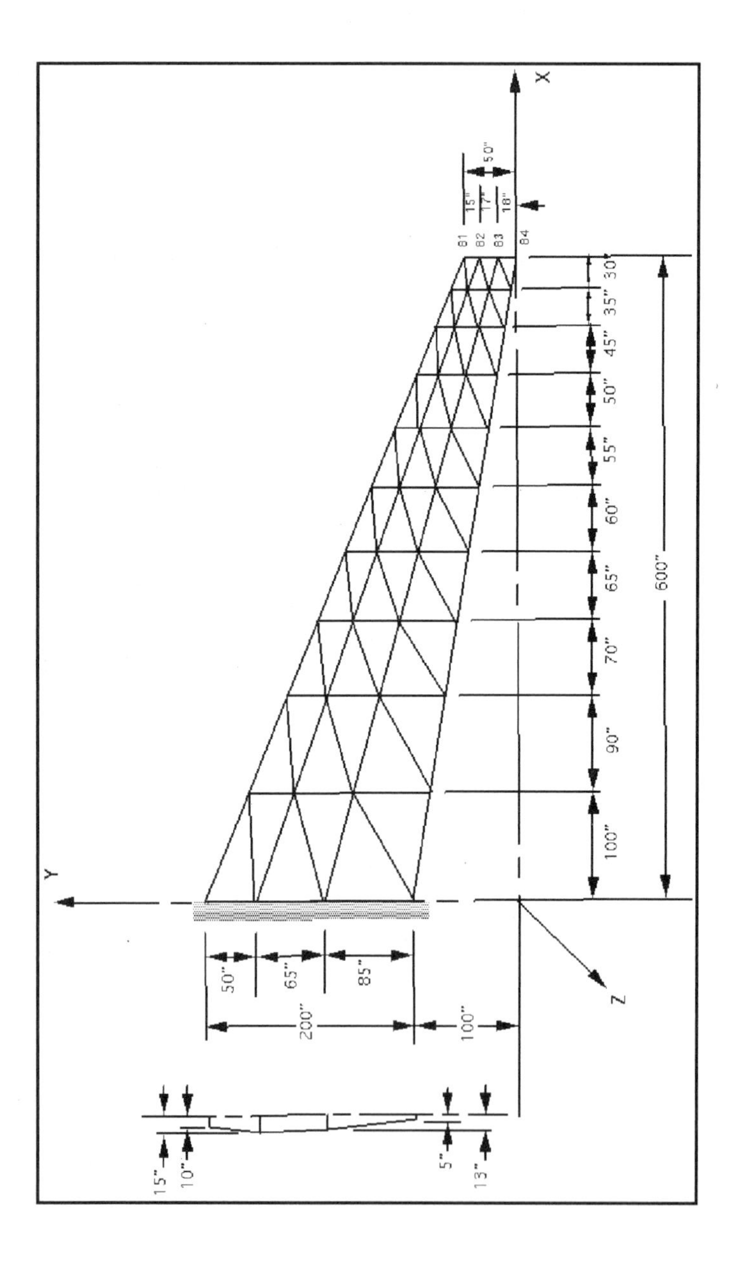

Figure 14.8. Wing model.

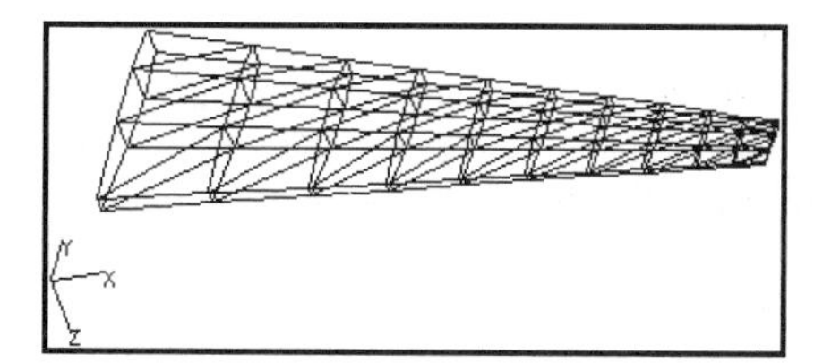

(a) Full Wing Model

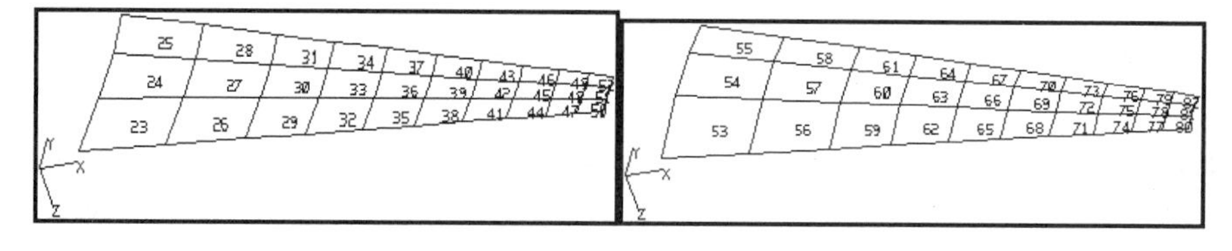

(b) Front Skin Panel Assembly

(c) Back Skin Panel Assembly

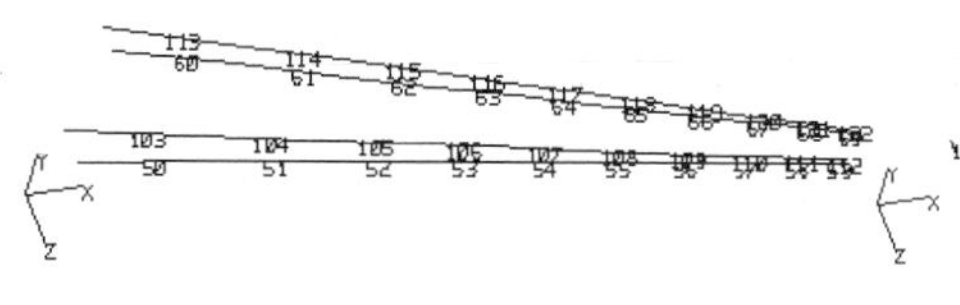

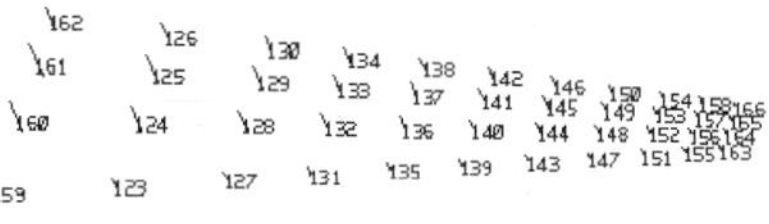

(d) Longitudinal Spar Cap Assembly

(e) Vertical Spar Cap Assembly

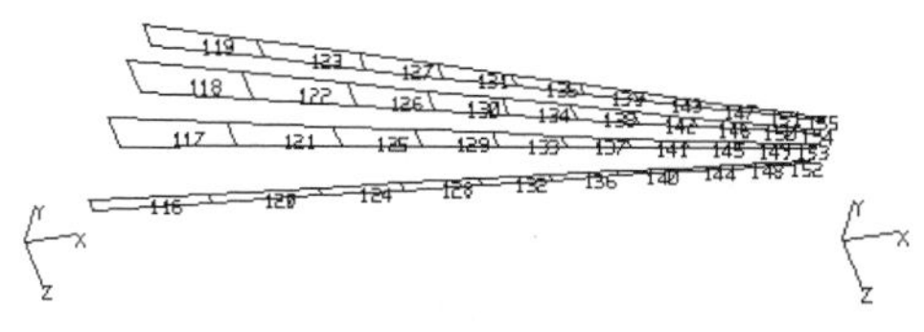

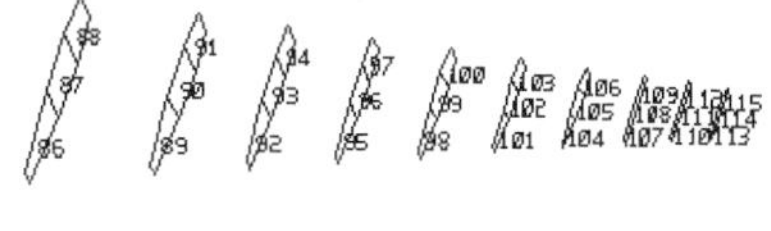

(f) Horizontal Shear Panel Assembly

(g) Vertical Shear Panel Assembly

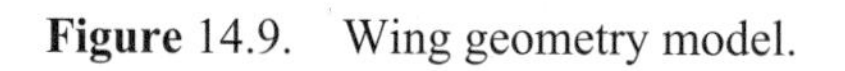

Figure 14.9. Wing geometry model.

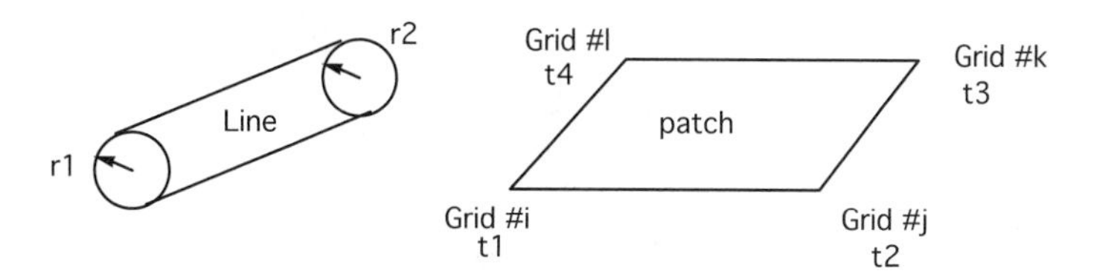

Figure 14.10. Design parameterization for line and surface design components.

Finite Element Modeling

NASTRAN [100] truss element ROD and membrane element TRIA3 are used for analysis. There are 88 nodes and 274 elements (84 truss elements, and 70 quadrilateral and 120 triangular membrane elements) in the wing FE model, as shown in Fig. 14.9(a).

The following load and boundary conditions are applied to the wing:

- Node displacements in the *x*-, *y*-, and *z*-directions at the wing's root (where it connects to the fuselage) are restricted.

- A set of equivalent nodal forces that simulate a uniform pressure loading of 0.556 psi is applied to the front skin panel.

The wing is made of aluminum, and has the following properties: Young's modulus $E = 10.6 \times 10^6$ psi, Shear modulus $G = 4.0 \times 10^6$ psi, Poisson's ratio $\nu = 0.3$, and mass density $\rho = 0.284$ lb/in^3.

Design Parameterization

The line design entities of the wing model are parameterized as uniform circular sections, with a longitudinal spar cap area of 0.02 in^2 and a vertical spar cap area of 0.2 in^2. For each line design entity in the longitudinal spar cap, the radius r1 at the first end point is defined as a free design parameter with a value of 0.07979 in, and the radius r2 at the other end is proportional to r1 with a ratio of 1.0.

Table 14.9. Design parameterization of wing model.

dp ID	Line ID	Surface ID	Assembly	Current Value
1	50–59		LSPAR CAPS	0.079788 in
2	60–69		LSPAR CAPS	0.079788 in
3	103–112		LSPAR CAPS	0.079788 in
4	113–122		LSPAR CAPS	0.079788 in
5		23–25	FRONT PANEL	0.2 in
6		26–28	FRONT PANEL	0.2 in
7		29–31	FRONT PANEL	0.2 in
8		32–34	FRONT PANEL	0.2 in
9		53–55	BACK PANEL	0.2 in
10		56–58	BACK PANEL	0.2 in
11		59–61	BACK PANEL	0.2 in
12		62–64	BACK PANEL	0.2 in
13		35–37	FRONT PANEL	0.1 in
14		38–40	FRONT PANEL	0.1 in
15		41–43	FRONT PANEL	0.1 in
16		44–46	FRONT PANEL	0.1 in
17		47–49	FRONT PANEL	0.1 in
18		50–52	FRONT PANEL	0.1 in
19		65–67	BACK PANEL	0.1 in
20		68–70	BACK PANEL	0.1 in
21		71–73	BACK PANEL	0.1 in
22		74–76	BACK PANEL	0.1 in
23		77–79	BACK PANEL	0.1 in
24		80–82	BACK PANEL	0.1 in
fixed	123–166		VSPAR CAPS	0.25231 in
fixed		86–155	SHEAR PANEL	0.2 in

For each line in the vertical spar caps, the radius at each end is fixed with a value of 0.2523 in. For each surface (patch) design entity in the front and back skin panels, a uniform thickness is defined as the design parameter, with values of 0.2 in for the first half (close to the fuselage) and 0.1 in for the second half (close to the tip). For each surface in the shear panel, thickness is fixed with the value 0.2 in. Figure 14.10 shows the design parameters for the line and surface design components. Table 14.9 summarizes the locations and the current values of design parameters.

Finite Element Analysis

NASTRAN was used to perform static and modal analyses of the wing model. Figure 14.11 shows the deformed shape of the wing obtained from static analysis, illustrating that maximum deformation occurs at the wing tip.

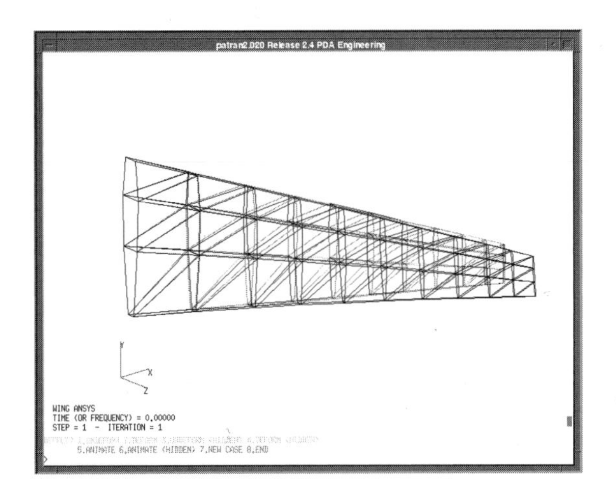

Figure 14.11. Deformed shape of wing.

Table 14.10. Definition of performance measures.

Perf. ID	Performance Measure	Description	Value
1	Volume		34280.5in^3
2	Displacement1	node 86, z-direction	21.772563 in
3	Displacement2	node 44, z-direction	20.414493 in
4	Stress1	element 5, von Mises, average	8565.6318 psi
5	Stress2	element 11, von Mises, average	14227.3626 psi
6	Stress3	element 12, von Mises, average	13457.1935 psi
7	Stress4	element 26, von Mises, average	14426.3480 psi
8	Stress5	element 25, von Mises, average	14381.3538 psi
9	Stress6	element 45, von Mises, average	7502.6018 psi
10	Stress7	element 18, von Mises, average	8977.6010 psi
11	Stress8	element 33, von Mises, average	13505.3928 psi
12	Stress9	element 34, von Mises, average	13138.3520 psi
13	Stress10	element 54, von Mises, average	11701.8567 psi
14	Stress11	element 53, von Mises, average	12248.5369 psi
15	Stress12	element 72, von Mises, average	4411.7140 psi
16	Natural Frequency	first mode	1316.8003 Hz

Definition of Performance Measure

For the purposes of this design problem, the followings are defined as performance measures: mass, volume, displacements, stresses, and natural frequency. These measures and their values (obtained from NASTRAN) are listed in Table 14.10.

Design Sensitivity Analysis

The design sensitivity coefficient of the volume can be explicitly obtained from the surface area of the panels and from the radius of the truss. The adjoint variable method is used to compute the design sensitivity coefficients of the displacement and stress. The design sensitivity coefficient of the natural frequency does not require a reanalysis procedure. Presuming that the natural frequency is a simple eigenvalue, only numerical integration is needed to compute its design sensitivity. Table 14.11 shows the design sensitivity coefficients for those performance measures obtained using NASTRAN.

The accuracy of the sensitivity results was verified using the forward finite difference method with a 1% perturbation, and the verification results are summarized in Table 14.12, with good agreement.

Table 14.11. Design sensitivity coefficient.

Performance Measure	DP1	DP5	DP8	DP9
Volume	312.281	18817.7	8270.58	18817.7
Displacement1	–0.145868	–9.45529	–3.62596	–9.45199
Displacement2	–0.134197	–9.27717	–3.30270	–9.27403
Stress1	–80.3410	–31835.4	–19.9950	4133.88
Stress2	–135.840	–46481.9	1.18237	4643.15
Stress3	–126.147	–46112.4	–6.90316	4162.21
Stress4	–167.943	–46399.5	33.1162	5283.78
Stress5	–164.474	–46134.4	23.0779	5040.19
Stress6	–139.459	–29665.6	31.1451	10908.2
Stress7	–94.8019	–5273.58	–75.6904	–111.351
Stress8	–141.091	–877.670	200.565	–1626.36
Stress9	–136.284	–2060.38	134.403	–1772.15
Stress10	–145.493	2816.03	69.9826	–1097.58
Stress11	–131.811	11.2227	3506.02	4.70456
Stress12	–128.232	–76.7234	–1395.53	–522.521
Natural Frequency	–22.0639	651.778	25.1018	588.544

Table 14.12. Verification of design sensitivity information.

Perf. Measure	$\Psi(b)$	$\Psi(b+\delta b)$	$\delta\Psi$	Ψ'	$\Psi'/\Delta\Psi \times 100\%$
Volume	34280.5	34318.1	37.6	37.63	100.1
Displacement1	21.7726	21.7538	–0.0188	–0.0189	100.56
Displacement2	20.4145	20.3961	–0.0184	–0.01855	100.8
Stress1	8565.6	8502.4	–63.2	–63.67	100.7
Stress2	14227.0	14135.0	–92.0	–92.96	101.0
Stress3	13457.0	13366.0	–91.0	–92.22	101.3
Stress4	14426.0	14334.0	–92.0	–92.9	101.3
Stress5	14381.0	14290.0	–91.0	–92.27	101.4
Stress6	7502.6	7443.7	–58.9	–59.33	100.7
Stress7	8977.6	8967.1	–10.5	–10.55	100.4
Stress8	13505.1	13503.4	–1.7	–1.75	103.0
Stress9	13138.0	13134.0	–4.0	–4.12	103.0
Stress10	11702.0	11707.5	5.5	5.63	102.4
Stress11	12248.5	12255.5	7.0	7.01	100.1
Stress12	4411.7	4408.9	–2.8	–2.79	99.6
Frequency	1316.8	1318.0	1.2664	1.2771	100.8

$\delta b_5 = 0.01$ (1%).

Definition of Cost and Constraints

Table 14.13 shows the cost and constraint functions (and their corresponding upper or lower bounds) used to characterize the wing design. Note that at the current design the average von Mises stress and the z-displacement at node 86 violate the design constraints, making the design infeasible. The side constraints are defined for all design parameters with corresponding upper and lower bounds, as listed in Table 14.14.

Trade-off Analysis

Based on the cost and constraint definitions, a trade-off analysis is performed to find the best design direction. Because the current design is infeasible, a constraint correction algorithm [96] is selected for the trade-off study. A quadratic programming (QP) subproblem is formed, and it is sent to the QP solver to find the best design direction. Table 14.15 shows the design direction obtained from the QP solver.

Table 14.13. Cost and constraint function of wing model.

Function	Description	Lower Bound	Current Design	Upper Bound
Cost	Volume		34280.507	
Constraint1	Stress4		14426.347	14000.0
Constraint2	Stress5		14381.353	14000.0
Constraint3	Displacement1		21.772	20.0
Constraint4	Natural Frequency	1300.0	1316.8	

Table 14.14. Side constraints definitions of wing model.

DP ID	Lower Bound	Current Value	Upper Bound
1	0.05	0.07979	0.1
2	0.05	0.07979	0.1
3	0.05	0.07979	0.1
4	0.05	0.07979	0.1
5	0.01	0.2	1.0
6	0.01	0.2	1.0
7	0.01	0.2	1.0
8	0.01	0.2	1.0
9	0.01	0.1	1.0
10	0.01	0.1	1.0
11	0.01	0.1	1.0
12	0.01	0.1	1.0
13	0.01	0.1	1.0
14	0.01	0.1	1.0
15	0.01	0.2	1.0
16	0.01	0.2	1.0
17	0.01	0.2	1.0
18	0.01	0.2	1.0
19	0.01	0.1	1.0
20	0.01	0.1	1.0
21	0.01	0.1	1.0
22	0.01	0.1	1.0
23	0.01	0.1	1.0
24	0.01	0.1	1.0

Table 14.15. Design direction obtained from trade-off analysis.

DP ID	Current Value	Direction	Perturbation	%
1	0.07979	0.007181	0.000179	0.225
2	0.07979	0.003549	0.000088	0.111
3	0.07979	0.007176	0.000179	0.225
4	0.07979	0.003544	0.000088	0.111
5	0.2	0.4654	0.011637	5.818
6	0.2	0.3450	0.008627	4.313
7	0.2	0.2547	0.006367	3.183
8	0.2	0.1785	0.004462	2.231
9	0.1	0.4653	0.011632	11.63
10	0.1	0.3448	0.008620	8.620
11	0.1	0.2546	0.006365	6.365
12	0.1	0.1784	0.004460	4.460
13	0.1	0.2162	0.005405	5.405
14	0.1	0.1255	0.003137	3.137
15	0.2	0.06501	0.001625	0.812
16	0.2	0.02947	0.000736	0.368
17	0.2	0.01219	0.000304	0.152
18	0.2	0.005198	0.000129	0.065
19	0.1	0.2160	0.0054	5.40
20	0.1	0.1253	0.003132	3.132
21	0.1	0.06492	0.001623	1.623
22	0.1	0.02941	0.000735	0.735
23	0.1	0.01216	0.000303	0.303
24	0.1	0.005184	0.0001296	0.129

Table 14.16. What-if results and verifications.

Function	Description	Current Value	Predicted Value	FEA Results	Accuracy
Cost	Volume	34280.507	35301.818	35301.814	100.00
Constraint1	Stress4	14426.348	13929.413	13946.372	103.53
Constraint2	Stress5	14381.354	13887.869	13904.728	103.53
Constraint3	Displacement1	21.773	21.265	21.278	97.34
Constraint4	Frequency	1316.80	1334.503	1334.503	100.23

What-if Study

A what-if study is performed following the design direction suggested by the trade-off study, listed in Table 14.16, and using a step size of 0.025 in. The approximate cost and constraint values calculated using both the design sensitivity coefficients and design perturbation are listed in Table 14.16. A finite element analysis, performed at the perturbed design, verifies that the approximations are accurate. These results are summarized in Table 14.16.

14.2 Shape Design Applications

The shape design problem is closely related to the geometric modeler, since a design change means remodeling of the structural geometry. It is therefore very important to connect the design program to the geometric modeler. In this section, commercial geometric modelers, such as PATRAN and Pro/Engineer [101], are used to represent the structural geometry and to make an analysis model during the seamless design optimization process.

14.2.1 Design of a Three-Dimensional Clevis

Physical Model

A redesigned clevis model [102], shown in Fig. 14.12, is utilized to demonstrate the shape design procedure of a structural problem. This clevis connects a tow bar to the top eye hook of the tracked vehicle. When the tracked vehicle is subjected to certain maneuvers, interference develops between the clevis and the hookup. In order to eliminate such interference, a chamfer was placed on the lower surface of the clevis in order to increase the tow bar's vertical motion, as shown in Fig. 14.12. However, the redesigned clevis failed in the field test, due to the increased stress concentration on its lower surface.

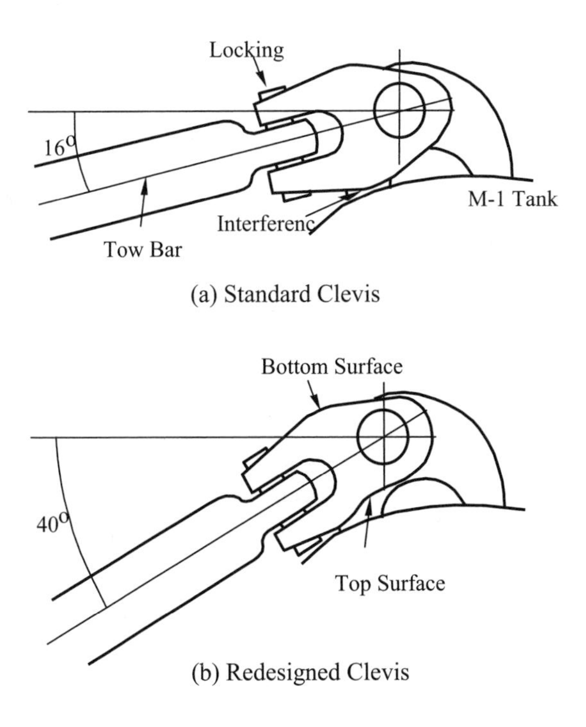

Figure 14.12. Three-dimensional clevis.

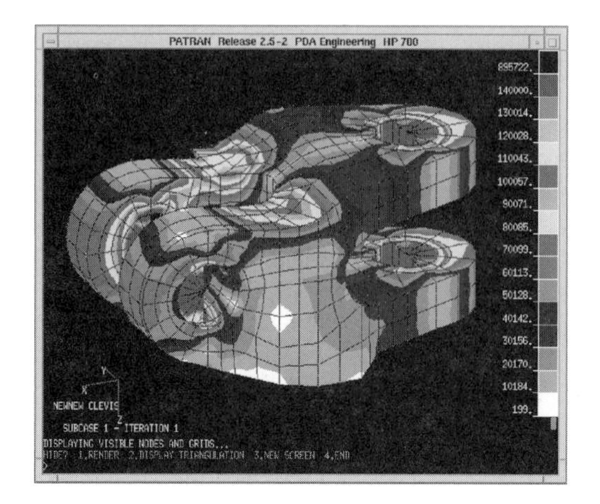

Figure 14.13. Von Mises stress distribution of redesigned clevis.

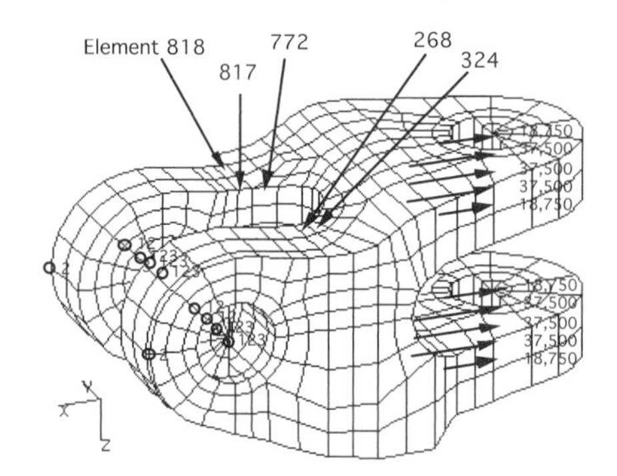

Figure 14.14. Finite element model and boundary conditions.

Finite Element Model

Finite element analysis results confirmed the higher surface concentration of stress, as illustrated in Fig. 14.13, under the boundary conditions that were applied at the four half pins, as shown in Fig. 14.14. Note that the four high stresses are artificial, due to the application of boundary conditions. The material used is SAE 1045 with the following properties: Young's modulus $E = 10.5 \times 10^6$ psi, Poisson's ratio $\nu = 0.3$, and yielding stress 185 ksi. The finite element model has 928 20-node isoparametric elements (ANSYS STIF95) and 5369 nodes, with 16,000 degrees of freedom.

Design Model Definition

The objective in redesigning the clevis is to reduce stress values below 75% of the material yield stress by altering the five boundary surfaces, without significantly increasing the weight. The five design boundary surfaces are front, rear, front slot, rear slot, and bottom, as shown in Fig. 14.15.

A geometric patch format is chosen to parameterize the five design boundary surfaces. For the front surface, y-coordinates of the seventeen grid points are linked as five independent shape design parameters, as illustrated in Fig. 14.15(a). C^0-continuity is retained by linking design parameters at the grids of interpatch boundaries, in addition to the fact that tangent vectors and twist vectors are invariant. For the rear surface, y-coordinates are allowed to vary, while, x-coordinates are allowed to vary for the two slots. For the bottom surface, z-coordinates of grid points are defined as shape design parameters. Consequently, 25 independent shape design parameters are defined for the five surfaces in order to parameterize the clevis.

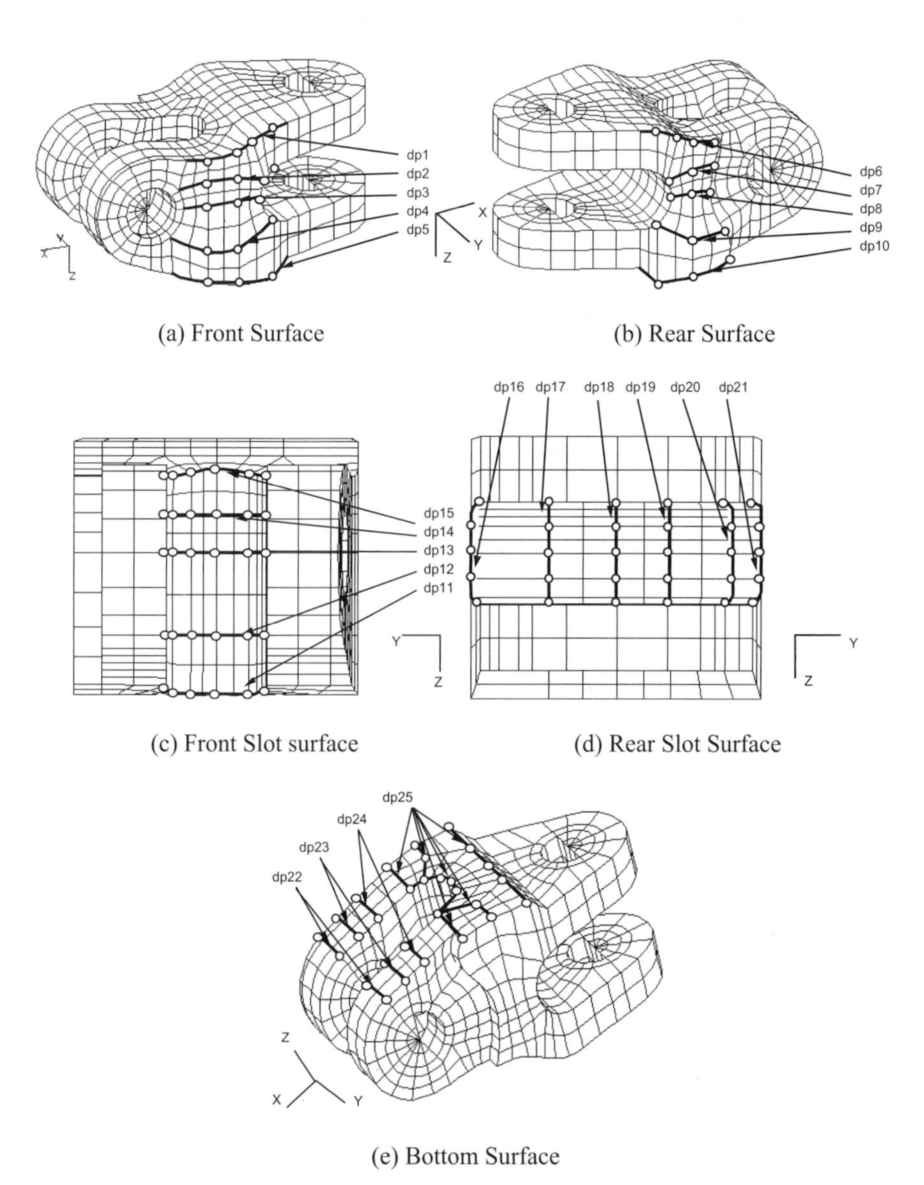

(a) Front Surface

(b) Rear Surface

(c) Front Slot surface

(d) Rear Slot Surface

(e) Bottom Surface

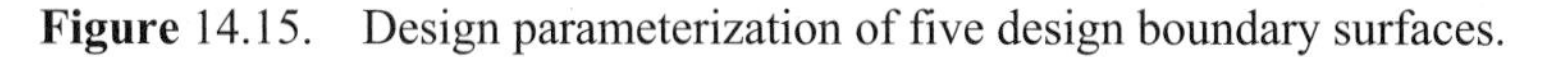

Figure 14.15. Design parameterization of five design boundary surfaces.

Table 14.17. Cost and (selected) constraint function definitions.

Cost/Constraint	Element	Current Values	Upper Bound	Status
Cost (volume)		89.7099 in^3		
Constraint 669	817	122.3435 ksi	138.75 ksi	inactive
Constraint 460	268	132.4939 ksi	138.75 ksi	inactive
Constraint 462	324	132.8683 ksi	138.75 ksi	inactive
Constraint 550	772	141.2288 ksi	138.75 ksi	violated
Constraint 672	818	146.0990 ksi	138.75 ksi	violated

The volume and maximum von Mises stress at the integration point of each finite element are defined as performance measures, excluding those artificially high stresses in the elements surrounding the pins due to the boundary conditions. There are 668 stress performance measures and 1 volume performance measure that have been defined. The maximum stresses are found at elements 772 and 818, as shown in Figs. 14.13 and 14.14. The volume is chosen as the cost function, with a value of 89.71 in^3 at the initial design. These 668 stress performance measures are defined as constraints with an upper bound of 138.75 ksi ($0.75 \times \sigma_y$). The five finite elements with high stresses shown in Figs. 14.13 and 14.14 are listed in Table 14.17 with their stress values at the initial design.

Design Sensitivity Analysis and Results Display

The design velocity field is computed using the boundary displacement method, while, the sensitivity coefficients are calculated using the direct differentiation method. The design sensitivity coefficient of stress at element 772 is displayed using color plots. Figure 14.16(a) shows that if on the front surface design parameter dp1 is moved a unit of magnitude in the negative *y*-direction to increase the thickness of the front side, the stress in element 772 is decreased by 6078 psi. Design parameter dp2 similarly decreases the stress by a lesser amount. The other design parameters are not influential. Figure 14.16(b) shows that perturbing design parameter dp6 a unit magnitude in the positive *y*-direction causes the rear surface to be thicker, decreasing the stress concentration 5208 psi. The other design parameters defined on the rear surface similarly decrease the stress, but by a lesser amount. Figure 14.16(c) shows that design parameter dp14 has the most influence on stress performance. Perturbing design parameter dp14 a unit magnitude in the *x*-direction makes the upper front slot thicker, and decreases the stress concentration 19,581 psi. Design parameter dp15 has the opposite behavior: moving it in the negative *x*-direction will reduce stress at element 772. For the rear slot surface, moving the design parameter dp19 one unit of magnitude in the negative *x*-direction decreases the stress concentration 13,524 psi, as shown in Fig. 14.16(d). The other design parameters defined on the rear slot surface decrease the stress, but by a lesser amount.

Among the five geometric features, the design perturbations on the bottom surface have the least influence on the stress concentration, as shown in Fig. 14.16(e). However, it is interesting to note that instead of adding material to the top surface, moving the grids inward reduces the stress concentration. This is because the overall stress field is redistributed when material is cut from the bottom surface, such that stress concentration is reduced.

In summary, the most influential design parameters are located on the two slots. The closer the grids are to the stress concentration area, the more influential they are. Moreover, top surface variation has little influence.

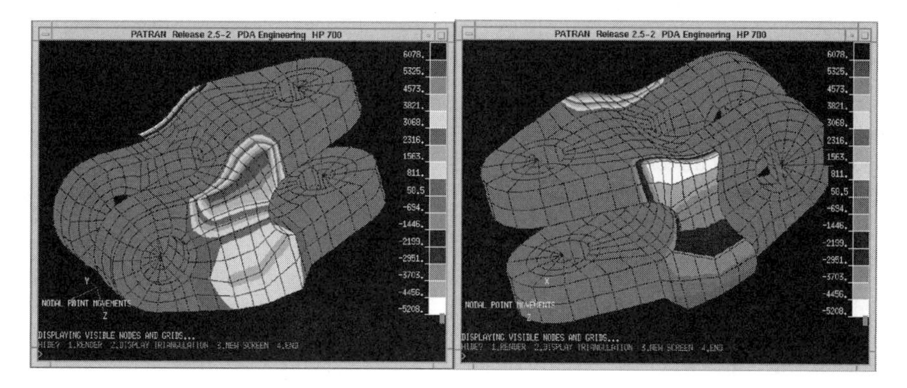

(a) Sensitivity for Front Surface (b) Sensitivity for Rear Surface

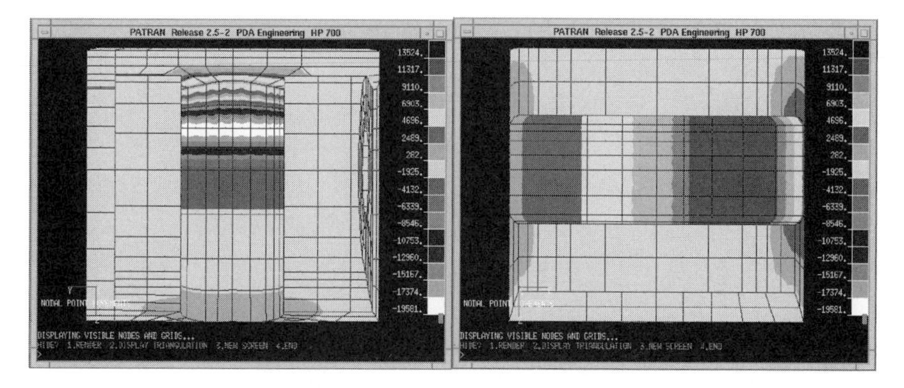

(c) Sensitivity for Front Slot Surface (d) Sensitivity for Rear Slot Surface

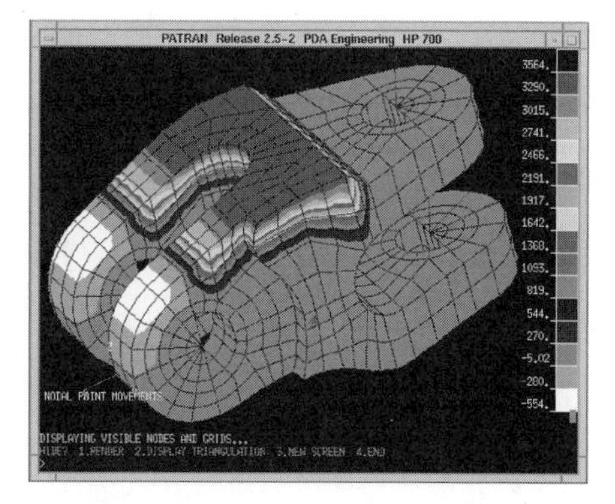

(e) Sensitivity for Bottom Surface

Figure 14.16. Contour display of design sensitivity of stress at element 772.

The design sensitivity coefficient of stress in element 772 is also displayed in the bar chart shown in Fig. 14.17, while the sensitivity coefficients of high stress at element 818 are displayed in a bar chart shown in Fig 14.18. The steepest descent directions of the two stress constraints are found to be in conflict, for example, dp6.

Design Trade-off and What-if Study

Since some design directions for the two high stresses are in conflict, a trade-off study is performed to search for the best design direction using the constraint correction option [96]. The design direction is found and a what-if study is performed using the direction and a step size of 0.5 in as the design perturbation.

Based on the design perturbation, new cost and constraint function values are predicted by the what-if study. As listed in Table 14.18, the results show that stress measures are improved and the cost function is reduced by 0.9%. The what-if results are shown in Fig. 14.19.

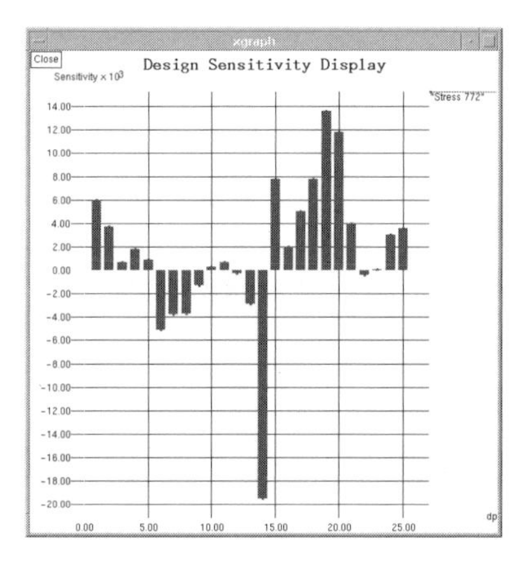

Figure 14.17. Bar chart of design sensitivity of stress at element 772.

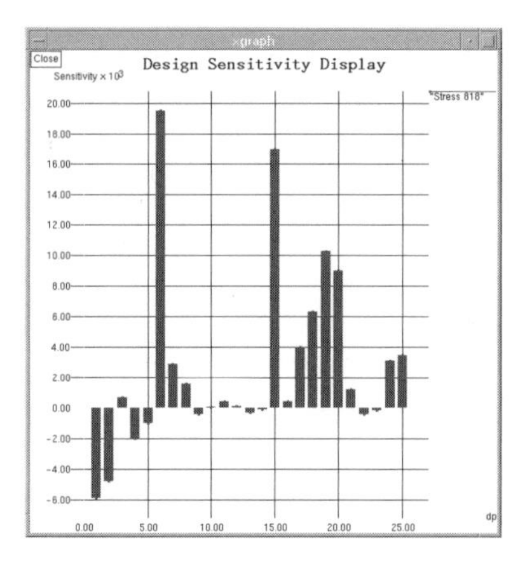

Figure 14.18. Bar Chart of deign sensitivity of stress at element 818.

Table 14.18. What-if study results of cost and selected constraints.

Cost/Constraint	Element	Current Values	Predicted Value	% Change	FEA Results
Cost (volume)		89.7099 in^3	88.8996 in^3	–0.90	88.8949 in^3
Constraint 669	817	122.3435 ksi	113.1767 ksi	–7.5	115.3268 ksi
Constraint 460	268	132.4939 ksi	125.3103 ksi	–5.4	126.5164 ksi
Constraint 462	324	132.8683 ksi	125.6632 ksi	–5.4	125.8293 ksi
Constraint 550	772	141.2288 ksi	135.8741 ksi	–3.8	136.5004 ksi
Constraint 672	818	146.0990 ksi	130.2237 ksi	–10.9	129.4148 ksi

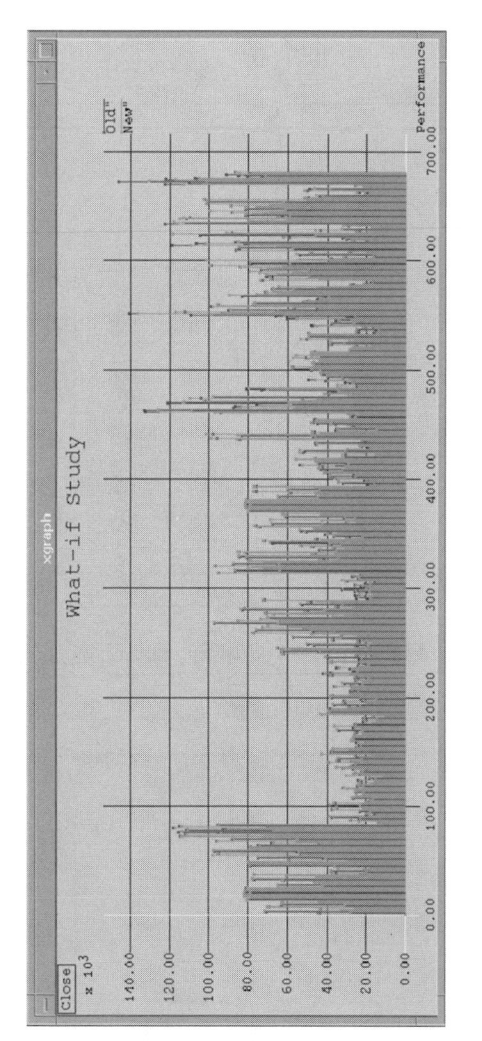

Figure 14.19. Bar chart of what-if results.

Interactive Design Iterations

Four interactive design iterations are carried out using the trade-off and what-if studies of the interactive design mode. The design history is summarized in Table 14.19 and the history of cost and selected constraint functions are shown in Figs. 14.20 and 14.21, respectively.

Table 14.19. Design iteration history.

Design Iteration	Cost Function (Volume in^3)	Constraint 550 (Stress 772 ksi)	Constraint 672 (Stress 818 ksi)
Initial Design	89.7	146.099	141.229
1	88.9	129.415	136.500
2	81.9	125.200	135.474
3	79.1	122.285	135.200
Optimal Design	77.7	120.463	135.122
Reduction	13.3%	17.6%	4.3%

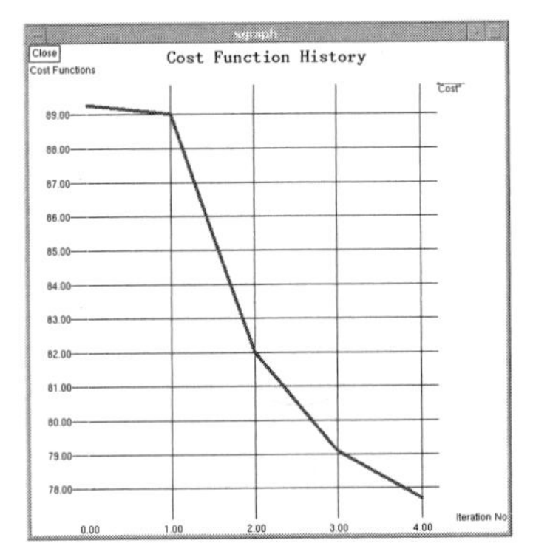

Figure 14.20. Cost function history.

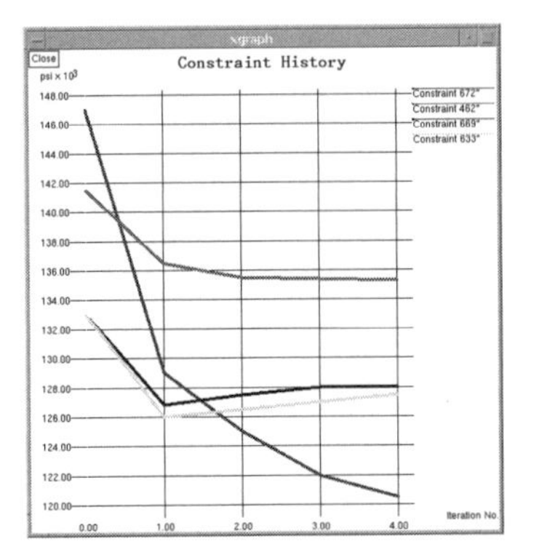

Figure 14.21. (Selected) constraint function history.

Clevis shape at the final design is displayed in Fig 14.22. As shown in the figures, the rear surface becomes thinner and the bottom surface moves upward, suggesting more symmetric design. The front slot becomes an "S" shape, as shown in Fig. 14.12, when projected on x-z plane. No significant changes are found in the front surface and rear slots. The stress contour shown in Fig. 14.23 reveals that the stress concentration at elements 772 and 818 has been reduced and that stress increases near the bottom. However, this increase is not critical. The overall stress distribution throughout the clevis is more even.

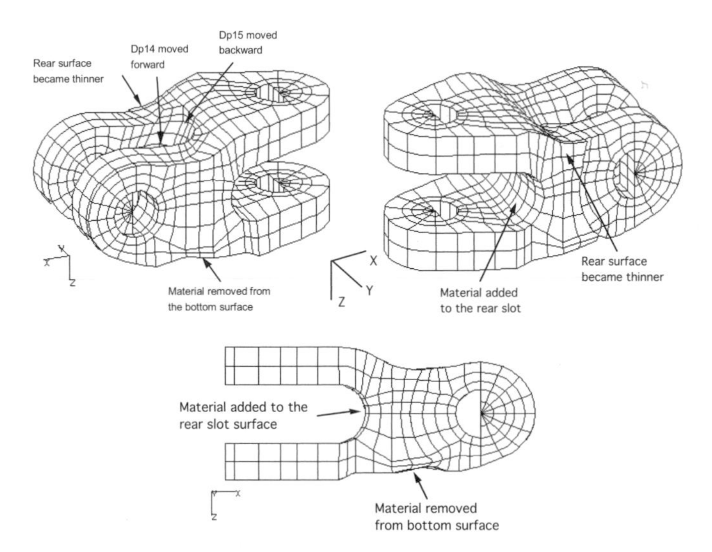

Figure 14.22. Clevis shape at final design.

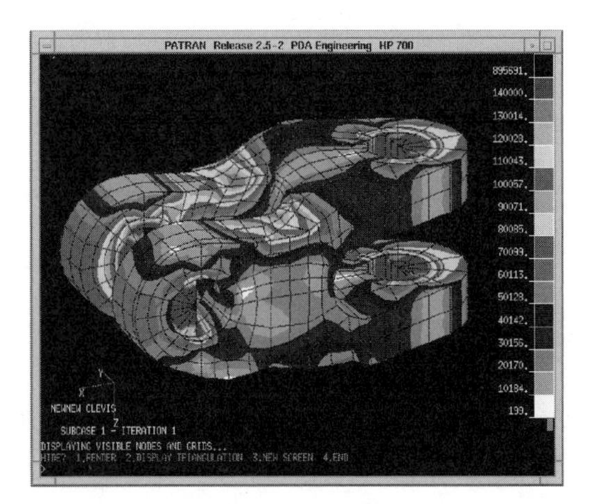

Figure 14.23. Stress contour of clevis at final design.

14.2.2 Design of Turbine Blade

Such high-speed rotating objects as blades, flywheels, and rotors are very common components of turbo-machinery, and include turbines, fans, and compressors. With a rotating object, inertia forces are generated in radial and tangential directions due to angular velocity and angular acceleration, respectively. Two types of design changes contribute to the variation of the inertia force for a rotating object: object translation and object shape change. For a pure translation type of design change, the inertia force varies due to a change in the length of its position vector (rotation radius). In a shape design change, the inertia force varies due to mass redistribution and/or a change in the rotation radius. Note that for a rotating object both geometric shape and inertia force variation affect the structural behavior. In this section, the design optimization procedure of a turbine blade is described.

A turbine blade is inserted into a slot in a disc mounted on a rotating shaft, as shown in Fig. 14.24. The blade can be separated into four major parts: the airfoil, platform, shank, and dovetail. Due to shaft rotation, fluid pressure is applied to the surface of the airfoil, and a centrifugal force is applied to the whole blade structure. After analyzing the blade model, centrifugal force is identified to be the dominant contribution to the blade's structural deformation; therefore, fluid pressure is ignored in finite element analysis and design sensitivity analysis. Moreover, from finite element analysis it is found that the platform does not significantly contribute to the blade's structural behavior, and is thus removed in the modeling process. Note that the shank sustains the majority of stress flow that moves from the dovetail to the airfoil, due to rotation.

Shape of the airfoil is determined by performing an aerodynamics analysis, which is not considered to change in this study. However, the shape of the shank and dovetail can be modified to improve the blade's structural performance.

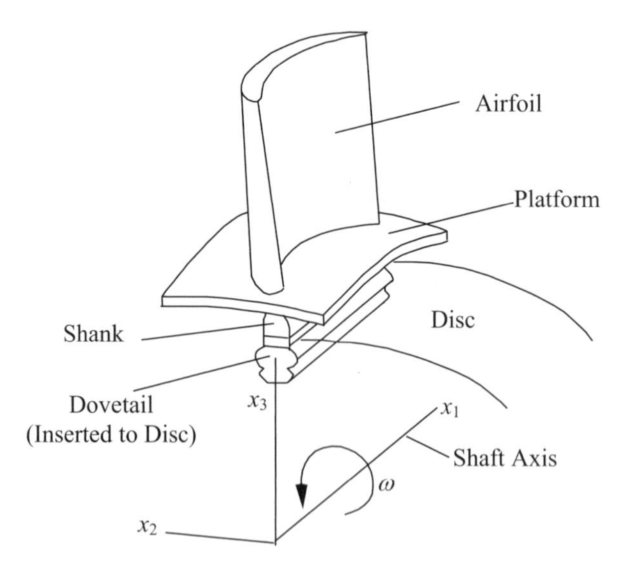

Figure 14.24. Turbine blade physical model.

Structural Modeling

The blade finite element model is created using the geometric modeler PATRAN [42]. To extract geometric data from CAD, the IGES format is used to translate the wireframe model from the CAD tool to PATRAN, as shown in Fig. 14.25(a). Note that the wireframe model accurately describes the airfoil and the dovetail. However, the straight lines that connect the airfoil and the dovetail do not provide an appropriate model for forming the shank. Instead of straight lines, it is necessary to place smooth surfaces between the airfoil and the dovetail, with C^0- and C^1-continuity across the upper and lower interfaces, to create an interface both between the airfoil and the shank, and between the shank and the dovetail.

When creating the airfoil and the dovetail in the wireframe model, unnecessary curves and points are removed, and a number of patches are created using parametric curves. To create the shank, appropriate geometric patches are found that can connect the airfoil and the dovetail, with specified tangent vectors across interfaces to maintain C^0- and C^1-continuity. Once the geometric patches are created for the shank, hyperpatches are formed using the surrounding patches.

After the geometric model is completed, finite elements are meshed in each hyperpatch to generate the finite element model shown in Fig. 14.25(b). The PATRAN blade model has 315 3-D 20-node elements (ANSYS STIF95) and 2388 nodes. The material properties are Young's Modulus $E = 2.99938 \times 10^7$, Poison's ratio $\nu = 0.29$, and mass density $\rho = 7.317313 \times 10^{-4}$. For boundary conditions, the node displacements are fixed on two sides of the dovetail in all directions, and a constant angular velocity $\omega =$ 1,570.8 rad/sec (15,000 rpm) is applied in the x_1-direction to create centrifugal force throughout the blade model, which is a body force type of loading.

Finite element analysis is carried out using the ANSYS code. From the von Mises stress contour shown in Fig. 14.26, stress is found to be high in the shank. However, in the dovetail and the airfoil, stress levels are comparatively low. Consequently, the objective of designing the turbine blade is to relieve the stress concentration in the shank by changing its shape.

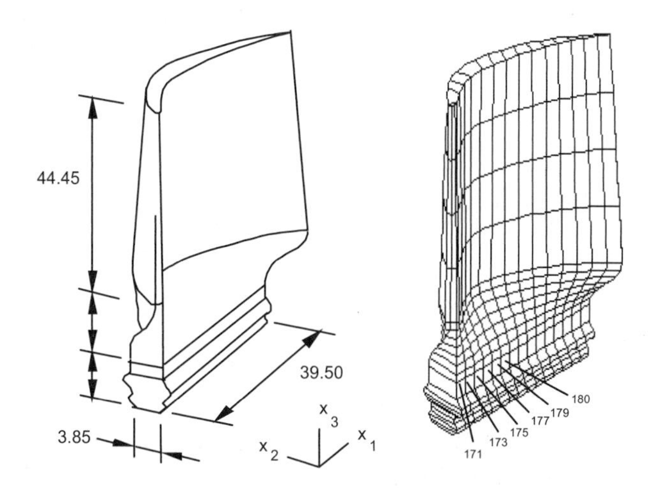

(a) Key Dimensions (Unit: mm) (b) Finite Element Model

Figure 14.25. Turbine blade model.

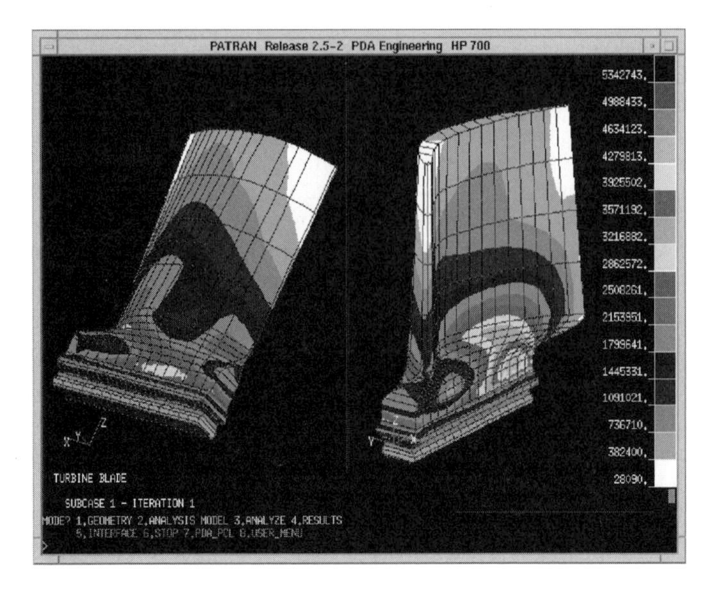

Figure 14.26. Von Mises stress contour of turbine blade model.

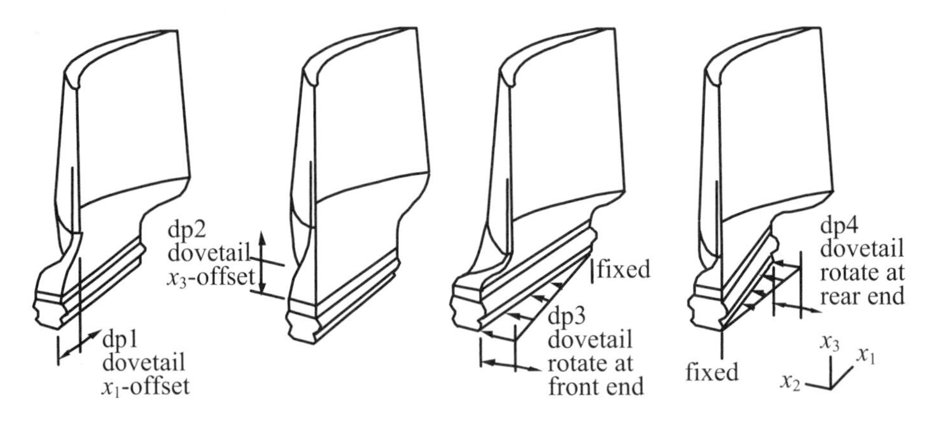

Figure 14.27. Design parameterization for blade.

Design Parameterization

Four design parameters, shown in Fig. 14.27, are defined in the model: offset of the dovetail in the x_1- and x_3-directions, and rotations of the dovetail about the axes that are parallel to the x_3-axis. The four design parameters characterize the design changes by repositioning the dovetail. Thus, the dovetail maintains a constant ffal shape and a straight boundary, even though the shape of the shank itself changes. The shank's shape change and the translation of the dovetail both affect the structural behavior of the blade, since mass is redistributed due to these changes. The effect of the design change due to the dovetail x_2-offset is equivalent to moving both dp3 and dp4 the same degree.

Table 14.20. Gauss stress performance measures over 4.0 MPa.

Element	Gauss Point	Stress
171	2	0.48195E+07
173	13	0.42270E+07
175	6	0.41918E+07
177	6	0.41596E+07
179	6	0.41024E+07
180	6	0.40169E+07

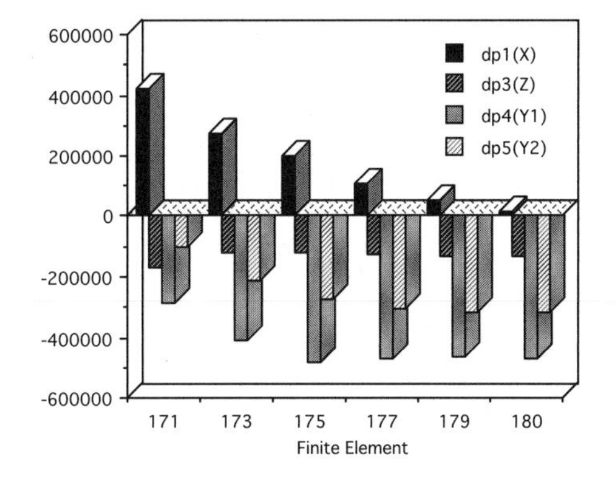

Figure 14.28. Stress sensitivity displayed in bar chart.

Performance Measures

Turbine blade design emphasizes the reduction of stress concentration in the shank area by changing the shank shape, and by repositioning the dovetail without significantly increasing its mass. Since the blade mass and volume differ only by a factor of mass density ρ, volume is taken into account. Consequently, structural volume and high stress measures (over 4.0 MPa) are considered as performance measures. At the current design, structural volume is 15268.304 mm^3, and high stresses occur in six elements, as listed in Table 14.20. Locations of these elements in the blade are shown in Fig. 14.25(b).

Design Sensitivity Analysis

Isoparametric mapping and direct differentiation methods are used to evaluate velocity fields and sensitivity coefficients, respectively. Accuracy of the sensitivity coefficients can be verified using the what-if study and the finite element analysis results described below.

For stress performance measures, design sensitivity coefficients for the six high stress measures are displayed using a bar chart (Fig. 14.28). The chart shows that increasing design parameter dp1, that is, moving the dovetail in the x_1-direction, increases the stress measures, and the effect decreases in the x_1-direction. However, increasing the other three design parameters decreases the stress measures. Among the four design parameters, dp3, the dovetail rotation at the front end, generally has the largest effect, while dp2, the dovetail x_3-offset, has the least effect. To reduce stress at element 171 (the highest stress

Table 14.21. Design sensitivity coefficients and design perturbation.

dpid	Sensitivity (171)	Current Values	Perturbation
1	419459.8	–1.975	–0.768
2	–169430.9	–5.715	0.310
3	–287500.8	–4.292	0.526
4	–104484.8	–4.292	0.191

Table 14.22. What-if results and verification.

Performance	Current Value	Predicted Value	% change	FEA Results	Accuracy %
Volume	15268.3041	15182.7667	–0.56	15182.7573	100.0
Stress 171	4819509.93	4273413.35	–11.33	4289771.25	97.0
Stress 173	4226974.28	3722203.13	–11.94	3734589.77	97.5
Stress 175	4191807.41	3695253.89	–11.85	3704051.12	98.2
Stress 177	4159635.85	3732537.12	–10.27	3740478.96	98.1
Stress 179	4102384.61	3714012.11	–9.49	3723099.27	97.7
Stress 180	4016851.13	3657743.19	–8.94	3666848.63	97.5

measure), x_1-offset, dp1, has the largest effect. For example, moving the dovetail 1.0mm in the x_1-direction will increase the stress by around 0.42 MPa. Rotation at the rear end, dp4, has the least effect, as documented in Table 14.21. The trend is the same for the other stress performance measures, but with differing degrees of effect.

What-if Study

A design improvement that focuses on reducing the highest stress performance measure, identified in element 171, is carried out using the what-if design step. A steepest descent design direction for the stress in element 171 and a step size of 1.0 mm, which yields the design perturbations shown in Table 14.21, have been selected to perform the what-if study. Prediction of volume and stress measures using design sensitivity coefficients is listed in Table 14.22. Accuracy of the predicted performance measures are verified to be excellent using the FEA results obtained for the perturbed design. Reduction of stress performance measures due to such design change is displayed in Fig. 14.29.

From Fig. 14.29, stresses at the six elements are reduced by approximately 10% due to the design change. However, blade volume is reduced by 0.56%. Such a design change not only lowers stress concentration at the shank, but also reduces the structural volume. Based on such a design change, the finite element mesh is updated for the blade, using the velocity field and design perturbation.

The predicted performance measures and finite element mesh update can be obtained within a few workstation CPU minutes, which is extremely efficient compared with FEA methods at the perturbed design. Moreover, the design engineer can efficiently try out several design changes, without going through the modeling and FEA process. The element von Mises stress contours at current and perturbed designs are also displayed in Fig. 14.30 for stress in all elements. Sensitivity coefficients are computed for all stress measures using the direct differentiation method. Note that the contour shown in Fig. 14.30 is obtained from what-if prediction, which has been verified using the FEA results.

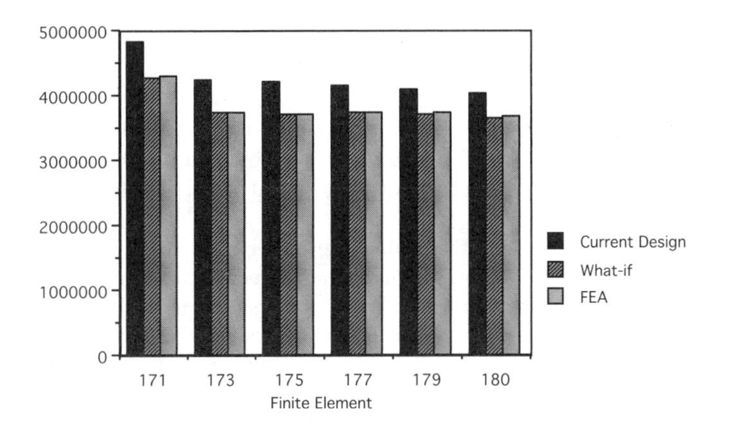

Figure 14.29. Stress what-if study results displayed in bar chart.

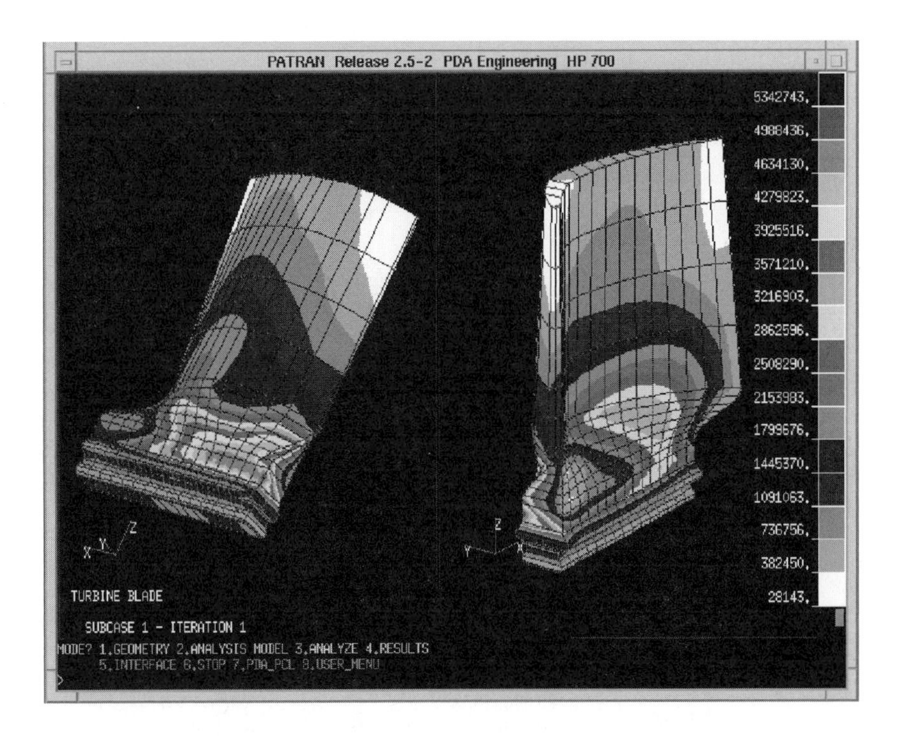

Figure 14.30. Von Mises stress contour at current and perturbed designs.

As shown in Fig. 14.30, the highest stress in the shank decreases from approximately 4.8 to 4.2 MPa. High stresses at the other five elements also decrease. However, stress distribution for the rest of the region remains nearly unchanged due to the design perturbation, which is desirable.

Design Optimization

The objective of the turbine blade design is to minimize the mass of the blade and to keep stress below a certain limit. The objective is achieved by repositioning the dovetail, and therefore, changing the shape of the shank, which is characterized by the four design parameters defined in Fig. 14.27.

Since the mass and volume of the blade differ only by a factor of mass density ρ, the volume is considered as the cost function to be minimized. At the initial design, structural volume is 15268.30 mm^3. The von Mises stress at element Gauss points of all 315 finite elements are defined as constraints with an upper bound of 4.5 MPa. The 10 finite elements that have stress values over 4.1 MPa in the design iterations are shown in Fig. 14.31, and their stress values at initial design are listed in Table 14.23. Also, the x_3-displacement of nodes at the top surface of the airfoil (77 nodes) and the x_1-displacement of nodes at the front and rear corners of the airfoil (8 nodes) must be less than 5 mm to maintain clearance between blade and engine casing, as shown in Fig. 14.31. Thus, in total, 400 design constraints have been defined.

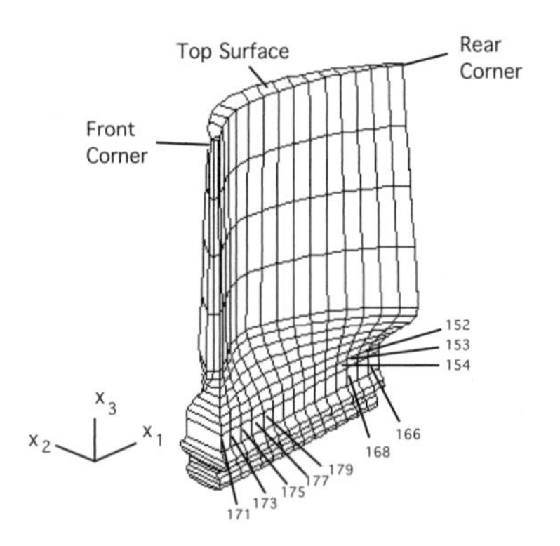

Figure 14.31. Locations of displacement constraints and high stresses.

Table 14.23. Selected gauss stress measures at initial design.

Element	Stress	Upper Bound	Status
171	4.81950993	4.5	violated
173	4.22697428	4.5	inactive
175	4.19180741	4.5	inactive
177	4.15963585	4.5	inactive
179	4.10238461	4.5	inactive
168	2.74871275	4.5	inactive
166	2.15874325	4.5	inactive
154	2.71213725	4.5	inactive
153	2.33576175	4.5	inactive
152	2.22792225	4.5	inactive

Table 14.24. Side constraint definition.

Parameter	Lower Bound	Current Value	Upper Bound
dp1 (x_1-movement)	–5.0	0.0	5.0
dp2 (x_3-movement)	–4.0	0.0	4.0
dp3 (front rotation)	–5.0	0.0	5.0
dp4 (rear rotation)	–5.0	0.0	5.0

As shown in Table 14.23, the stress on element 171 is larger than 4.5 MPa; therefore, the initial design is infeasible. Side constraints are defined for the four design parameters, as listed in Table 14.24, where design parameters 3 and 4 are restricted in the range of –5 to 5 mm, to prevent the dovetail rotation angle from falling outside the ±15° range. Design parameter 2 is restricted to –4 to 4 mm to prevent a large volume change in the disc, although the disc itself is not being considered in the study. Moreover, the bounds defined for the first design parameter ensure the alignment of the airfoil in the blade.

Design optimization of the turbine blade is performed using ANSYS and the modified feasible direction method in DOT, together with the sensitivity analysis and design model update programs.

Optimum design of the turbine blade is obtained in three iterations, with 15 FEAs and three design sensitivity analyses. Boundary displacement and direct differentiation methods are employed to compute the velocity field and sensitivity coefficients, respectively. The design velocity field computed in the first iteration is used for all subsequent design iterations, while the velocity field and design perturbation are used to update the finite element mesh for the new design.

The optimization histories for cost, constraint, and design parameters are shown in Figs. 14.32 – 34. Figure 14.32 shows that the cost function starts at 15268.30 mm^3 and decreases to 14501.83 mm^3 at the first design iteration. During the last two iterations the cost function converges to an optimum criterion defined in DOT. Note that at the initial (infeasible) design cost reduction is possible since the design search direction that corrects the constraint violation (stress at element 171) also reduces the cost (contributed by design parameter dp2). The sensitivity coefficients shown in Table 14.25 illustrate this behavior. The cost is reduced further until it achieves a minimum of 14159.43 mm^3. It is noted that the displacement constraints are inactive throughout the design iterations.

At optimum, all stresses are below the upper bound, as illustrated in Fig. 14.33 where the history of the high stresses at the ten elements is given. Stress at element 171 is initially larger than the upper bound, and is reduced to 3.316 MPa at the optimum design. The highest stress at the optimum design is 3.579 MPa, found at element 179. A comparison of stress at element 179 at optimum to stress at element 171 at the initial design shows that stress has been reduced by 25.7%, and stress in the shank is distributed more evenly at the optimum design.

Figure 14.34 and Table 14.26 show the design parameter history. All design parameters start from value 0.0, and the second design parameter becomes active at optimum to reduce the blade volume. At optimum, the dovetail rotates 6.1° to adjust stress distribution in the shank, which is within the ±15° limit. The blade finite element models at the initial and optimum designs are shown in Fig. 14.35.

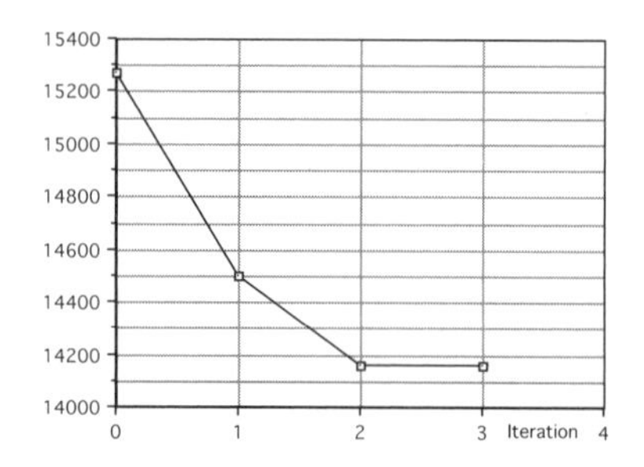

Figure 14.32. Cost function history.

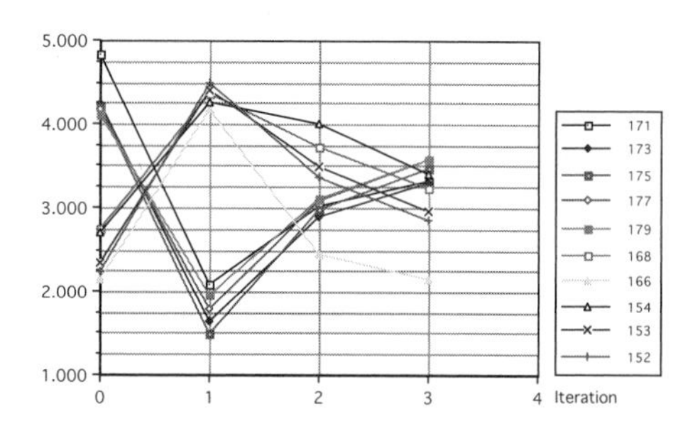

Figure 14.33. Stress constraint function history (selected).

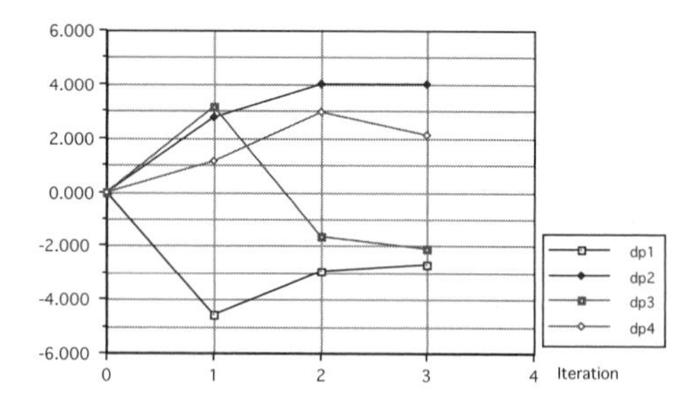

Figure 14.34. Design parameter history.

Table 14.25. Design sensitivity and search direction at initial design.

Direction	dp1	dp2	dp3	dp4
Cost Sensitivity	–.337623E–03	–.276670E+03	0.869761E+00	–.870733E+00
Stress (171) Sensitivity	0.419460E+06	–.169431E+06	–.287501E+06	–.104485E+06
Search Direction	–.100000E+01	0.607480E+00	0.684767E+00	0.249735E+00

Table 14.26. Design parameter history.

Iteration	dp1	dp2	dp3	dp4
Initial Design	0.0	0.0	0.0	0.0
Iteration 1	–4.569	2.775	3.129	1.140
Iteration 2	–2.947	4.000	–1.632	2.943
Iteration 3 (optimum)	–2.664	4.000	–2.121	2.118

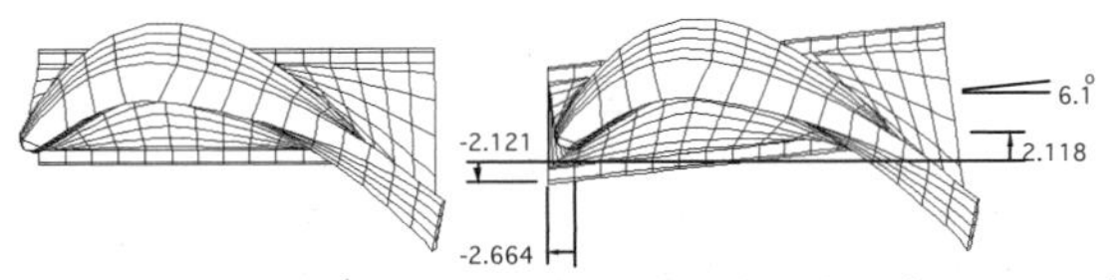

(a) Top View (Initial Design) (b) Top View (Optimum Design)

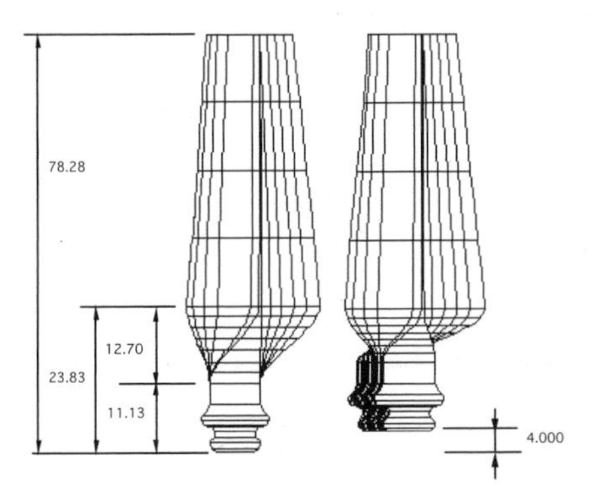

(c) Front View (Initial Design) (d) Front View (Optimum Design)

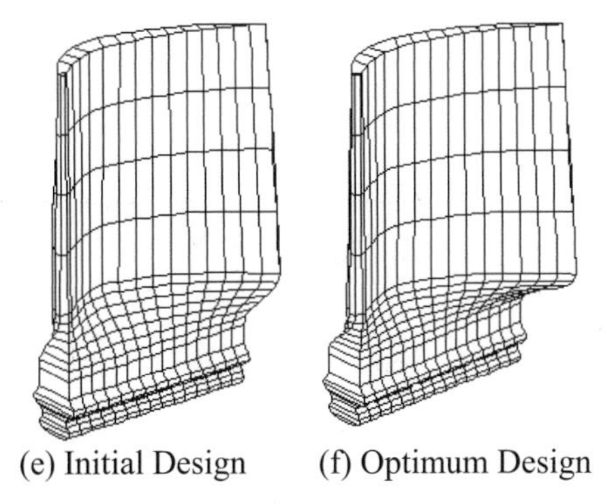

(e) Initial Design (f) Optimum Design

Figure 14.35. Blade models at initial and optimum designs.

14.3 Configuration Design Applications

14.3.1 Design Crane Structure

An 18-bar crane structure, treated in [103], is used as an example of configuration design optimization. The initial geometry and loading conditions are shown in Fig. 14.36. The element cross-sectional areas are linked as follows: $A_1 = A_4 = A_8 = A_{12} = A_{16} = 10.71$ in^2, $A_2 = A_6 = A_{10} = A_{14} = A_{18} = 15.19$ in^2, $A_3 = A_7 = A_{11} = A_{15} = 1.94$ in^2, and $A_5 = A_9 = A_{13} = A_{17} = 5.19$ in^2. Young's modulus is $E = 1.0 \times 10^7$ psi. Finite element analysis is performed using ANSYS 2-D truss element STIF1. There are 18 truss elements, 11 nodal points, and 18 degrees of freedom. Each finite element is treated as a single design component. The perturbed design, shown in Fig. 14.37, is defined so that nodal points 3, 5, 7, and 9 move upward in the x_2 direction by 4", 3", 2", and 1", respectively. In this example, even though the sizing design can be easily treated, only the configuration design variable is considered. In the perturbed design, both the shape (i.e., the length) and the orientation of a truss design component may be changed. A linear shape design velocity field that is consistent with the truss displacement field is used for each truss design component.

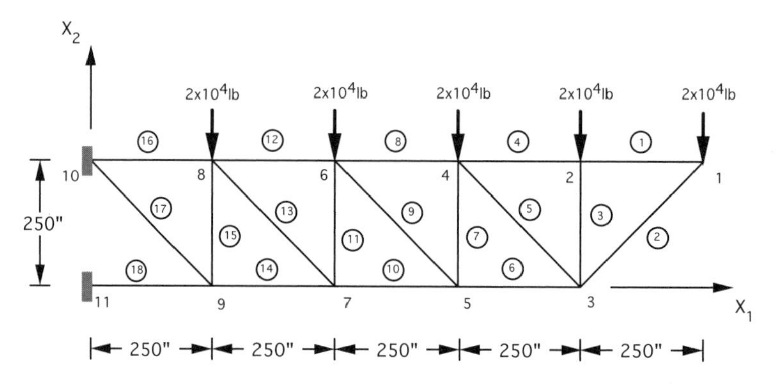

Figure 14.36. Eighteen-bar crane structure.

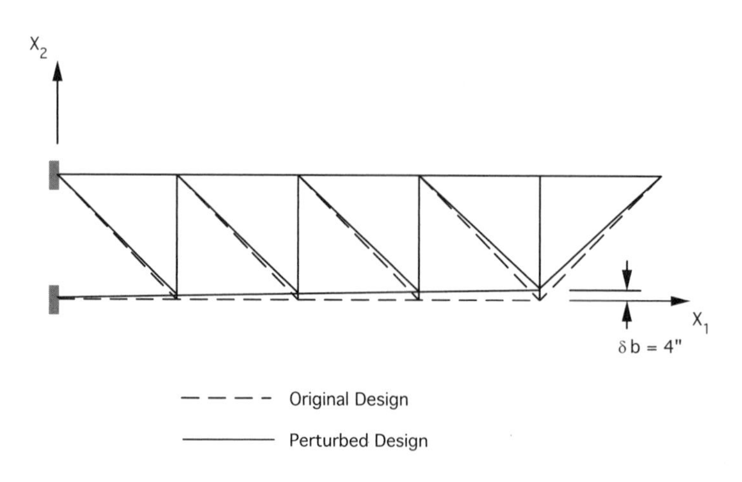

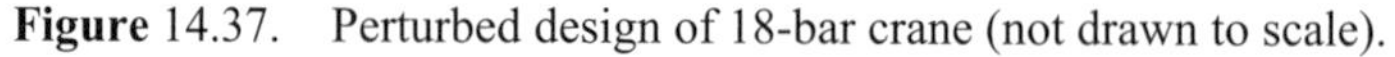

Figure 14.37. Perturbed design of 18-bar crane (not drawn to scale).

Table 14.27. Configuration design sensitivity results of a crane.

Node/ Elem.	Dir./ Stress	$\Psi(b-\delta b)$	$\Psi(b+\delta b)$	$\Delta\Psi(b)$	$\Psi'(b)$	Ratio %
1	x_1	0.97318E+0	0.98775E+0	0.14567E–1	0.14566E–1	100.0
1	x_2	–0.13967E+2	–0.13804E+2	0.16388E+0	0.16391E+0	100.0
1	σ_{axial}	0.18380E+4	0.18978E+4	0.59773E+2	0.59757E+2	100.0
2	σ_{axial}	–0.18474E+4	–0.18772E+4	–0.29803E+2	–0.29793E+2	100.0
3	σ_{axial}	0.00000E+0	0.00000E+0	0.00000E+0	0.00000E+0	0.0
4	σ_{axial}	0.18380E+4	0.18978E+4	0.59773E+2	0.59757E+2	100.0
5	σ_{axial}	0.10878E+5	0.10922E+5	0.43607E+2	0.43598E+2	100.0
6	σ_{axial}	–0.39032E+4	–0.39980E+4	–0.94814E+2	–0.94799E+2	100.0
7	σ_{axial}	–0.20741E+5	–0.20493E+5	0.24746E+3	0.24742E+3	100.0
8	σ_{axial}	0.55358E+4	0.56703E+4	0.13447E+3	0.13445E+3	100.0
9	σ_{axial}	0.16382E+5	0.16316E+5	–0.65411E+2	–0.65397E+2	100.0
10	σ_{axial}	–0.78373E+4	–0.79637E+4	–0.12641E+3	–0.12640E+3	100.0
11	σ_{axial}	–0.31173E+5	–0.30678E+5	0.49488E+3	0.49485E+3	100.0
12	σ_{axial}	0.11116E+5	0.11295E+5	0.17928E+3	0.17927E+3	100.0
13	σ_{axial}	0.21929E+5	0.21667E+5	–0.26160E+3	–0.26159E+3	100.0
14	σ_{axial}	–0.13114E+5	–0.13220E+5	–0.10533E+3	–0.10533E+3	100.0
15	σ_{axial}	–0.41648E+5	–0.40823E+5	0.82476E+3	0.82474E+3	100.0
16	σ_{axial}	0.18600E+5	0.18749E+5	0.14940E+3	0.14939E+3	100.0
17	σ_{axial}	0.27521E+5	0.26976E+5	–0.54498E+3	–0.54498E+3	100.0
18	σ_{axial}	–0.19750E+5	–0.19750E+5	–0.40018E–9	0.29559E–11	–0.7

Based on the orientation change of the design component, derivatives of the orientation design velocity are computed. Configuration design sensitivity results for the displacement and axial stress performance measures are presented in Table 14.27. In Table 14.27, $\Psi(b-\Delta b)$ is the value of the performance measure at the backward perturbed design; $\Psi(b+\Delta b)$ is the value at the forward perturbed design; $\Delta\Psi(b)$ is the change from the central finite difference method; and $\Psi'(b)$ is the predicted change from the design sensitivity result. The ratio between the central finite difference and the predicted change is given in the last column, with 100% meaning complete agreement.

For both displacement and axial stress performance measures, excellent agreement is obtained between the predicted sensitivity results and the central finite difference, except for stress on element 18 where the amount of change is very small. Note that the axial stress on element 3 is zero in the initial configuration. Although the contribution of element 3 vanishes in the current design, because its effect may become significant in a different configuration, it is not eliminated.

As shown in Chapter 7, configuration design sensitivity expressions are composed of two parts: shape variation and orientation change. It is useful to know what percentage of the predicted sensitivity results can be attributed to shape effects and what percentage to orientation effects. The contribution of shape and orientation effects on the configuration design sensitivity analysis of the crane structure is shown in Table 14.28. In Table 14.28, $\Psi'(b)$ is the predicted change in the performance measure, and is the same as the sixth column in Table 14.27. $\Psi'_{sh}(b)$ is the amount contributed from the shape effect, and $\Psi'_{or}(b)$ is the contribution from the orientation effect. The percentage of each contribution is calculated by $[|\Psi'_{sh}|/(|\Psi'_{sh}| + |\Psi'_{or}|) \times 100]\%$ for the shape effect, and $[|\Psi'_{or}|/(|\Psi'_{sh}| + |\Psi'_{sh}|) \times 100]\%$ for the orientation effect. Table 14.28 shows that the contribution from the orientation effect is dominant for all selected performance measures. For the displacement at node 1 in the x_2-direction, both the shape and orientation effect are

Table 14.28. Contributions of shape and orientation effects on configuration design sensitivity analysis of a crane.

Node/ Elem.	Dir./ Stress	$\Psi'(b)$	Shape effect $\Psi'_{sh}(b)$	%	Orientation effect $\Psi'_{or}(b)$	%
1	x_1	0.14566E–1	–0.82598E–16	0.00	0.14566E–1	100.00
1	x_2	0.16391E+0	0.67178E–01	40.99	0.96728E–1	59.01
1	σ_{axial}	0.59757E+2	–0.18720E–14	0.00	0.59757E+2	100.00
2	σ_{axial}	–0.29793E+2	0.10658E–13	0.00	–0.29793E+2	100.00
3	σ_{axial}	0.00000E+0	0.00000E+00	0.00	0.00000E+0	0.00
4	σ_{axial}	0.59757E+2	–0.17513E–12	0.00	0.59757E+2	100.00
5	σ_{axial}	0.43598E+2	–0.12920E–11	0.00	0.43598E+2	100.00
6	σ_{axial}	–0.94799E+2	0.41862E–12	0.00	–0.94799E+2	100.00
7	σ_{axial}	0.24742E+3	0.20224E–11	0.00	0.24742E+3	100.00
8	σ_{axial}	0.13445E+3	–0.57849E–12	0.00	0.13445E+3	100.00
9	σ_{axial}	–0.65397E+2	–0.14520E–11	0.00	–0.65397E+2	100.00
10	σ_{axial}	–0.12640E+3	0.70407E–12	0.00	–0.12640E+3	100.00
11	σ_{axial}	0.49485E+3	0.25956E–11	0.00	0.49485E+3	100.00
12	σ_{axial}	0.17927E+3	–0.10081E–11	0.00	0.17927E+3	100.00
13	σ_{axial}	–0.26159E+3	–0.14963E–11	0.00	–0.26159E+3	100.00
14	σ_{axial}	–0.10533E+3	0.10888E–11	0.00	–0.10533E+3	100.00
15	σ_{axial}	0.82474E+3	0.26381E–11	0.00	0.82474E+3	100.00
16	σ_{axial}	0.14939E+3	–0.14689E–11	0.00	0.14939E+3	100.00
17	σ_{axial}	–0.54498E+3	–0.13774E–11	0.00	–0.54498E+3	100.00
18	σ_{axial}	0.29559E–11	0.14618E–11	48.32	0.15632E–11	51.68

important, although the latter is more important than the former. The sensitivity results indicate that in the crane of the structure displacement and axial stress performance measures will be most affected by varying the orientations of components.

Design Optimization

A nonlinear programming algorithm is used to iteratively optimize the layout of a crane structure, using the configuration design sensitivity results. The minimum weight of the structure is considered, with constraints on the element stress. The coordinates of nodal points are treated as design parameters. The objective of the crane structure design may be stated as follows: subject to geometric limitations, design the layout of the structural members such that the weight of the crane will be minimized, and none of the structural members will yield under a loading condition. To achieve this goal, the optimal design problem can be formulated:

$$\begin{aligned} &\text{minimize} && f(\boldsymbol{b}) \\ &\text{subject to} && \sigma \le \sigma_0 \\ & && \boldsymbol{b}^L \le \boldsymbol{b} \le \boldsymbol{b}^U, \end{aligned} \tag{14.1}$$

where $f(\boldsymbol{b})$ is the objective function, $\boldsymbol{b}$ is the design parameter vector, σ is the axial stress, σ_0 is the given allowable stress, and $\boldsymbol{b}^L$ and $\boldsymbol{b}^U$ are the geometric limitations.

The objective function, which is the weight of the structure, can be defined as

$$f(\boldsymbol{b}) = \sum_{i=1}^{NM} \rho A_i L_i(\boldsymbol{b}), \tag{14.2}$$

where NM is the number of structural members, ρ is the material density, A_i is the cross-sectional area, and $L_i(\boldsymbol{b})$ is the length of the ith member. In (14.2), $L_i(\boldsymbol{b})$ is a function of design parameter **b,** which indicates the position of grid points. Material density is given as ρ = 0.1 lb/in^3 and cross-sectional areas are defined as constant. The x_1 and x_2 coordinates of nodal points 3, 5, 7, and 9 are treated as the design parameter, that is, $\boldsymbol{b} = [x_1(3), x_2(3), x_1(5), x_2(5), x_1(7), x_2(7), x_1(9), x_2(9)]^T$. The lower bound for each design parameter is –10 in, while the upper bound is 1500 in. The axial stress performance measure is defined for all members in the structure. There are 18 stress constraints. The allowable stress σ_0 is given as 20 ksi.

The design problem in (14.1) can be solved using a gradient-based mathematical programming algorithm. The linearization method LINRM [104] is used to solve the optimal design problem, and is based on a recursive quadratic programming algorithm. The concept behind such an algorithm is that instead of solving the optimality criteria, a small perturbation of the design parameter is determined at each iteration to reduce the objective function and to eliminate constraint violations. A more detailed discussion of this algorithm can be found in [104]. When the objective function cannot be further reduced by changing the design parameters, it is said that the program has converged and that an optimum solution has been obtained. It has been proved [105] that a local optimum solution is obtained when the L^2-norm of the design perturbation becomes zero.

The results of the optimal design of a crane structure are summarized in Table 14.29. The weight of the crane is reduced 9.7%, from 4322.77 to 3904.90 lb. In addition, large constraint violations have been removed, indicating that a much better design has been obtained with a significant saving of material. The convergence histories of the objective function and the L^2-norm of the design perturbation are shown in Figs. 14.38 and 14.39.

A slow convergence is shown near the optimal point in Figs. 14.38 and 14.39, using LINRM. About 8.2% of the cost reduction is achieved in the first seven iterations. Between the 8th and 75th iteration, only 1.4% of the cost reduction is obtained. After the 75th iteration, the value of the objective function stays almost the same. This type of slow convergence has been observed in other studies of configuration design optimization [106]. Hansen and Vanderplaats [103] presented an optimum configuration of the same crane structure, using an approximate structural analysis that is based on a first order Taylor expansion of the member forces. They obtained an optimum solution in eight

Table 14.29. Numerical results of the optimal design of a crane.

Initial design	Final design	LINRM	Hansen [103]
$x_1(3)$*	1000.00	920.80	881.42
$x_2(3)$	0.00	202.36	178.76
$x_1(5)$	750.00	647.28	628.90
$x_2(5)$	0.00	151.20	124.92
$x_1(7)$	500.00	428.17	390.54
$x_2(7)$	0.00	83.11	66.79
$x_1(9)$	250.00	305.07	313.16
$x_2(9)$	0.00	42.28	45.03
Cost f	4322.77 lb	3904.90 lb	3906.8 lb
L^2–norm of gradient	54.13692	0.27161	–
No. of active constraints	1.00	7	–
Max. constraint violation	1.577319	0.42802×10^{-4}	–
No. of iteration		95	8

*Note: Coordinates are in inches.

iterations using DOT [98]. The results obtained in [103] are listed in Table 14.29 and compared with those obtained in this section using LINRM. At optimum, the values of the objective function are nearly identical. In addition, the optimum configurations are similar, as seen by comparing Figs. 14.40(b) and 14.40(e). The configurations obtained from LINRM at the 7th, 20th, and 95th iterations are shown in Figs. 14.40(c), 14.40(d), and 14.40(e), respectively. Although the configuration at the 20th iteration is close to the final shape, the program takes another 75 iterations to arrive at the final design, as shown in Figs. 14.40(d) and 14.40(e). The final result tends to have a sharp tip with evenly distributed cross members.

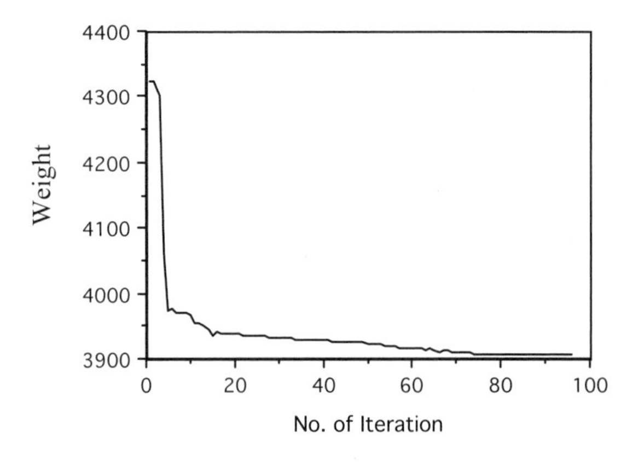

Figure 14.38. Convergence history of objective function.

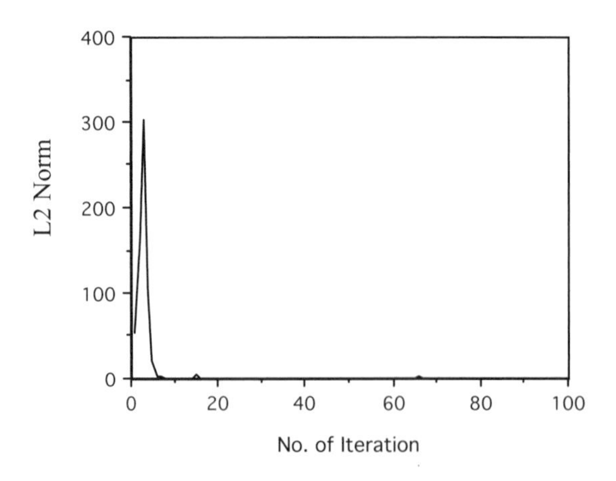

Figure 14.39. Convergence history of L^2-norm of design perturbation.

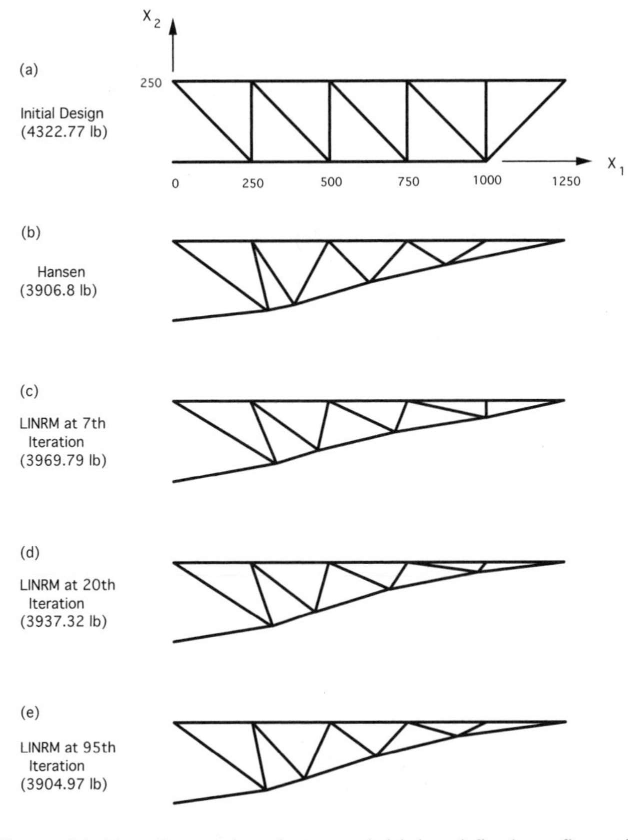

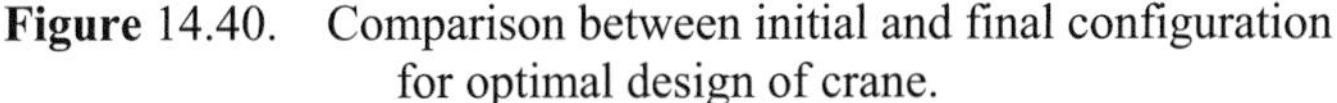

Figure 14.40. Comparison between initial and final configuration for optimal design of crane.

14.4 Nonlinear Design Applications

The computational analysis of a nonlinear structural problem is a challenging enough area without discussing design optimization. Since the cost of structural analysis is very expensive, design optimization requires a large amount of computational time. The main issue in the design of a nonlinear structural problem is the accuracy for faster convergence in optimization, and efficiency of design sensitivity information. In this section, the design applications of nonlinear structural problems are introduced.

The design sensitivity procedure of a nonlinear structural problem is carried out at each time step of the response analysis. Since design sensitivity analysis uses the same stiffness matrix as response analysis, this matrix has to be saved at each time step. Consequently, a tremendous storage capacity is required if the design sensitivity analysis is implemented out of the response analysis code. It is much more efficient if the design sensitivity code is implemented within the response analysis code. For this reason, numerical examples in this section are carried out using a public domain code or an analysis code that is developed for the purpose.

14.4.1 Design of Windshield Wiper

In this section an automotive windshield wiper is optimized to reduce the structural volume and concentration of stress, while maintaining a contact force between the wiper and glass. The meshfree method [48] is used for structural analysis and design sensitivity analysis purposes. The continuum-based structural and sensitivity formulations are discretized by reproducing the kernel particle method (RKPM), where the structural domain is represented using a finite number of particles. RKPM introduces a modified kernel function by imposing a reproducing condition such that the estimates of displacement variables exactly reproduce polynomials up to a certain degree. Unlike a conventional FEA method, which may present difficulties for large deformation analysis and shape optimization, a solution using RKPM is much less sensitive to mesh distortion.

Figure 14.41 shows the geometry of the windshield blade and glass, with the discretized particles of RKPM. Since there is a significant difference in stiffness between rubber and glass, the glass can be assumed to be rigid compared with the rubber material. For convenience, the rigid wall is approximated as a straight line, and a vertical geometry of the rigid wall is added to ensure a smooth deformation of the contact region. The upper part of the blade is supported by a steel slab. Material constants and contact parameters are shown in Table 14.30.

As the rigid wall moves to the left, the edge of the blade is in contact with the glass, which is modeled as flexible-rigid body contact. Blade deformation is large enough that the first wing of the blade makes contact with the second wing, which is modeled as a flexible-flexible body contact. In addition, the second wing makes contact with the steel slab, and this contact is modeled as a flexible-rigid body contact. The function of the thin neck is to generate flexibility so that blade direction can easily be reversed when the blade changes direction. The role of the wing is to supply enough contact force at the tip. Figure 14.42 shows a von Mises stress contour plot with the deformed geometry at the final configuration. Stress concentration is found at the neck and the edge due to the bending effect.

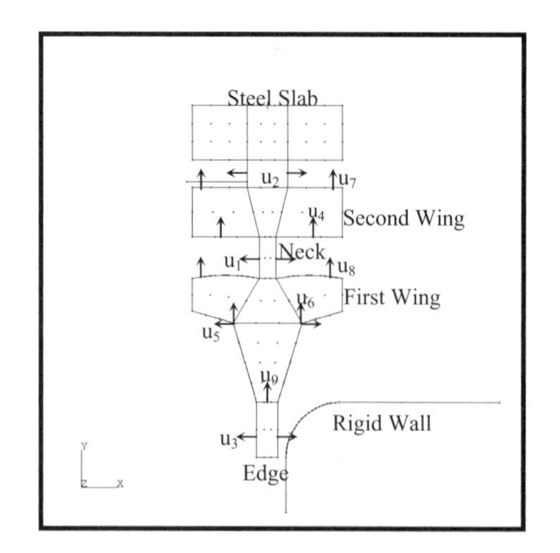

Figure 14.41. Geometry and design parameters of windshield wiper.

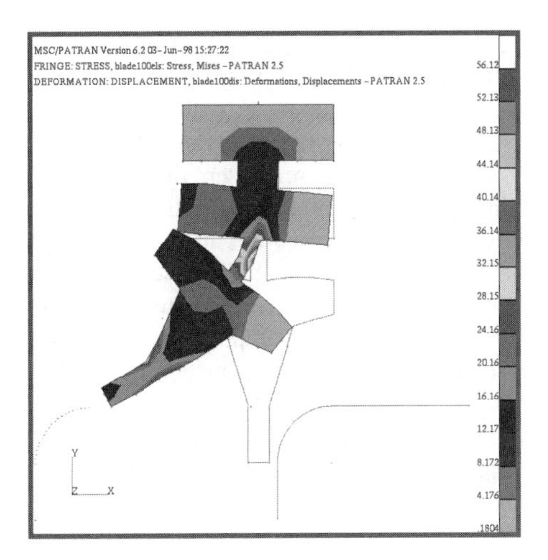

Figure 14.42. Von Mises stress contour plot of windshield wiper.

Table 14.30. Material constants and contact parameters.

D_{10} = 80 kPa	D_{01} = 20 kPa	K = 80 MPa
μ = 0.15	$\omega_n = 10^7$	$\omega_t = 10^6$

Table 14.31. Sensitivity results and comparison with finite difference method.

Performance	ΔΨ	Ψ′	(ΔΨ/Ψ′) × 100%
u_1			
Area	.28406E-5	.28406E-5	100.00
vm_{53}	.19984E-3	.19984E-3	100.00
vm_{54}	.28588E-3	.28588E-3	100.00
F_{cy}	.55399E-5	.55399E-5	100.00
u_3			
Area	.68663E-5	.68663E-5	100.00
vm_{53}	.19410E-3	.19410E-3	100.00
vm_{54}	.68832E-4	.68832E-4	100.00
F_{cy}	.43976E-4	.43976E-4	100.00
u_5			
Area	.33000E-5	.33000E-5	100.00
vm_{53}	.22762E-4	.22762E-4	100.00
vm_{54}	.77289E-5	.77286E-5	100.00
F_{cy}	.62356E-5	.62355E-5	100.00
u_6			
Area	-.24000E-5	-.24000E-5	100.00
vm_{53}	-.16452E-4	-.16452E-4	100.00
vm_{54}	.36694E-5	.36690E-5	100.01
F_{cy}	-.26072E-5	-.26072E-5	100.00

The geometry of the structure is parameterized using nine shape design variables, as shown in Fig. 14.41. The design velocity field at the boundary is obtained first by perturbing the boundary curve corresponding to the design variable, and the domain design velocity field is computed using the isoparametric mapping method presented in Chapter 13. Four performance measures are chosen: the total area of the windshield wiper, two von Mises stresses in the neck region, and the y-directional contact force at the edge.

The design sensitivity analysis is carried out at each converged load step to compute the material derivative of the displacement. The sensitivities of other performance measures are computed at the final converged load step using the material derivative of the displacement. The cost of sensitivity computation is about 4% of that of response analysis per design parameter, which is quite efficient compared with the finite difference method. The accuracy of the sensitivity is compared with the forward finite difference results for a perturbation size of $\tau = 10^{-6}$. Table 14.31 provides data on the accuracy of the sensitivity results. The second column $\Delta\Psi$ denotes finite difference results, and the third column represents the change in the performance measure from the proposed method. Excellent sensitivity results are obtained.

The objective of this design problem is to reduce the amount of windshield material by changing its shape. Stress concentration should be reduced, while the contact force at the tip should be increased. The design optimization problem is

$$\begin{aligned} &\text{mininize} && Area(39) \\ &\text{subject to} && \sigma_{53}(75), \sigma_{54}(45), \sigma_{76}(32), \sigma_{84}(34) \leq 55 \\ &&& F_{y128}(5) \geq 5.5 \\ &&& -0.2 \leq u_i \leq 0.2 \; i = 1,3,7,8 \\ &&& -0.3 \leq u_i \leq 0.3 \; i = 2,4 \\ &&& -0.6 \leq u_i \leq 0.6 \; i = 5,6 \\ &&& -0.1 \leq u_i \leq 0.1 \; i = 9, \end{aligned} \tag{14.3}$$

where the values given within parentheses are the initial design values. Design optimization is carried out using the sequential quadratic programming (SQP) method.

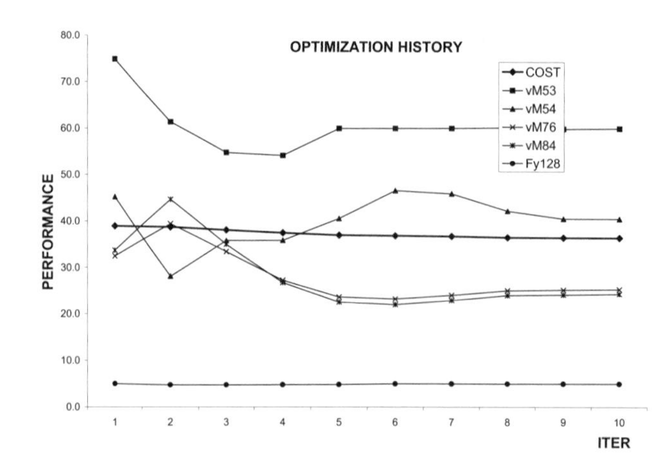

Figure 14.43. Optimization history of windshield wiper contact problem.

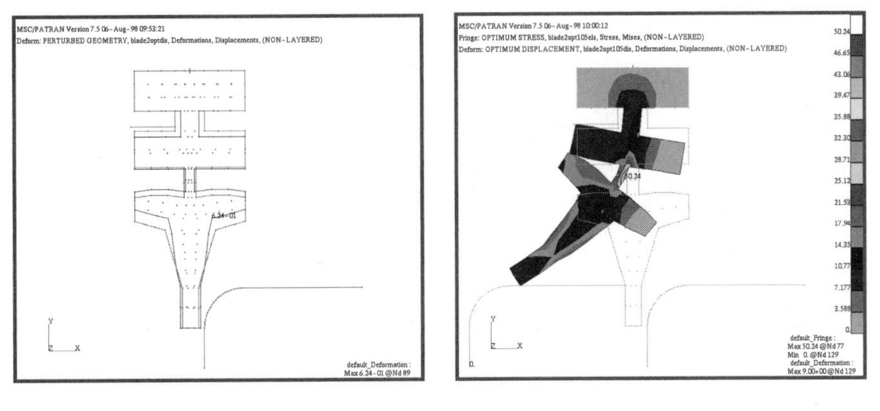

(a) Optimum Shape (b) Analysis Result

Figure 14.44. Optimized shape design and analysis result.

The performance values are supplied to the optimizer by using the nonlinear analysis method (RKPM), and the sensitivity coefficients are provided by using the methods presented in Chapter 10. Optimization is converged after 10 iterations, and Fig. 14.43 shows the history of the cost and constraint functions. The cost function, which is the total area, is reduced by 3.5%. Figure 14.44 shows the optimized shape and analysis results. Edge thickness has been increased, increasing the contact force, while neck thickness has been reduced, decreasing the bending stress in that region. The thickness of the first wing is reduced to the lower bound since the level of the stress is relatively low.

14.4.2 Configuration Design of Vehicle A–Pillar

An automotive A-pillar model, shown in Fig. 14.45, is in need of a design that has less deflection at the junction, and with some specified stress limitations when it is subjected to an oblique impact loading. Each design component is modeled using six beam elements.

The objective function, which is the total deflection at the junction of the three components at the final load step, can be defined as

$$f(\boldsymbol{b}) = \sqrt{\sum_{i=1}^{3} (z_i)_m^2} \quad \text{at node } m = 1 \text{ and load step } n = 100, \tag{14.4}$$

where n = 100 is the final load step. Constraint functions are given for the structure's mass, shear stress, and effective plastic strain, as

$$\begin{aligned} &mass \le 1.6 \\ &mass \ge 1.4 \\ &\tau^{13} \le 1.0 \times 10^{9} \\ &\varepsilon_{eff}^{1} \le 6.5 \times 10^{-2} \\ &\varepsilon_{eff}^{7} \le 6.5 \times 10^{-2} \end{aligned} \tag{14.5}$$

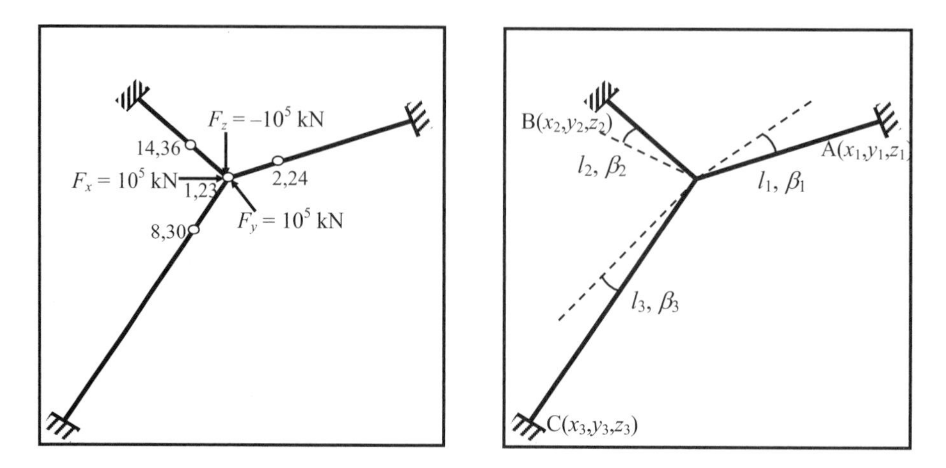

Figure 14.45. A-pillar model and design variables for optimum design.

Table 14.32. Design variables and their bounds (unit: m).

Design Variable	Lower Bounds	Initial Design	Upper Bounds
1 (x1)	0.594018	0.600000	0.603000
2 (y1)	–0.060200	0.000000	0.060200
3 (x2)	–0.060200	0.000000	0.060200
4 (y2)	0.594018	0.600000	0.603000
5 (x3)	–0.829064	–0.600000	0.180701
6 (z3)	–0.829064	–0.600000	0.180701

Design variable vector $\boldsymbol{b}$ is defined as the Cartesian coordinates of points A(x_1, y_1, z_1), B(x_2, y_2, z_2), and C(x_3, y_3, z_3), as shown in Fig. 14.45. It is converted to the lengths and orientation angles of three design components during the configuration design sensitivity analysis and optimization process, because the orientation changes can be expressed more accurately in an angular coordinate system than in a Cartesian one. Note that even though the length and Euler angles of each component are used as design variables during the configuration design sensitivity analysis and optimization process, the design parameter consists of the global coordinates of each component.

The design velocity field of each finite element is computed using the initial and perturbed geometry of three components that are assumed to remain straight after design perturbation. The initial design and their bounds are defined in terms of their global coordinates in Table 14.32.

After a total of 55 iterations, the optimum design is found. Optimum design results of the A-pillar model are summarized in Table 14.33. The total displacement at node 1, which is the objective function, is significantly reduced from 10.010 to 2.718 cm. In addition, constraint violations have been removed in the optimum design. A-pillar members tend to be configured perpendicular to one another at the optimum design, as shown in Fig. 14.46. The transient displacement at node 1 is shown in Fig. 14.47. All structural constraints (3, 4, and 5) at specified time steps, as well as the mass requirement (constraints 1 and 2) are satisfied. The deformed shape of the original and optimum design is shown in Fig. 14.48. To resist vertical deflection, the inclined member is rotated in the vertical direction. To satisfy the third constraint, the shear stress is decreased in element 13, primarily by rotating the member, as shown in Fig. 14.49. In addition, the effective plastic strains in elements 1 and 7 are changed according to the specified criteria, as shown in Fig. 14.50. The optimization history of the objective function is shown in Fig. 14.51.

Table 14.33. Results for optimum design of A-pillar.

	Number	Lower Bound	Initial	Optimum	Upper bound
Objective function			1.00966e−1	2.71771e−2	
Constraints	1		−1.34379e−3	−1.19170e−1	
	2		−1.41321e−1	−6.65710e−3	
	3		−1.09789e−1	−1.09390e+0	
	4		3.89037e−1	−1.57446e−1	
	5		−2.16917e−1	−1.98013e−2	
Design variables	1	5.94018e−1	6.00000e−1	6.00704e−1	6.03000e−1
	2	−6.01996e−2	0.00000e+0	−4.14010e−2	6.01996e−2
	3	−6.01996e−2	0.00000e+0	4.11542e−2	6.01996e−2
	4	5.94018e−1	6.00000e−1	5.95579e−1	6.03000e−1
	5	−8.29064e−1	−6.00000e−1	−1.46342e−3	1.80701e−1
	6	−8.29064e−1	−6.00000e−1	−6.07691e−1	1.80701e−1

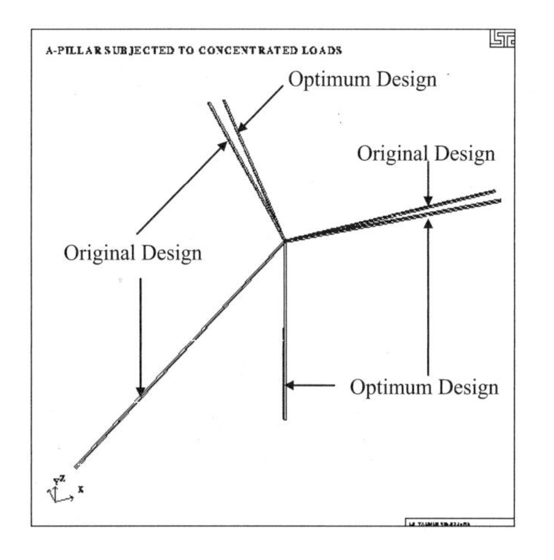

Figure 14.46. Initial and optimum design of A-pillar.

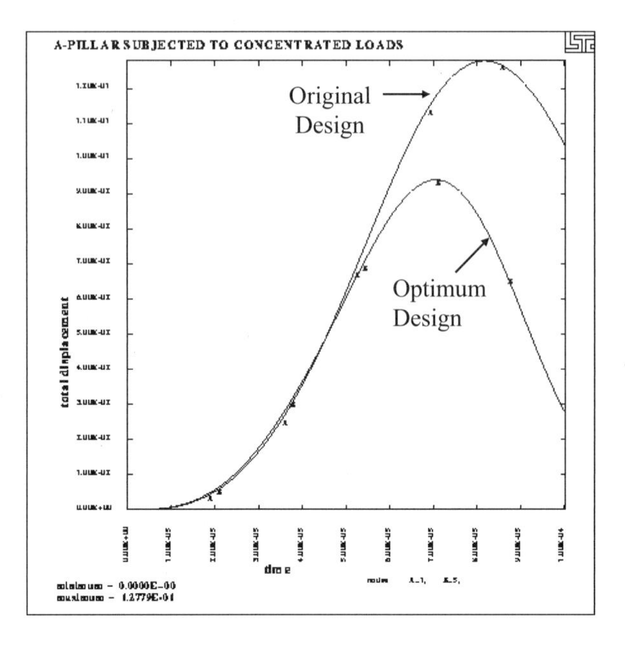

Figure 14.47 Total displacement at node 1.

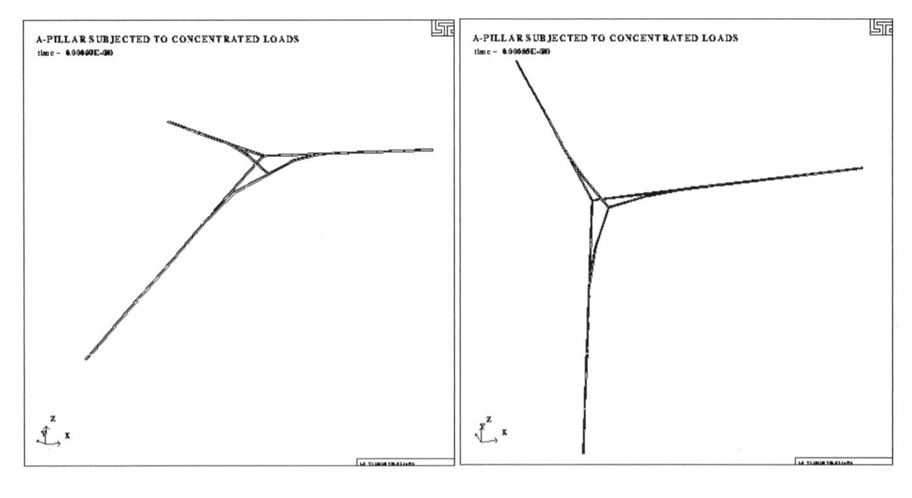

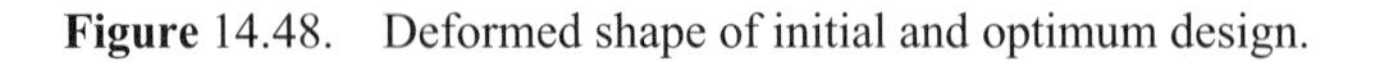
(a) Initial Design (b) Optimum Design

Figure 14.48. Deformed shape of initial and optimum design.

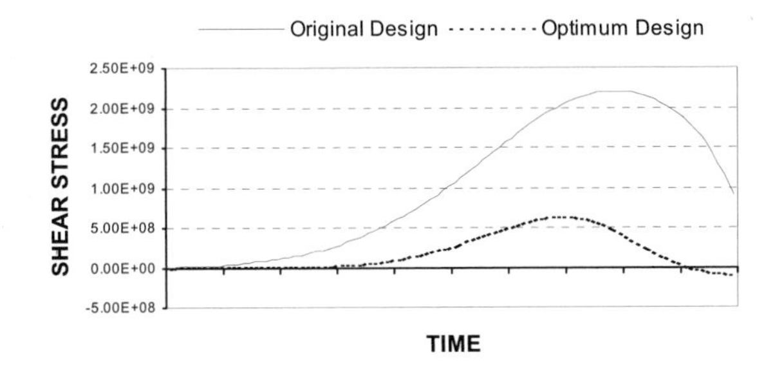

Figure 14.49. Shear stress at element 13.

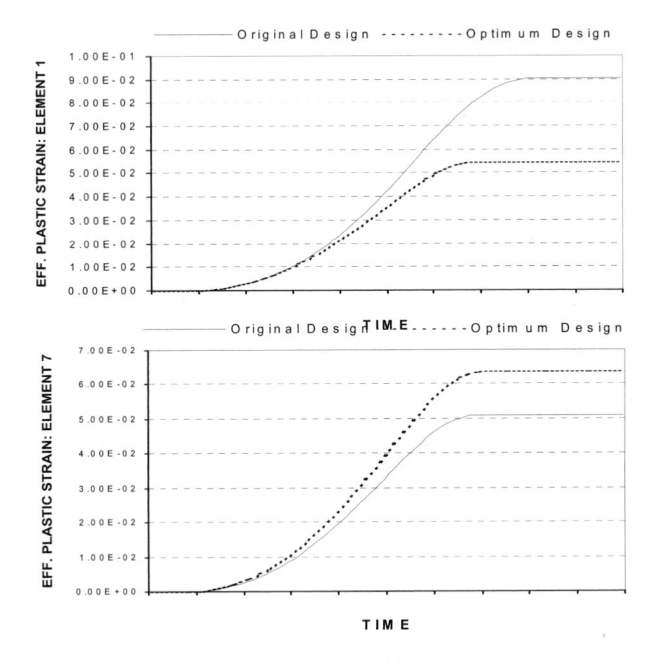

Figure 14.50. Effective plastic strains at elements 1 and 7.

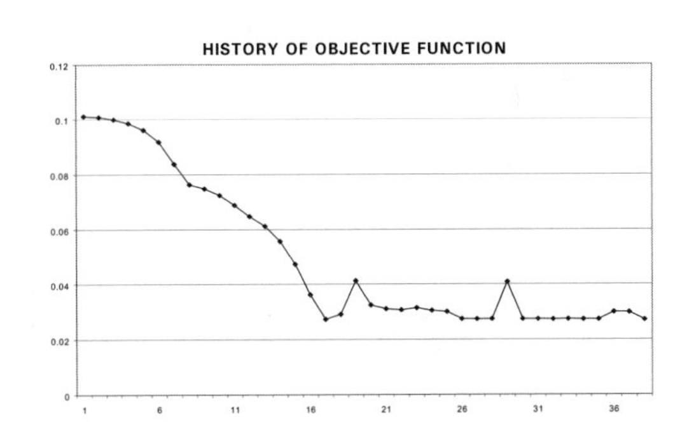

Figure 14.51. Optimization history of objective function.

14.4.3 Stability Design of Vehicle Passenger Compartment Frame

Consider a vehicle passenger compartment modeled as a planar, closed frame structure that projects the side-frame assembly of an automotive body on the vertical plane, as shown in Fig. 14.52 [107]. This finite element model is composed of 45 equivalent beam members. The moment of inertia (around the axis perpendicular to the x_1-x_2 plane) and the cross-sectional area are equal to the sum of the corresponding members of both side frames. Nodal point coordinates are given in Table 14.34. The six flexible joints at nodes 5, 11, 17, 26, 33, and 38 in Fig. 14.52 correspond to the six major joint assemblies in each side frame of the passenger compartment. An important aspect pertaining to the flexibility of a body joint is its nonlinear moment-rotation (M-θ) relationship. The M-θ relationship of joints is given as a piecewise linear curve, as shown in Fig. 14.53.

To determine the load condition, consider a frontal collision with a barrier as shown in Fig. 14.52. As shown in Fig. 14.54, during the deceleration time from a certain speed to zero, part of a vehicle's kinetic energy is dissipated by the work done by plastic deformation of the front portion of the automobile passenger compartment. It is assumed in this example that a nonstructural weight of 1,045 kg (75% of the vehicle sprung mass) and a structural frame weight are imposed on the frame in a horizontal direction during the deceleration time from t_1 to t_2, as shown in Fig. 14.54. Out of the nonstructural mass, 35% of the sprung mass is distributed over the bottom rail as nodal mass, while 65% is distributed over the other nodal points. The engine/transmission mass is distributed at nodes 1, 4, and 5. The mass distribution at each nodal point is shown in Table 14.35.

The design variables are five cross-sectional areas of the beam members. The element groups for each design variable are shown in Table 14.36. For a fixed cross-sectional beam, the moment of inertia can be uniquely defined as $I = \alpha A^2$, where α is a constant that depends on the cross-section type and A is the cross-sectional area. In this example, a rectangular hollow cross section is used with two fixed geometric ratios between width b and height h, and between width b and thickness t. In Table 14.37, the properties of three rectangular cross-sectional types are given with corresponding design variable numbers. The initial design vector is $[100, 100, 100, 100, 100]^T$.

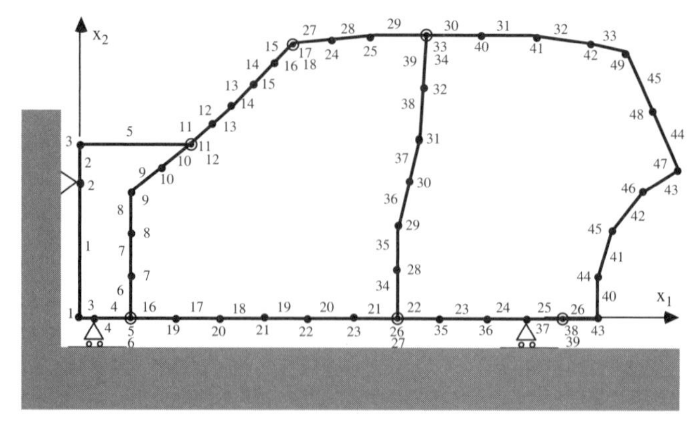

Figure 14.52. Vehicle passenger compartment frame.

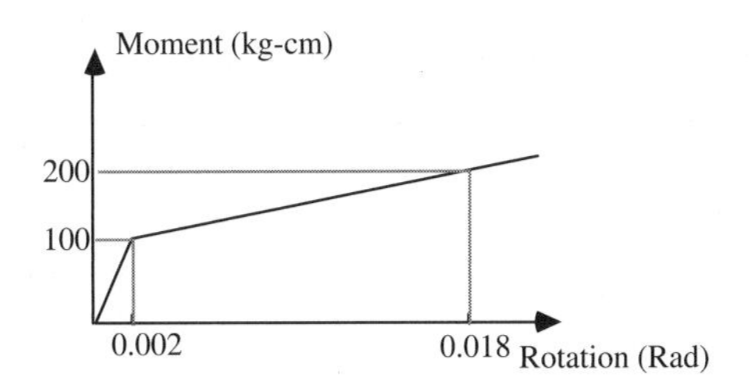

Figure 14.53. Moment-rotational relation of joints.

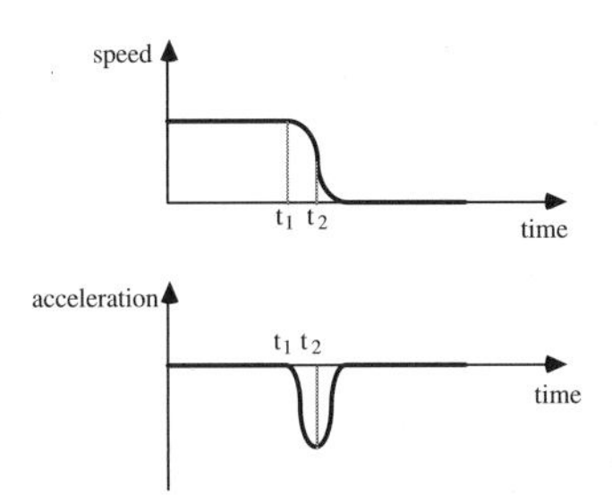

Figure 14.54. Velocity and acceleration diagram during vehicle collision.

Table 14.34. Coordinates of nodal points for vehicle passenger compartment frame (unit : cm).

Node	x_1	x_2	Node	x_1	x_2
1	0.0000	0.0000	2	0.0000	42.750
3	0.0000	54.625	4	4.8750	0.0000
5	17.625	0.0000	6	17.625	0.0000
7	17.625	13.250	8	17.625	26.500
9	17.625	39.750	10	28.000	47.188
11	38.375	54.625	12	38.375	54.625
13	45.438	61.063	14	52.500	67.500
15	59.250	74.125	16	66.000	80.750
17	72.750	87.375	18	72.750	87.375
19	32.625	0.0000	20	47.625	0.0000
21	62.625	0.0000	22	77.625	0.0000
23	92.625	0.0000	24	85.313	88.438
25	97.875	89.500	26	107.63	0.0000
27	107.63	0.0000	28	107.63	14.750
29	107.63	29.500	30	111.38	43.188
31	115.13	56.875	32	116.19	73.188
33	117.25	89.500	34	117.25	89.500
35	122.63	0.0000	36	137.63	0.0000
37	152.63	0.0000	38	163.63	0.0000
39	163.63	0.0000	40	135.81	89.500
41	154.38	89.500	42	170.00	86.875
43	174.63	0.0000	44	174.63	13.250
45	179.50	27.375	46	189.63	40.125
47	201.50	47.625	48	193.56	65.938
49	185.63	84.250			

Table 14.35. Nodal mass distribution for vehicle compartment frame.

Node	Mass	Node	Mass	Node	Mass
1	61.005	2	31.046	3	28.567
4	132.54	5	53.243	6	53.243
7	15.063	8	15.063	9	15.063
10	14.514	11	18.159	12	18.159
13	10.868	14	10.868	15	10.751
16	10.751	17	5.3800	18	5.3800
19	21.266	20	21.266	21	21.266
22	21.266	23	21.266	24	14.337
25	14.337	26	10.633	27	10.633
28	16.768	29	16.768	30	16.131
31	16.313	32	18.581	33	20.306
34	20.306	35	21.266	36	21.266
37	18.424	38	7.8010	39	7.8010
40	21.099	41	21.099	42	18.012
43	7.8010	44	14.965	45	16.885
46	18.512	47	15.964	48	22.687
49	20.355				

* The gravitational acceleration (9.8 m/sec^2).

Table 14.36. Design variable linking for vehicle compartment frame.

Design	Elements Linked	Number of Element
A_1	6–15	10
A_2	16–26	11
A_3	27–33	7
A_4	34–39	6
A_5	40–45	6

Table 14.37. Properties of rectangular hollow cross sections.

Cross Section #	Design Id.	h/b	t/b	$c = I/A^2$
1	A_1, A_4, A_5	1.5	0.075	0.8609
2	A_2	0.5	0.05	0.2866
3	A_3	1.0	0.05	0.7939

The design variables are five cross-sectional areas of the beam members. The element groups for each design variable are shown in Table 14.36. For a fixed cross-sectional beam, the moment of inertia can be uniquely defined as $I = \alpha A^2$, where α is a constant that depends on the cross-section type and A is the cross-sectional area. In this example, a rectangular hollow cross section is used with two fixed geometric ratios between width b and height h, and between width b and thickness t. In Table 14.37, the properties of three rectangular cross-sectional types are given with corresponding design variable numbers. The initial design vector is $[100, 100, 100, 100, 100]^T$.

For structural analysis purposes, STIF3 (two-dimensional beam element) is used for beam members, STIF39 (nonlinear force-deflection element) is used for flexible joints, and STIF 21 (general mass) of ANSYS is used for nodal masses. An aluminum alloy with Young's modulus of $E = 7.4 \times 10^5$ kg/cm^2 is used as the linear elastic material of the beam members. The specific weight of 2.69 gram/cm^3 is used as the self-weight of beam members. Using the incremental analysis method of ANSYS for the initial design, the critical deceleration d_{cr} is found to be between 3.41278 and 3.41279g, where g is the gravitational acceleration. In other words, the vehicle frame buckles at the deceleration $d_{cr} = 3.41279g$.

At several deceleration levels between $d = 3.0g$ and $d = 3.41278g$, design sensitivity coefficients of the critical deceleration are evaluated. In structural analysis of the frame, inertia loads of structural masses are imposed with the applied deceleration. In particular, the inertia loads due to the self-weight of the frame are proportional to the cross-sectional areas. In other words, the load linear form ℓ_u in Section 8.1.2 depends on design variables (cross-sectional areas). The design sensitivity coefficients with a uniform design are presented in the last column of Table 14.38. The second and third columns represent the estimated critical load factor (lowest eigenvalue) and the design sensitivity coefficient of the critical load factor. The design sensitivity of the critical deceleration in the last column is the multiplication of the applied deceleration d/g in the first column and the design sensitivity of the actual critical load factor $\partial\beta/\partial A_{\text{uniform}}$ in the third column.

The design sensitivity results are verified using the finite difference method. Design sensitivity coefficients with uniform design at the applied deceleration level $d = 3.41278g$ are compared with the results using the finite difference method for three design perturbations: 10, 1, and 0.1%. In Table 14.39, $\Delta d_{cr}/g$ is the finite difference of the critical deceleration divided by g, and d'_{cr}/g is the predicted change by the proposed design sensitivity method for the corresponding design perturbation. The agreements between Δd_{cr} and d'_{cr} in the last column of Table 14.39 approach 101.4% with a 0.1% design perturbation. This rate of convergence indicates that results from the finite difference method agree with these results.

Table 14.38. Design sensitivity of critical deceleration for vehicle compartment frame.

d/g	t_ζ	$\partial\beta/\partial A_{\text{uniform}}$	$\partial d_{cr}/\partial A_{\text{uniform}}$
3.00000	2.2622	0.018158	0.054474
3.40000	1.1532	0.018110	0.061574
3.41000	1.0694	0.018103	0.061731
3.41200	1.0362	0.018100	0.061757
3.41270	1.0117	0.018097	0.061759
3.41278	1.0024	0.018096	0.061758

Table 14.39. Verification of design sensitivity of critical deceleration using the finite difference method for vehicle compartment frame (applied deceleration = 3.41278g).

Area	Perturbation	d_{cr}/g	$\Delta d_{cr}/g$	d'_{cr}/g	$d'_{cr}/\Delta d_{cr}$
100.0	—	3.41278	–	–	–
110.0	10.0%	4.00490	0.59211	0.61758D–0	104.3%
101.0	1.0%	3.47290	0.06012	0.61758D–1	102.7%
100.1	0.1%	3.41888	0.00609	0.61758D–2	101.4%

The design sensitivity coefficient of the critical deceleration factor β at the final deceleration $d = 3.41278g$ is [0.22141E–5, 0.13299E–1, 0.10683E–2, 0.83548E–4, 0.36434E–2 $]^T$. At the initial design the cross-sectional area A_2 of the bottom rail members (elements 16–26) is most effective for controlling the buckling behavior. The cross-section area A_2 of the windshield pillar (A-pillar) has no significant effect on the critical deceleration.

Design Optimization

Design optimization of the vehicle passenger compartment frame with stability constraints is described for the critical load with respect to the sizing design variables. For nonlinear structural systems, the system can collapse before the final load level is reached during the incremental analysis. In this case, the design must be improved to make the structure stable before other constraints can be imposed in the design optimization process. With an active set strategy in the optimization algorithm, design sensitivity of the critical load is required when the system collapses, or is near its critical limits.

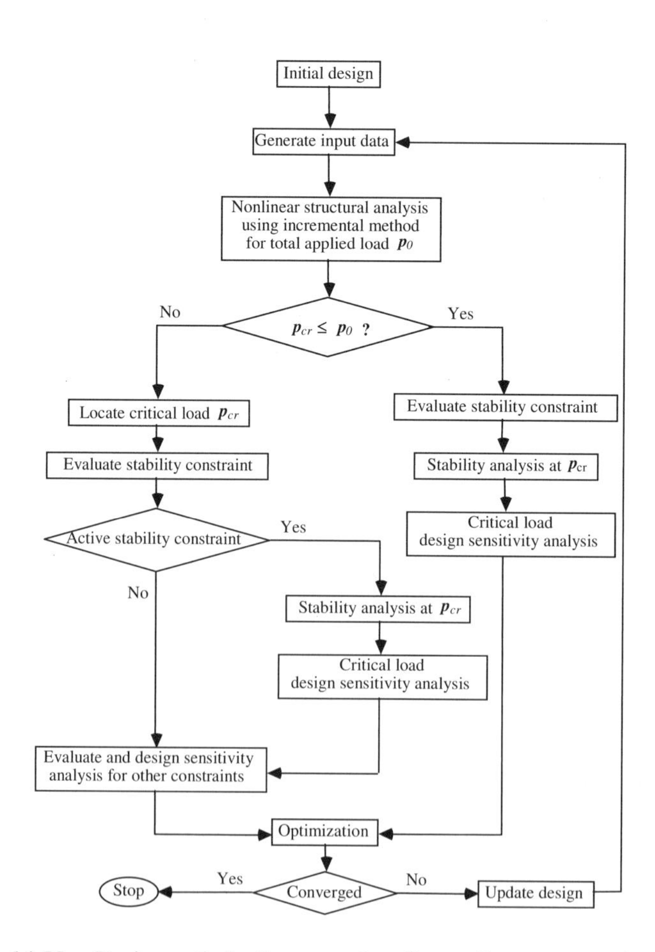

Figure 14.55. Design optimization procedure for nonlinear structural systems.

For numerical implementation with an established finite element analysis code, the computational procedure is given in Fig. 14.55. The nonlinear structural system is first analyzed with an initial design to obtain structural responses for the total applied load $\boldsymbol{p}_0$. An incremental analysis method is carried out to locate the critical load $\boldsymbol{p}_{cr}$, in case the system collapses. Near the critical limit point, a small loading step (e.g., 1% of the total load) should be used to find an accurate estimate. When the system buckles before the total applied load is reached ($\boldsymbol{p}_{cr} \leq \boldsymbol{p}_0$), an optimization problem with only the stability constraint is solved to make the structure stable. Otherwise, $\boldsymbol{p}_0 < \boldsymbol{p}_{cr}$, and all constraints are considered in the optimization process. To locate the critical load in the case of $\boldsymbol{p}_{cr} < \boldsymbol{p}_0$, the incremental analysis scheme is continued until it collapses. At the critical limit point, the stability constraint is checked to determine whether it is active. If it is, then the stability analysis is carried out to obtain the eigenvector associated with the lowest eigenvalue. The above steps are repeated until an optimum is found.

The vehicle passenger compartment frame in Fig. 14.52, which has undergone design sensitivity analysis using the method presented in Section 9.1.2, is now considered for design optimization. The moment-rotation curve of the six flexible joints is piecewise linear, as shown in Fig. 14.53. For the loading condition, the inertia load of the member self-mass and the nodal mass distribution (given in Table 14.35) are imposed, with deceleration in the x_2-direction.

During vehicle frontal impact, the average deceleration level of the passenger compartment frame is $20g$ [108]. The goal of the optimization problem is to minimize the total volume of the vehicle frame shown in Fig. 14.52. Only the stability constraint is considered: that the critical deceleration level of the compartment frame is $20g$ in the direction of the x_2-axis. The optimization problem can be mathematically formulated as

$$\begin{aligned} &\text{minimize} \quad && f = \sum_{i=1}^{5} L_i A_i \\ &\text{subject to} \quad && 1 - \frac{d_{cr}}{20g} = 0, \end{aligned} \tag{14.6}$$

where L_i is the length, and A_i is the cross-sectional area of the ith design group, as defined in Table 14.36. The configuration of the frame is fixed with L_1 = 112.767 cm, L_2 = 157.000 cm, L_3 = 113.403 cm, L_4 = 90.578 cm, and L_5 = 98.436 cm. The initial design variables are A_i = 100 cm^2, i = 1~5.

The stability analysis is carried out at every design iteration, since the stability constraint is imposed as an equality constraint. The Gradient Projection Method [99] is used for the nonlinear optimization problem.

The final optimization results are summarized in Tables 14.40 and 14.41, and Fig. 14.56. The design optimization converges after 23 iterations. The detailed cost and stability constraint history of the optimization process is also given in Table 14.40. The history of the cost shown in Fig. 14.56 reveals a slow convergence, which is a characteristic of the Gradient Projection Method.

In Table 14.41, the histories of five design variables are shown. At the optimum design given in the last row of Table 14.41, the cross-sectional area of the rocker panel members (the second design group with elements 16–26 in Fig. 14.52) increased the most (356.33 cm^2), while the cross-sectional area of the windshield members (the first design group with elements 6–15 in Fig. 14.52) decreased the most (58.24 cm^2). This trend agrees with the automotive vehicle body model, which has small cross sections at the windshield members and large cross sections at the bottom rocker members.

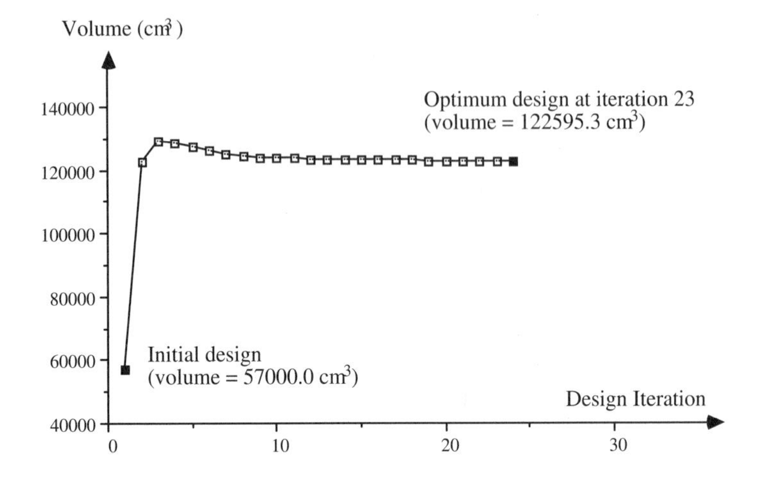

Figure 14.56. History of cost for vehicle compartment frame optimization.

Table 14.40. Cost and constraint history of vehicle compartment frame optimization.

Iteration	$d_{cr}/g^{\S}$	Violation	Cost (cm^3)	Cost Reduction
1	3.415	0.82925	57000.0	
2	16.15	0.1925	122467.4	65467.4
3	19.55	0.0225	129063.8	6596.4
4	19.895	0.00525	128512.7	–551.1
5	19.95	0.0025	127408.9	–1103.8
6	19.995	0.00025	126267.3	–1141.6
7	20.005	–0.00025	125351.0	–916.3
8	20.005	–0.00025	124658.0	–693.0
9	20.005	–0.00025	124146.4	–511.6
10	20.0075	–0.000375	123777.8	–368.6
11	20.0115	–0.000575	123676.0	–101.8
12	20.0015	–0.000075	123618.7	–57.3
13	20.0015	–0.000075	123563.2	–55.5
14	20.00175	–0.0000875	123512.1	–51.1
15	20.00175	–0.0000875	123461.0	–51.1
16	20.00125	–0.0000625	123417.4	–43.6
17	20.00175	–0.0000875	123201.2	–216.2
18	20.00025	–0.0000125	123046.9	–154.3
19	19.99925	0.0000375	122810.0	–236.9
20	19.9935	0.000325	122707.2	–102.8
21	19.9955	0.000225	122655.3	–51.9
22	19.9975	0.000125	122626.9	–28.4
23	19.9985	0.000075	122591.4	–35.5
24	19.9985	0.000075	122595.3	–3.9

§g is the magnitude of the gravitational acceleration.

Table 14.41. Design variable history of vehicle compartment frame optimization.

Iteration	A_1 (cm^2)	A_2 (cm^2)	A_3 (cm^2)	A_4 (cm^2)	A_5 (cm^2)
1	100.00	100.00	100.00	100.00	100.00
2	100.05	438.31	127.17	102.12	192.68
3	94.79	431.47	163.65	114.60	223.43
4	88.10	422.86	167.76	117.54	231.82
5	81.77	414.49	169.86	119.19	237.26
6	76.13	406.74	171.39	120.47	241.54
7	71.43	399.87	172.96	121.85	245.47
8	67.69	393.88	174.56	123.29	249.12
9	64.81	388.68	176.08	124.70	252.46
10	62.70	384.18	177.47	126.05	255.48
11	62.33	383.32	177.62	126.21	255.93
12	62.04	382.54	177.85	126.44	256.44
13	61.77	381.78	178.07	126.67	256.95
14	61.51	381.05	178.29	126.89	257.44
15	61.26	380.32	178.50	127.11	257.93
16	61.04	379.63	178.71	127.32	258.40
17	60.00	376.26	179.70	128.36	260.69
18	59.31	373.36	180.55	129.28	262.72
19	58.40	368.32	181.98	130.89	266.29
20	58.16	364.69	182.98	132.11	269.05
21	58.13	362.08	183.64	133.01	271.16
22	58.16	360.19	184.07	133.65	272.76
23	58.23	357.49	184.60	134.57	275.20
24	58.24	356.33	184.76	134.98	276.52

14.5 Fatigue and Durability Design Applications

14.5.1 Design of Engine Exhaust Manifold

A ground vehicle's engine exhaust manifold is used to demonstrate a fatigue life analysis and the sensitivity calculation method. This component of the engine is subjected to a large thermal load due to the high temperature of exhaust gas, in addition to a variable, dynamic mechanical load. The combination of these loads significantly affects the fatigue life of the manifold.

Thermoelastic Model

Reports from the automotive industry show that the exhaust manifold under investigation cracks during testing. To simulate the test process, dynamic analysis and FEA are employed. The analysis procedures are decoupled into thermal analysis, transient stress analysis, and life prediction.

The manifold is made of cast iron, with an average thermal conductivity of $k = 0.0026$ W/mm/K [109]. The average temperature of the exhaust gas is 1100 ^{0}C. The temperature of the cooling outer gas is 23 ^{0}C. Figure 14.57 shows the thermal model with its thermal boundary conditions. The thermal load is assumed to be convection only, so the entire surface boundary is treated as Γ_θ^2. The experimental data were not available for this manifold, so the convection film coefficients are computed using approximate formulas obtained from Bejan [110]. The film coefficient on the inner surface is $h_{\text{int}} = 1.148 \times 10^{-5}$

W/mm^2/^{0}C, while on the outer surface it is $h_{ext} = 2.148 \times 10^{-6}$ W/mm^2/^{0}C. The output from the thermal analysis is the temperature field, which is computed using finite element analysis.

For the elastic model, the dynamic load is simulated in an interval of 1.453 sec, which is one block, using a sinusoidal function of magnitude between 0 and 1. The load is applied at the junction between the exhaust manifold and the muffler, as shown in Fig. 14.57. Elastic material properties are dependent on the temperature (Table 14.42) and linear interpolation is used. The temperature field is applied as a thermal load. The peak dynamic load is 720 N. The expansion coefficient is $\alpha = 1.25 \times 10^{-5}$/^{0}C and the stress-free temperature (reference temperature) is $T^0 = 23$ ^{0}C.

Structural Analysis

The ANSYS finite element model [28] contains 8188 nodes and 4782 elements. Among these, 3685 are solid 8-node elements, and 1097 are shell 4-node elements. Some of the elements are degenerated as wedges (solid) and triangles (shell). The same finite element model is used for thermal analysis and elasticity analysis. For the thermal problem, the convective loads are applied on the inner and outer surface of the manifold and on the interface between the engine cylinders and the exhaust manifold. For the elasticity problem, displacements in the z-direction at the nodes on the interface with the engine are imposed as zero. To avoid rigid body movement, two nodes have additional constraints. The dynamic load is applied at the nodes near the bolts that connect the muffler to the exhaust manifold. The force from the attachment bolts to the engine is neglected.

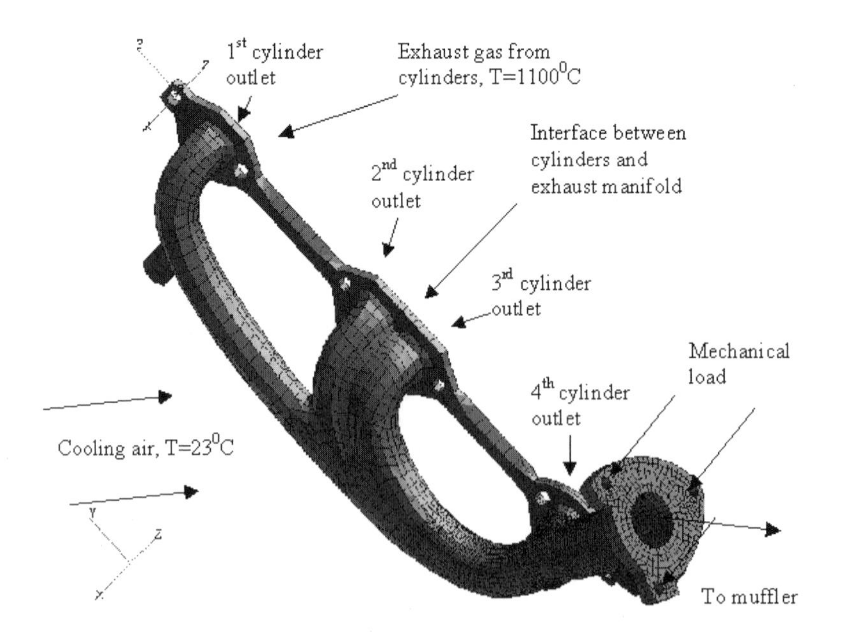

Figure 14.57. Thermoelastic model.

Table 14.42. Elastic material properties.

Temperature (Celsius)	Young's Modulus (N/mm^2)	Poisson's Ratio
150	165000	0.300
200	153000	0.280
300	150000	0.270
400	148000	0.260
500	141000	0.250
550	137000	0.245
600	132000	0.240
700	123000	0.230
750	117000	0.220
816	109000	0.210
1010	95000	0.200

The temperature field obtained using the ANSYS thermal analysis is applied as the thermal load for the elasticity analysis. Thermal analysis contains a single load, corresponding to the highest temperature of the exhaust gas, while elasticity analysis contains two loads, one corresponding to the maximum mechanical load applied in the negative x-direction, and the second corresponding to the maximum load applied in the positive x-direction. The thermal load is applied in both load cases because the stiffness matrix corresponding to the bilinear form contains terms that depend on the temperature. The total dynamic stress tensor is obtained from

$$\sigma_{ij}(z,\theta,t) = \varphi_1(t)\sigma_{ij}(z_1,\theta) + \varphi_2(t)\sigma_{ij}(z_2,\theta), \tag{14.7}$$

where z_1 and z_2 are the displacement fields for the two loads under consideration, and $\varphi_1(t)$ and $\varphi_2(t)$ are sinusoidal functions with interval of 1.453, shifted by one half cycle each other.

Fatigue life is calculated using a von Mises equivalent strain method that is implemented in DRAW [111]. The material properties for the corresponding uniaxial model are fatigue strength coefficient 807 N/mm^2, fatigue strength exponent –0.08, fatigue ductility coefficient 0.29, and fatigue ductility exponent –0.60. For the local elastic-plastic effect, the Ramberg-Osgood equation is involved [112], with cyclic strength coefficient 800 N/mm^2, and cyclic strain hardening exponent 0.12.

It is interesting to note that the lowest fatigue life does not appear in the region where the highest stress (and the peak load) occurs. Similar observations have been made for components with strictly mechanical loads [85]. Such behavior is due to the fact that fatigue life depends on the variation of stresses in time, as well as on their absolute magnitude at peak load. A fatigue life contour for the initial design is shown in Fig. 14.58 (for convenience, a decimal logarithm of the fatigue life has been used). Table 14.43 contains the fatigue life at the most critical nodes, which are located in the area with lowest fatigue life in Fig. 14.58. This area is enlarged in Fig. 14.59. The results show that the lowest life is 7.46625×10^6 blocks in the initial design. With one block being 1.453 seconds, this corresponds to one year, assuming that average use is eight hours per day, seven days per week.

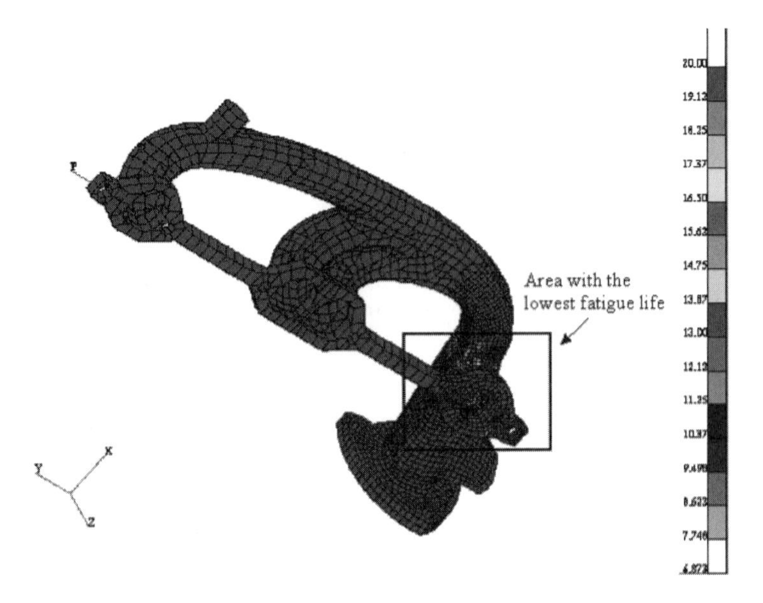

Figure 14.58. Initial fatigue life contour.

Table 14.43. Initial life at most critical nodes.

Node ID	Life	Node ID	Life	Node ID	Life
3331	7.466250e+06	3282	3.377298e+07	3325	6.284750e+08
3549	8.953949e+06	3552	4.960476e+07	3544	7.609912e+08
3546	1.364464e+07	3480	1.827871e+08	3766	1.022470e+09
3266	1.415534e+07	3110	1.945110e+08	3743	1.050892e+09
3329	1.610933e+07	3323	3.394617e+08	3691	1.306688e+09
3333	3.039790e+07	3554	5.659472e+08	3510	1.321799e+09
3548	3.328448e+07	3270	6.114220e+08	3771	4.231992e+09

Design Parameterization
Most nodes with a low fatigue life are concentrated in the area between the fourth cylinder outlet and the junction with the muffler. This area is parameterized in an attempt to increase the life of the component after the proper design changes. A geometric model is created using MSC/PATRAN [42]. Eight shape design parameters are selected for this problem, as shown in Fig. 14.59. Design parameters 1–4 characterize the inner and outer cross sections of two cylindrical cross sections of the manifold (AA and BB); design parameters 5–8 characterize the inner and outer cross section of the junction area between the fourth runner and the main exhaust pipe (CC). Note that bi-cubic patches are employed to represent the boundary geometry of the manifold segment, shown in Fig. 14.59. Eight corresponding design velocity fields are computed using the isoparametric mapping method presented in Chapter 13.

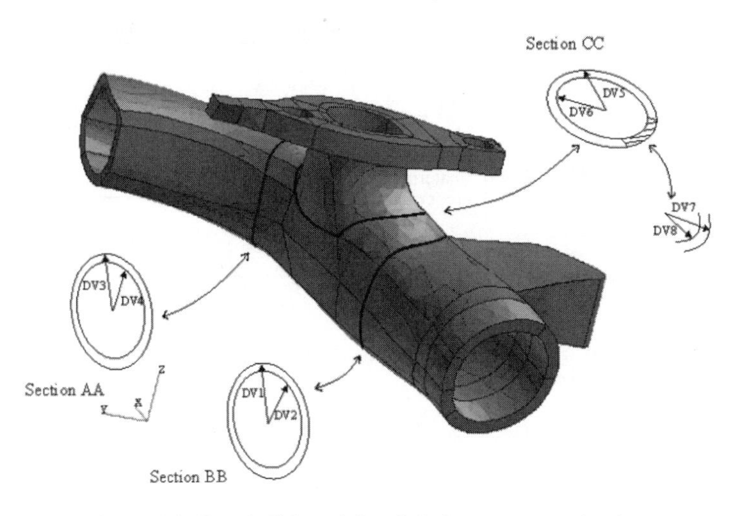

Figure 14.59. Solid model and design parameterization.

Design Sensitivity Analysis

Figure 14.60 shows the dependency of the fatigue life on design variables. Figure 14.61 shows the computational flowchart of the fatigue life design sensitivity analysis. An analytical approach for fatigue life sensitivity analysis is difficult, if not impossible, since the fatigue life is obtained after performing peak-valley editing and rainflow counting procedures. Thus, the finite difference method is used to compute the design sensitivity of the component fatigue life as shown in Fig. 14.61, using the predicted stress history obtained from the continuum shape design sensitivity analysis of thermoelastic problem presented in Section 6.5.3.

The design sensitivity coefficients of stress components are computed with respect to the eight design parameters using the direct differentiation method. Once the sensitivity of the transient stress history is obtained using the design sensitivity analysis method described in Section 6.6, increments of stress history can be obtained by

$$\Delta\sigma_{ij}(z,\theta,t) = \frac{d}{d\boldsymbol{b}}\left(\sigma_{ij}(z,\theta,t)\right)\Delta\boldsymbol{b}, \tag{14.8}$$

where $\Delta\boldsymbol{b}$ is the design perturbation. The design perturbation $\Delta\boldsymbol{b}$ must be small for linear approximation of the fatigue life. On the other hand, in numerical calculation, $\Delta\boldsymbol{b}$ cannot be too small since it may introduce numerical noise.

The transient stress history of the perturbed design can be approximated by

$$\sigma_{ij}(\boldsymbol{b}+\Delta\boldsymbol{b}) = \sigma_{ij}(\boldsymbol{b}) + \Delta\sigma_{ij}(z,\theta,t), \tag{14.9}$$

where σ_{ij} is the component of stress tensor. The perturbed temperature field is also needed for life computation of the component. This perturbed temperature field is easier to compute, because the temperature field does not depend on time t, it depends only on the design perturbation $\Delta\boldsymbol{b}$ as

$$\theta(\boldsymbol{b}+\Delta\boldsymbol{b}) = \theta(\boldsymbol{b}) + \frac{d\theta}{d\boldsymbol{b}}\Delta\boldsymbol{b}. \tag{14.10}$$

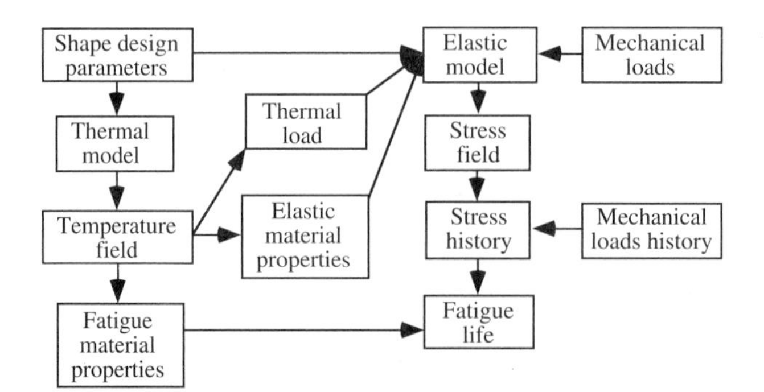

Figure 14.60. Dependence of fatigue life on shape design parameters.

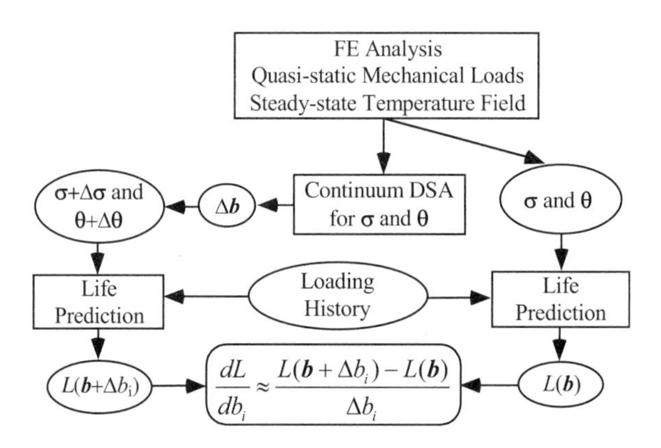

Figure 14.61. Flowchart for design sensitivity analysis of fatigue life.

The new transient stress history is then used to calculate the fatigue life $L(\boldsymbol{b}+\Delta\boldsymbol{b})$ of the component at the perturbed design. The design sensitivity coefficient of the component fatigue life with respect to the design parameter b_i can be obtained from

$$\frac{dL}{db_i} \approx \frac{L(\boldsymbol{b}+\Delta b_i) - L(\boldsymbol{b})}{\Delta b_i}. \tag{14.11}$$

Note that (14.8) through (14.11) must be evaluated for each design parameter.

To determine the design sensitivity of the fatigue life, the hybrid method is employed. Accuracy is verified at those nodes with a low fatigue life, by using the overall finite difference. In Table 14.44, $L(\boldsymbol{b})$ and $L(\boldsymbol{b} + \Delta b_5)$ are fatigue lives at the initial and the perturbed design, with a 0.01 mm perturbation of design parameter dp5; ΔL is the perturbation obtained from the overall finite difference method; L' is the predicted perturbation using design sensitivity $L' = dL/db_5 \times \Delta b_5$; and $\Delta L/L'$ represents the accuracy of the design sensitivity. A value close to 100% for the latter indicates accurate design sensitivity.

Table 14.44. Verifications of design sensitivity coefficients of life.

Node ID	$L(\mathbf{b} + \delta b_5)$	$L(\mathbf{b})$	$L(\mathbf{b} + \delta b_5) - L(\mathbf{b})$	dL/db_5	$\delta L/L'$%	Change %
3331	.7721602E+07	.7466250E+07	.25535E+06	.25536E+08	99.9	3.42
3549	.9520163E+07	.8953949E+07	.56621E+06	.57732E+08	101.9	6.32
3546	.1417754E+08	.1364464E+08	.53291E+06	.53291E+08	100.0	3.90
3266	.1459943E+08	.1415534E+08	.44409E+06	.46629E+08	104.9	3.13
3329	.1713074E+08	.1610933E+08	.10214E+07	.10436E+09	102.1	6.34
3333	.3097522E+08	.3039790E+08	.57732E+06	.57732E+08	100.0	1.89
3548	.3528288E+08	.3328448E+08	.19984E+07	.20428E+09	102.2	6.00
3282	.3372857E+08	.3377298E+08	–.44409E+05	–.44409E+07	100.0	–.13
3552	.5146994E+08	.4960476E+08	.18652E+07	.19540E+09	104.7	3.76
3480	.1852740E+09	.1827871E+09	.24869E+07	.24869E+09	100.0	1.36

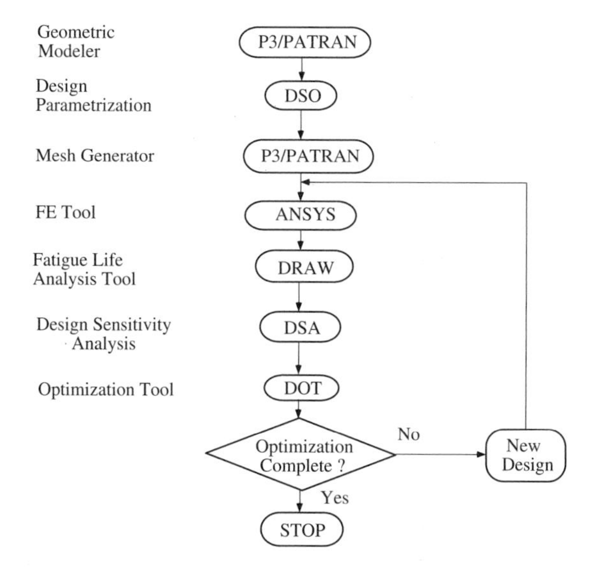

Figure 14.62. Optimization flowchart.

Design Optimization

The shape of the exhaust manifold is optimized using the defined shape design parameters. The objective is to minimize the volume, while increasing durability from one year to six years. At the initial design, the structural volume is 5.24409×10^5 mm^3. Fatigue lives at the most critical 100 nodes are defined as constraints, with a lower bound of 4.628×10^7 blocks, or 6 years of service life. At the initial design, the lowest fatigue life is found at node 3331 with 7.46625×10^6 blocks until failure, which corresponds to 1.03 years of service life. Design optimization is performed using ANSYS as the FEA code, the Design Optimization Tool (DOT) for optimization, and DRAW for the fatigue life computation. P3/PATRAN is used as the preprocessor and postprocessor. The modified feasible direction method of DOT is employed for optimization. Figure 14.62 shows the computational flowchart.

The initial design was infeasible, since eight constraints were violated. A feasible design was obtained after six iterations. The cost function was further reduced, while still keeping the design feasible after the sixth iteration. Fourteen iterations were successfully completed, with 134 finite element analyses and 14 design sensitivity computations. After 14 iterations, ANSYS finite element analysis could not be further performed without re-meshing, which was not pursued. The optimization history of the cost, design parameters and first nine normalized constraints are shown in Figs. 14.63 and 14.64. Table 14.45 shows the cost and fatigue lives at the initial and optimum designs, and the percentage of change.

It can be noted that fatigue life is significantly improved, with little change in volume. During the optimization process, the fatigue life was computed at 100 nodes, for which fatigue lives were low at the initial design. At the optimum design, fatigue lives are computed again to verify results for all nodes in the model. The lowest fatigue life at the final design is found at node 3329 with 4.622968 $\times$ 10^7 blocks until failure, which corresponds to 6.4 years of service life. The life contour plot of the optimum design is shown in Fig. 14.65 (a decimal logarithm of the fatigue life is displayed).

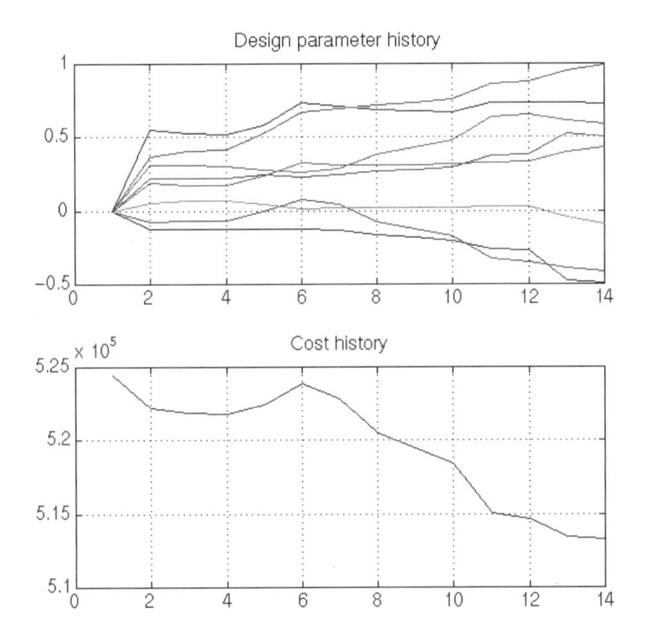

Figure 14.63. Design parameters history and cost history.

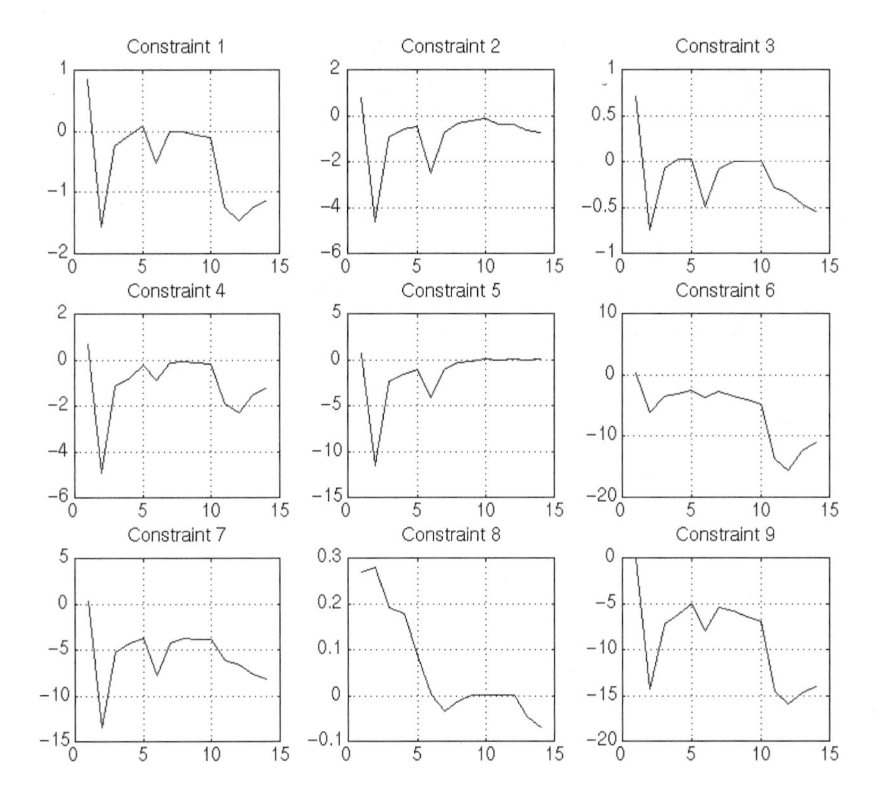

Figure 14.64. History of the first nine constraints (normalized).

Table 14.45. Cost and first nine constraints at initial and optimum design.

Function	Description	Initial design	Optimum design	% changes
Cost	Volume	524409 mm^3	513278 mm^3	–2.122
Constraint 1	Life, node 3331	7.466250e+06	9.920953e+07	1228.7
Constraint 2	Life, node 3549	8.953949e+06	8.077982e+07	802.1
Constraint 3	Life, node 3546	1.364464e+07	7.118750e+07	421.6
Constraint 4	Life, node 3266	1.415534e+07	1.040056e+08	634.7
Constraint 5	Life, node 3329	1.610933e+07	4.622968e+07	186.9
Constraint 6	Life, node 3333	3.039790e+07	5.623945e+08	1750.0
Constraint 7	Life, node 3548	3.328448e+07	4.259703e+08	1179.8
Constraint 8	Life, node 3282	3.377298e+07	4.951594e+07	46.61
Constraint 9	Life, node 3552	4.960476e+07	6.966871e+08	1304.5

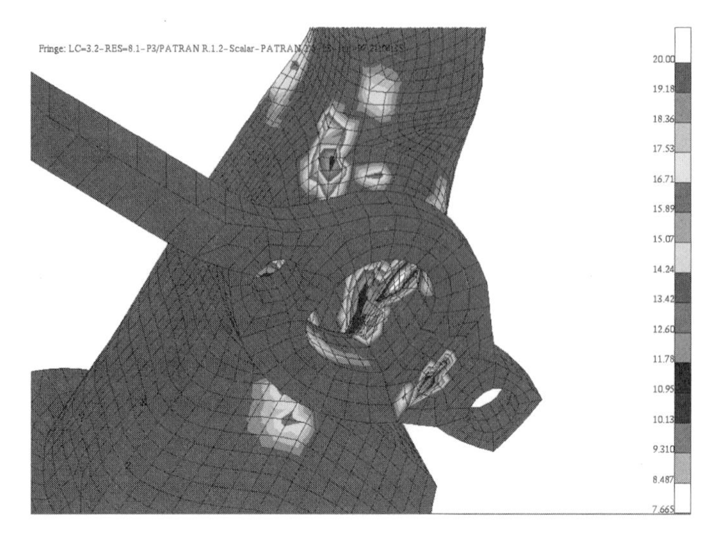

Figure 14.65. Fatigue life contour in critical area at optimum design.

References

[1] Bathe, K.-J., *Finite Element Procedures in Engineering Analysis*. 1996, Englewood Cliffs, NJ: Prentice-Hall.

[2] Bathe, K.-J., E. Ramm, and E.L. Wolson, *Finite Element Formulations for Large Deformation Dynamic Analysis*. International Journal of Numerical Methods in Engineering, 1975, **9**:353–386.

[3] Mallet, R.H. and P.V. Marcal. *Finite Element Analysis of Nonlinear Structures*. In *Journal of the Structural Division, ASCE*, 1968.

[4] Brendel, B. and E. Ramm, *Linear and Nonlinear Stability Analysis of Cylindrical Shells*. Computers and Structures, 1980, **12**:549–558.

[5] Rammerstofer, F.G., *Jump Phenomena Associated with the Stability of Geometrically Nonlinear Structures*. In *Recent Advances in Nonlinear Computational Mechanics,* E. Hinton, D.R.J. Owen, and C. Taylor, ed. 1982, UK: University College of Swansea, pp. 119–153.

[6] Borri, C. and H.W. Hufendiek, *Geometrically Nonlinear Behavior of Space Beam Structures*. Journal of Structural Mechanics, 1985, **13**1–26.

[7] Marcal, P.V., *Instability Analysis Using the Incremental Stiffness Matrices*. In *Lectures on Finite Element Methods in Continuum Mechanics*, J.T. Oden and E.R.A. Oliveria, ed. 1973, Huntsville: University of Alabama in Huntsville, pp. 545–561.

[8] Bathe, K.-J. and E.N. Dvorkin, *On the Automatic Solution of Nonlinear Finite Element Equations*. Computers and Structures, 1983, **17**:871–879.

[9] Ciarlet, P.G., *Lectures on Three-Dimensional Elasticity*. 1983, Berlin: Springer-Verlag.

[10] Novozhilov, V.V., *Foundations of the Nonlinear Theory of Elasticity*. 1953, Rochester, NY: Graylock Press.

[11] Brush, O.B. and B.O. Almroth, *Buckling of Bars, Plates, and Shells*. 1975, New York, NY: McGraw-Hill.

[12] Oden, J.T. and N. Kikuchi, *Finite Element Methods for Constrained Problems in Elasticity*. International Journal for Numerical Methods in Engineering, 1982, **18**:701–725.

[13] Malkas, D.S. and T.J.R. Hughes, *Mixed Finite Element Methods-Reduced and Selective Integration Techniques: A Unification of Concept*. Computer Methods in Applied Mechanics and Engineering, 1978, **15**:63–81.

[14] Sussman, T. and K.J. Bathe, *A Finite Element Formulation for Nonlinear Incompressible Elastic and Inelastic Analysis*. Computers and Structures, 1987, **26**:357–409.

[15] Chen, J.S., C.T. Wu, and C. Pan, *A Pressure Projection Method for Nearly Incompressible Rubber Hyperelasticity, Part I: Theory and Part II: Application.* ASME Journal of Applied Mechanics, 1996, **63**:862–876.

[16] Hughes, T.J.R., *The Finite Element Method*. 1987, Englewood Cliffs, NJ: Prentice-Hall.

[17] Carey, G.F. and J.T. Oden, *Finite Elements, A Second Course. Vol II*. 1983, Englewood Cliffs, NJ: Prentice-Hall.

[18] Simo, J.C. and S. Govindjee, *Nonlinear B-Stability and Symmetric Preserving Return Mapping Algorithms for Plasticity and Viscoplasticity*. International Journal for Numerical Methods in Engineering, 1991, **31**:151–176.

[19] Simo, J.C. and R.L. Taylor, *Consistent Tangent Operator for Rate-Independent*

Elastoplasticity. Computer Methods in Applied Mechanics and Engineering, 1985, **48**:101–118.

[20] Hughes, T.J.R. and J. Winget, *Finite Rotation Effects in Numerical Integration of Rate Constitutive Equations Arising in Large-Deformation Analysis.* International Journal for Numerical Methods in Engineering, 1980, **15**:1862–1867.

[21] Lee, E.H., *Elastic-Plastic Deformation at Finite Strains.* Journal of Applied Mechanics, 1969, **36**:1–6.

[22] Simo, J.C., *Algorithms for Static and Dynamic Multiplicative Plasticity That Preserve the Classical Return Mapping Schemes of the Infinitesimal Theory.* Computer Methods in Applied Mechanics and Engineering, 1992, **99**:61–112.

[23] Kim, N.H., Y.H. Park, and K.K. Choi, *An Optimization of a Hyperelastic Structure with Multibody Contact Using Continuum-Based Shape Design Sensitivity Analysis.* Structural Optimization, 2001, **21**:196–208.

[24] Duvaut, G. and J.L. Lions, *Inequalities in Mechanics and Physics.* 1976, Berlin: Spring-Verlag.

[25] Kikuchi, N. and J.T. Oden, *Contact Problems in Elasticity: A Study of Variational Inequalities and Finite Element Method.* 1988, Philadelphia: SIAM.

[26] Wriggers, P., T.V. Van, and E. Stein, *Finite Element Formulation of Large Deformation Impact-Contact Problems with Friction.* Computers and Structures, 1990, **37**:319–331.

[27] Simo, J.C. and R.L. Taylor, *Quasi-compressible Finite Elasticity in Principal Stretches. Continuum Basis and Numerical Algorithms.* Computer Methods in Applied Mechanics and Engineering, 1991, **85**:273–310.

[28] DeSalvo, G.J. and J.A. Swanson, *ANSYS Engineering Analysis System, User's Manual Vol. I and II.* 1989, Houston, PA: Swanson Analysis Systems Inc.

[29] Kamat, M.P., N.S. Khot, and V.B. Venkayya, *Optimization of Shallow Trusses Against Limit Point Instability.* AIAA Journal, 1984, **22**(3):403–408.

[30] Wu, C.C. and J.S. Arora, *Design Sensitivity Analysis and Optimization of Nonlinear Structural Response Using Incremental Procedure.* AIAA Journal, 1986, **25**(8):1118–1125.

[31] Santos, J.L.T. and K.K. Choi, *Design Sensitivity Analysis of Non-linear Structural Systems Part II: Numerical Method.* International Journal for Numerical Methods in Engineering, 1988, **26**(9):2097–2114.

[32] Choi, K.K. and J.L.T. Santos, *Design Sensitivity Analysis of Nonlinear Structural System. Part I: Theory.* International Journal for Numerical Methods in Engineering, 1987, **24**:2039–2055.

[33] Gallagher, R.H., ed. *Finite Element Representations for Thin Shell Instability Analysis.* In Buckling of Structures, IUTAM Symposium, 1974, B. Bundiansky ed. 1976, New York: Springer-Verlag.

[34] Vidal, C.A. and R.B. Haber, *Design Sensitivity Analysis for Rate-Independent Elastoplasticity.* Computer Methods in Applied Mechanics and Engineering, 1993, **107**:393–431.

[35] Hughes, T.J.R. and J.E. Marsden, *Mathematical Foundations of Elasticity.* 1983, Englewood Cliffs, NJ: Prentice-Hall.

[36] Kim, N.H., K.K. Choi, and J.S. Chen, *Structural Optimization of Finite Deformation Elastoplasticity Using Continuum-Based Shape Design Sensitivity Formulation.* Computers and Structures, 2001, **79**(20–21):1959–1976.

[37] Haug, E.J., K.K. Choi, and V. Komkov, *Design Sensitivity Analysis of Structural Systems.* 1986, London: Academic Press.

[38] Santos, J.L.T. and K.K. Choi, *Shape Design Sensitivity Analysis of Nonlinear Structural System.* Structural Optmization, 1992, **4**:23–35.

[39] Cho, S. and K.K. Choi, *Design Sensitivity Analysis and Optimization of Nonlinear*

Transient Dynamics Part II: Configuration Design. International Journal of Numerical Methods in Engineering, 2000, **48**(3):375–399.

[40] Hibbit, Karlsson, and Sorensoen, *ABAQUS Users' Manual.* 1989, Providence, RI: Hibbit, Karlsson & Sorensoen Inc.

[41] Choi, K.K. and K.H. Chang, *A Study of Design Velocity Field Computation for Shape Optimal Design.* Finite Elements in Analysis and Design, 1994, **15**:317–341.

[42] MacNeal-Schwendler, *MSC/PATRAN User's Manual.* Vol. 70. 2000, Los Angeles: MacNeal Schwendler Corp.

[43] Kleiber, M., *Shape and Non-shape Structural Sensitivity Analysis for Problems with Any Material and Kinematic Non-linearity.* Computer Methods in Applied Mechanics and Engineering, 1993, **108**:73–97.

[44] Zhang, Q., S. Mukherjee, and A. Chandra, *Shape Design Sensitivity Analysis for Geometrically and Materially Nonlinear Problems by the Boundary Element Method.* International Journal of Solids and Structures, 1992, **29**:2503–2525.

[45] Dutta, A., *A Study of Implicit and Explicit Methods for Design Sensitivity Analysis of Nonlinear Dynamic Systems.* In *Department of Civil and Environment Engineering.* 1996, Iowa City, IA: University of Iowa.

[46] Simo, J.C., *On the Computational Significance of the Intermediate Configuration and Hyperelastic Stress Relations in Finite Deformation Elastoplasticity.* Mechanics of Materials, 1985, **4**:439–451.

[47] Miehe, C., E. Stein, and W. Wagner, *Associative Multiplicative Elastoplasticity: Formulation and Aspects of the Numerical Implementation Including Stability Analysis.* Computers and Structures, 1994, **52**:969–978.

[48] Liu, W.K., S. Jun, and Y.F. Zhang, *Reproducing Kernel Particle Methods.* Internaltional Journal for Numerical Method in Fluids, 1995, **20**:1081–1106.

[49] Choi, K.K. and N.H. Kim, *Design Optimization of Springback in a Deepdrawing Process.* AIAA Journal, 2002, **40**(1):147–153.

[50] Rousselet, B., *Quelques Résultats en Optimisation de Domains.* 1982, Nice, France: Université de Nice.

[51] Timoshenko, S.P. and J. N. Goodier, *Theory of Elasticity.* 1951, New York: McGraw-Hill.

[52] Reissner, E., *The Effect of Transverse Shear Deformation on the Bending of Elastic Plates.* Journal of Applied Mechanics, 1945, **12**:A69–A77.

[53] Kim, N.H., et al., *Meshfree Analysis and Design Sensitivity Analysis for Shell Structures.* International Journal for Numerical Methods in Engineering, 2002, **53**:2087–2116.

[54] Haftka, R.T. and R.V. Grandhi, *Structural Shape Optimization—A Survey.* Computer Methods in Applied Mechanics and Engineering, 1986, **57**:91–106.

[55] Ding, Y., *Shape Optimization of Structures—A Literature Survey.* Computers and Structures, 1986, **24**(6):985–1004.

[56] Bhavikatti, S.S. and C.V. Ramakrishnan, *Optimum Shape Design of Rotating Disks.* Computers and Structures, 1980, **11**:397–401.

[57] Prasad, B. and J.F. Emerson, *Optimal Structural Remodeling of Multi-objective Systems.* Computers and Structures, 1984, **18**(4):619–628.

[58] Kristensen, E.S. and N.F. Madsen, *On the Optimum Shape of Fillets in Plates Subjected to Multiple in-Plane Loading Cases.* International Journal for Numerical Methods in Engineering, 1976, **10**:1007–1019.

[59] Pedersen, P. and C.L. Laursen, *Design for Minimum Stress Concentration by Finite Elements and Linear Programming.* Journal of Structural Mechanics, 1982–83, **10**:375–391.

[60] Yang, R.J. and K.K. Choi, *Accuracy of Finite Element Based Shape Design*

Sensitivity Analysis. Journal of Structural Mechanics, 1985, **13**:223–239.

[61] Luchi, M.L., A. Poggialini, and F. Persiani, *An Interactive Optimization Procedure Applied to the Design of Gas Turbine Discs*. Computers and Structures, 1980, **11**:629–637.

[62] Weck, M. and P. Steinke, *An Efficient Technique in Shape Optimization*. Journal of Structural Mechanics, 1983–4, **11**:433–449.

[63] Braibant, V., C. Fleury, and P. Beckers, *Shape Optimal Design: An Approach Matching CAD and Optimization Concepts*. 1983, Liege, Belgium: Aerospace Laboratory of the University of Liege.

[64] Braibant, V. and C. Fleury, *Shape Optimal Design Using B-Spline*. Computer Methods in Applied Mechanics and Engineering, 1984, **44**:247–267.

[65] Yao, T.M. and K.K. Choi, *3-D Shape Optimal Design and Automatic Finite Element Regridding*. International Journal for Numerical Methods in Engineering, 1989, **28**:369–384.

[66] Yao, T.M. and K.K. Choi, *Shape Optimal Design of an Arch Dam*. ASME Journal of Structural Engineering, 1989, **115**(9):2401–2405.

[67] Choi, K.K. and T.M. Yao, *3-D Shape Modeling and Automatic Regridding in Shape Design Sensitivity Analysis*. In *Sensitivity Analysis in Engineering*. 1987, NASA Conference Publication 2457, Washington, DC: National Aeronautics and Space Administration, pp. 329–345.

[68] Mortenson, M.E., *Geometric Modeling*. 1985, New York: Wiley.

[69] Prezmieniecki, J.S., *Theory of Matrix Structural Analysis*. 1968, New York: McGraw-Hill.

[70] Shames, I.H. and C.L. Dym, *Energy and Finite Element Methods in Structural Mechanics*. 1985, New York: McGraw-Hill.

[71] Timoshenko, S.P. and S. Woinowsky-Krieger, *Theory of Plates and Shells*. 2nd ed. 1959, New York: McGraw-Hill.

[72] Cook, R.D., *Concepts and Applications of Finite Element Analysis*. 1981, New York: Wiley.

[73] Zienkiewicz, O.C., *The Finite Element Method*. 1977, New York: McGraw-Hill.

[74] Godse, M.M., E.J. Haug, and K.K. Choi, *A Parametric Design Methodology for Concurrent Engineering*. 1991, Center for Computer-Aided Design, Iowa City, IA: University of Iowa.

[75] Yang, R.J. and M.J. Fiedler, *Design Modeling for Large-Scale Three-Dimensional Shape Optimization Problems*. ASME Computers in Engineering, 1987, pp. 177–182.

[76] Kodiyalam, S., V. Kumar, and P.M. Finnigan, *A Constructive Solid Geometry Approach to Three-Dimensional Shape Optimization*. AIAA Journal, 1992, **30**(5): 1408–1415.

[77] Botkin, M.E., *Shape Optimization of Plate and Shell Structures*. AIAA Journal, 1982, **20**(2):268–273.

[78] Fleury, C. and V. Braibant. *Shape Optimal Design—A Performing CAD Oriented Formulation*. In *25th AIAA SDM Conference*. 1984, Palm Springs, CA.

[79] Yang, R.J. and M.E. Botkin, *A Modular Approach for Three-Dimensional Shape Optimization*. AIAA Journal, 1987, **25**(3):492–497.

[80] Imam, M.H., *Three-Dimensional Shape Optimization*. International Journal for Numerical Methods in Engineering, 1982, **18**:661–673.

[81] Choi, K.K., *Shape Design Sensitivity Analysis and Optimal Design of Structural Systems*, in *Computed Aided Optimal Design*. C.A.M. Soares, ed. 1987, Heidelberg: Springer-Verlag, pp. 439–492.

[82] Rajan, S.D. and A.D. Belegundu, *Shape optimization Approach Using Fictitious Loads*. AIAA Journal, 1989, **27**(1):102–107.

[83] Belegundu, A.D. and S.D. Rajan, *A Shape Optimization Approach Based on Natural Design Variables and Shape Functions.* Computer Methods in Applied Mechanics and Engineering, 1988, **66**:89–106.

[84] Zhang, S. and A.D. Belegundu, *A System Approach for Generating Velocity Fields in Shape Optimization.* In *Optimization of Large Scale Structural Systems,* 1991, Berchtesgaden, Germany: NATO Advanced Study Institute.

[85] Chang, K.H. and K.K. Choi, *A Geometry-Based Parameterization Method for Shape Design of Elastic Solids.* Mechanics of Structures and Machines, 1992, **20**(2):215–252.

[86] Yang, R.J., *A Three-Dimensional Shape Optimization—SHOP3D.* Computers and Structures, 1989, **31**(6):881–890.

[87] Botkin, M.E., *Shape Design Modeling Using Fully Automatic Three-Dimensional Mesh Generator.* Finite Elements in Analysis and Design, 1991, **10**:165–181.

[88] Kikuchi, N., *Adaptive Grid-Design Methods for Finite Element Analysis.* Computer Methods in Applied Mechanics and Engineering, 1986, **55**:129–160.

[89] Winslow, A.M., *Numerical Solution of the Quasilinear Poisson Equation in a Nonuniform Triagular Mesh.* Journal of Computational Physics, 1977, **1**:149–172.

[90] Tortorelli, D.A., *A Geometric Representation Scheme Suitable for Shape Optimization.* Mechanics of Structures and Machines, 1993, **21**(1).

[91] Hou, J.W. and J.S. Cheen, *On the Design Velocity Field in the Domain and Boundary Methods for Shape Optimization.* AIAA Paper, 1988, **88-2338**:1032–1040.

[92] Rajan, S.D. and L. Gani. *A Comparison of Natural and Geometric Approaches for Shape Optimal Design.* In *AIAA 31st SDM Conference.* 1990, Long Beach, CA.

[93] Chang, K.H. and K.K. Choi. *Shape Design Sensitivity Analysis and What-if Tool for 3-D Design Applications.* In *Concurrent Engineering Tools and Technologies For Mechanical System Design.* 1992, Iowa City, IA: NATO-Army-NASA Advanced Study Institute.

[94] Chang, K.H., et al., *Design Sensitivity Analysis and Optimization Tool (DSO) for Shape Design Applications.* Computing Systems in Engineering, 1995, **6**(2):151–175.

[95] Chang, K.H. and K.K. Choi, *Design Sensitivity Analysis and Optimization Tool (DSO) Example Manual.* 1994, Iowa City, IA: Center for Computer-Aided Design, College of Engineering, University of Iowa.

[96] Arora, J.S., *Introduction to Optimum Design.* 1989, New York: McGraw-Hill.

[97] AEA-Industrial-Technology, *Harwell Laboratory.* 1990, Oxfordshire, England.

[98] Vanderplaats, G.N. and S.R. Hansen, *DOT User's Manual.* 1990, Goleta, CA: VMA Engineering.

[99] Haug, E.J. and J.S. Arora, *Applied Optimal Design.* 1979, New York: Wiley.

[100] MacNeal-Schwendler, *MSC/NASTRAN User's Manual Vols. I and II.* Vol. 70. 2000, Los Angeles, CA: MacNeal Schwendler Corp.

[101] Parametric, T.C., *Pro/ENGINEER Fundamentals.* 1995, Waltham, MA: Parametric Technology Corporation.

[102] Peterson, S. and T.A. Stone, *Finite Element Analysis of M88 Tow Bar Clevis.* 1987, TACOM Report.

[103] Hansen, S.R. and G.N. Vanderplaats. *An Approximation Method for Configuration Optimization of Trusses.* In *AIAA/ASME/ASCE/AHS 29th Structures, Structural Dynamics and Materials Conference.* 1988.

[104] Choi, K.K., et al., *Pshenichny's Linearization Method for Mechanical System Optimization.* ASME Journal of Mechanical Design, 1983, **105**(1):97–103.

[105] Pshenichny, B.N. and Y.M. Danilin, *Numerical Methods in External Problems.* 1978, Moscow: MIR Publishers.

[106] Imai, K., *Configuration Optimization of Trusses by the Multiplier Method.* 1978, Los Angeles, CA: Mechanics and Structures Department, School of Engineering and Applied Science, University of California.

[107] Chang, D.C. and C.-M. Ni, *Plastic Deformation Analysis.* In *Modern Automotive Structural Analysis*, M.M. Kamal and J.A. Wolf, ed. 1981, New York: Van Nostrand Reinhold.

[108] Kamal, M.M. and K.-H. Lin, *Collision Simulation.* In *Modern Automotive Structural Analysis*, M.M. Kamal and J.A. Wolf, ed. 1981, New York: Van Nostrand Reinhold.

[109] Arpaci, V.S., *Conduction Heat Transfer*. 1966, Reading, MA: Addison-Wesley.

[110] Bejan, A., *Convection Heat Transfer*. 1984, New York: Wiley.

[111] Tang, J. and K.K. Choi, *Durability and Reliability Analysis Workspace (DRAW) User's Reference*. 1994, Iowa City, IA: Center for Computer-Aided Design, College of Engineering, University of Iowa.

[112] Bannantine, J.A., *Fundamentals of Metal Fatigue Analysis*. 1990, Englewood Cliffs, NJ: Prentice-Hall.

Index

Mechanical Engineering Series *(continued from page ii)*

M. Kaviany, **Principles of Convective Heat Transfer, 2nd ed.**

M. Kaviany, **Principles of Heat Transfer in Porous Media, 2nd ed.**

E.N. Kuznetsov, **Underconstrained Structural Systems**

P. Ladevèze, **Nonlinear Computational Structural Mechanics: New Approaches and Non-Incremental Methods of Calculation**

P. Ladevèze and J.-P. Pelle, **Mastering Calculations in Linear and Nonlinear Mechanics**

A. Lawrence, **Modern Inertial Technology: Navigation, Guidance, and Control, 2nd ed.**

R.A. Layton, **Principles of Analytical System Dynamics**

F.F. Ling, W.M. Lai, D.A. Lucca, **Fundamentals of Surface Mechanics: With Applications, 2nd ed.**

C.V. Madhusudana, **Thermal Contact Conductance**

D.P. Miannay, **Fracture Mechanics**

D.P. Miannay, **Time-Dependent Fracture Mechanics**

D.K. Miu, **Mechatronics: Electromechanics and Contromechanics**

D. Post, B. Han, and P. Ifju, **High Sensitivity Moiré:** Experimental Analysis for Mechanics and Materials

F.P. Rimrott, **Introductory Attitude Dynamics**

S.S. Sadhal, P.S. Ayyaswamy, and J.N. Chung, **Transport Phenomena with Drops and Bubbles**

A.A. Shabana, **Theory of Vibration: An Introduction, 2nd ed.**

A.A. Shabana, **Theory of Vibration: Discrete and Continuous Systems, 2nd ed.**